Lecture Notes in Mathematics

Edited by A. Dold and B. Eckmann

1432

K. Ambos-Spies G.H. Müller
G.E. Sacks (Eds.)

Recursion Theory Week

Proceedings of a Conference held in
Oberwolfach, FRG, March 19–25, 1989

Springer-Verlag
Berlin Heidelberg New York London
Paris Tokyo Hong Kong Barcelona

Editors

Klaus Ambos-Spies
Gert H. Müller
Mathematisches Institut, Universität Heidelberg
Im Neuenheimer Feld 288, 6900 Heidelberg, Federal Republic of Germany

Gerald E. Sacks
Department of Mathematics, Harvard University
and Massachusetts Institute of Technology
One Oxford Street, Cambridge, MA 02138, USA

Mathematics Subject Classification (1980): 03Dxx, 03Exx, 68Qxx

ISBN 3-540-52772-9 Springer-Verlag Berlin Heidelberg New York
ISBN 0-387-52772-9 Springer-Verlag New York Berlin Heidelberg

Printing and binding: Druckhaus Beltz, Hemsbach/Bergstr.
2146/3140-543210 – Printed on acid-free paper

PREFACE

In continuation of a first Recursion Theory Week 1984 (see the Lecture Notes in Mathematics, Vol. 1441 (1985)), another such week took place in the Mathematisches Forschungsinstitut Oberwolfach from March 19 to March 25, 1989. Not only the formal talks but also many fruitful discussions and conversations contributed to the success of the meeting. All of us enjoyed the atmosphere and the facilities of the Forschungsinstitut and we express here our warm thanks to its director Professor Martin Barner and to its staff, once again.

The papers contained in this volume provide a global view of the recent progress in the vast area of Recursion Theory as understood today. Unfortunately not all the papers given at the conference could be published here in order to meet the deadline we had set ourselves, taking into account the length of the usual refereeing process.

The editors express their gratitude to all of the participants for their respective contributions: in discussions, in the talks given and in preparing the papers.

We acknowledge with gratitude the financial help of the Division of Logic, Methodology and Philosophy of Science that we received both for the first conference in 1984, and also this time. In both cases it was used to cover the expenses of longdistance flights for some of the participants.

K. Ambos-Spies (Heidelberg, FRG)
G.H. Müller (Heidelberg, FRG)
G.E. Sacks (Cambridge, Mass., USA)

March 1990

TABLE OF CONTENTS

Honest Polynomial Reductions and Exptally Sets

Klaus Ambos-Spies* Steven Homer† Dongping Yang‡

1 Introduction

This paper is a continuation and extension of recent research on the properties of honest polynomial reductions. It generalizes many of the previous results and addresses several central questions raised in the original work concerning minimal degrees. Honest polynomial time reductions were first considered in the papers of Homer [H-85,H-87]. The main focus there was on the existence of minimal degrees for honest polynomial Turing reductions. The original results were greatly simplified and strengthened in Homer-Long [LH-87] and particularly in Ambos-Spies [AS-87,AS-89]. In these later papers the role of *EXPTALLY* sets for this theory was seen to play a central role, as the sets of minimal degree constructed were all *EXPTALLY* sets. *EXPTALLY* sets are extremely sparse and have properties which allow many of the constructions and results to be simplified.

Two types of theorems were prominent in the previous papers. First, theorems were proved showing that various sets could not be minimal with

*Mathematisches Institut, Universität Heidelberg, Heidelberg, West Germany

†Boston University, Computer Science Department, Boston MA, USA 02215. Supported in part by National Science Foundation Grants MIP-8608137 and CCR-8814339 and a Fulbright-Hays Research Fellowship. This research was done while the author was a Guest Professor at the Mathematisches Institut, Heidelberg University.

‡Institute of Software, Academia Sinica, Beijing, China 100080. This research was supported by a grant of the Stiftung Volkswagenwerk. The research was done while the author visited the Mathematisches Institut, Heidelberg University.

respect to honest reductions. These included the recursive sets, r.e. complete sets and P-immune sets. Second, assuming $P = NP$, the existence of minimal set was shown. (A set $B \notin P$ is minimal if the only sets below B are in P.) The first such sets were minimal for honest Turing reductions and were recursive in O''. In Ambos-Spies[AS-87] these results were improved to apply to honest many-one reductions as well and to construct minimal sets which are r.e. It was previously shown that no recursive set can be minimal.

Two central questions emerged from this study,
(1). Is the existence of minimal degrees dependent on whether $P = NP$? In particular, does $P \neq NP$ imply that minimal degrees do not exist?
(2). Assuming $P = NP$ we can prove that minimal degrees exist. How complicated must such degrees be ?

We address both of these questions in this paper. In Section 2 we present a generalized and unified version of many of these earlier results. In particular we consider properties of the honest polynomial time degrees and their structure under various reductions. The main result concerns the structure of these degrees below a fixed $EXPTALLY$ set. Assuming $P = NP$, we prove that any such structure is a Boolean algebra and that, for some nonrecursive set A, the Boolean algebra has atoms and hence in particular, minimal degrees.

We then turn to the two questions above. For (1) we prove that for some honest reducibilities, the question is oracle dependent. We construct an oracle relative to which $P \neq NP$ and yet there exist minimal degrees. On the other hand we construct an oracle relative to which no minimal degrees exist. The first results hold for all of the honest reducibilities, the second for all but honest Turing reducibility.

For (2) we apply our results on the structure of degrees below $EXPTALLY$ sets together with some recent results of Downey [Do-89] and Jockusch [Jo-89]. We characterize, in terms of the high and low hierarchy for Turing degrees, which degrees can contain minimal sets and conclude that minimal degrees can be low_2 but not low_1.

2 Definitions and Notation

The central concepts of this paper are honest polynomial time reductions. These reductions differ from the usual polynomial reductions in the strings they are allowed to query. Because of their time bounds, polynomial reductions can only query strings which are at most a polynomial longer than their input. Honest reductions, in addition, are allowed only to query strings which are no more than polynomially shorter than the input as well. In other words, honest polynomial time reductions are strongly polynomial in the sense that, for a given input, the reduction procedure queries only strings in a polynomial sized interval about the length of the input.

All of the various polynomial reductions have honest analogues. For our purposes though, it is sufficient to consider three types of reductions: many-one, Turing and one-truth-table. The honest many-one reduction is unusual in that we allow our reduction functions to have the special values + and −. These indicate the input element is in or out of the set, respectively. The reason for these special cases is that we want to allow our many-one reductions to sometimes determine an answer without necessarily querying the oracle. The + and − outputs permit such determinations and yield a more natural and reasonable theory than the one gotten without them. (For an investigation of part of the more traditional many-one theory see Ambos-Spies [AS-89] and Downey, Gasarch, Homer and Moses [DGHM-89].)

In this paper all of the reductions, degrees, etc. we use are polynomial. We often omit explicit mention of this fact in out notation.

Definition:

(1). A total function f is (polynomially) honest if there is a polynomial q such that for all x, $q(|f(x)|) \geq |x|$.

(2). A set A is honest many-one polynomial time reducible to B ($A \leq^h_m B$) if there is an honest, polynomial time computable function $f : \Sigma^* \Longrightarrow \Sigma^* \cup \{+,-\}$, such that $x \in A \iff f(x) = +$ or $f(x) \in B$.

(3). A set A is honest 1-truth-table polynomial time reducible to a set B ($A \leq^h_{1-tt} B$) if there is an honest, polynomial time computable function g and a polynomial time computable function h such that for all x, $A(x) = h(x, B(g(x)))$.

(4). A set A is honest polynomial time Turing reducible to B ($A \leq^h_T B$) if there is a polynomial time bounded oracle Turing machine M and a polynomial q such that for any input x, $A(x) = M^B(x)$ and any string y queried about B by $M^B(x)$ is such that $q(|y|) \geq |x|$.

(5). For $r \in \{m, 1-tt, T\}$, A is $\leq^h_r$-equivalent to B ($A \equiv^h_r B$) if $A \leq^h_r B$ and $B \leq^h_r A$. The $\leq^h_r$-degree of a set A, denoted $deg^h_r(A)$ is $\{B : B \equiv^h_r A\}$. $DEG^h_r(\leq A) = \{deg^h_r(B) : B \leq^h_r A\}$.

(6). For $r \in \{m, 1-tt, T\}$, a set $A \notin P$ is said to be $\leq_r^h$-minimal if $\forall B(B \leq_r^h A \Longrightarrow B \in P$ or $A \leq_r^h B)$.

Several of the results in Section 3 bear upon the structure $< DEG_r^h(\leq A), \leq_r^h >$:= the $\leq_r^h$-degrees of sets $B \leq_r^h A$ partially ordered by the $\leq_r^h$ relation.

We fix enumerations M_i^S of polynomial time bounded, polynomially honest oracle Turing machines. For all i, we let p_i be a polynomial which witnesses both the polynomial time bound and the honesty of M_i^S for all oracle sets S. We think of M_i^S as the ith honest polynomial time bounded Turing reduction to a set S.

When dealing with relativizations we will often be considering relativized honest polynomial reductions. The reductions considered will again be one of the three types defined above, but in each case the reduction procedure will have access, in the usual way, to an oracle set. We will denote these relativized reductions by $\leq_r^{h,A}$ where A is the oracle set and $r \in \{m, 1-tt, T\}$.

Subproblems play an important role in our discussion. A set B is a subproblem of A if there is a set $F \in P$ such that $B = A \cap F$. Note that if B is a subproblem of A then $B \leq_m^h A$. $SUB(A)$ denotes the collection of all subproblems of the set A.

We assume familiarity with the low hierarchy from recursion theory (see Soare [So-87]). We say a degree (or a set) is *low* if it (or its degree) is in the class low_1.

3 EXPTALLY Sets

In this section we examine the structure of the honest polynomial degrees of *EXPTALLY* sets. We will see that these sets have a simple degree structure and that the results obtained generalize earlier work on honest reducibilities by Ambos-Spies and Homer.

We first define *EXPTALLY* sets and examine some properties of these sets as they relate to honest reductions.

Definition:

(i) $\delta(0) = 0$ and $\delta(n+1) = 2^{\delta(n)}$

(ii) $E = \{0^{\delta(n)} : n \geq 0\}$

(iii) A is *exptally* if $A \subseteq E$.

EXPTALLY is the class of all exptally sets.

The next few propositions show that sets which are reducible to $EXPTALLY$ sets are reducible via simple reductions.

Lemma 1 (1) Let A, B, C be given such that $A \in EXPTALLY$, $C \leq_T^h B$ and $B \leq_T^h A$. Then $C \leq_{1\text{-}tt}^h B$.
(2) If moreover, $C \leq_m^h A$ and $B \leq_m^h A$ then $C \leq_m^h B$.

Proof: Fix polynomially honest, polynomial time bounded oracle machines M_0 and M_1 such that $C = M_0^B$ and $B = M_1^A$. Note that, as $A \in EXPTALLY$, for all sufficiently long x there is at most one string of the form $0^{\delta(n)}$ which can be queried by M_1 on input x. It follows immediately that $B \leq_{1\text{-}tt}^h A$, say via reduction (g_1, h_1). (See [AS-87] for details).

We will now convert M_0^B to an $\leq_{1\text{-}tt}^h$-reduction by eliminating non-relevant oracle queries using the reduction (g_1, h_1). Call y relevant if $g_1(y) = 0^{\delta(n)}$ for some n and $h_1(y, 0) \neq h_1(y, 1)$. Note that for *non*relevant y, the query "$y \in B$?" is trivial since $B(y) = h_1(y, 0)$. Hence, without loss of generality we may assume that M_0^B queries only relevant strings. Moreover, for all sufficiently large x, namely those satisfying $p_1(p_0(|x|)) < 2^{p_1^{-1}(p_0^{-1}(|x|))}$, if y_1, y_2 are relevant queries of M_0^B on input x, then $g_1(y_1) = g_1(y_2)$, whence

$$B(y_2) = \begin{cases} B(y_1) & \text{if } h_1(y_1, 1) = h_1(y_2, 1) \\ 1 - B(y_1) & \text{if } h(y_1, 1) \neq h_1(y_2, 1) \end{cases}$$

So, in the reduction $M_0^B(x)$, all but the first relevant query can be eliminated, hence $C \leq_{1\text{-}tt}^h B$.

For a proof of (2), assume $C \leq_m^h A$ via f_0, $B \leq_m^h A$ via f_1, and $C \leq_{1\text{-}tt}^h B$ via (g, h). Then an honest reduction function f witnessing $C \leq_m^h B$ is defined by distinguishing the following polynomially decidable cases.

Case 1: $f_0(x) \notin E$

Then define

$$f(x) = \begin{cases} + & \text{if } f_0(x) = + \\ - & \text{otherwise} \end{cases}$$

Case 2: $f_0(x) \in E$

Case 2.1: $f_1(g(x)) = f_0(x) = 0^{\delta(n)}$

Let $f(x) = g(x)$.

Case 2.2: $f_1(g(x)) \neq f_0(x)$

Case 2.2.1: $f_1(g(x)) \notin E$.
Let

$$f(x) = \begin{cases} + & \text{if } (f_1(g(x)) = + \text{ and } h(g(x),1) = 1) \text{ or } (f_1(g(x)) \neq + \text{ and } h(g(x),0) = 1) \\ - & \text{otherwise} \end{cases}$$

Case 2.2.2: $f_1(g(x)) \in E$
Case 2.2.2.1: $f_1(g(x)) = 0^{\delta(n)} (= f_0(x))$
Let $f(x) = g(x)$.
Case 2.2.2.2: Otherwise
In this case, $f_1(g(x)) = 0^{\delta(m)} \neq 0^{\delta(n)} = f_0(x)$, for some m. This can happen only for finitely many strings x. Hence we define,

$$f(x) = \begin{cases} + & \text{if } x \in C \\ - & \text{otherwise} \end{cases}$$

■

Lemma 2 Let A and B be sets such that $B \leq^h_{1\text{-}tt} A$. There is a set D such that $D \equiv^h_{1\text{-}tt} B$ and $D \leq^h_m A$.

Proof: Fix an $\leq^h_{1\text{-}tt}$-reduction (g, h) of B to A. Then it is easy to check that
$D = \{x \in B : h(x,0) \leq h(x,1)\} \cup \{x \in \overline{B} : h(x,0) > h(x,1)\}$ has the desired properties. ■

The next Theorem shows that the structure of the honest degrees below an *EXPTALLY* set does not depend on which honest reducibility is considered.

Theorem 3 For $A \in EXPTALLY$,
$< DEG^h_T(\leq A), \leq^h_T > = < DEG^h_{1\text{-}tt}(\leq A), \leq^h_{1\text{-}tt} > \cong < DEG^h_m(\leq A), \leq^h_m >$.

Proof: The first equality is immediate by Lemma 1(1). For the second part, by Lemma 2 there is a total function $\Phi : \{B : B \leq^h_{1\text{-}tt} A\} \rightarrow \{B : B \leq^h_m A\}$ such that $\Phi(B) \equiv^h_{1\text{-}tt} B$. Then $B \leq^h_{1\text{-}tt} D \Longrightarrow \Phi(B) \leq^h_m \Phi(D)$, by the last part of Lemma 1. This Φ induces a homomorphism $\phi : < DEG^h_{1\text{-}tt}(\leq A), \leq^h_{1\text{-}tt} > \rightarrow < DEG^h_m(\leq A), \leq^h_m >$ (namely $\phi(deg^h_{1\text{-}tt}(B) := deg^h_m(\Phi(B))$. By definition of Φ, ϕ is one-to-one. Finally, if $B \leq^h_m A$, then by definition of Φ and by Lemma 1(2), $B \equiv^h_m \Phi(B)$, hence ϕ is onto. ■

Lemma 4 Let A be $EXPTALLY$. Then $< DEG^h_m(\leq A), \leq^h_m >$ is a distributive lattice.

Proof: Since the $\leq^h_m$ degrees form a distributive upper semi-lattice (see [AS-89]), it suffices to show that, for any $b_0, b_1 \in DEG^h_m$, $b_0 \wedge b_1$ exists. So fix B_i, f_i such that $B_i \leq^h_m A$ via f_i ($i = \{0,1\}$). We need to show there is a set D such that $D \leq^h_m B_i$ and $\forall C(C \leq^h_m B_i \implies C \leq^h_m D)$. Let $D = \{< x_0, x_1 >: f_0(x_0) = f_1(x_1) \in A\}$.
Then $D \leq^h_m B_i$ via

$$g_i(< x_0, x_1 >) = \begin{cases} x_i & \text{if } f_0(x_0) = f_1(x_1) \in E \\ - & \text{otherwise.} \end{cases}$$

Finally assume $C \leq^h_m B_i$ via h_i. Then $x \in C \iff h_i(x) \in B_i$ or $h_i(x) = + \iff f_i(h_i(x)) \in A$ or $f_i(h_i(x)) = +$ or $h_i(x) = +$ ($i = \{0,1\}$).
Moreover there is a number n such that for any string x of length $\geq n$, $f_0(h_0(x)), f_1(h_1(x)) \in E \implies f_0(h_0(x)) = f_1(h_1(x))$. Hence $C \leq^h_m D$ via k where for x with $|x| \geq n$,

$$k(x) = \begin{cases} + & \text{if } \exists i \leq 1 (h_i(x) = + \text{ or } f_i(h_i(x)) = +) \\ < h_0(x), h_1(x) > & \text{if } f_0(h_0(x)) = f_1(h_1(x)) \in E \\ - & \text{otherwise} \end{cases}$$

and $k(x) = C(x)$ for $|x| < n$. ∎

We next consider the structure of all subproblems of an $EXPTALLY$ set under $\leq^h_m$ reductions.

Lemma 5 (a). For any sets A, B_0, B_1 such that $B_0, B_1 \in P$:
(i). $B_0 \subseteq B_1 \implies deg^h_m(A \cap B_0) \leq^h_m deg^h_m(A \cap B_1)$
(ii). $deg^h_m(A \cap B_0) \vee deg^h_m(A \cap B_1) = deg^h_m(A \cap (B_0 \cup B_1))$
(iii). If $A \in EXPTALLY$ then
$deg^h_m(A \cap B_0) \wedge deg^h_m(A \cap B_1) = deg^h_m(A \cap (B_0 \cap B_1))$.
(b). For $A \in EXPTALLY$,
$< SUB(A), \leq^h_m >$ is a sublattice of $< DEG^h_m(\leq A), \leq^h_m >$
(c). For $A \in EXPTALLY$, $< SUB(A), \leq^h_m > \cong < SUB(E - A), \leq^h_m >$.

Proof: (a). (i) and (ii) are straightforward. To prove (iii), by (i) it suffices to show $(C \leq^h_m A \cap B_0$, and $C \leq^h_m A \cap B_1) \implies C \leq^h_m A \cap (B_0 \cap B_1)$. So assume $C \leq_m^h A \cap B_i$ via f_i. Then there is a number n such that for all x with $|x| \geq n$, $f_0(x), f_1(x) \in E \Longrightarrow f_0(x) = f_1(x)$. Hence $C \leq_m^h A \cap (B_0 \cap B_1)$ via f with

$$f(x) = \begin{cases} + & \text{if } |x| < n \text{ and } x \in C \text{ or } f_0(x) = + \text{ or } f_1(x) = + \\ f_0(x) & \text{if } f_0(x) = f_1(x) \in E \\ - & \text{otherwise} \end{cases}$$

(b). is immediate from (a).
(c). If $A \subseteq E$, $B_0, B_1 \in P$ and $A \cap B_0 \leq_m^h A \cap B_1$ via f then $(E - A) \cap B_0 \leq_m^h (E - A) \cap B_1$ via

$$g(x) = \begin{cases} \text{f(x)} & \text{if } x \in E \cap B_0 \text{ and } f(x) \in E \cap B_1 \\ + & \text{if } x \in E \cap B_0 \text{ and } f(x) \in \{-\} \cup (\Sigma^* - (E \cap B_1)) \\ - & \text{otherwise} \end{cases}$$

■

Lemma 6 For $A \in EXPTALLY$ with $A \notin P$, $< SUB(A), \leq_m^h >$is a Boolean algebra.

Proof: By the previous two lemmas, $< SUB(A), \leq_m^h >$is a distributive lattice. Hence it suffices to show that $< SUB(A), \leq_m^h >$is complemented. But, by parts (a)(ii) and (a)(iii) of Lemma 5, $deg_m^h(A \cap \overline{B})$ is the complement of $deg_m^h(A \cap B)$. ■

Lemma 7 Let $A \notin P$ be a recursive *EXPTALLY* set. Then $< SUB(A), \leq_m^h >$is the countable atomless Boolean algebra.

Proof: Obviously $SUB(A)$ is countable. Hence, by Lemma 6, it suffices to show that if $A \cap B \notin P$ there is a $D \in P$ such that $\emptyset <_m^h A \cap D <_m^h A \cap B$. The existence of such D has been shown by Ladner [La-75]. ■

In contrast to Lemma 7, we will see that all finite Boolean algebras can also be represented by the honest polynomial many-one degrees of subproblems of nonrecursive *EXPTALLY* sets.

Theorem 8 For any $k \geq 1$ there is an r.e. set $A \in EXPTALLY$ such that $< SUB(A), \leq_m^h >$is the k-atom Boolean algebra.

For the proof we will need to extend the recursion theoretic concept of cohesiveness to arbitrary complexity classes.

Definition: Let C be any class of sets. A set A is C-cohesive if
(i). A is infinite, and (ii). $\forall C \in C$ ($A \cap C$ is finite or $A \cap \overline{C}$ is finite).

For C we will consider the classes r.e., prim (= the primitive recursive sets), and P. Clearly A r.e.-cohesive $\Longrightarrow A$ prim-cohesive $\Longrightarrow A$ P-cohesive. Cohesive set have been well-studied in recursion theory. Martin [Ma-66] has shown: For any high r.e. set A there is an r.e. set B such that $A \equiv_T B$ and $\overline{B}$ is r.e.-cohesive.

In fact, all co-r.e. r.e.-cohesive sets have high degree. Whether every r.e.-cohesive set below $0'$ is high is an open question (see Soare [So-87]). Non-high prim-cohesive sets have been obtained by Jockusch. In fact, he has shown that there is a *low*$_2$ prim-cohesive set [Jo-89].

On the other hand, by Breitbart's splitting theorem [Br-78], no recursive set is P-cohesive. In fact, Downey's proof that no low set is $\leq_T^h$-minimal shows that no low set is P-cohesive[Do-89].

To apply these cohesiveness results we need the following definition.

Definition: For any set A define
$\mathrm{ET}(A) = \{0^{\delta(n)} : z_n \in A\}$ (Here z_n is the nth string in the canonical enumeration.)
$ET_{k,i}(A) = \{0^{\delta(kn+i)} : z_n \in A\}$
$ET^k(A) = \{0^{\delta(kn+i)} : z_n \in A, 0 \leq i < k\} = \bigcup_{0 \leq i < k} ET_{k,i}(A)$
The following proposition is straightforward.

Proposition 9 (a). [AS-87] A P-cohesive $\Longrightarrow \|< SUB(A), \leq_m^h >\| = 2$.
(b). A prim-cohesive $\Longrightarrow ET_{k,i}(A)$ prim-cohesive.

Lemma 10 Let A be prim-cohesive. Then $< SUB(ET^k(A)), \leq_m^h >$ is the k-atom Boolean algebra.

Proof: By Lemma 6, $< SUB(ET^k(A)), \leq_m^h >$ is a Boolean algebra, so it suffices to show that there are degrees $a_0, \ldots, a_{k-1} \in SUB(ET^k(A))$ such that
(i). $i \neq j \Longrightarrow a_i$ and a_j are incomparable with respect to $\leq_m^h$.

(ii). $\forall b \in SUB(ET^k(A)) \exists \alpha \subseteq \{0,..,k-1\}(b = \bigvee_{i\in\alpha} a_i)$.
Let $a_i = deg_m^h(ET_{k,i}(A))$ and $E_{k,i} = \{0^{\delta(kn+i)} : n \geq 0\} \in P$. Then $ET_{k,i}(A) = ET^k(A) \cap E_{k,i}$ and $a_i \in SUB(ET^k(A))$.

For a proof of (i) assume, for a contradiction, that $i \neq j$ and $a_i \leq_m^h a_j$. Then by Lemma 5(iii) $ET_{k,i}(A) \equiv_m^h ET_{k,i} \cap ET_{k,j} = \emptyset$. Hence $ET_{k,i}(A) \in P$. Since $ET_{k,i}(A)$ is prim-cohesive, this contradicts Breidbart's result [Br-78] that no low set is P-cohesive.

Finally, for a proof of (ii) fix a subproblem $ET^k(A) \cap B$ of $ET^k(A)$ and let $b = deg_m^h(ET^k(A) \cap B)$. Then, since the P sets $E_{k,0}, ..., E_{k,k-1}$ partition E, $ET^k(A) \cap B \equiv_m^h \bigoplus_{i=0}^{k-1}(ET^k(A) \cap E_{k,i} \cap B) = \bigoplus_{i=0}^{k-1} ET_{k,i}(A) \cap B$.
By Proposition 9(b) and since A prim-cohesive $\Longrightarrow$ A P-cohesive, $ET_{k,i}(A)$ is P-cohesive, whence either
$ET_{k,i}(A) \cap B \equiv_m^h ET_{k,i}(A)$, or $ET_{k,i}(A) \cap B \equiv_m^h \emptyset$. Now for
$\alpha = \{i : (0 \leq i < k) \wedge ET_{k,i}(A) \cap B \equiv_m^h ET_{k,i}(A)\}$, $ET^k(A) \cap B = \bigoplus_{i\in\alpha} ET_{k,i}(A)$, that is $b = \bigvee_{i\in\alpha} a_i$.

We are now ready to prove Theorem 8.

Proof of Theorem 8: By Martin's Theorem on cohesive sets, there is an r.e. set B such that $\overline{B}$ is r.e.-cohesive, and hence prim-cohesive. Let $C = ET^k(\overline{B})$. Then, by Lemma 10, $< SUB(C), \leq_m^h >$ is the k-atom Boolean algebra. Moreover, $A := E - C$ is r.e. and, by Lemma 5(c), $< SUB(A), \leq_m^h > \cong < SUB(C), \leq_m^h >$. ∎

As observed by Ambos-Spies [AS-87], $P = NP$ implies that every set A is subproblem complete under $\leq_m^h$. That is, $B \leq_m^h A \Longrightarrow B \equiv_m^h A \cap C$, for some $C \in P$. Hence, in this case, $< DEG_m^h(\leq A), \leq_m^h > = < SUB(A), \leq_m^h >$. So, assuming $P = NP$, Theorem 8 extends to the structure $< DEG_m^h(\leq A), \leq_m^h >$ and therefore, by Theorem 3, to $< DEG_T^h(\leq A), \leq_T^h >$.

The following lemma pins down the points at which the assumption $P = NP$ is used to prove this extension of Theorem 8. This analysis will be crucial for the relativization results of the next sections.

First some notation.
Let $f : \Sigma^* \longrightarrow \Sigma^* \cup \{+, -\}$ be honest and P-computable. Then, $range_E(f) := \{f(x) : x \in \Sigma^*\} \cap E$ and $IG_E(f) := \{< x, 0^{\delta(n)} >: \exists y(f(xy) = 0^{\delta(n)})\}$.

By the honesty of f, the existential quantifier in the definition of $IG_E(f)$ is polynomially bounded, whence $IG_E(f) \in NP$. Moreover, $range_E(f) \leq_m^h IG_E(f)$ This implies that $range_E(f) \in NP$ as well.

Lemma 11 Let A, B, and f be given such that $A \subseteq E$ and $B \leq_m^h A$ via f. Then
(i). $B \leq_m^h A \cap range_E(f)$
(ii). $A \cap range_E(f) \leq_m^{h,I} B$, where $I = IG_E(f)$.
In particular, if $P = NP$ then $range_E(f) \in P$ and $B \equiv_m^h A \cap range_E(f)$.

Proof: (i). $B \leq_m^h A \cap range_E(f)$ via f.
(ii). A honest, P^I-computable function $g : \Sigma^* \longrightarrow \Sigma^* \cup \{+,-\}$ which reduces $A \cap range_E(f)$ to B is defined by,

$$g(x) = \begin{cases} - & \text{if } x \notin range_E(f) \\ y & \text{if } x \in range_E(f) \text{ and } y \text{ is the first string such that } f(y) = x. \end{cases}$$

Note that $x \in range_E(f)$ is P^I-decidable since $range_E(f) \leq_m^p I$, and if $x \in range_E(f)$ then a standard prefix search with oracle I yields the desired string y.

The last part of the proof follows from (i) and (ii) since $P = NP$ implies that $range_E(f) \in P$ and $I \in P$. ∎

Using the previous results we now prove a theorem characterizing the structure of $< DEG_r^h(\leq A), \leq_r^h >$ for *EXPTALLY* sets A. This can be viewed as generalizing several of the original results concerning the structure of honest reductions.

Theorem 12 Assume $P = NP$. Let $r \in \{m, 1-tt, T\}$.
(a). For any *EXPTALLY* set $A \notin P$, $< DEG_r^h(\leq A), \leq_r^h >$ is a Boolean algebra.
(b). For any **recursive** *EXPTALLY* set $A \notin P$, $< DEG_r^h(\leq A), \leq_r^h >$ is the countable atomless Boolean algebra. In particular, for any recursive *EXPTALLY* sets $A_0, A_1 \notin P$, $< deg_r^h(\leq A_0), \leq_r^h > \cong < deg_r^h(\leq A_1), \leq_r^h >$.
(c). For any $k \geq 1$, there is an r.e. (*EXPTALLY*) set A_k such that $< deg_r^h(\leq A_k), \leq_r^h >$ is the k-atom Boolean algebra. In fact, any high r.e. Turing degree contains such a set.
(d). For any $k \geq 1$, there is a low_2 *EXPTALLY* set A_k such that $< deg_r^h(\leq A_k), \leq_r^h >$ is the k-atom Boolean algebra.

Proof: By Theorem 3 and Lemma 11,
$< DEG_r^h(\leq A), \leq_r^h > \cong < SUB(A), \leq_m^h >$ for any *EXPTALLY* set A. Hence (a), (b) and the first part of (c) are immediate by Lemma 6, Lemma 7 and

Theorem 8, respectively. The second part of (c) and (d) require straightforward extensions of Theorem 8 based on the corresponding results of Martin [Ma-66] and Jockusch [Jo-89] on cohesive sets. ∎

Theorem 12 unifies and extends the results on $\leq_r^h$-minimal sets in [AS-87, Ho-87, LH-87]. Recall that a set A is $\leq_r^h$-minimal if $A \notin P$ and $\forall B(B \leq_r^h A \Longrightarrow B \in P$ or $B \equiv_r^h A)$. Obviously A is $\leq_r^h$-minimal if $< deg_r^h(\leq A), \leq_r^h >$ is the 1-atom Boolean algebra. Thus, parts (c) and (d) subsume results on $\leq_r^h$-minimal sets.

Homer showed that $\leq_r^h$-minmal sets cannot be recursive. Downey ([Do-89]) has extended this by showing that no low_1 set is $\leq_r^h$-minimal. By part (d) of our Theorem this result is optimal with respect to the jump hierarchy (assuming $P = NP$). It refutes a conjecture of Downey that the $\leq_r^h$-minimal sets must be high.

$\leq_r^h$-minimal sets demonstrate a difference in the algebraic structures of the $\leq_r^h$-degrees and the $\leq_r^p$-degrees (assuming $P = NP$). By the above observation, this difference does not occur in the recursive sets. Ambos-Spies [AS-89] has proved a difference on the recursive sets, assuming $P = NP$. He has shown that for any recursive set $A \notin P$, $< deg_r^p(\leq A), \leq_r^p >$ is not complemented. Further, he constructed a recursive set $A \notin P$ such that (if $P = NP$), $< deg_r^h(\leq A), \leq_r^h >$ is complemented. This latter result is subsumed by our Theorem 12(b) which proves this for arbitrary $EXPTALLY$ sets.

The second part of (b) shows that, assuming $P = NP$, the complexity of $< DEG_r^h(\leq A), \leq_r^h >$ is not, in general, related to the complexity of A, since we can find $EXPTALLY$ sets of arbitrarily high complexity.

4 An Oracle Relative to which $P \neq NP$ and There Exist $\leq_r^h$–Minimal Sets

There have been a number of different constructions of sets of minimal degree for honest polynomial reductions. All of these have used the assumption that $P = NP$. One crucial question in this study is whether this assumption is necessary and in particular whether the existence of any of these $\leq_r^h$–minimal sets is equivalent to $P = NP$. Here we give a negative answer to these questions relative to some oracle. We do this by constructing a set A such that relative to A, $P \neq NP$ and yet Theorem 12(c) above still holds, so in particular there exist $\leq_r^h$ minimal sets. In fact, as we will show later, our oracle is such that parts (a) and (b) of Theorem 12 fail.

The following Lemma is the key to our construction.

Definition: For any function $f : \mathcal{N} \Longrightarrow \mathcal{N}$ let
$CL_f(E) = \{x \in \Sigma^* : \exists n(|x| \leq \delta(n) < f(|x|) \text{ or } \delta(n) \leq |x| < f(\delta(n)))\}$.
A set $A \subseteq \Sigma^*$ is called E-close if there is a polynomial p such that $A \subseteq CL_p(E)$. Note that $CL_p(E) \in P$ for any polynomial p.

Lemma 13 There is a recursive set A such that $P^A \neq NP^A$ and for any E-close set $B \in NP^A, B \in P^A$.

Proof: The proof combines the strategies of Baker Gill and Solovay [BGS-75] for constructing oracles B and C such that $P^B = NP^B$ and $P^C \neq NP^C$. Both strategies consist of a sequence of "local steps". By alternating the two strategies in an appropriate way, we can construct an oracle A which collapses NP^A to P^A on certain intervals so that we can satisfy our condition for E-close sets, but still does not totally collapse NP^A to P^A.

Recall that for any oracle X we have an NP^X complete set
$K^X = \{< e, x, 0^n >: N_e^X \text{ accepts } x \text{ in less than } n \text{ steps}\}$. Here N_e^X is the eth nondeterministic oracle Turing machine in some standard enumeration. See [BGS-75] for a proof of this fact. Now note that if $B \in NP^X$ is E-close, say $B \subseteq CL_q(E)$, then B is $\leq_m^h$-reducible to some E-close subproblem of K^X. Namely, if r is a polynomial which is both the time bound of the reduction f which reduces B to K^X as well as witnessing the honesty of f, then for any polynomial $s(n) \geq \max(q(r(n)), r(q(n)))$, $B \leq_m^h K^X \cap CL_s(E)$ via

$$g(x) = \begin{cases} f(x) & \text{if } x \in CL_q(E) \\ - & \text{otherwise} \end{cases}$$

Hence, to satisfy the second part of the Lemma it suffices to meet the requirements

$Q_e : K^A \cap CL_{p_e}(E) \in P^A$, for all $e \geq 0$, where $p_e(n) = n^e + e$.

To meet Q_e, we let K^A look (almost) like A in $CL_{p_e}(E)$. To be more precise, we will ensure,

$Q'_e : \exists n_e \forall x(|x| > n_e \rightarrow (K^A \cap CL_{p_e}(E))(x) = (A \cap CL_{p_e}(E))(x))$.

Clearly, Q'_e implies Q_e.

To ensure that $P^A \neq NP^A$ we define the NP^A set
$L^A = \{O^n : \exists x(x \in A \text{ and } |x| = n)\}$. Our construction will guarantee that L^A is not in P^A. To this end we have requirements
$R_e : L^A \neq M_e^A$, for all $e \geq 0$.
Recall that M_e^A is the eth deterministic polynomial time oracle Turing machine and has honesty and time bound p_e. As is customary M_e^A is identified with the characteristic function of the set of strings it accepts.

Since Q'_e completely determines $A(x)$ for all $x \in CL_{p_e}(E)$ of length greater than n_e, we have to choose the numbers n_e so that for any polynomial p there are infinitely many intervals $\{x : n \leq |x| \leq p(n)\}$ on which the Q'_e requirements do not determine the value of $A(x)$. We can then define A on these intervals to meet the R_e requirements.

To this end let $h(0) = 0$ and $h(n+1) = \max\{e : p_e(p_e(p_e(\delta(n)))) < \delta(n+1)\}$. It is easy to check that h is recursive, nondecreasing and unbounded.

Now let $lb(0) = \delta(0) = ub(0)$ and
$lb(n+1) = \max\{m : p_{h(n+1)}(m) \leq \delta(n+1)\}$, $ub(n+1) = p_{h(n+1)}(\delta(n+1))$, and let $U = \{x : \exists n > 0(lb(n) < |x| < ub(n))\}$. Then U is recursive,
(*) $CL_{p_e}(E) - U$ is finite for all e,
and
(**) $\forall e \forall n \exists m \geq n(\{x : m \leq |x| \leq p_e(m)\} \cap U = \emptyset)$.

Now by (*), $K^A \cap U = A \cap U$ will guarantee that all Q'_e requirements are met, while (**) leaves room for meeting the diagonalization requirements R_e.

We can now describe the construction of A. At stage s of the construction $A(x)$ is defined for strings of length s. Simultaneously with A we define a recursive set S containing the indices of those R_e requirements which have been satisfied. Let A_s and S_s denote the strings and numbers put into A and S, respectively, by the end of stage s. Note that by construction $A_s = \{x : x \in A \text{ and } |x| \leq s\} =: A^{\leq s}$. Initially we set $A_0 = S_0 = \emptyset$.

Stage $s+1$:

Case 1: $0^{s+1} \in U$

Then let $A(x) = K^{A_s}(x)$ for all strings x of length $s+1$ and set $S_{s+1} = S_s$.

Case 2: $0^{s+1} \notin U$ and $0^s \in U$.

Then fix the least number e such that $e \notin S_s, p_e(s+1) < 2^{s+1}$ and $\{x : s+1 \leq |x| \leq p_e(s+1)\} \cap U = \emptyset$. (If no such number e exists then let $A(x) = 0$ for all x of length $s+1$ and $S_{s+1} = S_s$.) Let $M_e^{A_s}(0^{s+1}) = i \in \{0,1\}$. If $i = 0$ then choose the least string y of length $s+1$ not queried by $M_e^{A_s}$ on input 0^{s+1} and let $A_{s+1} = A_s \cup \{y\}$. Otherwise let $A_{s+1} = A_s$. In either case let $S_{s+1} = S_s \cup \{e\}$.

Case 3: $0^{s+1} \notin U$ and $0^s \notin U$

Then let $A_{s+1} = A_s$ and $S_{s+1} = S_s$.

This completes the construction.

To show that A has the desired properties we first observe that the construction is effective and that during the construction we fix increasingly long initial segments of A, hence A is recursive. Moreover, by the definition of K^X, $K^A(x) = K^{A^{<|x|}}(x)$, so Case 1 of the construction implies that $K^A \cap U = A \cap U$. From this we get the second part of the Lemma. Finally, to show that $P^A \neq NP^A$ holds, it suffices to show that all requirements R_e are met. For a contradiction assume that R_e is not satisfied. There are two cases.

Case 1: $e \in S$

Fix s such that $e \in S_{s+1} - S_s$. Then by construction $L^A(0^{s+1}) = L^{A_{s+1}}(0^{s+1}) \neq M_e^{A_s}(0^{s+1}) = M_e^{A_{s+1}}(0^{s+1})$. Moreover, Case 3 of the construction applies to all stages t with $s+1 < t \leq p_e(s+1)$. It follows that $A^{\leq p_e(s+1)} = A_{p_e(s+1)} = A_{s+1}$, and so $M_e^A(0^{s+1}) = M_e^{A_{s+1}}(0^{s+1}) \neq L^A(0^{s+1})$, a contradiction.

Case 2: $e \notin S$

In this case there is a stage t such that

(i) $\forall e' \leq e(e' \in S \Longrightarrow e' \in S_t)$, and

(ii) $\forall s \geq t(p_e(s) < 2^s)$.

By (**), and since U is infinite, there is a stage $s+1 > t$ such that $0^s \in U$ and $\{x : s+1 \leq |x| \leq p_e(s+1)\} \cap U = \emptyset$. Hence by case 2 of the construction e or a smaller e' will enter S at stage $s+1$, i.e. $S_s \cap \{0, 1, ..., e\} \neq S_{s+1} \cap \{0, 1, .., e\}$ contrary to (i).

This completes the proof of the Lemma. ∎

Using the lemma we can now prove the main theorem of this section.

Theorem 14 There is a recursive set A such that

(i) $P^A \neq NP^A$,

and for any $r \in \{m, 1-tt, T\}$ the following hold,

(ii) if $B \notin P^A$ is *EXPTALLY* , then $< DEG_r^{h,A}(\leq B), \leq_r^{h,A} >$ is a Boolean algebra.

(iii) If $B \notin P^A$ is *EXPTALLY* and recursive, then

$< DEG_r^{h,A}(\leq B), \leq_r^{h,A} >$ is the atomless countably infinite Boolean algebra.

(iv) For any $k \geq 1$ there is an r.e (*EXPTALLY*) set A_k such that

$< DEG^{h,A}(\leq A_k), \leq_r^{h,A} >$ is the k-atom Boolean algebra. In particular, there is an r.e. $(EXPTALLY)$ set A_1 such that A_1 is $\leq_r^{h,A}$-minimal.

Note: The rest of theorem 12 holds here as well.

Proof: Let A as in the previous Lemma. Then (i) holds. For the other three properties we first note that the results in Section 3 relativize. Hence it is sufficient to replace the application of Lemma 11 (which used the $P = NP$ assumption) in the proof of Theorem 12 with an application of the following Lemma.

Lemma 15 Let A be as in Lemma 13 and let C, B, f be such that $B \subseteq E$ and $C \leq_m^{h,A} B$ via f. Then $C \equiv_m^{h,A} B \cap D$ for some $D \in P^A$.

Proof: As in the absolute case, $range_E(f)$ and $IG_E(f)$ are in NP^A. Moreover, these sets are E-close, whence in fact $range_E(f)$ and $IG_E(f)$ are in P^A. So we can argue that $C \equiv_m^{h,A} B \cap D$ for $D = range_E(f)$ as in the proof of Lemma 11. ∎

5 An Oracle Relative to Which There are No Minimal Degrees

The results of the previous sections might lead one to believe that sets of minimal degree exist outright and we can eliminate the $P = NP$ assumption which, to this point, has been needed in their proof. While this may be the case, we show here that this is not true relative to all oracles. Namely, we construct an oracle relative to which no set is $\leq_m^h$-minimal. Hence the question of the existence of $\leq_m^h$-minimal sets is oracle dependent.

Theorem 16 There is a recursive set A such that no set B is $\leq_m^{h,A}$-minimal.

Proof: Fix a recursive enumeration $\{f_n^A\}$ of P^A-time functions mapping Σ^* to $\Sigma^* \cup \{+, -\}$. A will be constructed to satisfy the following requirements.

$$P_n : \forall x(|x| = n)\ [\exists y(|y| = |x|^2 \text{ and } xy \in A)]$$

$$N_n : \exists s_n \forall x(|x| \geq s_n)(|f_n^A(x)| \geq |x|^2 + |x| \Longrightarrow f_n^A(x) \notin A)$$

Claim: If a recursive set A satisfies all P_n and N_n then no set B is $\leq_m^{h,A}$–minimal

Proof: Fix $B \subseteq \Sigma^*$. Without loss of generality we can assume B is nonrecursive. (If B is recursive use Ladner's splitting Theorem, which relativizes.) Let $C = \{xy : |y| = |x|^2 \text{ and } xy \in A \text{ and } x \in B\}$.
(1) $C \leq_m^{h,A} B$
Proof: $z \in C \iff z = xy$ and $|x|^2 = y$ and $xy \in A$ and $x \in B$. So the reduction is,

$$g(z) = \begin{cases} x & \text{if } z = xy, |x|^2 = |y| \text{ and } z \in A \\ - & \text{otherwise.} \end{cases} \qquad \blacksquare$$

(2)$B \leq_m C$ (Here $\leq_m$ is the recursive many-one reduction.)
Proof: To check if $x \in B$, consider all y with $|y| = |x|^2$. Find the least such y with $xy \in A$ (such exist by $P_{|x|}$). Then $x \in B \iff xy \in C$. ∎
Hence, by (2) and since B is not recursive, C is nonrecursive. As A is recursive, C is not in P^A.
(3) $B \not\leq_m^{h,A} C$
Proof: Assume $B \leq_m^{h,A} C$ via f_n^A for a contradiction.
Then $x \in B \iff f_n^A(x) \in C \cup \{+\}$.
Case 1: $\exists x(|x| \geq s_n, x \in B$ and $|f_n^A(x)| \geq |x|^2 + |x|)$. Then by N_n, $f_n^A(x) \notin A$. This then implies $f_n^A(x) \notin C \cup \{+\}$, a contradiction.
Case 2: $\forall x(|x| \geq s_n$ and $x \in B \Longrightarrow |f_n^A(x)| < |x|^2 + |x|.)$
In this case we show that B is recursive via the following algorithm.

Input x.
If $|x| < s_n$, use table look-up to see if $x \in B$.
If $|x| \geq s_n$ then compute $f_n^A(x)$.
If $f_n^A(x) \in \{+,-\}$ then output $f_n^A(x)$.
If $|f_n^A(x)| \geq |x|^2 + |x|$ then output $-$.
If $|f_n^A(x)| < |x|^2 + |x|$, see if $f_n^A(x) = uv$ with $|u|^2 = |v|$.
If there are no such u, v, output $-$.
If there exists such u, v, then $x \in B \iff uv \in A$ and $u \in B$.
If $uv \notin A$ then output $-$ (Recall that A is recursive.),
otherwise input u to the algorithm.

Now, note that if u is found as in the algorithm, $|u| < |x|$. So the algorithm is repeated at most $|x|$ times on input x. It is straightforward to check that the algorithm always halts and accepts B, a contradiction. ∎

By (1), (2), and (3) we see that $C <_m^{h,A} B$ and $C \notin P^A$ and so B is not minimal.

We are left to construct A satisfying P_n and N_n, for all n. Let $h : \mathcal{N} \Longrightarrow \mathcal{N}$ be a nondecreasing, unbounded function such that $\Sigma_{m \leq n}(\Sigma_{i<h(m)} 2^m p_i(m) + 2^m) < 2^{n^2}$. (Recall that p_i is the time bound for f_i.)

At stage n we determine all elements of A of length less than $(n+1)^2 + (n+1)$. For each x with $|x| = n$ we add exactly one string xy with $|x|^2 = |y|$, to A. Simultaneously we define a restraint set R_n of strings which we never allow to enter A.

Initially $A_0 = R_0 = \emptyset$.
Stage n: (Given $A_{n-1} = A^{<n^2+n}$ and R_{n-1}.)
Let $R_n = \{x : |x| \geq n^2 + n$ and $\exists i < h(n) \exists y (|y| = n$ and x is used negatively in $f_i^{A_{n-1}}(y)$ or $f_i^{A_{n-1}}(y) = x\} \cup R_{n-1}$.
For each x of length n, find the least y, $|y| = |x|^2$ such that $xy \notin R_n$ and add xy to A.
This ends the construction.

Note that $||R_n - R_{n-1}|| \leq \Sigma_{i<h(n)} 2^n + p_i(n) 2^n$ and so $||R_n|| \leq \Sigma_{m \leq n} \Sigma_{i<h(m)} 2^m + p_i(m) 2^m$, hence, by the definition of h, $||R_n|| < 2^{n^2}$. For x with $|x| = n$, there are 2^{n^2} strings y with $|y| = |x|^2$. So some xy with $|y| = |x|^2$ will not be in R_n and so can be added to A. Thus A satisfies P_n.

To see that A satisfies N_n, set s_n = the least stage such that $h(s_n) > n$. Then for any y with $|y| \geq s_n$ and any $s \geq |y|$ we have $f_n^{A_{s-1}}(y) = f_n^A(y)$ and $|f_n^{A_{s-1}}(y)| \geq |y|^2 + |y| \Longrightarrow f_n^{A_{s-1}}(y) \notin A$, since $R_n \cap A = \emptyset$. ∎

It is important to note that while this proof extends to some other honest polynomial time reducibilities, it does not seem to apply to $\leq_T^h$. The key difference seems to be that the proof works only for non-adaptive reducibilities, while $\leq_T^h$ is adaptive. In particular, a very similar argument suffices to construct an oracle relative to which there are no minimal sets with respect to $\leq_{tt}^h$. For $\leq_T^h$ we can show the weaker result that there is an oracle relative to which no **tally** set is minimal. Again the proof is along the lines of the above argument.

Finally, the proof of the Theorem can be modified to show that relative to the same oracle A constructed there, parts (a) and (b) of Theorem 12 fail.

Corollary 17 There is a recursive set A such that for any set $B \notin NP^A$, $< DEG_m^{h,A}(\leq B), \leq_m^{h,A}>$ does not form a Boolean algebra.

Note that for any $EXPTALLY\, B$, the conclusion holds for $< DEG_T^{h,A}(\leq B), \leq_T^{h,A}>$ as well by the relativized version of Theorem 3.
Proof: Fix A as constructed in the proof of the last Theorem and let B be any set not in NP^A. Let $C = \{xy : |y| = |x|^2 \wedge xy \in A \wedge x \in B\}$.
Then, as shown in the above proof,
(1). $C \leq_m^{h,A} B$ via a function f such that for any x, $|f(x)| + |f(x)|^2 = |x|$ or $f(x) \in \{+, -\}$.
Moreover, note that,
(2). $C \notin P^A$ since, by requirements P_n and by the definition of C, $x \in B \iff \exists y(|y| = |x|^2 \wedge xy \in C)$. Hence $C \in P^A$ would imply $B \in NP^A$, contrary to assumption.

We will show that for any set D,
(3) $B \leq_m^{h,A} C \oplus D \Longrightarrow \exists E \notin P^A(E \leq_m^{h,A} C, D)$.
One can easily check that (1) and (3) impy that the element $deg_m^{h,A}(C)$ of $DEG_m^{h,A}(\leq B)$ has no complement in $< DEG_m^{h,A}(\leq B), \leq_m^{h,A} >$, whence the latter is not a Boolean algebra.

For a proof of (3), fix D such that $B \leq_m^{h,A} C \oplus D$, say via h. By (1), choose f such that $C \leq_m^{h,A} B$ via f and $|f(x)| + |f(x)|^2 = |x|$ if $f(x) \notin \{+, -\}$. Let $F = \{xy : |y| = |x|^2 \wedge xy \in A \wedge h(f(xy)) \in 1\Sigma^*\}$ and $E = C \cap F$. Note that E contains the elements of C which are reduced to the D side of $C \oplus D$ in the reduction $C \leq_m^{h,A} C \oplus D$ via $h \circ f$. Obviously $F \in P^A$ and so $E \leq_m^{h,A} C$ via

$$k_0(z) = \begin{cases} z & \text{if } z \in F \\ - & \text{otherwise} \end{cases}$$

Moreover, $E \leq_m^{h,A} D$ via

$$k_1(z) = \begin{cases} w & \text{if } z \in F \text{ and } h(f(xy)) = 1w \\ - & \text{otherwise} \end{cases}$$

It remains to show that $E \notin P^A$. Since, by (2), $C \notin P^A$, it suffices to show that (4) $C \leq_m^A E$. (Note that here the reduction is not required to be honest.)
Proof of (4): Let $G = \{xy : |y| = |x|^2 \wedge xy \in A \wedge h(f(xy)) \in 0\Sigma^*\}$.
Note that $G \in P^A$ and $C \equiv_m^A (C \cap G) \oplus (C \cap F)$. (Namely, for $z \notin F \cup G, z \in$

$C \iff z = xy$, for some xy such that $|y| = |x|^2, xy \in A$ and $h(f(xy)) = +$ (where $h(+) = +$).)

Hence it suffices to show $C \cap G \leq_m^A E$. Let

$$j(x) = \begin{cases} w & \text{if } h(x) = 0w \\ - & \text{otherwise} \end{cases}$$

and

$$k(z) = \begin{cases} j(f(z)) & \text{if } z \in G \\ - & \text{otherwise} \end{cases}$$

Then (5) $C \cap G \leq_m^{h,A} C$ via k. By the requirements N_n, there is a number s such that $|x| \geq s \Longrightarrow |j(x)| < |x|^2 + |x|$ or $j(x) \notin A$. Since $|z| = |f(z)|^2 + |f(z)|$, this implies $|z| \geq s^2 + s$ which implies $|k(z)| < |z|$ or $k(z) \notin A$. Hence there is a number n such that for any string z,

(6) $n \leq |z| \Longrightarrow |k^n(z)| < s^2 + s$ or $k^n(z) \notin A$.

Now $C \cap G \leq_m^A E$ as follows. Without loss of generality it suffices to compute $(C \cap G)(z)$ for $z \in G$ and $|z| \geq s^2 + s$. Let n be as in (6). Then $|k^n(z)| < s^2 + s$ or $k^n(z) \notin G$ (Note that $G \subseteq A$). By the assumption on z, $n > 0$. From (5) it follows that $(C \cap G)(z) = (C \cap G)(k^{n-1}(z))$, and $k^{n-1}(z) \in C \cap G \iff k^n(z) \in C$ or $k^n(z) = +$. Then $(C \cap G)(z)$ can be computed according to the following cases.

(i). $k^n(z) = - \Longrightarrow (C \cap G)(z) = 0$.
(ii). $k^n(z) = + \Longrightarrow (C \cap G)(z) = 1$.
(iii). $k^n(z) \in \Sigma^* - A \Longrightarrow k^n(z) \notin C \Longrightarrow (C \cap G)(z) = 0$.
(iv). $k^n(z) \in A - (F \cup G) \Longrightarrow (C \cap G)(z) = C(k^n(z)) = (C \cap (A - (F \cup G)))(k^n(z)))$ and $C \cap (A - (F \cup G)) \in P^A$.
(v). $k^n(z) \in A \cap F \Longrightarrow (C \cap G)(z) = (C \cap F)(k^n(z)) = E(k^n(z))$.
(vi). $k^n(z) \in A \cap G \Longrightarrow |k^n(z)| < s^2 + s$. So $(C \cap G)(z) = C(k^n(z))$ can be computed by table look-up. ∎

6 Conclusions and Open Problems

Progress has been made in resolving two of the major open problems concerning the existence of honest minimal degrees. Assuming $P = NP$, the determination of which Turing degrees contain minimal sets has been essentially resolved. One small question left is to decide exactly which Turing degrees which are not low_1 contain minimal sets.

The question of the relationship between the existence of minimal degrees and the $P = NP$ problem appears considerably more difficult. For $\leq_m^h$ or for $\leq_{tt}^h$ or for tally sets with respect to $\leq_T^h$ we have shown that this question is oracle dependent. While oracle dependence does not settle this relationship, it indicates that there are considerable difficulties in settling it. One remaining problem is the construction of an oracle relative to which no $\leq_T^h$-degree is minimal.

References

[AS-87] K. Ambos-Spies, *Honest Polynomial Reducibilities, Recursively Enumerable Sets, and the P=?NP Problem*, Second Annual Structure in Complexity Theory Conference, 60-68.

[AS-89] K. Ambos-Spies, *Honest Polynomial Time Reducibilities and the P=?NP Problem*, JCSS 39, 250-281.

[BGS-75] T. Baker, J. Gill and R. Solovay, *Relativizations of the P=?NP Question*, SIAM Journal of Computer Science 1, 305–322.

[Br-78] S.I. Breidbart, *On Splitting Recursive Sets*, JCSS 17, 56-64.

[Do-89] R. Downey, *On Computational Complexity and Honest Polynomial Degrees*, Theoretical Computer Science, to appear.

[DM-89] R. Downey and M. Moses, *On the Structure of Honest Poly-m-degrees*, Manuscript.

[DGHM-89] R. Downey, W. Gasarch, S. Homer and M. Moses, *On Honest Polynomial Reductions, Relativizations, and P = NP*, Fourth Annual Structure in Complexity Theory Conference, 196-207.

[GH-87] W.I. Gasarch and S. Homer, *Recursion Theoretic Properties of Minimal Honest Polynomial Degrees*, University of Maryland Computer Science Department, TR 1803. Also Boston University TR 87-005.

[Ho-85] S. Homer, *Minimal Polynomial Degrees of Non-Recursive Sets*, Lecture Notes in Mathematics No. 1141 (Recursion Theory Week), 193–202.

[Ho-87] S. Homer, *Minimal Degrees for Polynomial Reducibilities*, Journal of the Association of Computing Machinery 34, 480–491.

[Jo-89] C. Jockusch, personal communication.

[La-75] R. Ladner, *On the Structure of Polynomial Time Reducibility*, Journal of the Assoc. for Computing Mach. 22, No. 1, 155–171.

[LH-87] T.J. Long and S. Homer, *Honest Polynomial Degrees and P=?NP*, Theoretical Computer Science 51, 265–280.

[Ma-66] D.A. Martin, *Classes of Recursively Enumerable Sets and Degrees of Unsolvability*, Z. Math. Logik Grundlag. Math. 12, 295-310.

[So-87] R.I. Soare, *Recursively Enumerable Sets and Degrees*, Springer Verlag (Ω Series), Berlin and New York, 1987.

ON THE STRUCTURE OF DEGREES BELOW O'

Marat M. Arslanov
Kazan State University, Lenin str.
18,420087, Kazan, USSR

We shall examine the Boolean algebra of sets which is generated by the recursively enumerable (r.e.) sets. A set $A \subseteq \omega$ is called n-r.e. if $A = \lim_s A_s$ for some recursive sequence $\{A_s\}_{s\in\omega}$ such that for all x $A_0(x) = 0$ and card $\{s: A_s(x) \neq A_{s+1}(x)\} \leq n$. A set A is called ω-r.e. if for some recursive function f and for all x card $\{s: A_s(x) \neq A_{s+1}(x)\} \leq f(x)$ (the set A is f-r.e.). Obviously, the only 0-r.e. set is $\emptyset$, the 1-r.e. sets are the usual r.e. sets, and the 2-r.e. sets are the differences of r.e. sets (d-r.e.sets). For $1 \leq \alpha \leq \omega$ a Turing degree is called α-r.e. if it contains an α-r.e. set; it is called properly α-r.e. if it is α-r.e. but not β-r.e. for any $\beta < \alpha$.

Our notation and terminology are mostly standard. They are summarized below.

Weshall write $\mathcal{D}_\alpha$, $1 \leq \alpha \leq \omega$, for the upper semilattice of (Turing) degrees which contain α-r.e.-sets, partially ordered by the relation $\leq$ of Turing reducibility. For degrees $\underset{\sim}{a}$ and $\underset{\sim}{b}$, let $\underset{\sim}{a} < \underset{\sim}{b}$ mean that $\underset{\sim}{a} \leq \underset{\sim}{b}$ and $\underset{\sim}{b} \nleq \underset{\sim}{a}$; let $\underset{\sim}{a} \cup \underset{\sim}{b}$ be the least upper bound of the degrees $\underset{\sim}{a}$ and $\underset{\sim}{b}$. $\{W_x\}$, $\{\varphi_x\}$ and $\{\varphi_x^A\}$ denote the standard enumerations of all r.e. sets, all one-place partial-recursive functions, and partial-recursive functions with oracle A, respectively. For a set A, $A(x) = 1$ if $x \in A$, and $A(x) = 0$ if $x \notin A$; $A\upharpoonright x = \{A(0), \ldots, A(x)\}$; $\bar{A} = \omega - A$; $A \oplus B = \{2x : x \in A\} \cup \{2x+1 : x \in B\}$; deg(A) is the degree of A. If $\deg(A) \leq \deg(B)$ then we write $A \leq_T B$. $\varphi_{x,s}(t)$ is defined and equals to $\varphi_x(t)$ if the value of $\varphi_x(t)$ is computable in s steps by an effective procedure for computing the values of φ_x. Otherwise $\varphi_{x,t}(t)$ is not defined. For a r.e. set A we write A_s for the value of A at the end of stage s. The use of a computation $\varphi_e^A(x)$ (denoted by use(A;e,x)) is 1 plus the largest number from oracle A used in the computation if $\varphi_e^A(x)\downarrow$; and 0 otherwise (likewise for use(A;e,x,s), the use at stage s).

Let $\langle x_1, x_2\rangle$ denote some fixed recursive pairing function, i.e.

a 1-1 map from $\omega \times \omega$ onto ω ; for $n > 2$ let $\langle x_1, \dots, x_n \rangle = \langle\langle \dots \langle x_1, x_2 \rangle, \dots \rangle, x_n \rangle$. Let π_1, π_2 be recursive functions such that for all x $\langle \pi_1(x), \pi_2(x) \rangle = x$.

We begin with the following result which asserts that $\mathcal{D}_f = \mathcal{D}_g$ for unbounded recursive functions f and g.

Proposition 1. Let f,g be recursive functions, A is f-r.e. and rang(g) is infinite. Then there exists a set $B \equiv_T A$ such that B is g-r.e.

Proof. Let $A_1 : \omega^2 \to \omega$ be a recursive function such that $A(e) = \lim_k A_1(k,e)$ and $\text{card}\{k : A_1(k,e) \neq A_1(k+1,e)\} \leq f(e)$ for all e. Define a strictly increasing recursive function h by $h(0) = \mu y\{f(0) < g(y)\}$ and $h(e) = \mu y\{h(e-1) < y \& g(y) \geq f(e)\}$ for $e > 0$. Let $B = h(A)$. Clearly, $A \equiv_T B$.

To see that B is g-r.e. we define the recursive function $B_1 : \omega^2 \to \omega$ as follows:

$$B_1(s,e) = \begin{cases} 1 & \text{if } \exists z < e(h(z) = e \ \& \ A_1(s,z) = 1), \\ 0 & \text{otherwise} \end{cases}$$

It is easy to see that B_1 is recursive and for all $e \in \omega$ $B(e) = \lim_s B_1(s,e)$. Note that if $\forall z < x\ (h(z) \neq x)$ then

$$\text{card}\{s : B_1(s,x) \neq B_1(s+1,x)\} = 0 \leq g(x),$$

and if $\exists z < x(h(z) = x)$ then

$$\text{card}\{s : B_1(s,x) \neq B_1(s+1,x)\} \leq \text{card}\{s : A_1(s,z) \neq A_1(s+1,z)\}$$
$$\leq f(z) \leq g(h(z)) = g(x).$$

Thus, $\text{card}\{s : B_1(s,x) \neq B_1(s+1,x)\} \leq g(x)$ for all $x \in \omega$.

Before presenting our other results, we survey the related results known to us.

Cooper[5] has shown that properly n-r.e. degrees do exist for every n, $1 < n < \omega$. Lachlan,Hay and Lerman (unpublished) have proved that for all m,n, $0 < n, m < \omega$, and any n-r.e. degree $\underset{\sim}{a} > \underset{\sim}{0}$ there is an m-r.e. degree $\underset{\sim}{b}$ such that $\underset{\sim}{0} < \underset{\sim}{b} < \underset{\sim}{a}$.

In [1,3] we have proved following related results.

1) For all n, $1 < n < \omega$, there exist n-r.e. degrees $\underset{\sim}{a} < \underset{\sim}{b}$ such that there is no (n-1)-r.e. degree $\underset{\sim}{c}$ between them ([1, Theorem 7] and [3,Theorem 7-XIV]).

2) For all n, $1 < n < \omega$, there exists an n-r.e. degree $\underset{\sim}{a}$ which

is not (n-1)-REA (a degree $\underset{\sim}{a}$ is n-REA [10] if either n=1 and $\underset{\sim}{a}$ is r.e. or $n > 1$ and $\underset{\sim}{a}$ is r.e. in some (n-1)-REA degree $\underset{\sim}{b} \leq \underset{\sim}{a}$). It is easy to see (see, for example, [10]) that avery n-r.e. degree is n-REA.

3) Every d-r.e. degree $\underset{\sim}{d} > \underset{\sim}{0}$ can be cupped to $\underset{\sim}{0}'$ in the d-r.e. degrees (i.e. there is a d-r.e. degree $\underset{\sim}{a} < \underset{\sim}{0}'$ such that $\underset{\sim}{d} \cup \underset{\sim}{a} = \underset{\sim}{0}'$) ([2,Theorem 1] and [4,Theorem 2]). This result contrasts with the result by Yates and Cooper [6] that there exists an r.e. degree $\underset{\sim}{a}$ with $\underset{\sim}{0} < \underset{\sim}{a} < \underset{\sim}{0}'$ such that no r.e. degree $\underset{\sim}{b} < \underset{\sim}{0}'$ cups to $\underset{\sim}{0}'$.

4) For every r.e. degree $\underset{\sim}{a} < \underset{\sim}{0}'$ there exists a properly d-r.e. degree $\underset{\sim}{b}$ such that $\underset{\sim}{a} < \underset{\sim}{b} < \underset{\sim}{0}'$ ([2,Theorem 2] and [4,Teorem 4]).

Note that in [8] authors have noted that 1) has also been proved by Hay and Lerman (unpublished); 2) was later proved independently by Jockush and Shore [10,Theorem 1.7] along the same lines. Note that they also proved the existence of 2-REA degrees below $\underset{\sim}{0}'$ which are not even ω-r.e.

Sacks [13] has proved that any r.e. degree $\underset{\sim}{b} > \underset{\sim}{0}$ splits over $\underset{\sim}{0}$. Robinson [12] has shown that any r.e. degree $\underset{\sim}{b} > \underset{\sim}{0}$ splits over any low r.e. degree $\underset{\sim}{c} < \underset{\sim}{b}$. Lachlan [11] has proved that this is not true in general when $\underset{\sim}{c}$ is not low. Moreover, Harrington (unpublished) has shown that $\underset{\sim}{0}'$ does not splits over some r.e. degree $\underset{\sim}{b} < \underset{\sim}{0}'$. Studying this question for n-r.e. degrees, we have shown in Theorem 4 that $\underset{\sim}{0}'$ splits by ω-r.e. degrees over any r.e. degree $\underset{\sim}{b} < \underset{\sim}{0}'$ (i.e. for any r.e. degree $\underset{\sim}{b} < \underset{\sim}{0}'$ there exist ω-r.e. degrees $\underset{\sim}{c}_1, \underset{\sim}{c}_2$ such that $\underset{\sim}{b} < \underset{\sim}{c}_1 < \underset{\sim}{0}'$, $\underset{\sim}{b} < \underset{\sim}{c}_2 < \underset{\sim}{0}'$ and $\underset{\sim}{c}_1 \cup \underset{\sim}{c}_2 = \underset{\sim}{0}'$).
Note that in [2,Theorem 1] we sketched a proof of more strong version of this theorem when the degrees $\underset{\sim}{c}_1$ and $\underset{\sim}{c}_2$ are d-r.e. But, unfortunately, the detailed construction leads to a weaker result stated in Theorem 4.

The proof of Theorem 4 uses the basic approach devised in [2]. In Proposition 2 we describe a preliminary construction which plays an important role in the proof.

Proposition 2. Let H be any coinfinite r.e. set which is not hypersimple, and let f be an recursive function which majorize $H = \{ h_0 < h_1 < \ldots \}$, i.e. for any $x \geq 0$ $f(x) \geq h_x$. Then there exist r.e. sets H_1 and H_2 such that $H_2 \subseteq H_1$, $H_1 - H_2 = \{ h_0, \langle h_0, h_1 \rangle, \langle h_0, h_1, h_2 \rangle, \ldots \}$ and the following conditions are satisfied:

a) $H_1 - H_2 \equiv_T H$;

b) $H_1 - H_2$ is retraceable;

c) $H_1 - H_2$ is not hyperimmune. Moreover, we can suppose that the disjoint strong array $D_0, D_1, D_2, \ldots$ witnessed by non-hyperimmunity of $H_1 - H_2$ has following additional properties:

c1) For every $t \geqslant 0$ card $\{(H_1-H_2) \cap D_t\} = 1$;

c2) $\forall t \forall x \forall y \{x \in (H_1-H_2) \cap D_t \ \&\ y \in D_t - H_1 \rightarrow x < y$.

Proof. We first give an informal construction. Fix an enumeration H_s $s \in \omega$ of H and simultaneously enumerate integers h_0, $\langle h_0, h_0+1 \rangle$, $\langle h_0, h_0+1, h_0+2 \rangle, \ldots$ into H_1 until some $h_0+i, i > 0$, has enumerated in H. Then transfer all integers $\langle h_0, h_0+1, \ldots, h_0+i \rangle, \langle h_0, h_0+1, \ldots h_0+i, h_0+i+1 \rangle, \ldots$, which to this moment have been enumerated in H_1, from H_1 to H_2 and enumerate beginning from this moment integers $\langle h_0, h_0+1, \ldots, h_0+i-1, h_0+i+1 \rangle, \langle h_0, h_0+1, \ldots, h_0+i-1, h_0+i+1, h_0+i+2 \rangle$, ... into H_1 until some other integer h_0+j has appeared in H. Then again transfer all integers $\langle h_0, \ldots, h_0+j, \ldots \rangle$ from H_1 to H_2 and so on.

Construction of H_1 and H_2. Define $H_{1,0} = \{h_0\}$, $H_{2,0} = \emptyset$ and for $s \geqslant 0$ $H_{2,s+1} = H_{2,s} \cup \{t: t \in H_{1,s} \ \&\ \exists i_1 < \ldots < i_r (t = \langle h_0, i_1, \ldots, i_r \rangle \ \&$ $\exists j (1 \leq j \leq r \ \&\ i_j \in H_{s+1}))\}$.

Let y be a greatest element of $H_{1,s} - H_{2,s+1}$ and $y = \langle i_0, i_1, \ldots, i_r \rangle$, where $h = i_0 < i_1 \ldots < i_r$. Define $H_{1,s+1} = H_{1,s} \cup \{\langle y, t \rangle$,where $t = \mu z \{z > i_r \ \&\ z \notin H_{s+1}\}$.

Let $H_1 = \lim_s H_{1,s}$ and $H_2 = \lim_s H_{2,s}$.

The construction immediately yields $H_1 - H_2 = \{h_0, \langle h_0, h_1 \rangle, \langle h_0, h_1, h_2 \rangle, \ldots\}$. Hence, $H_1 - H_2 \equiv_T H$ and $H_1 - H_2$ is retraceable. To prove that $H_1 - H_2$ is not hyperimmune suppose that g is recursive function majorizing $\bar{H}$ and consider the following disjoint strong array:

$D_0 = \{h_0\}, D_1 = \{\langle h_0, x \rangle : h_0 < x \leq g(1)\}$, $D_2 = \{\langle h_0, x, y \rangle : h_0 < x < y \leqslant g(2)\}, \ldots$

Clearly, $\forall n \{D_n \cap (H_1 - H_2) = \{\langle h_0, h_1, \ldots, h_n \rangle\}$. To prove c2) suppose that $\langle x_0, x_1, \ldots, x_t \rangle \in D_t - H_1$. By construction we have $x_0 = h_0$, $x_1 = h_{i_1}, \ldots, x_t = h_{i_t}$ for some $0 < i_1 < \ldots < i_t$. Therefore, $\langle h_0, h_1, \ldots, h_t \rangle < \langle x_0, x_1, \ldots, x_t \rangle$.

In Theorem 4 we construct r.e. sets A,B,C,D to satisfy various requirements. Among these requirements we have the condition $\emptyset' \leqslant_T (A-B) \oplus (C-D)$. It would be convenient to meet requirements of this form using the construction described in Proposition 2. The proof of the following theorem,which is of independent interest, may help to understand this idea.

Theorem 3. For any $n > 1$ there exist properly n-r.e. degrees $\underset{\sim}{a}$ and $\underset{\sim}{b}$ such that $\underset{\sim}{a} \cup \underset{\sim}{b} = \underset{\sim}{0}'$ and $\underset{\sim}{a}'=\underset{\sim}{0}', \underset{\sim}{b}'=\underset{\sim}{0}'$.

Proof. We prove the level 2 version. The proof of the theorem for all n is similar.The proof use a finite injury priority argument.

We recursively enumerate A,B,C,D, $B \subseteq A, D \subseteq C$, to meet for all e requirements:

$$R^1_{e=\langle i,j,x\rangle}: A-B \neq \varphi_i^{W_x} \vee W_x \neq \varphi_j^{A-B} ;$$

$$R^2_{e=\langle i,j,x\rangle}: C-D \neq \varphi_i^{W_x} \vee W_x \neq \varphi_j^{C-D}$$

(to ensure that deg(A-B) and deg(C-D) non r.e.).

$$R^3_e : \exists^\infty s(\varphi_{e,s}^{A_s-B_s} (e)\downarrow) \to \varphi_e^{A-B} (e)\downarrow ;$$

$$R^4_e : \exists^\infty s(\varphi_{e,s}^{C_s-D_s} (e)\downarrow) \to \varphi_e^{C-D} (e)\downarrow$$

(the usual lowness requirements).

$$R^5 : (A-B)\cap (C-D) =\{ \cup_{x\in I} D_x \} - H_2 \text{ for some infinite set } I;$$

$$R^6 : \forall x \{(A-B)\cap (C-D) \cap D_x \neq \emptyset \to (A-B) \cap (C-D) \cap (H_1-H_2) \cap D_x \neq \emptyset$$

(The sets H_1, H_2 and the strong array $\{D_x\}$ are defined in Proposition 2 for the case $H \equiv_T \emptyset'$.)

Last two requirements ensure that $\emptyset' \equiv_T H \equiv_T H_1-H_2 \leq_T (A-B) \cap (C-D)$. Indeed, to test whether $e \in H_1-H_2$ we find a number x such that $F =_{def} (A-B) \cap (C-D) \cap D_x \neq \emptyset$ and for every $t \in D_x$ we have $\pi_2(t) > e$ (remember $t = \langle \pi_1(t), \pi_2(t) \rangle$). If $\hat{e}$ is the least number of F then by R^6 and points c1) and c2) of Proposition 2 we have $\hat{e} \in H_1-H_2$. Therefore, $\hat{e} = \langle h_0, h_1, \dots, h_x \rangle$, where $h_x > e$ and $H_1-H_2=\{ h_0, h_1, \dots \}$.Clearly, $e \in H_1-H_2$ iff $e=h_k$ for some $0 \leq k \leq x$.

In satisfying requirements R^1_e, R^2_e we shall put some numbers into A, B,C,D. So that this will not interfere with the satisfaction of requirements R^5 and R^6, we fix infinite disjoint recursive sets R_1, R_2 such that $(\cup_{n=0}^{\infty} D_n) \cap (R_1 \cup R_2) = \emptyset$, and we use numbers R_i to satisfy requirement R_i, $1 \leq i \leq 2$.

To satisfy requirements R^1_e and R^2_e we use the well-known Cooper's method (see,for example, [3,Theorem 2-1]).To satisfy R^1_e, $e=\langle i,j,x\rangle$, we proceed as follows: we successively choose numbers $a \in R_1$ which have not yet been put into A (and hence not into B), and we wait for a stage s such that for some least u and v

$$A_s-B_s(a) = \varphi_{i,s}^{W_{x,s}\restriction u}(a) \ \& \ W_{x,s}\restriction u = \varphi_{j,s}^{A_s-B_s\restriction v}\restriction u$$

(If this never happens then R_e^1 is satisfied). At the first such step s we put a into A_{s+1} and restrain with priority R_e^1 $A-B\restriction v$ from other strategies from now on. Suppose that at some step $s'>s$ we again have for some u' and v'

$$A_{s'}-B_{s'}(a) = \varphi_{i,s'}^{W_{x,s'}\restriction u'}(a) \ \& \ W_{x,s'}\restriction u' = \varphi_{j,s'}^{A_{s'}-B_{s'}\restriction v'}\restriction u'$$

(If this never happens then R_e^1 is satisfied). We transfer a to B and restrain with priority R_e^1 $A-B\restriction v'$. Now R_e^1 is satisfied because

$$\varphi_j^{A-B}\restriction u = \varphi_{j,s}^{A_s-B_s}\restriction u = W_{x,s}\restriction u \subsetneq W_x\restriction u .$$

Requirement R_e^2 is satisfied in an analogies way.

To satisfy requirements R_e^3 and R_e^4 we use the usual "lowness strategy": to satisfy R_e^3 we attempt to restrain with priority R_e^3 from other strategies any elements $x \leq use(A_s-B_s,e,e,s)$. If $\varphi_{e,s}^{A_s-B_s}(e)\downarrow$, and R_e^3 succeeds in preventing any $x \leq use(A_s-B_s,e,e,s)$ from later entering A-B, then $A-B\restriction u = A_s-B_s\restriction u$, so $\varphi_e^{A-B}(e)\downarrow$.

To meet requirements R_e^1,R_e^2,R_e^3,R_e^4 $e\in\omega$ simultaneously we carry out the usual finite injury priority argument. The priority ranking of the requirements is $R_0^1,R_0^2,R_0^3,R_0^4,R_1^1,R_1^2,R_1^3,R_1^4,\ldots$

We now need to include the strategy for R^5 and R^6. To meet requirements R^5 and R^6 it suffices from time to time to put some sets D_x (whose elements are not restrained by any requirements) into A and C, and as the elements of D_x within H_2 are enumerated, to enumerate them either in B or in D.

Here the main difficulty is that we may first put elements of some D_x into A and C, then restrain A-B,C-D from other strategies by some R_e^i so that this restraint includes D_x, and after that we may need to enumerate some elements of D_x in B or in D.

To overcome this we include to strategies for R_e^i the following crucial change. As elements of any D_x restrained by the requirement R_0^1 within H_2 are enumerated, we enumerate them in B (but not in D) may be injuring requirements $R_e^2,R_e^4,e\in\omega$, of lower priority. Obviously, this strategy allow eventually to satisfy R_0^1. Further, as elements of any D_x, restrained by the requirement R_0^2 and does not restrained by R_0^1, within H_2 are enumerated, we enumerate them in D (but not in B) may be injuring requirements R_e^1, R_e^3, $e \in \omega$, of lower priority. This strategy allow eventually to satisfy R_0^1, and R_0^2 and so on.

The method used here can be adapted to prove the following theorem.

Theorem 4. Let A be an r.e. set and $A <_T \emptyset'$. Then there exist ω-r.e. sets B, C such that $A <_T B <_T \emptyset'$, $A <_T C <_T \emptyset'$ and $B \oplus C \equiv_T \emptyset'$.

Proof. We construct ω-r.e. sets B and C to meet for all e requirements:

R_e^1: $K \neq \varphi_e^{A \oplus B}$;

R_e^2: $K \neq \varphi_e^{A \oplus C}$;

R^3: $B \cap C = \{\bigcup_{x \in I} D_x\} - H_2$ for some infinite set I;

R^4: $\forall x \{B \cap C \cap D_x \neq \emptyset \rightarrow B \cap C \cap (H_1 - H_2) \cap D_x \neq \emptyset\}$.

Here K is creative set. The sets H_1, H_2 and the strong array $\{D_x\}$ have been defined in preceding theorem.

The requirements R^3 and R^4 guarantee $\emptyset' \leq_T B \oplus C \leq_T (A \oplus B) \oplus (A \oplus C)$ (see the proof of Theorem 3). It follows from R_e^1, R_e^2, $e \in \omega$, that $A \oplus B <_T \emptyset'$ and $A \oplus C <_T \emptyset'$.

To satisfy requirements R_e^3 and R_e^4, $e \in \omega$, we use the Sacks' agreement method. For instance, for R_e^1 at any stage $s > 0$ we define

$$l(e,s) = \max\{z: K_s \upharpoonright z = \varphi_e^{A_s \oplus B_s} \upharpoonright z\},$$

$$r(e,s) = \max\{use(A_s \oplus B_s, e, x, s): x \leq l(e,t)\}$$

and restrain $A \oplus B \upharpoonright r(e,s)$ with priority R_e^1 from other strategies from now on. If the whole construction is finite injury and $K \nleq_T A$ then this strategy ensure $K \neq \varphi_e^{A \oplus B}$ (see [14, Theorem VII.3.1]). We use the same strategy for R_e^2.

We satisfy requirements R^3 and R^4 by the same method which was used in the proof of Theorem 3: we put from time to time the elements of some sets D_x (whose elements are not restrained by "negative" requirements R_e^1, R_e^2) into B and C, and as elements of D_x within H_2 are enumerated, we take them away from B (if the elements are not restrained by negative requirements), or (otherwise) take them away either from B or from C depending on priorities of restraints which contain the elements.

But now we have a new difficulty. One may have an element x such that, for instance, $\varphi_0^{A \oplus B}(x)$ or $\varphi_0^{A \oplus B}(x)\downarrow$ & $\varphi_0^{A \oplus B}(x) \neq K(x)$, but $\exists^\infty s(\varphi_{0,s}^{A_s \oplus B_s}(x) = K_s(x))$ and it may force to restrain with priority R_0^1 numbers infinitely many times preventing satisfaction of R_0^2 (the elements of $B_s \cap C_s \cap D_x \cap H_{2,s}$ are forced to be taken away every time from C). Of course, there must be stages when the restraints of R_0^1 drop back, and at these stages s_1 we can correct the situation: the elements of $B_s \cap C_s \cap D_{s_2} \cap H_{2,s}$ restrained by "wrong restraints" of R_0^1 and restrained by R_0^2 we may put again into C and take them away from B. But now the situation may be repeated with R_0^2 instead of R_0^1 : there are wrong restraints of R_0^2 which force elements of $B_s \cap C_s \cap D_x \cap H_{2,s}$ to be taken away continuously from C and so on.

We can try to remove this obstacle using the following change in the construction:

a_0) Elements of $D_x \cap B_s \cap C_s \cap H_{2,s}$ restrained by R_0^1 we take away every time from C;

b_0) If at some stage $s_1 > s$ we discover that some elements of $D_x \cap B_s \cap C_s \cap H_{2,s}$ were taken away from C by wrong restraints of R_0^1 and they are restrained by R_0^2, then put them into C and take away from B;

c_0) If at some stage $s_2 > s_1$ we discover that some of these elements were taken away from B and were put into C by wrong restraints of R_0^2 and if they are restrained by R_1^1 and does not restrained by R_0^1 , then we again put them into B and take away from C and so on.

a_1) Elements of $D_x \cap B_s \cap C_s \cap H_{2,s}$ restrained by R_0^2 but not by R_0^1 we take away every time from B (may be injuring requirements of lower priority) and so on .

Clearly this strategy eventually satisfies all requirements R_e^1, R_e^2 $e \in \omega$,but the sets B and C are not necessarily ω-r.e.Now the following crucial cange in the construction allows to override this last obstacle:

The case a_0) is unchanged. In the case b_0) we transfer an element from B to C only if it does not belong to first set D_x which was put into B and C.In the case c_0) we transfer an element only if it does not belong to first two sets D_x which were put into B and C and so on.

This change of the strategy cannot eventually hinder to satisfy requirements $R_e^1, R_e^2, e \in \omega$ .FOR instance, the first enumerated in $B \cap C$

set D_x, which now "stay" on B, may prevent from satistying R_0^2, but after a stage s big enough the requirement will be satisfied. Clearly, the sets B and C are now ω-r.e.

We want to close with a few open questions. Let $\mathcal{D}_{Fin} = \bigcup_{n\in\omega} \mathcal{D}_n$. It is clear that $\mathcal{D}_{Fin} \subseteq \mathcal{D}_\omega$, $\mathcal{D}_\omega - \mathcal{D}_{Fin} \neq \emptyset$ and $\mathcal{D}_{Fin}$ is upper semilattice.

Q1) (Density problem) For all degrees $\underset{\sim}{a} < \underset{\sim}{b}$, $\underset{\sim}{a}, \underset{\sim}{b} \in \mathcal{D}_{Fin}$, does there exists a degree $\underset{\sim}{c} \in \mathcal{D}_{Fin}$ such that $\underset{\sim}{a} < \underset{\sim}{c} < \underset{\sim}{b}$?

Note that recently Harrington, Lachlan, Lempp and Soare [9] have proved that the density problem fails in the d-r.e. degrees. Cooper [7] have showed that the low_2 n-r.e. degrees are dense.

Q2) ("Monster" problem for $\mathcal{D}_{Fin}$) Fpr all n-r.e. degrees $\underset{\sim}{a} < \underset{\sim}{b}$, $1 \leq n < \omega$, do there exist degrees degrees $\underset{\sim}{c}_0, \underset{\sim}{c}_1 \in \mathcal{D}_{Fin}$ such that $\underset{\sim}{a} < \underset{\sim}{c}_0 < \underset{\sim}{b}$, $\underset{\sim}{a} < \underset{\sim}{c}_1 < \underset{\sim}{b}$ and $\underset{\sim}{c}_0 \cup \underset{\sim}{c}_1 = \underset{\sim}{b}$?

Q3) The same question for $\underset{\sim}{b} = \underset{\sim}{0}'$, $\underset{\sim}{b} = \underset{\sim}{0}''$.

Q4) Whether Th $(\langle \mathcal{D}_n, \leq \rangle)$, Th $(\langle \mathcal{D}_m, \leq \rangle)$, Th $(\langle \mathcal{D}_{Fin}, \leq \rangle)$ for $m \neq n$, $m > 1$, $n > 1$ are pairwise distinct?

We have noted above that Th $(\langle \mathcal{D}_{r.e.}, \leq \rangle)$ and Th $(\langle \mathcal{D}_n, \leq \rangle)$ are distinct for $1 < n \leq \omega$.

REFERENCES

1 . M.M.Arslanov, An hierarchy of degrees of unsolvability, Prob. Meth. Cyb. 18(1982),10-17.

2 . M.M.Arslanov, Structural properties of the degrees below 0', Sov.Math.Dokl.,N.S. 283, 2(1985), 270-273.

3 . M.M.Arslanov, Recursively Enumerable Sets and Degrees of Unsolvability, Kazan Univ. Press, Kazan, USSR, 1986.

4. M.M.Arslanov, On the upper semilattice of Turing degrees below 0', Sov.Math. 7(1988), 27-33.

5 . S.B.Cooper, Degrees of Unsolvability, Ph.D.Thesis, Leicester University, Leicester, 1971.

6 . S.B.Cooper, On a theorem of C.E.M.Yates, Handwritten notes, 1974.

7. S.B.Cooper, The density of the low_2 n-r.e. degrees, to appear.

8 . R.L.Epstein,R.Haas,R.L.Kramer, Hierarchies of sets and degrees below 0', Lect.Not.Math. 859(1981), 32-48.

9. L.Harrington,A.H.Lachlan, S.Lempp,R.I.Soare, On the nondensity of the d-r.e.degrees, to appear.

10. C.G.Jockush,R.A.Shore, Pseudo-jump operators 11: Transfinite iteration, hierarchies and minimal covers, J.Symb.Log.49(1984), 1205-1236.

11.A.H.Lachlan, A recursively enumerable degree which will not split over all lesser ones, Ann.Math.Log. 9(1975), 307-365.

12. R.M.Robinson, Jump restricted interpolation in the recursively enumerable degrees, Ann.Math. 93(1971), 586-596.

13. G.E.Sacks, On the degrees less than 0', Ann.Math. 77(1963), 211-231.

14. R.I.Soare, Recursively Enumerable Sets and Degrees, Springer-Verlag, Berlin,New York,London,1987.

Positive Solutions to Post's Problem

C. T. Chong and K. J. Mourad

National University of Singapore

Post's problem asks whether there exists an incomplete recursively enumerable (r.e.) set which is not recursive. This problem has been solved positively for ω (the Friedberg Muchnik Theorem), for all admissible ordinals (Sacks and Simpson [9]), for all inadmissible ordinals β such that β^* is a β-regular cardinal (Friedman [4]), for all inadmissible β whose Σ_1-cofinality is at least the Σ_1-projectum β^* (Friedman [4]), and for all models of fragments of arithmetic which satisfy Σ_1 induction (Simpson, see [8]). Negative solutions to Post's problem have been obtained by Friedman for various inadmissible ordinals (under pointwise reducibility) [4], and it is known that under GCH, the relativized version of Post's problem has a negative solution for singular cardinals of uncountable cofinality (Friedman [6]).

Let $\mathcal{M}$ be a structure which supports a theory of computation. In a general setting, a set $B \subseteq \mathcal{M}$ is (pointwise) recursive in a set $A \subseteq \mathcal{M}$ if there is a reduction procedure Φ_e such that $\Phi_e(A) = B$. This means, writing W_e for the eth $\mathcal{M}$-r.e. set and K_c and K_d for the cth and dth $\mathcal{M}$-finite sets respectively, that one has

(1) $$x \in B \longleftrightarrow (\exists c)(\exists d)[(x,1,c,d) \in W_e \;\&\; K_c \subset A \;\&\; K_d \subset \bar{A}];$$

and

(2) $$x \notin B \longleftrightarrow (\exists c)(\exists d)[(x,0,c,d) \in W_e \;\&\; K_c \subset A \;\&\; K_d \subset \bar{A}].$$

When A and B are $\mathcal{M}$-r.e., the sensitiveness of the solution of Post's problem to the relation $K_c \subset B$ in (1) and (2) becomes especially acute. An r.e. set B is *tame* if for every $\mathcal{M}$-finite set K, $K \subset B$ if and only if there is a σ such that $K \subset B^\sigma$, where B^σ is the $\mathcal{M}$-finite subset of B enumerated by stage σ. In α recursion theory, it is a basic result that every α-r.e. set is tame. On the other hand, there exist inadmissible ordinals β in which the only tame r.e. sets are those β-recursive in the empty set $\emptyset$. If $\mathcal{M}$ is a model of some fragment of Peano arithmetic, then Σ_1 collection is the minimal requirement for every $\mathcal{M}$-r.e. set to be tame. For a positive solution to Post's problem, the device of ingeneous combinatorial techniques in a priority construction (to ensure that every requirement is eventually satisfied) relies in an essential way on the tameness of the r.e. sets constructed. On the other hand, negative solutions exist in cases of ordinals where non-trivial tame r.e. sets do not exist (for example, when $\beta = \aleph_{\omega_1}^L \cdot \omega$ and when $\mathcal{M} = (L_{\aleph_{\omega_1}^L}, \emptyset')$, in which all tame $\mathcal{M}$-r.e. sets (i.e. sets which are $\Sigma_1(\emptyset')$ over $\aleph_{\omega_1}^L$) are recursive in $\emptyset'$).

While (1) and (2) are arguably the most obvious definition to be adopted for the notion of reducibility in a general setting (although the relation 'B is Δ_1 in A' would be the most faithful preservation), it is in our view not necessarily the most natural for r.e. sets or relativized r.e. sets. Indeed in classical recursion theory, there is a special appeal of r.e. sets characterized by the dynamism of growth in stages. The intuition that any growth on a finite initial segment must be completed in finitely many stages is in turn derived from the tameness of these sets. This unique feature is carried over to reducibility considerations. Hence in (1) and (2) the relation $K_c \subset A$ is equivalent to (and is indeed treated as) $K_c \subset A^s$ for some natural number s. This *dynamic* aspect of reducibility, which is independent of the tameness of r.e. sets, has not hitherto been studied in generalized recursion theory. We believe that it deserves to be investigated since it is one which, for r.e. sets and relativized r.e. sets, is very close to the heart of the classical intuition, and since it provides another angle of looking at Post's problem that is very natural. In particular, it is an interesting problem to see whether positive solutions exist for a notion of reducibility which emphasizes the dynamism of r.e. sets, not only for cases where negative solutions are known for (1) and (2), but for all limit ordinals. Ultimately, one wants to know whether a generalization of the Friedberg-Muchnik Theorem holds (for all limit ordinals) under this reducibility.

Let β be a limit ordinal. Then S_β is a rudimentarily closed structure. Let $C \subset S_\beta$. Given sets A, $B \subset S_\beta$ which are β-r.e. in C, we say that B is *weakly* $d\beta$-recursive in A (relative to C) if there is an e and an enumeration $\{A^\sigma\}$ (using C) such that for all x,

$$x \in B \longleftrightarrow (\exists\sigma)(\exists c)(\exists d)[(x,1,c,d) \in W_e \;\&\; K_c \subset A^\sigma \;\&\; K_d \subset \bar{A}]; \tag{3}$$

and

$$x \notin B \longleftrightarrow (\exists\sigma)(\exists c)(\exists d)[(x,0,c,d) \in W_e \;\&\; K_c \subset A^\sigma \;\&\; K_d \subset \bar{A}]. \tag{4}$$

We denote this relation by $B \leq^C_{wd\beta} A$. If $C = \emptyset$, we write $B \leq_{wd\beta} A$ instead. Clearly (3) and (4) imply (1) and (2) respectively (note that in general, the truth of (3) and (4) depend on the enumeration of A and B relative to C). In this paper we prove that Post's problem has a positive solution under weak $d\beta$ reducibility for all ordinals : .

THEOREM 1. *Let α be admissible. Then there exist sets A and B which are r.e. in and above $\emptyset'$ such that $A \not\leq^{\emptyset'}_{wd\alpha} B$ and $B \not\leq^{\emptyset'}_{wd\alpha} A$.*

THEOREM 2. *Let β be a limit ordinal. Then there exist β-r.e. sets A and B such that $A \not\leq_{wd\beta} B$ and $B \not\leq_{wd\beta} A$.*

We will first prove two theorems which handle the most difficult cases.

THEOREM 3. *Let α be an admissible ordinal which is a limit of α-cardinals such that the Σ_2 cofinality of α is less than α. Then there exist A and B which are Σ_1 over and above $\emptyset'$ such that $A \not\leq^{\emptyset'}_{w\,d\alpha} B$ and $B \not\leq^{\emptyset'}_{w\,d\alpha} A$.*

In particular, a positive solution to (relativized) Post's problem exists for $\alpha = \aleph^L_{\omega_1}$ under weak $d\alpha$ reducibility, in contrast to [6] for α-degrees, and a positive solution exists for $\aleph^L_\omega$, a case which remains open for $\aleph^L_\omega$-degrees.

The reader will observe from the proof in the next section that the following can be derived in a similar manner:

COROLLARY. *If $\alpha = \aleph_\omega$ denotes the real ωth cardinal, and if $F : \omega \to \aleph_\omega$ is a cofinal function, then there exist A and B which are r./e. relative to F such that $B \not\leq^F_{w\,d\alpha} A$ and $A \not\leq^F_{w\,d\alpha} B$.*

Let β^* denote the Σ_1 projectum of β. If β is inadmissible, then $\beta^* < \beta$ is a β cardinal. Friedman [3] showed that if β^* is a regular β cardinal, then Post's problem has a positive solution (under (1) and (2)). Some positive and negative results have been obtained for the case when β^* is a singular β-cardinal. We show that under weak $d\beta$ reducibility, positive solutions exist for the remaining case:

THEOREM 4. *Let β be inadmissible such that β^* is a singular β cardinal, and such that the Σ_1 cofinality of β is less than β^*. Then there exist β-r.e. sets A and B such that $A \not\leq_{w\,d\beta} B$ and $B \not\leq_{w\,d\beta} A$.*

The proofs of these results originate from our construction of a pair of $\mathcal{M}$-r.e. sets having incomparable Turing degrees (the Friedberg-Muchnik Theorem), for every model $\mathcal{M}$ of a fragment of Peano arithmetic which satisfies the Σ_1 collection scheme (Chong and Mourad [1]). A novelty of this construction is that it does not use any priority argument. This approach is flexible enough to allow one to sidetrack some of the basic difficulties in priority argument of dealing with r.e. sets which are not tame. In [2] we exploit this technique further by contructing a pair of incomparable $\aleph^L_\omega$-degrees lying between $\emptyset'$ and $\emptyset''$.

Proof of Theorem 3

Let α be a limit of α-cardinals with the property that α is not Σ_2 admissible. Let $\kappa < \alpha$ be the Σ_2-cofinality of α, and let $F : \kappa \to \alpha$ be a $\Sigma_2(L_\alpha)$ increasing cofinal function such that $F(0) = 0$ and $F(\rho)$ is an infinite successor α-cardinal greater than κ for each $0 < \rho < \kappa$.

DEFINITION. *An interval I is an α-finite set of the form $\{x|F(\rho) \leq x < \gamma\}$ for some $\rho < \kappa$ and $\gamma < F(\rho+1)$. We call γ the cut of I.*

We construct sets A and B which are r.e. in $\emptyset'$ and which are unions of unboundedly many intervals.

LEMMA 1. *If $A \subset \alpha$ is a union of unboundedly many intervals, then $\emptyset' \leq_\alpha A$.*

PROOF: We may take $\emptyset'$ to be the set of α-cardinals. To decide if $x \in \emptyset'$, let $y > x$ be in A such that y and x do not belong to the same interval. Then by assumption the beginning of the interval containing y is an α-cardinal $F(\rho)$, and so by Σ_1-stability, x is an α-cardinal if and only if it is a cardinal in L_y ($y \geq F(\rho)$).

Let $[x,y) = \{z|x \leq z < y\}$. We first set once and for all $A \cap [F(2\rho), F(2\rho+1)) = [F(2\rho), F(2\rho+1))$ and $B \cap [F(2\rho+1), F(2\rho+2)) = [F(2\rho+1), F(2\rho+2))$, for every $\rho < \kappa$, Let $A^{<\sigma} = \cup_{\tau<\sigma} A^\tau$ and $B^{<\sigma} = \cup_{\tau<\sigma} B^\tau$ ($\sigma < \kappa$) be an α-finite unions of intervals.

We say that x is in block ρ if $x \in [F(\rho), F(\rho+1))$.

The construction proceeds in κ many stages, beginning with $A^0 = B^0 = \emptyset$. At even stage σ, assume by induction that $A^{<\sigma}$ and $B^{<\sigma}$ are α-finite unions of intervals contained in $F(\sigma+1)$. More specifically, let the restriction of $A^{<\sigma}$ to block ρ be denoted $I_\rho^{<\sigma}$, for all $\rho < \kappa$. Let $\rho < \kappa$ be fixed. Suppose that e is in block ρ. In the following consider ν, $2\rho+1 \leq \nu \leq \sigma+1$ to be an odd ordinal. For each x in block 2ρ, each α-finite sequence $\mathbf{c} = (c_\varsigma)$, where $\varsigma < 2\rho$ is an odd ordinal such that $c_\varsigma >$ the cut of $I_\varsigma^{<\sigma}$ lies in block ς, define $(z_{e,\nu,\mathbf{c}}(x))$ to be the least sequence $(z_\nu(x))$ (if it exists) such that $z_\nu(x)$ is in block ν and greater than or equal to the cut of $I_\nu^{<\sigma}$, with the property that no positive condition K_c contained in $A^{<\sigma}$ and negative condition K_d contained in $\cup_\varsigma[c_\varsigma, F(\varsigma+1)) \cup (\cup_\nu[z_\nu(x), F(\nu+1))$ yields $(x, 0, c, d) \in W_e^\sigma$. For each ρ and each $\mathbf{c}$, the set

$$K = \{(e, x, \rho, \mathbf{c})|e \in \text{block}(\rho) \ \& \ (z_{e,\nu,\mathbf{c}}(x)) \text{ is defined}\}$$

is α-finite (indeed $F(2\rho+1)$-finite by the assumption that F is mapped only to successor cardinals) and (uniformly) α-recursive in $\emptyset'$ (as a function of ρ and $\mathbf{c}$). Let $\hat{z}_\nu^\rho < F(\nu+1)$ be the least ordinal greater than $z_{e,\nu,\mathbf{c}}(x)$ for $(e, x, \rho, \mathbf{c})$ in K (guaranteed by the regularity of $F(\nu+1)$).

Again by the regularity of $F(\nu+1)$ and the fact that $F(\nu+1) > \kappa$, we have a $(\hat{z}_\nu)$ such that $\hat{z}_\nu^\rho < \hat{z}_\nu < F(\nu+1)$ for each ν. Let $I_\nu^\sigma = [F(\nu), \hat{z}_\nu)$ (recall our condition that ν is an odd ordinal such that $2\rho+1 \leq \nu \leq \sigma+1$). Set $A^\sigma = \cup_{\nu \leq \sigma+1} I_\nu^\sigma$. Let

$$x_{e,\mathbf{c}} = \text{least } x \text{ such that } (z_{e,\nu\mathbf{c}}(x)) \text{ is not defined.}$$

Let $K^* = \{x_{e,\mathbf{c}}|x_{e,\mathbf{c}} \in \text{ block } 2\rho \text{ exists}\}$, and enumerate into B all y such that $y \leq x_{e,\mathbf{c}}$ for some $x_{e,\mathbf{c}} \in K^*$. Note that at the end of stage σ, the set of elements of B in block 2ρ is an interval bounded in $F(2\rho+1)$ (by the fact that $F(2\rho+1)$ is a successor cardinal).

At odd stage σ, we interchange the roles of A and B in the construction, with the understanding that for e in block ρ, we consider x in block $2\rho+1$, and $\nu < \sigma+1$ is now an even ordinal greater than or equal to $2\rho+2$. Similarly, $\varsigma < 2\rho+1$ is now an even ordinal.

This ends the construction at stage σ.

Since the function F is recursive in $\emptyset'$, we see that A and B are r.e. in $\emptyset'$. Furthermore, by Lemma 1, these two sets compute $\emptyset'$. We verify that they are *wdα*-incomparable relative to $\emptyset'$.

Suppose that $B \leq_{wd\alpha} A$ via Φ_e. Assume that e is in block ρ. Let b be the cut of $B \cap [F(2\rho), F(2\rho+1))$. Choose a $\sigma < \kappa$ such that there exist $K_c \subset A^{<\sigma}$, $K_d \subset \bar{A}$ such that $(b, 0, c, d) \in W_e^\sigma$. For $\varsigma < 2\rho$ an odd ordinal, let $c_\varsigma = \min(K_d \cap [F(\varsigma), F(\varsigma+1))$. Let $\mathbf{c} = (c_\varsigma)$.

Now at stage σ, if $(z_{e,\nu,\mathbf{c}}(b))$ is defined, then the construction ensures, by the choice of $(\hat{z}_\nu)$, that no $K_{c'} \subset A^{<\sigma}$, $K_{d'} \subset \bar{A}^\sigma$ yields $(b, 0, c', d') \in W_e^\sigma$, contradicting our choice of σ and the assumption that $\Phi_e^\sigma(A; b) = B(b) = 0$. Hence $(z_{e,\nu,\mathbf{c}}(b))$ is not defined. But then this implies that at stage σ, $x_{e,\mathbf{c}} \leq b$ exists. Now it is not possible that $x_{e,\mathbf{c}} < b$, since otherwise we must have $\Phi_e(A; x_{e,\mathbf{c}}) = B(x_{e,\mathbf{c}}) = 1$, which is not possible by the definition of $x_{e,\mathbf{c}}$. Hence $x_{e,\mathbf{c}} = b$. But here we always enumerate $x_{e,\mathbf{c}} = b$ in B, a contradiction (since $\Phi_e(A; b) = 0$). Thus B is not weakly *dα*-recursive in A relative to $\emptyset'$.

Proof of Theorem 4

Let $f : S_\beta \to \beta^*$ be a β-recursive injection. Let κ be the $\Sigma_1(S_\beta)$ cofinality of β and let κ^* be the $\Sigma_1(S_\beta)$ cofinality of β^*. The hypothesis here is that $\kappa < \beta^*$ (i.e. strongly inadmissible). Since β^* is a singular β-cardinal, there is a β-recursive, increasing and cofinal function $F : \kappa^* \to \beta^*$ such that $F(0) = 0$ and $F(\rho) > \kappa$ is a successor β-cardinal for every $\rho < \kappa^*$.

We construct β-r.e. sets A, $B \subset \beta^*$ in κ many stages. We declare firstly that $A \cap [F(2\rho), F(2\rho+1)) = [F(2\rho), F(2\rho+1))$, and $B \cap [F(2\rho+1), F(2\rho+2)) = [F(2\rho+1), F(2\rho+2))$. Let $g : \kappa \to \beta$ be an increasing, cofinal β-recursive function such that $g(0) > \beta^*$.

Let σ be even. Suppose that $A^{<\sigma}$ and $B^{<\sigma}$ are β-finite unions of intervals contained in β^*. We say that x is in block ρ if $F(\rho) \leq x < F(\rho+1)$. Define $I_\nu^{<\sigma}$ as before. Assuming that the cut of each I_ν^τ, $\tau < \sigma < \kappa$, is less than $F(\nu+1)$ (for ν an odd ordinal), we have by the fact that $F(1) > \kappa$ and the regularity of $F(\nu+1)$ that the same conclusion holds for $I_\nu^{<\sigma}$.

Consider $f[g(\sigma)] \subset \beta^*$. Let $\rho < \kappa^*$ be fixed. Suppose that $e < \sigma$ and $f(g(e))$ is in block ρ. In the following consider ν, $2\rho+1 \leq \nu < \kappa$ to be an odd ordinal. For each x in block 2ρ, each β-finite sequence $\mathbf{c} = (c_\varsigma)$, where $\varsigma < 2\rho$ is an odd ordinal such that $c_\varsigma >$ the cut of $I_\varsigma^{<\sigma}$ and lies in block ς, define $(z_{e,\nu,\mathbf{c}}(x))$ to be the least sequence $(z_\nu(x))$ (if it exists) such that $z_\nu(x)$ is in block ν and greater than or equal to the cut of $I_\nu^{<\sigma}$, with the property that no positive condition K_c contained in $A^{<\sigma}$

and negative condition K_d contained in $\cup_\varsigma[c_\varsigma, F(\varsigma+1)) \cup (\cup_\nu[z_\nu(x), F(\nu+1))$ yields $(x, 0, c, d) \in W_e^{g(\sigma)}$. Now the set

$$K = \{(e, x, \rho, \mathbf{c}) | f(g(e)) \in \text{block}(\rho) \ \& \ (z_{e,\nu,\mathbf{c}}(x)) \text{ is defined}\}$$

is β-finite. By the choice of F (which is mapped only to successor β-cardinals), K is bounded in $F(2\rho+1)$ and so by β-stability is in fact $F(2\rho+1)$-finite. Let $\hat{z}_\nu^\rho < F(\nu+1)$ be the least ordinal greater than $z_{e,\nu,\mathbf{c}}(x)$ for $(e, x, \rho, \mathbf{c})$ in K (guaranteed by the regularity of $F(\nu+1)$).

Now by the regularity of $F(\nu+1)$ and by the fact that $F(\nu+1) > \kappa$, we have a $(\hat{z}_\nu)$ such that $\hat{z}_\nu^\rho < \hat{z}_\nu < F(\nu+1)$ for each ν. Let $I_\nu^\sigma = [F(\nu), \hat{z}_\nu)$ (recall our condition that ν is an odd ordinal such that $2\rho+1 \leq \nu < \kappa$). Set $A^\sigma = \cup_{\nu \leq \sigma+1} I_\nu^\sigma$. Let

$$x_{e,\mathbf{c}} = \text{least } x \text{ such that } (z_{e,\nu\mathbf{c}}(x)) \text{ is not defined.}$$

Let $K^* = \{x_{e,\mathbf{c}} | x_{e,\mathbf{c}} \in \text{ block } 2\rho \text{ exists}\}$, and enumerate into B all y such that $y \leq x_{e,\mathbf{c}}$ for some $x_{e,\mathbf{c}} \in K^*$. Note that at the end of stage σ, the set of elements of B in block 2ρ is an interval bounded in $F(2\rho+1)$ (by the fact that $F(2\rho+1)$ is a successor β-cardinal).

At odd stage σ, we interchange the roles of A and B in the construction as in the admissible case. This ends the construction at stage σ.

Since F is β-recursive, we see that A and B are r.e. We now verify that they are $wd\beta$-incomparable.

Suppose that $B \leq_{wd\beta} A$ via Φ_e. Assume that $f(g(e))$ is in block ρ. Let b be the cut of $B \cap [F(2\rho), F(2\rho+1))$. Choose a $\sigma < \kappa$ such that $e < \sigma$, and there exist $K_c \subset A^{<\sigma}$, $K_d \subset \bar{A}$ such that $(b, 0, c, d) \in W_e^{g(\sigma)}$. For $\varsigma < 2\rho$ an odd ordinal, let $c_\varsigma = \min(K_d \cap [F(\varsigma), F(\varsigma+1))$. Let $\mathbf{c} = (c_\varsigma)$.

Now at stage σ, if $(z_{e,\nu,\mathbf{c}}(b))$ is defined, then the construction ensures, by the choice of $(\hat{z}_\nu)$, that no $K_{c'} \subset A^{<\sigma}$, $K_{d'} \subset \bar{A}^\sigma$ yields $(b, 0, c', d') \in W_e^{g(\sigma)}$, contradicting our choice of σ and the assumption that $\Phi_e^\sigma(A; b) = B(b) = 0$. Hence $(z_{e,\nu,\mathbf{c}}(b))$ is not defined. But then this implies that at stage σ, $x_{e,\mathbf{c}} \leq b$ exists. Now it is not possible that $x_{e,\mathbf{c}} < b$, since otherwise we must have $\Phi_e(A; x_{e,\mathbf{c}}) = B(x_{e,\mathbf{c}}) = 1$, which is not possible by the definition of $x_{e,\mathbf{c}}$. Hence $x_{e,\mathbf{c}} = b$. But here we always enumerate $x_{e,\mathbf{c}} = b$ in B, a contradiction (since $\Phi_e(A; b) = 0$). Thus B is not weakly $d\beta$-recursive in A.

Observe that the argument in Theorem 4 can be adapted to show that in the inadmissible structure $(L_\alpha, \emptyset')$, where α is admissible and $\Sigma_2(L_\alpha)$ cofinality $< \Sigma_2(L_\alpha)$ projectum $=$ a singular $(L_\alpha, \emptyset')$-cardinal, there exist A and B which are r.e. in and above $\emptyset'$ which are not weakly $d\alpha$ comparable. The key observation here is that if α is admissible, then the $\Sigma_2(L_\alpha)$ cofinality function and the $\Sigma_2(L_\alpha)$ projection function are both Σ_1 in $\emptyset'$. We are then in the realm of inadmissible recursion theory.

Post's Problem under Dynamic Reducibility

We now prove that positive solutions exist for all α and β under weak d-reducibility. Let α be an admissible ordinal.

PROOF OF THEOREM 1: If the Σ_2 cofinality of α is at least equal to the Σ_2 projectum of α, then in this case Post's problem under pointwise reducibility has a positive solution above $\emptyset'$ (relativizing Friedman's proof in [4] for the structure $(L_\alpha, \emptyset')$). If $\Sigma_2(L_\alpha)$ cofinality $< \Sigma_2(L_\alpha)$ projectum, and the latter is a regular $(L_\alpha, \emptyset')$-cardinal, then the argument in Friedman [3] (apropriately relativized) applies.

Next suppose that $\Sigma_2(L_\alpha)$ cofinality $< \Sigma_2(L_\alpha)$ projectum $= \alpha$. Then α is a limit of α-cardinals, and Theorem 3 gives a positive solution to the problem. Finally suppose that $\Sigma_2(L_\alpha)$ cofinality $< \Sigma_2(L_\alpha)$ projectum is less than α and is a singular $(L_\alpha, \emptyset')$-cardinal. Apply the relativized version of Theorem 4 on the structure $(L_\alpha, \emptyset')$ to get the desired sets A and B.

Now let β be a limit ordinal.

PROOF OF THEOREM 2: If $\Sigma_1(S_\beta)$ cofinality $\geq \beta^*$, or if β^* is a regular β-cardinal, then β-r.e. sets A and B which are pointwise incomparable exist ([3], [4]). Hence we need to consider only the case when β^* is a singular β-cardinal greater than $\Sigma_1(S_\beta)$ cofinality. Theorem 4 then implies the theorem.

Let A and B be β-r.e. in C. A notion closely related to that of weak $d\beta$-reducibility is $d\beta$-reducibility (both relative to C), where β-finite sets K of the form $\exists\sigma(K \subset B^\sigma\}$ and $K \subset \bar{B}$, instead of single elements (of B say), are computed (from A) via dynamic enumeration, for any enumeration of A relative to C. Denote this relation by $B \leq_{d\beta} A$. It is easy to see that $\leq_{d\beta}$ is transitive.

We say that B is *finitely* β-recursive in A (written $B \leq_{f\beta} A$) if, for every x, the positive and negative conditions about A used to compute $B(x)$ are finite sets. It is worthwhile to note that in Lemma 1 we actually have $\emptyset' \leq_{f\alpha} A$. Friedman [7] has shown that for all limit ordinals β, there exist β-r.e. sets A and B such that neither is finitely β-recursive in the other. Clearly

$$\leq_{f\beta} \Rightarrow \leq_{wd\beta} \Rightarrow \leq_{w\beta}$$

for pointwise reduction, and

$$\leq_{d\beta} \Rightarrow \leq_\beta$$

for β-finite set reduction. Maass [10] has generalized the notion of finite reducibility to that of I-finite reducibility where only small (relative to the cofinality) computations are considered. He shows the existence of incomparable r.e. sets under this reducibility. Thus Theorem 2 can be regarded as a strengthening of the results of Friedman and Maass. In [2] we provide a detailed analysis of various notions of reducibility and their significance in the solutions of Post's problem.

References

[1] C. T. Chong and K. J. Mourad, The Friedberg-Muchnik Theorem without Σ_1 induction, *in preparation*

[2] C. T. Chong and K. J. Mourad, Post's problem and singularity, *in preparation*

[3] S. D. Friedman, Post's problem without admissibility, *Advances in Math.* **35** (1980), 30–49

[4] S. D. Friedman, β recursion theory, *Trans. Amer. Math. Soc.* **255** (197 9), 173–200

[5] S. D. Friedman, Negative solutions to Post's problem I, in: *Generalized Recursion Theory II*, North-Holland, 1978, 127–134

[6] S. D. Friedman, Negative solutions to Post's problem II, *Annals Math.* **113** (1981), 25–43

[7] S. D. Friedman, An introduction to β recursion theory, in: *Generalized Recursion Theory II*, North-Holland, 1978, 111–126

[8] M. Mytilinaios, Finite injury and Σ_1 induction, *J. Symbolic Logic* **54** (1989), 38–49

[9] G. E. Sacks and S. G. Simpson, The α-finite injury method, *Annals Math. Logic* **4** (1972), 343–368

[10] W. Maass, Recursively invariant β- recursion theory,*Annals Math. Logic* **21** (1981), 27–73

The metamathematics of Fraïssé's order type conjecture

P. Clote[1]
Department of Computer Science, Boston College

Abstract. A *well ordering* has the property that any non-empty subset has a *minimum* element. In [Girard 87], J.-Y. Girard proved that arithmetical comprehension is equivalent to a kind of ordinal exponentiation axiom, which states that *if X is well ordered then 2^X is well ordered*, where 2^X is the collection of finite sequences drawn from X under reverse lexicographic order. A *well partial ordering* has the property that any non-empty subset has finitely many *minimal* elements. In this paper, we attempt to generalize Girard's result and similarly characterize stronger comprehension axioms by such "exponentiation" axioms. In pursuing the Friedman-Simpson program of analyzing the proof theoretic "content" of mathematical theorems, we analyze the proof theoretic complexity of certain results in the theory of well quasi-orderings (wqo). The principal results are that for a fixed non-negative integer n,

(i) ATR_0 proves that the collection S_n of countable scattered linear orderings at level n of the Hausdorff hierarchy is better quasi ordered (bqo), and

(ii) ATR_0 proves that " *if α is an ordinal and Q is bqo then Q^α is bqo*".

We conjecture that the techniques introduced will eventually allow a proof in ATR_0 of Fraïssé's order type conjecture (proved by R. Laver) which states that the collection L of all countable linear orderings is wqo under embeddability.

Introduction. In 1947, R. Fraïssé [Fraïssé 48] conjectured that the collection **L** of all countable linear orderings is well quasi-ordered under embeddability. In [Laver 71], R. Laver proved this theorem and the even stronger result that **L** (and even the collection of all countable unions of scattered linear orderings) is better quasi-ordered (bqo) under embeddability. Laver's proof appears to require Π_2^1-transfinite induction. In this paper, we prove preliminary results which suggest that H. Friedman's system ATR_0 of second order arithmetic, also known as "predicative analysis", is sufficiently strong to prove Fraïssé's order type conjecture. Our key idea is to introduce an ordinal approximation α-wqo to the notion of bqo and to show a trade-off relation of the form: *Q is ω^{n+m}-wqo* iff *Q^{ω^n} is ω^m-wqo*. This proof does not use a "minimal bad sequence" argument and can be formalized in ATR_0. We believe that these tradeoff techniques can be employed to establish the conjecture that *if Q is bqo and α is an ordinal, then $Q^{Z\alpha}$ is bqo*. The provability of Fraïssé's conjecture in ATR_0 then would follow immediately, using the result proved in [Clote 88] that ATR_0 proves that any

1 Research partially supported by NSF grant # DCR-8606165.
AMS subject classification 03F35, 03F15, 03D55.
Key words and phrases: Fraïssé's order type conjecture, scattered linear ordering, arithmetic transfinite recursion.
Statement: This article is the final version and will not be submitted for publication elsewhere.

scattered linear ordering is embeddable in $\mathbf{Z}_\alpha$ for some α.

H. Friedman has shown that ATR_0 is equivalent to *comparability of well orderings*[2]. It is conjectured that ATR_0 is equivalent to the seemingly weaker statement that *the collection of well orderings is well quasi-ordered.* Because scattered linear orderings are so similar to well orderings[3], it is not inconceivable that one can similarly show that Fraïssé's conjecture implies ATR_0 and hence establish the conjecture that

Conjecture. Over ACA_0, ATR_0 equivalent with the statement that $\mathbf{L}$ is bqo (or $\mathbf{L}$ is wqo).

In light of the recent result by Friedman-Robertson-Seymour that ATR_0 is equivalent to the statement that *the collection of all finite graphs is wqo (bqo) under the relation of being a graph minor,* we believe that establishing the above conjecture would be of interest. We now turn to definitions and background material.

§1. Definitions and preliminary notions.

We refer the reader to [Simpson 85] for basic definitions and facts concerning the following finitely axiomatizable subsystems of second order arithmetic: *recursive comprehension* RCA_0, *weak König's lemma* WKL_0, *arithmetic comprehension* ACA_0, *arithmetic transfinite recursion* ATR_0, and $\Pi_1{}^1$*-comprehension* $\Pi_1{}^1\text{-}CA_0$. The principal axiom of ATR_0 is the statement

$$(X \text{ is well ordered}) \longrightarrow \exists Y \forall y \forall n[(\langle y,n\rangle \in Y \longleftrightarrow \Phi(y,\{\langle x,m\rangle : m <_X n \wedge \langle x,m\rangle \in Y\})],$$

where Φ is a first order formula possibly having free second order parameters other than Y, and the principal axiom of ACA_0 (resp. $\Pi_1{}^1\text{-}CA_0$) is

$\exists X \forall n(n \in X \longleftrightarrow \Theta(n,Y))$, where Θ is a first order (resp. $\Pi_1{}^1$) formula possibly having free second order parameters other than X.

We recall the following useful characterization of ATR_0.

Theorem 1. ([Simpson]) Over the base theory RCA_0, the following are equivalent:

[2] See [Fried 75] and [Fried 76] for a precise statement and the forthcoming monograph [Simp] for the first proof to appear in the literature.

[3] Recall that scattered linear orderings are built up inductively in the Hausdorff hierarchy. See [Clote 88] for a proof theoretic analysis of the Hausdorff decomposition theorem and of Hausdorff's characterization that the scattered linear orderings are exactly those appearing in the Hausdorff hierarchy.

1. ATR_0.
2. The schema

$$\forall i(\exists \text{ at most one } X)\Phi(i,X) \longrightarrow \exists Z \forall i(i \in Z \longleftrightarrow \exists X \Phi(i,X))$$

where $\Phi(i,X)$ is a first order formula possibly having free second order parameters other than Z.

3. For any sequence $<T_i : i \in N>$, if $\forall i(T_i$ has at most one infinite branch) then $\exists Z \forall i(i \in Z \longleftrightarrow T_i$ has an infinite branch).

A *well quasi-ordering* or *wqo* Q is a reflexive, transitive relation such that for any function $f:N \longrightarrow Fld(Q)$, there exist $i < j$ for which $f(i) \leq_Q f(j)$. We sometimes write $a \leq b \bmod Q$ instead of $a \leq_Q b$. The collection of equivalence classes mod Q of a wqo Q, where $a \approx b \bmod Q$ iff $a \leq_Q b$ and $b \leq_Q a$, forms a *well partial ordering* or *wpo* . Well partial orderings are a generalizations of ordinals. For instance, any non-empty subset A of a wpo Q has a finite set B of all "minimal" elements where $\forall a \in A \exists b \in B(b \leq_Q a)$. See [Kruskal 72] for an introductory survey of wqo theory.

Throughout this paper, all orderings, ordinals, etc. considered will be presumed to be *countable*, hence we often tacitly identify the field Fld(L) of an infinite ordering L with the natural numbers N. It is important then to distinguish between $x < y$ (as integers) and $x <_L y$ (in the ordering L). A linear ordering L is *scattered* if there is no order-preserving embedding of the rationals Q into L. In 1908, F. Hausdorff [Haus 08] gave a structure theorem for scattered linear orderings. Let 0 denote the empty linear ordering and 1 denote the linear ordering consisting of a single point. Following [Ros 82], the class of countable *very discrete* linear orderings is defined as follows, where **Z** denotes the set of positive and negative integers:

(i) $0, 1 \in VD_0$.
(ii) for $i \in \mathbf{Z}$, if $L_i \in \bigcup\{VD_\beta : \beta < \alpha\}$ then $\Sigma\{L_i : i \in \mathbf{Z}\} \in VD_\alpha$.

The class VD of countable very discrete linear orderings is taken to be $\bigcup \{VD_\alpha : \alpha < \omega_1\}$. If $L \in VD$ then the VD-rank of L, denoted by $r_{VD}(L)$, is the smallest ordinal α for which $L \in VD_\alpha$.

Definition. Powers of **Z**.
(i) $\mathbf{Z}^0 = 1$.
(ii) $\mathbf{Z}^{\alpha+1} = \mathbf{Z}^\alpha \cdot \omega^* + \mathbf{Z}^\alpha + \mathbf{Z}^\alpha \cdot \omega$
(iii) $\mathbf{Z}^\lambda = (\Sigma\{\mathbf{Z}^\alpha \cdot \omega : \alpha < \lambda\})^* + 1 + (\Sigma\{\mathbf{Z}^\alpha \cdot \omega : \alpha < \lambda\})$ for limit ordinals λ.

From work in [Clote 88], it follows that

Theorem 2. (ATR_0) If L is a scattered linear ordering, then $\exists X \exists i \exists f($ X is a well ordering $\wedge\ i \in$ Fld(X) $\wedge$ f is order preserving mapping from L into $\mathbf{Z}^i$).

If Q is a wqo, then the collection $Q^{<\omega}$ of finite sequences of elements of Q is a quasi-ordering under embeddability; i.e. for $s,t \in Q^{<\omega}$, $s \le t \bmod Q^{<\omega}$ iff there exists order preserving map $h : \{0,\dots,lh(s)-1\} \longrightarrow \{0,\dots,lh(t)-1\}$ for which $s(i) \le t(h(i)) \bmod Q$ for all $i < lh(s)$. The collection $\mathbf{P}_f(Q)$ of finite subsets of Q is a quasi-ordering under embeddability; i.e. for S,T finite subsets of Fld(Q), $S \le T \bmod \mathbf{P}_f(Q)$ iff there exists a map $h : S \longrightarrow T$ for which $q \le h(q) \bmod Q$ for all elements q of S.

The following shows the equivalence between arithmetic comprehension and the well-known Higman theorem.

Theorem 3. Over RCA_0, the following are equivalent[4]:
(i) ACA_0
(ii) $\forall Q$ (Q wqo $\longrightarrow Q^{<\omega}$ wqo).
(iii) $\forall Q$ (Q wqo $\longrightarrow \mathbf{P}_f(Q)$ wqo).
Proof sketch. See [Simpson 88] for unexplained terminology.
(i) ==> (ii) In [Girard 87] pp. 299-310, J.-Y. Girard showed that over RCA_0, ACA_0 is equivalent to the statement $\forall A,B($ A,B are well orderings $\longrightarrow (1+A)^B$ is a well ordering) and equivalent to the statement $\forall B($ B is well a orderings= $\longrightarrow 2^B$ is a well ordering). Here, for linear orderings A,B, the field of $(1+A)^B$ consists of all finite sequences $\langle\langle b_0,a_0\rangle,\dots,\langle b_{n-1},a_{n-1}\rangle\rangle$ where $a_i \in$ Fld(A), $b_i \in$ Fld(B) and $b_0 >_B \dots >_B b_{n-1}$. The linear ordering relation of $(1+A)^B$ is defined by

$$\langle\langle b_0,a_0\rangle,\dots,\langle b_{n-1},a_{n-1}\rangle\rangle \le \langle\langle b'_0,a'_0\rangle,\dots,\langle b'_{m-1},a'_{n=m-1}\rangle\rangle \bmod (1+A)^B$$

iff

$$(n \le m \wedge \forall i < n(b_i = b'_i \wedge a_i = a'_i)) \vee (\exists i < \min\{n,m\}((b_i <_B b'_i \vee (b_i = b'_i \wedge a_i <_A a'_i)) \wedge \forall j < i(a_j = a'_j \wedge b_j = b'_j)).$$

Sublemma 4.8 of [Simpson 88] states that the following is provable in RCA_0[5]. Let A be a countable partial ordering. If there exists a reification of A by the well ordering α, then there exists a *reification* of $A^{<\omega}$ by $\omega^{\omega^{\alpha+1}}$. Now if Q is wqo, then the tree T of bad finite sequences is well founded. Using ACA_0, the Kleene-Brouwer linearization L of T is well ordered. Thus Q has a reification by the well ordering L. Then $Q^{<\omega}$ has a

4 S.G. Simpson has indicated that, as well, both he and H. Friedman independently noticed the equivalence between (1) and (2).

5 See Lemma 5.2 of [SchSimp85].

reification by the well ordering $L\omega^{\omega}$. Thus $Q^{<\omega}$ is wqo.

(ii) ==> (iii) Obvious.

(iii) ==> (i) We show that over RCA_0,

$$\forall Q\,(\,Q \text{ wqo} \longrightarrow P_f(Q) \text{ wqo}) \text{ implies } \forall L(\,L \text{ is well ordered} \longrightarrow \omega^L \text{ is well ordered}).$$

By Girard's result, this clearly suffices. Suppose that L is well ordered, but that ω^L is not. Let $\{\alpha_m : m < \omega\}$ be a strictly decreasing sequence, where $\alpha_m = \omega^{\alpha_{m,0}} + \ldots + \omega^{\alpha_{m,n(m)}}$ and $\alpha_{m,0} \geq \alpha_{m,1} \geq \ldots \geq \alpha_{m,n(m)}$. Let $h(\alpha_m) = \{\alpha_{m,0}, \ldots, \alpha_{m,n(m)}\}$. Then $h(\alpha_m) \leq h(\alpha_k)$ implies that $\alpha_m \leq \alpha_k$. It follows that $P_f(Q)$ is not wqo, a contradiction. ◆

Remark. In [Higman 52] G. Higman proved (ii) and in [Erdös-Rado 52] P. Erdös-R. Rado proved (iii) and stated that independently Higman, B.H. Neumann (unpublished) and they found a proof of (ii).

For specific ordinal numbers like ω^{ω}, one says that ω^{ω} is well ordered if there are no infinite decreasing sequences of codes of smaller ordinals in the primitive recursive set of codes of ordinals smaller than ω^{ω}. The following theorem is a slight generalization of a result in [Simpson 88].

Theorem 4. Over WKL_0, the following are equivalent:
(i) $\forall n\, \forall Q(\,Q \text{ wqo} \longrightarrow Q^n \text{ wqo})$
(ii) $\forall \alpha(\,\alpha$ well ordering $\longrightarrow \alpha^{\omega}$ well ordering).

Proof sketch.
(i) ==> (ii) Suppose that α is a well ordering, but that α^{ω} is not. Let $\{\alpha_m : m < \omega\}$ be a strictly decreasing sequence, where $\alpha_m = \alpha^{k_{m,0}} + \ldots + \alpha^{k_{m,n(m)}}$ with $\omega > k_{m,0} \geq k_{m,1} \geq \ldots \geq k_{m,n(m)}$. Let $h(\alpha_m) = \langle k_{m,0}, \ldots, k_{m,n(m)} \rangle$. Then $h(\alpha_m) \leq h(\alpha_k)$ implies that $\alpha_m \leq \alpha_k$.
(ii) ==> (i) Let Q be wqo. By WKL_0, let Q^* be any linearization of Q. By formalizing the argument of [Fraïssé 86] p. 105 within RCA_0, Q^* is a well ordering. Thus Q has a reification by Q^*. By the argument of Sublemma 4.7 in [Simpson 88], Q^m has a reification by the well ordering $(Q^*)^m$ and hence Q^m is wqo. ◆

Let $Q = \{0,\ldots,n\text{-}1\}$ be quasi-ordered by the identity relation; i.e. $i \leq j \bmod Q$ iff $i=j$. Let N be quasi-ordered by the usual less than or equals relation; for $i,j \in N$, $i \leq j \bmod N$ iff $i \leq j$.

Theorem 5. [Simpson 88] Over RCA_0, the following are equivalent:
(i) $\forall n\, (\{0,\ldots,n\text{-}1\}^{<\omega} \text{ wqo})$.

(ii) $\omega^{\omega^{\omega}}$ is a well ordering.
(iii) $N^{<\omega}$ wqo.

The equivalence with (iii) is not stated, but follows easily from a minor adjustment to Simpson's proof. First, one proves the equivalence between (i*) and (ii), where (i*) is (i) except that $Q = \{0,...,n-1\}$ is quasi-ordered by the less than or equals relation. In [Simpson 88], it is shown that (ii) implies (iii), which then clearly implies (i*). ♠

§2. Auxilliary results.

With Higman's result, attempts were subsequently made to prove that *if Q is wqo then Q^{α} is wqo*, where the α-length sequences drawn from Q are quasi-ordered under embeddability. However, Rado gave an example of a wqo Q such that Q^{ω} is not wqo. He took Q to be $(N^2, \leq^*)$ where $(x,y) \leq^* (x',y')$ iff

(i) $x = x'$ and $y \leq y'$, or
(ii) $x' > \max(x,y)$.

Then Q is a wqo, but Q^{ω} is not wqo, for take $F : \omega \longrightarrow Q^{\omega}$ to be defined by $F(n) = <(n,0),(n,1),(n,2),...>$. In [Nash-Will 65] Nash-Williams later showed that desired result is true if one takes *restricted* transfinite sequences, where restricted means that the cardinality of the image of the sequence is finite. Realizing that the notion of wqo was not a sufficiently strong notion, in [Nash-Will 68], Nash-Williams introduced the stronger notion of *better quasi-ordered* (bqo) and by a difficult combinatorial argument, showed that *if Q is bqo then Q^{α} is bqo*. By a different argument which we can formalize in ATR_0, we show that ATR_0 proves this latter result.

If Q is an ordered set then let $[Q]^{\alpha}$ denote the collection of strictly increasing α- sequences drawn from Q. Similarly, $[Q]^{<\alpha}$ is the union of $[Q]^{\beta}$ for $\beta < \alpha$. If $s \in [Q]^{<\alpha}$ then lh(s) is the ordinal domain of s.

Definition. A *barrier* B is an infinite subset of the collection $[N]^{<\omega}$ of strictly increasing finite sequences of integers such that

(i) the *base* $\cup \{rng(s) : s \in B\}$ is infinite,
(ii) for any $f \in [N]^{\omega}$, there is an initial segment s of f belonging to B,
(iii) for all distinct $s,t \in B$, rng(s) is not contained in rng(t) and rng(t) is not contained in rng(s).

A *barrier'* satisfies the above definition with (iii') in place of (iii), where
(iii') for all distinct $s,t \in B$, s is not an intial segment of t and t is not an intial segment of s.

The notion of barrier was introduced by Nash-Williams, who in a series of articles, forged the subject and invented many of the combinatorial techniques ultimately used in R. Laver's resolution of Fraïssé's conjecture.

The related notion of barrier' was first introduced and investigated in [Clote 84].

Obviously, any barrier is a barrier', but not conversely :- for instance, $\{\langle 0,i\rangle : 0 < i \in \mathbf{N}\} \cup \{\langle m\rangle : 0 < m \in \mathbf{N}\}$ is a barrier' but not a barrier. It is known that every barrier is well-ordered lexicographically[6] and in [Assous 74] it is shown that the order types of the lexicographic orderings of barriers are exactly the ordinals $\omega^{\alpha}\cdot n$, where α is a countable ordinal and n is an integer. The same proof yields that any barrier' is well ordered lexicographically. We denote the order type of barrier (resp. barrier') B by $|B|$. For s,t belonging to barrier B, t is a *successor* of s if $s = \langle s_0,\ldots,s_n\rangle$ and $\langle s_1,\ldots,s_n\rangle$ is a proper initial segment of t. This is denoted by $s \angle t$. For s,t belonging to barrier' B, t is a *successor* of s if $s = \langle s_0,\ldots,s_n\rangle$ and either $\langle s_1,\ldots,s_n\rangle$ is a proper initial segment of t or for some $i \le n$, $\langle s_1,\ldots,s_n\rangle = t$. This will be denoted as well by $s \angle t$. We also say that s,t are successive elements of B. The *square* B^2 of barrier B is the collection of all s belonging to $[\mathbf{N}]^{<\omega}$ such that if $s = \langle s_0,\ldots,s_n\rangle$ then there exist $r = \langle s_0,\ldots,s_k\rangle$ and $t = \langle s_1,\ldots,s_n\rangle$ both belonging to B where $r \angle t$. It is easy to verify that $B^2 = \{\langle n\rangle\hat{\ }s : n \in$ base of B, $n <$ minimum of rng(s) $\}$ and hence if $|B| = \omega^{\alpha}$ then $|B^2| = \omega^{\alpha+1}$. The *square* B^2 of barrier' B is the collection of all s belonging to $[\mathbf{N}]^{<\omega}$ such that if $s = \langle s_0,\ldots,s_n\rangle$ then either (i) there exist $r = \langle s_0,\ldots,s_k\rangle$ and $t = \langle s_1,\ldots,s_n\rangle$ both belonging to B where $r \angle t$ or (ii) s belongs to B and for some $i \le n$, $\langle s_1,\ldots,s_i\rangle$ belongs to B. In this case, we write s as $r \# t$, so that $B^2 = \{s \# t : s,t \in B\}$. A mapping $F: B \longrightarrow Q$ from a barrier (resp. barrier') B into a quasi-ordering Q is *good* if there exist elements s,t belong to B such that $s \angle t$ and $F(s) \le F(t)$ mod Q. A mapping which is not good is called *bad*. A mapping F from B into Q is *perfect* if $s \angle t$ implies $F(s) \le F(t)$ mod Q for all s,t in B. The Nash-Williams' barrier theorem states that if B is a barrier and $F: B \longrightarrow \{0,\ldots,n\}$ then there is a subbarrier B' of B on which F is constant. This theorem is a generalization of Ramsey's theorem and is essentially equivalent to the Galvin-Prikry theorem for clopen sets. From results in [Clote 84] it is clear that ATR_0 proves the Nash-Williams' barrier theorem. It immediately follows from the barrier theorem that if $F:B \longrightarrow Q$ is a good then F restricted to some subbarrier B' of B is perfect. (Consider $G: B^2 \longrightarrow \{0,1\}$ by $G(s) = 0$ if $s = \langle s_0,\ldots s_k,\ldots,s_n\rangle$ and $\langle s_0,\ldots s_k\rangle$, $\langle s_1,\ldots s_n\rangle$ belong to B and $F(\langle s_0,\ldots s_k\rangle) \le F(\langle s_1,\ldots s_n\rangle)$.)

A quasi-ordering Q is a *better quasi-ordering* (bqo) if for all barriers B and mappings $F:B \longrightarrow Q$, F is good. A quasi-ordering Q is bqo' if every mapping from a barrier' into Q is good. By taking $B = \{\langle n\rangle : n \in \mathbf{N}\}$, it is clear that if Q is bqo (resp. bqo') then Q is wqo.

Conjecture. The set of all indices of recursive bqo's is Π_2^1-complete.

If $s:\alpha \longrightarrow Q$ and $t:\beta \longrightarrow Q$ then the sequence t is a *tail* of the sequence s if $\alpha = \gamma + \beta$ and for all $i < \beta$, $t(i) = s(\gamma+i)$. A sequence $s:\alpha \longrightarrow Q$ is *indecomposable* if s is embeddable in every non-empty tail of itself. It

6 See [Fraïssé 86] p. 66.

immediately follows from the definitions that the domain of an indecomposable sequence is an indecomposable ordinal, i.e. of the form ω^α. Following Fraïssé, we define a quasi-ordering Q to be α-bqo (resp. $<\alpha$-bqo, $\leq\alpha$-bqo) if for every sequence $s:\alpha \longrightarrow Q$ (resp. $s: \beta \longrightarrow Q$ where $\beta < \alpha$, $s: \beta \longrightarrow Q$ where $\beta \leq \alpha$) there is a non-empty tail t of s which is indecomposable.

We define the quasi-ordering Q to be α-wqo (resp. $<\alpha$-wqo, $\leq\alpha$-wqo) if every mapping F from any barrier B of order type α (resp. $<\alpha$, $\leq\alpha$) is good. For each countable additively indecomposable ordinal α, let B_α be the *canonical* barrier of order type α described on p. 382 of [Clote 84] (see also p. 12 of [Pouzet 72]).[7] We then define the quasi-ordering Q to be *canonically* α-wqo if for the canonical barrier B_α, every mapping $F: B_\alpha \longrightarrow Q$ is good.

Lemma 6. (ATR_0) Let B be any fixed barrier of order type α, where α is a countable additively indecomposable ordinal. Suppose that Q is a quasi-ordering and that it is the case that every mapping $F:B \longrightarrow Q$ is good. Then Q is $<\omega\cdot\alpha$-wqo.

Proof. Let B,α, Q satisfy the hypothesis of the lemma. Let C be an arbitrary barrier of order type stricly less than $\omega\cdot\alpha$. Suppose that $G: C \longrightarrow Q$. We must show that G is good. Let q_0 be an element not in Q and take Q' to be the quasi-ordering $Q \cup \{q_0\}$ where q_0 is taken to be incomparable with every element of Q. By hypothesis, every mapping $F: B \longrightarrow Q$ is good and so it is easy to see that every mapping $F: B \longrightarrow Q'$ is good. Now define $F: B \longrightarrow Q'$ by

$$F(s) = G(s \restriction n) \text{ if } s \restriction n \text{ is the unique initial segment of s belonging to C}$$

$$F(s) = q_0 \text{ otherwise.}$$

By hypothesis, F is good, so from the Nash-Williams' barrier theorem, there is a subbarrier B' of B on which F is perfect. An easy proof by induction on β shows that every subbarrier D' of a barrier D of additively indecomposable order type ω^β has order type ω^β. Thus $|B'| = \alpha = |B|$. Suppose that $F(B') = \{q_0\}$. Fix $s \in B'$ and let $\{n_{s,i} : i \in N\}$ be a listing of the elements of the base of B' which are greater than max rng(s). Then for all $i \in N$, $s \wedge <n_{s,i}>$ is an initial segment of a distinct element of C. This argument shows that there is an order preserving mapping from $\omega\cdot\alpha$ into C and so $|C| \geq \omega\cdot\alpha$. Thus $F(B') \neq \{q_0\}$ and so since q_0 is incomparable with every element of B, q_0 does not belong to F(B'). Hence there are s,t in B such that $s \angle t$ and $F(s) \leq F(t)$. This implies that G is good. ∎

[7] In [Clote 84], we described B_α of order type ω^α for every (not necessarily additively indecomposable) ordinal. For notational simplicity, we take B_α of order type α where α is additively indecomposable.

Corollary 7. For $1 \leq n < \omega$, within ACA_0 the following are proved equivalent:[8]

(i) Q is canonically ω^n-wqo

(ii) Q is ω^n-wqo

(iii) Q is $< \omega^{n+1}$-wqo

Proof. Note that ACA_0 is sufficiently strong to show the comparability of linear orderings of order type less than ω^ω (see [Clote 85]). The result follows from the previous lemma by noting that $\omega \cdot \omega^n = \omega^{n+1}$. ♠

Remark. We suspect but do not have a proof that the above corollary holds for all countable ordinals n, and that Q is canonically α-wqo iff Q is α-wqo for all additively indecomposable ordinals α. In any case, the lemma does yield that over ATR_0, canonical α-wqo implies $<\alpha$-wqo for any additively indecomposable ordinals α.

Lemma 8. (ATR_0) For α an additively indecomposable ordinal, if Q is $\leq\alpha$-bqo then $Q^{<\alpha}$ is wqo.

Proof. This is just Theorem 5.3 (1) on p. 222 of [Fraïssé 86], but we must use Theorem 3 to know that Higman's theorem (used in Fraïssé's proof) is provable in ATR_0.♠

Lemma 9. (ATR_0) For $\beta < \alpha$ both additively indecomposable ordinals, if Q is α-bqo then Q is β-bqo.

Proof. Let $s : \beta \longrightarrow Q$. We must show that s has an indecomposable tail. Without loss of generality, we may suppose that $\beta = \omega^\gamma$ is additively indecomposable, for otherwise write β in Cantor normal form $\omega^{\gamma_1} + ... + \omega^{\gamma_n}$ and consider the ω^{γ_n} tail of s.

Suppose first that $\alpha = \omega^\lambda$ is multiplicatively indecomposable and that $< \lambda_n : n < \omega >$ is a fundamental sequence for limit ordinal λ with the property, guaranteed by multiplicative indecomposability, that $\lambda_n + \gamma < \lambda_{n+1}$ for all $n < \omega$. Let $< \beta_n : n < \omega >$ be a fundamental sequence for β and let k(i) be the least n such that $i < \beta_n$.

Consider the sequence

$$\sum_{i < \beta} s(i).(\omega^{\lambda_{k(i)}} \cdot i)$$

(i.e.$\omega^{\lambda_{k(i)}} \cdot i$ many copies of s(i) followed by $\omega^{\lambda_{k(i+1)}} \cdot (i+1)$ many copies of s(i+1), etc.) This may be reindexed as a sequence $t : \alpha \longrightarrow Q$. By hypothesis, there is an indecomposable tail t' of t. Because for $i < j$ there are strictly more elements in the "s(i)-block" than the "s(j)-block", this gives rise to an indecomposable tail s' of s.

---

[8] The equivalence of these notions (though not in ACA_0) follows from Proposition 1 in [AP 82], together with results in [Fraïssé 86]. The statement of the result and proof technique is quite different.

Now suppose that $\alpha = \omega^{\lambda+n+1}$ and that Q is α-bqo and that $\beta = \omega^{\lambda+n}$ and that s: $\beta \longrightarrow$ Q. For each integer n, let $s^{(n)}$ be an $\omega^{\lambda+n}$ sequence of elements from Q defined by $s^{(n)}(i) = s(\omega^{\lambda}n + i)$ and consider the sequence

$$t = s + s^{(1)} + s^{(2)} + \ldots$$

Then by reindexing, t is an α sequence of elements in Q and so by hypothesis has an indecomposable tail t'. This gives rise to an indecomposable tail of s.

Lemma 10. (ATR_0) If X,Y $\in Q^{\alpha}$ and every initial segment I of X is embeddable in Y, then X is embeddable in Y.

Proof. By induction on $\beta < \alpha$, we define an order preserving map $f : \alpha \longrightarrow \alpha$ such that

$$X \restriction \beta \leq Y \restriction f(\beta) \bmod Q^{\alpha}.$$

Let $f(\beta)$ = least $\gamma < \alpha$ [$\gamma > \sup\{f(\zeta) : \zeta < \beta\} \wedge X(\beta) \leq Y(\gamma) \bmod Q$] else $f(\beta) = \alpha + 1$.
By arithmetic transfinite recursion, f is well-defined. If the image of f is not contained in α, then, using arithmetic transfinite recursion, find α_0, the least ordinal less than α such that $f(\alpha_0) = \alpha$. But then the initial segment $X \restriction (\alpha_0 + 1)$ cannot be embeddable in Y, a contradiction. Thus the image of f is contained in α, and f is an embedding of X into Y.

Lemma 11. (ATR_0) For α an additively indecomposable ordinal, if $Q^{<\alpha}$ is wqo then Q is α-bqo .
Proof. Suppose that Q is not α-bqo. Let $s : \alpha \longrightarrow Q$ be a sequence with no indecomposable tail and let $< \alpha_i : i \in N >$ be a fundamental sequence for the limit ordinal α. Let s_i be the tail of s defined by $s_i(\zeta) = s(\alpha_i + \zeta)$. We claim that

$$\forall i \forall j{\geq}i \; \exists \text{ initial segment } t_{i,j} \text{ of } s_i \; \exists\, k \geq j \text{ such that } t_{i,j} \text{ is not embeddable in } s_k.$$

Indeed, for otherwise some tail s_i would be indecomposable. Using Lemma 10 and arithmetic transfinite recursion, one can construct an array indexed by i of such initial segments $t_{i,j}$ such that $t_{i,j}$ is not embeddable in s_k where $k = k(i,j)$ and hence is not embeddable in any s_m for $m \geq k(i,j)$. Now define $r_0, r_1, \ldots, r_n, \ldots$ with $n \in$ N of elements of $Q^{<\alpha}$ with the property that if $i < j$ then r_i is not embeddable in r_j. But this contradicts the hypothesis that $Q^{<\alpha}$ is wqo. This argument comes fromTheorem 5.3(3) on p. 223 of [Fraïssé 86].

Corollary 12. (ATR_0) For α an additively indecomposable ordinal, Q is α-bqo iff Q is $\leq\alpha$-bqo iff $Q^{<\alpha}$ is wqo.

Lemma 13. (ATR_0) For α an additively indecomposable ordinal, if Q is $\leq\alpha$-bqo then Q is $\leq\alpha$-wqo.
Proof. One can easily formalize the proof due to M. Pouzet, see Lemma III.4.1 on p. 14 of [Pouzet 72] or Theorem 5.4 on p. 223 of [Fraïssé 86].

§2. Main results.

Lemma 14. (ATR_0) If α is an additively indecomposable ordinal and Q is (α+1)-wqo then Q^ω is α-wqo.

Proof. Suppose, in order to obtain a contradiction, that B is a barrier of order type α and that F: B $\longrightarrow$ Q^ω is a bad mapping. By a previous remark, the square B^2 of B is a barrier of order type $\alpha\cdot\omega$. Consider the mapping $G : B^2 \longrightarrow Q^{<\omega}$ defined by

G(s) = the least initial segment of F(r) not embeddable in F(t)

where $s = <s_0,...,s_k,...,s_n>$ and $r = <s_0,...,s_k>$ and $t = <s_1,...,s_n>$ with $r,t \in B$. Using ATR_0, by Lemma 10, one can define G. Furthermore, G is bad because F is bad. Note that the technique in sublemma 4.8 of [Simpson 88] suitable modified shows that

ACA_0 proves that *if R is α-wqo then $R^{<\omega}$ is α-wqo.*

This together with the hypothesis implies of the lemma implies that $Q^{<\omega}$ is (α+1)-wqo which then yields that G is good, a contradiction.♠

Lemma 15. For $1 \leq n,m < \omega$, there is a proof in ATR_0 that Q is ω^{n+m}-wqo iff Q^{ω^n} is ω^m-wqo.

Proof. We first sketch the direction from left to right. We show by meta-induction on n that if Q is ω^{n+m}-wqo then Q^{ω^n} is ω^m-wqo. For n = 0, this is by definition. Suppose the result true for n and assume that Q is $\omega^{(n+1)+m}$-wqo, so Q is ω^{n+m+1}-wqo. By Lemma 14, Q^ω is ω^{n+m}-wqo. By the meta-induction hypothesis, $(Q^\omega)^{\omega^n}$ is ω^m-wqo, and so $Q^{\omega^{n+1}}$ is ω^m-wqo.

We now sketch the direction from right to left. Suppose that Q^{ω^n} is ω^m-wqo and that B is a barrier of order type ω^{n+m} and that F : B $\longrightarrow$ Q. By Corollary 7, we may suppose without loss of generality that B is the canonical barrier $[N^{n+m}]$, the collection of all n+m strictly increasing sequences of integers. Let C be the canonical barrier $[N^m]$ and define $G : C \longrightarrow Q^{\omega^n}$ by letting $G(<x_1,...,x_m>)$ be the ω^n sequence of elements from Q given by the listing $\{F(<x_1,...,x_m,y_1,...,y_n>) : x_m < y_1 < ... < y_n\}$ where barrier elements from B are taken in lexicographic order. By hypothesis, G is good, hence there exist $x_1 < ... < x_{m+1}$ with $G(<x_1,...,x_m>) \leq G(<x_2,...,x_{m+1}>)$. Now the argument (due to Pouzet) in Theorem 5.4 on p. 223 of [Fraïssé 86] shows that F is good.♠

Theorem 16. For $1 \leq n < \omega$, there is a proof in ATR_0 that if Q is bqo then Q^{ω^n} is bqo.

Proof. If Q is bqo, then for all λ, Q is $\omega^{\lambda+n}$-wqo, so by Lemma 14, Q^{ω^n} is ω^λ-wqo and from Corollary 7

it follows that Q^{ω^n} is bqo.♠

Theorem 17. For $1 \leq n < \omega$, there is a proof in ATR_0 that the collection S_n of countable scattered linear orderings at level n of the Hausdorff hierarchy is better quasi ordered (bqo).

Proof. If Q is $\omega^{\lambda+1}$-wqo then by Lemma 14, Q^{ω} is ω^{λ}-wqo. To show that $Q^{\mathbf{Z}}$ is ω^{λ}-wqo, consider a mapping $F : B \longrightarrow Q^{\mathbf{Z}}$ where B is a barrier of order type ω^{λ}. Consider $G : B \longrightarrow Q^{\omega}$ defined by $G(s) = F(s) \mid \{n : n \in \mathbf{Z}, n \geq 0\}$. Since Q^{ω} is ω^{λ}-wqo, there is a subbarrier B' of B on which G is perfect. By previous remarks, $|B'| = |B|$. Now consider $H : B' \longrightarrow Q^{\omega}$ defined by

$$H(s) = F(s) \mid \{-n : n \in \mathbf{Z}, n < 0\}.$$

Since Q^{ω} is ω^{λ}-wqo, H is good and so the original function F is good. Repeating this argument shows that $Q^{\mathbf{Z}^n}$ is ω^{λ}-wqo and so $Q^{\mathbf{Z}^n}$ is bqo.

Now, the finite linear ordering $\{0,1\}$, where 0 and 1 are incomparable elements, is easily seen to be bqo. Thus $\{0,1\}^{\mathbf{Z}^n}$ is bqo and by Theorem 2, every scattered linear ordering at level n of the Hausdorff hierarchy is isomorphic to a sequence s: $\mathbf{Z}_n \longrightarrow \{0,1\}$, so we conclude the proof of the theorem.♠

Lemma 18. (ATR_0) For any quasi-ordering Q, Q is bqo iff Q is bqo'.

Proof. Clearly if Q is bqo', then Q is bqo. Suppose now that Q is bqo and that B is a barrier' and that $F : B \longrightarrow Q$. Without loss of generality, suppose that base(B) = N. Let $T_B \subseteq [\omega]^{<\omega}$ be the tree obtained from B by taking the closure with respect to initial segment, so

$$T_B = \{s \in [\omega]^{<\omega} : \exists\ t \in B\ (\text{ s is an initial segment of t})\}.$$

By arithmetical comprehension, T_B exists. Using arithmetical comprehension, it follows that the Kleene-Brouwer linearization L_B of T_B is well-ordered. Now use arithmetic transfinite recursion along the well ordering L_B to define barriers C_s for s in L_B. If $s \in B$, then take[9]

$$C_s = \{ \langle\rangle \}$$

else for $s \in L_B - B$, since s is a proper initial segment of an element of B, by (ii) of the definition of barrier', for all n > max(s), s^<n> belongs to L_B. By definition of the Kleene-Brouwer ordering, s^;<n> precedes s in L_B. Thus $C_{s\wedge<n>}$ is defined for all n>max(s). In this case, take

$$C_s = \bigcup_{n>\max(s)} \{<n>\} * C_{s\wedge<n>} * C_{s\wedge<n-1>} * \dots * C_{s\wedge<\max(s)+1>}$$

where for barriers' D,E we define $D*E = \{s\wedge t : s \in D, t \in E, \max(s) < \min(t)\}$. Thus in particular, $\{ \langle\rangle \} * E = E$. In [Clote 84] it is shown by an elementary combinatorial argument formalizable in ATR_0 that C_s is a barrier'. By construction it is clear that for each t in C_s there is a unique initial segment r of t belonging to B. Let $C = C_{\langle\rangle}$, where <> is the largest element in L_B. Define the induced map $G : C \longrightarrow Q$ by $G(s) = F(t)$ where t is the unique initial segment of s belonging to B. Clearly, G is good iff F is good. It follows that if Q is bqo

[9] In this case, C_s is technically not a barrier' because its base is not infinite.

then Q is bqo'. ∎

Lemma 19. (ATR_0) For any quasi-ordering Q, if Q is bqo' then for any ordinal α, Q^α is bqo'.

Proof. Assume the contrary in order to obtain a contradiction. Given a barrier' B and a bad mapping $F : B \longrightarrow Q^\alpha$, let $B_0 = B$. Given B_n let

$$B_{n+1} = \{s \in B_n : lh(F_n(s)) = 1\} \cup \{s \# t : s,t \in B_n \text{ and } lh(F_n(s)) > 1\}$$

define $F_{n+1} : B_{n+1} \longrightarrow Q^{<\alpha}$ by

$F_{n+1}(s) = F_n(s)$ if $s \in B_n$ and $lh(F_n(s)) = 1$ else

$F_{n+1}(s \# t)$ is the least initial segment of F(s) not embeddable in F(t).

Using arithmetic transfinite recursion and Lemma 12, the sequences $<F_n : n \in N>$, $<B_n : n \in N>$ are well defined. Now define

$$C = \{s \in \bigcup B_n : \text{if } s \in B_n \text{ then } lh(F_n(s)) = 1\}.$$

Claim. C is a barrier'.

Proof of claim. We first show (ii) in the definition of a barrier'. It is clear that base(B) = base(B_n) for all n. Given $f \in [\omega]^{<\omega}$, since each B_n is a barrier', let $s_n \in B_n$ satisfy $f \mid lh(s_n) = s_n$ for all n. But by construction, $lh(F_n(s_n)) > lh(F_{n+1}(s_{n+1}))$ for all n, thus producing an infinite decreasing sequence of ordinals, a contradiction. Hence (ii) holds. This implies that (i) holds. Toward establishing (iii'), suppose that $s,t \in C$ and s is a proper initial segment of t. Let n_0, n_1 be the least integer such that $s \in B_{n_0}$, $t \in B_{n_1}$. Let $m = \max\{n_0,n_1\}$. Then $s,t \in B_m$ and s is a proper initial segment of t. This contradicts the fact that B_m is a barrier'. Q.E.D. claim.

Now define $G : C \longrightarrow Q$ be defined by $G(s) = F_n(s_n)$ where n is such that $s_n = s$. By construction of the bad maps F_n, it follows that G is bad. But this contradicts the hypothesis that Q is bqo'. It follows that Q^α is bqo'. ∎

Theorem 20. (ATR_0) For any bqo Q and ordinal α, Q^α is bqo.

Proof. Immediate by the two previous lemmas. ∎

The result, *for any bqo Q and ordinal α, Q^α is bqo,* was first proved by Nash-Williams by different techniques which are not obviously formalizable in ATR_0. In [Nash-Will 65] Nash-Williams introduced the *barrier theorem* (essentially equivalent to the Galvin-Prikry theorem for clopen sets) and proved that *for any wqo Q and ordinal α, the collection of restricted α- sequences of Q are wqo.* Here restricted means that the cardinality of the image is finite. Since ATR_0 and the clopen Ramsey theorem are equivalent over ACA_0[10], it is natural to conjecture that ATR_0 is similarly equivalent to *for any bqo Q and ordinal α, Q^α is bqo* as well as to *for any wqo Q and ordinal α, the collection of restricted α- sequences of Q are wqo.* Note that Theorem

[10] proved in [FMS 82].

20 in a similar fashion yields a proof of Theorem 17). The techniques introduced in the last theorem may be sufficient to yield a proof of Fraïssé's conjecture in ATR_0 provided that one can a result of the form "if X,Y are scattered linear orderings both embeddable in $\mathbf{Z}^\alpha$ then recursively in 0^α (or close to that) one can determine whether X is embeddable in Y".

Acknowledgements. I would like to thank Leo Harrington for a conversation on this subject while at the Oberwolfach Recursion Theory Meeting.

References.

[AP 82] M.R. Assous and M. Pouzet, "Structures invariantes et classes de meilleurordre", Technical Report from Université Claude Bernard Lyon 1, 43, Bd du 11 novembre 1918 bat 101, 69622 Villeurbanne Cedex, February 1982, 10 pages.

[Clote 84] P. Clote, "A recursion theoretic analysis of the clopen Ramsey theorem", *Journal of Symbolic Logic* 49(2), 376-400 (1984).

[Clote 85] P. Clote, "Optimal bounds for ordinal comparison maps", *Archiv für math. Logik und Grundlagenforschung* **25**, 99-107 (1985).

[Clote 88] P. Clote, "Metamathematics of scattered linear orderings", to appear in *Archives for Mathematical Logic* (continuation of *Archiv für math. Logik und Grundlagenforschung*) .

[Erdös-Rado 52] P. Erdös-R. Rado, "Sets having divisor property, Solution to problem 4358", *Amer. Math. Monthly* **59** (1952), 255-257.

[Fraïssé 48] R. Fraïssé, "Sur la comparaison des types d'ordre", C.R. Acad. Sci., **226** (1948) Série A, 987-988 et 1,330-1,331.

[Fraïssé 86] R. Fraïssé, *Theory of Relations*, North Holland Publishing Co. (1986), 397 pages.

[Fried 75] H.M. Friedman, "Some systems of second order arithmetic and their use", *Proceedings of the International Congress of Mathematicians* in Vancouver 1974, (1975), 235-242.

[Fried 76] H.M. Friedman, "Systems of second order arithmetic with limited induction (Abstract), *J. Symbolic Logic* **41** (1976), 557-559.

[FMS 82] "A finite combinatorial principle which is equivalent to the 1-consistency of predicative analysis", *Logic Colloquium '80*, ed. G. Metakides, North-Holland, Amsterdam, (1982), 197-230.

[Girard 87] J.-Y. Girard, *Proof Theory and Logical Complexity*, Vol.1, in series *Studies in Proof Theory*, Bibliopolis Publishing Co., Naples, Italy (1987).

[Haus 08] F. Hausdorff, "Grundzüge einer Theorie der Geordneten Mengen", *Math. Ann.* **65** (1908), 435-505.

[Higman 52] G. Higman, "Ordering by divisibility in abstract algebras", *Proc. London Math. Soc.* (3) **2** (1952), 326-336.

[Kruskal 72] J. Kruskal, "The theory of well-quasi-ordering: a frequently discovered concept", *Jour. Comb. Theory* **13** (1972), 297-305.

[Laver 71] R. Laver, "On Fraïssé's order type conjecture", *Ann. of Math.* **93** (1971), 89-111.

[Nash-Will 65] C. St. Nash-Williams, "On well quasi-ordering transfinite sequences", *Proceedings Cambridge Phil. Soc.* **61**, 33-39 (1965).

[Nash-Will 68] C. St. Nash-Williams, "On better quasi-ordering transfinite sequences", *Proceedings Cambridge Phil. Soc.* **64**, 273-290 (1968).

[Pouzet 72] M. Pouzet, "Sur les prémeilleurordres", *Ann. Inst. Fourier, Grenoble* **22**(2), 1-20 (1972).

[Ros 82] J. Rosenstein, *Linear Orderings*, Volume **98** in series Pure and Applied Mathematics, eds. S. Eilenberg and H. Bass, Academic Press, Inc., (1982).

[Sacks] *Higher Recursion Theory*, forthcoming monograph. In preprint form as graduate course notes by D. MacQueen of a course by G.E. Sacks at M.I.T. in 1971-72.

[SchSimp85] "Ein in der reinen Zahlentheorie unbeweisbarer Satz über enliche Folgen von natürlichen Zahlen", *Archiv f. math. Logik und Grundl.* **25** (1985), 75-89.

[Simpson 85] S.G. Simpson, "Friedman's research on subsystems of second order arithmetic", *Harvey Friedman's Research on the Foundations of Mathematics*, L.A. Harrington et al. (eitors), North Holland Publishing Co. (1985), pp. 137-159.

[Simpson 88] S.G. Simpson, "Ordinal numbers and the Hilbert basis theorem", *Journal of Symbolic Logic*, **53** (1988), 961-974.

[Simpson] S.G. Simpson, Handwritten manuscript of sections from chapter, "Countable well orderings; analytic sets", from his forthcoming monograph on *Subsystems of Second Order Arithmetic.*

ENUMERATION REDUCIBILITY, NONDETERMINISTIC COMPUTATIONS AND RELATIVE COMPUTABILITY OF PARTIAL FUNCTIONS[1]

S. BARRY COOPER
School of Mathematics, University of Leeds
Leeds LS2 9JT, England

In a computation using auxiliary informational inputs one can think of the external resource making itself available in different ways. One way is via an oracle as in Turing reducibility, where information is supplied on demand without any time delay. Alternatively the Scott graph model for lambda calculus suggests a situation where new information, only some of it immediately related to the current computation, is constantly being generated (or *enumerated*) over a period of time in an order which is not under the control of the computer. For some purposes, such as in classifying the relative computability of total functions without any time restrictions, it makes no difference whether oracles or enumerations supply auxiliary informational inputs. But this is not generally the case in situations involving partially accessible information or time-bounds, where nondeterministic computations are involved. Clearly both models of computation (based on oracles or enumerations) have wide validity, although much more is known about the former via the rich and extensive theory of Turing computability. The purpose of this article is to survey the existing literature related to the latter with an emphasis on enumeration reducibility and its associated degree structure.

§1. Notions of relative computability for partial functions.

In practice a function may have values which are difficult or even impossible to compute according to natural criteria, in which case we are concerned with computability, or perhaps relative computability, of *partial* functions.

There is an immediate extension to partial functions of the Turing notion of relative computability in which unanswered questions to an oracle for a partial function results in an infinite wait and hence in nontermination of the computation. Let $\Phi_i, i \in \omega$, be the partial recursive functional corresponding to the ith Turing machine in some standard listing $\{Z_i\}_{i\in\omega}$ (taking Turing machines to be defined as in Davis [**Da58**] as consistent finite sets of quadruples), and let $\langle\cdot,\cdot\rangle$ be a standard coding of the pairs of numbers onto the numbers. We define $\Phi_i^g(x)$ for g a partial function as the output obtained from Z_i on input x with oracle graph$(g) = \{\langle x,y\rangle \mid g(x) = y\}$, where during the computation an applicable quadruple $q_iS_jq_kq_l$ is interpreted as leading to internal state q_k if the number $\langle m,n\rangle$ coded on the tape is in graph(g), to q_l if $\langle m,n'\rangle$ is in graph(g) for some $n' \neq n$, and nowhere otherwise (the computation gets stuck waiting for an undefined value of g).

[1] Preparation of this paper partially supported by S.E.R.C. research grant no. GRF/42003.

DEFINITION 1.1 (Sasso [Sas71], Skordev [Sk72]). We say that f is *T-reducible to g* (written $f \leq_T g$) iff $f = \Phi_i^g$ for some partial recursive functional Φ_i, $i \in \omega$. We write $f \equiv_T g$ iff $f \leq_T g \& g \leq_T f$.

$\equiv_T$ is easily seen to be an equivalence relation. We call the equivalence classes of partial functions under $\equiv_T$ the *T-degrees* and write $\mathcal{D}_T$ for the set of all T-degrees with the ordering $\leq$ induced by $\leq_T$. $\mathcal{D}_T$ is an upper semi-lattice (but by [Sas71] not a lattice, using an adaptation of the Kleene/Post argument [KP54] for $\mathcal{D}$) with least element $\mathbf{0}_T$ = the set of all p.r. functions; and there is an immediate natural embedding of the degrees of unsolvability $\mathcal{D}$ into $\mathcal{D}_T$ (onto $\mathcal{T}$, the T-degrees of total functions). Most of what is known about the structure of the T-degrees is due to Sasso and Casalegno. We give some of the more significant results:

THEOREM 1.2 (Sasso [Sas71]). *For any degree $\mathbf{a} \in \mathcal{T}$ there is a degree $\mathbf{b}$ minimal over $\mathbf{a}$.*

The existence of minimal covers in $\mathcal{T}$ for total degrees implies of course the existence of minimal T-degrees. On the other hand Sasso also shows that $\mathbf{0}_T$ is the only total T-degree with a *strong* minimal cover. All the known constructions of minimal covers, minimal T-degrees and initial segments of $\mathcal{D}_T$ depend strongly on the distribution of the $\mathbf{a}$-*semicharacteristic* degrees for $\mathbf{a} \in \mathcal{T}$.

DEFINITION 1.3. If $A \subseteq \omega$, the *semicharacteristic function* S_A for A is the function whose value is 1 on A and undefined elsewhere. A *semicharacteristic* T-degree is one containing a semicharacteristic function. We write $\mathcal{S}$ for the set of all semicharacteristic degrees. $\mathbf{b} \geq \mathbf{a}$ is $\mathbf{a}$-*semicharacteristic* iff $\mathbf{b} = \mathbf{a} \cup \mathbf{c}$ for some $\mathbf{c} \in \mathcal{S}$.

THEOREM 1.4 (Casalegno [Cas85]). *Every countable distributive lattice with a least element can be isomorphically embedded as an initial segment of (semicharacteristic) T-degrees.*

Using the Exact Pair Theorem for the T-degrees and the main result of [NS80] Casalegno is able to deduce that the first order theories of $\mathcal{D}$ and $\mathcal{D}_T$ are recursively isomorphic. However:

THEOREM 1.5 (Casalegno [Cas85]). *$\langle \mathcal{D}, \leq \rangle$ and $\langle \mathcal{D}_T, \leq \rangle$ are not elementarily equivalent.*

To see this, we need only notice that

(1) All minimal T-degrees are semicharacteristic,
(2) The join of two semicharacteristic degrees is semicharacteristic,
(3) No total degree other than $\mathbf{0}_T$ is semicharacteristic, and
(4) For each $\mathbf{a} \in \mathcal{D}_T$ there is a total $\mathbf{b} \geq \mathbf{a}$.

It follows that there is no upper cone in $\mathcal{D}_T$ of degrees which are the joins of two minimal degrees, in contrast to the situation in $\mathcal{D}$ (see [Co72]). The main result for the T-degrees below $\mathbf{0}'_T$ is the following:

THEOREM 1.6 (Sasso [**Sas73**]). *There is a co-r.e., nonrecursive $A \subseteq \omega$ such that S_A is of minimal T-degree.*

As mentioned above, the construction is simpler than the Spector [**Sp56**] minimal degree construction for the total degrees in that it suffices to make $\deg_T(S_A)$ minimal in $\boldsymbol{S}$ ($\boldsymbol{S}$ is a nontrivial ideal in $\mathcal{D}_T$). This means in effect that the problem is reduced to that of finding a minimal *pc-degree* (see the section on strong enumeration reducibilities below for the definition of ***partial conjunctive reducibility***).

The T-degrees have not been extensively studied, and there are a number of outstanding questions worth listing, mainly due to Sasso.

QUESTIONS 1.7.

(1) Since segments above a nontotal T-degree $\mathbf{a}$ cannot in general be considered in terms of segments of $\mathbf{a}$-semicharacteristic degrees, does there exist a minimal cover for a nontotal degree? Do all T-degrees have minimal covers? Are there minimal upper bounds for all countable ideals in $\mathcal{D}_T$? More generally, consider questions of homogeneity.
(2) Characterise the jumps of the minimal T-degrees (cf. [**Co73**]).
(3) Which total degrees have minimal predecessors (and in particular do all total degrees have minimal predecessors)?
(4) Examine the structure of the degrees of co-r.e. semicharacteristic functions.
(5) Are the total degrees definable in the structure of $\mathcal{D}_T$? Is the jump definable in $\mathcal{D}_T$? (cf. [**Cotal**])

Let $\{W_i\}_{i\in\omega}$, $\{D_i\}_{i\in\omega}$ be, respectively, standard listings of the recursively enumerable sets and the finite sets of numbers. If $D = D_i$, say, we write $\langle x, D\rangle$ for $\langle x, i\rangle$. A more general reducibility between partial functions is obtained from:

DEFINITION 1.8 (Friedberg and Rogers [**FR59**]). We say that $\Psi : 2^\omega \to 2^\omega$ is an ***enumeration operator*** (or ***e-operator***) iff for some r.e. set W

$$\Psi(B) = \{x \mid (\exists D)[\langle x, D\rangle \in W \,\&\, D \subseteq B]\},$$

each $B \subseteq \omega$. For any sets A, B define A ***is enumeration reducible to*** B (or A ***is e-reducible to*** B, written $A \leq_e B$) by

$$A \leq_e B \Leftrightarrow A = \Psi(B) \text{ for some e-operator } \Psi.$$

This definition formalises the notion of A being reducible to B if and only if there is a procedure such that given any enumeration of the members of B the procedure uniformly provides us with an enumeration of A. We notice that an e-operator can be thought of as being the union of a r.e. set of T-operators.

Writing $\mathrm{graph}(f) = \{\langle x, y\rangle \mid f(x) = y\}$, etc, we can define for partial functions f, g

$$f \leq_e g \Leftrightarrow \mathrm{graph}(f) \leq_e \mathrm{graph}(g).$$

We tend to identify f and $\mathrm{graph}(f)$, for instance by writing Ψ^g for the partial function f (possibly not single-valued) for which $\mathrm{graph}(f) = \Psi^{\mathrm{graph}(g)}$. The notion of *relative partial recursiveness* of partial functions introduced by Kleene in **[Kl52]** as an extension of the definition by means of systems of equations of the notion of relative recursiveness of total functions is equivalent to $\leq_e$ between graphs of partial functions. As for $\leq_T$ we have that $\leq_e$ is a transitive, reflexive reducibility, and yields a degree structure (the *enumeration* or *e-degrees*, written $\mathcal{D}_e$), with the induced ordering relation $\leq$ under which (see Rogers **[Rog67]**) $\mathcal{D}_e$ is an upper semi-lattice with a least element ($\mathbf{0}_e$ = the set of all r.e. sets). Using the above definition of $f \leq_e g$, we have a degree structure for the partial functions.

DEFINITION 1.9 (Myhill **[My61]**, Rogers **[Rog67]**). The *partial degree* $\mathbf{f}$ of f is $\{g \mid f \leq_e g \,\&\, g \leq_e f\}$, where we write $\mathbf{f} \leq \mathbf{g} \Leftrightarrow f \leq_e g$. We write $\mathcal{P}$ for the set of all partial degrees with the ordering $\leq$.

(Sasso, for example in **[Sas75]**, uses the term "partial degrees" interchangably for each of the three main degree structures, $\mathcal{D}_T$, $\mathcal{D}_e$ and $\mathcal{D}_{WT}$ restricted to the partial functions.)

There is an obvious isomorphism between the structure $\mathcal{P}$ of the partial degrees and $\langle \mathcal{D}_e, \leq\rangle$, which depends on the fact that every e-degree contains a *single-valued* set (that is, a set A such that $\langle m, n\rangle, \langle m, n'\rangle \in A \Rightarrow n = n'$, each $m, n, n' \in \omega$). Since $\leq_e$ and $\leq_T$ agree on the total functions, this gives a natural embedding (Myhill **[My61]**) of $\mathcal{D}$ into $\mathcal{D}_e$, and onto the *total* e-degrees (that is those e-degrees containing a *total* set A, meaning $A = \mathrm{graph}(f)$ for some total function f). If $\mathbf{a}$ is a Turing degree we sometimes write $\mathbf{a}_e$ for the corresponding total e-degree.

The distinction between $\leq_T$ and $\leq_e$ on the partial functions is best described in terms of Turing machines. If we allow the possibility of *inconsistent* quadruples $q_i S x q_k$, $q_i S y q_l$, $x q_k \neq y q_l$, for a Turing machine Z, we obtain the notion of a *nondeterministic oracle machine*. Z uses an oracle for a partial function g as before, but can nondeterministically pursue different computations, thought of as forming the *branches* of a *computation tree*.

DEFINITION 1.10. We say that f is *nondeterministically Turing computable from* g (written $f \leq_{NT} g$) iff there is a nondeterministic oracle-machine Z with oracle g such that

$$\mathrm{graph}(f) = \{\langle m, n\rangle \mid \text{on input } n \text{ some computation branch of } Z \text{ with oracle } g \text{ terminates with output } m\}.$$

Then:

THEOREM 1.11 (Cooper, Sasso, and McEvoy [**McE84**]). *For any partial functions* f, g, $f \leq_e g$ *if and only if* $f \leq_{NT} g$.

To illustrate the difference between $\leq_T$ and $\leq_e$, we notice (following Myhill [**My61**]) that if A is any nonrecursive set with characteristic function χ_A then

$$\chi_A \leq_e S_{A\oplus\bar{A}} \text{ but } \chi_A \not\leq_T S_{A\oplus\bar{A}}.$$

This is because on the one hand $\chi_A \leq_e S_{A\oplus\bar{A}}$ via the e-operator defined by $\{(\langle 1,n\rangle, \{\langle 1,2n\rangle\}) \mid n \in \omega\} \cup \{(\langle 0,n\rangle, \{\langle 1,2n+1\rangle\}) \mid n \in \omega\}$, and on the other hand $\chi_A \leq_T S_{A\oplus\bar{A}}$ would imply that $\chi_A \leq_T 1 =$ the constant function which takes value 1 everywhere, giving χ_A recursive.

Case [**Ca71**] pointed out an interesting consequence of the proof of Feferman's theorem in [**Fe57**] which says that every truth-table degree contains a first-order theory. Since the truth-table reduction in the proof is a positive reduction it follows that every e-degree contains a first-order theory. As a theory is axiomatisable if and only if it is effectively enumerable (Craig [**Cr53**]), this means that the enumeration degrees can be thought of as degrees of unaxiomatisability.

We examine the structure of the enumeration degrees in more detail below, but before that consider a notion of relative computability between partial functions intermediate between $\leq_T$ and $\leq_e$.

DEFINITION 1.12 (Myhill and Shepherdson [**MSh55**], Rogers [**Rog67**]). Let $\mathcal{P}$ be the set of all (single-valued) partial functions. We say that $\Theta : \mathcal{P} \to \mathcal{P}$ is a (*partial*) *recursive operator* iff there is an e-operator Ψ such that $\Theta^f = \Psi^f$ for all $f \in \mathcal{P}$ (such that $\Theta^f\downarrow$ or $\Psi^f \in \mathcal{P}$, respectively).

Recursive operators are essentially equivalent to the completely computable or compact functionals of partial functions defined by Davis in [**Da58**]. Following Sasso, we call the corresponding reducibility *weak Turing reducibility*, and write $f \leq_{WT} g$ if and only if $f = \Theta^g$ for some recursive operator. See Rogers [**Rog67**], p.281, for a proof of the fact (originally due to Myhill and Shepherdson) that $\leq_{WT}$ is strictly stronger than $\leq_e$. It is easy to see that e and WT computations are distinguished by the way consistency requirements are imposed on the different computational branches, the requirement being relative in the e case and absolute in the WT case. In the former we allow Ψ^g as long as Ψ^g is single-valued, whereas in the latter we must have Ψ^h single-valued for *all* $h \in \mathcal{P}$.

Many results concerning the e-degrees can be carried over to WT-degrees. For instance, it will follow from Gutteridge's theorem (below) that there are no minimal WT-degrees (so neither $\mathcal{D}_{WT}$ nor $\mathcal{D}_e$ are elementarily equivalent to $\mathcal{D}_T$ or $\mathcal{D}$). It is an open question whether $\mathcal{D}_{WT}$ and $\mathcal{D}_e$ are elementarily equivalent. As for the e-degrees the undecidability of the first-order theory of the WT-degrees is not yet known, although in [**He79a**] Hebeisen states that the "almost all" theory of the WT-degrees is decidable.

Little is known about the structure of the T- or WT-degrees within particular e-degrees.

Definition 1.13. Let $\leq_r$, $\leq_{r'}$ be two reflexive, transitive reducibilities with $\leq_r \subseteq \leq_{r'}$. We say that an r'-degree **a** is *r-contiguous* iff **a** contains exactly one r-degree.

Rozinas [**Roz74**] showed that any nonzero WT-contiguous e-degree is *quasi-minimal* (that is, bounds no nonzero total degrees), and Hebeisen [**He79b**] demonstrated the abundance of these WT-contiguous degrees.

See [**Sas75**] for further information comparing $\leq_e$, $\leq_{WT}$ and $\leq_T$.

§2. The Scott graph model for lambda calculus and generalised enumeration operators.

In this section we give a brief account of how the enumeration operators can be used to provide a countable version of the graph model for λ-calculus. In so doing, we offer further evidence that enumeration reducibility is *the* fundamental, general concept of relative computability in as much as the nature of the computable universe is intimately bound up with the set of enumeration operators. Most of this material originally appeared in Scott [**Sc75a**] and [**Sc75b**] ([**Sc75b**] is a shorter version of [**Sc75a**]). See Odifreddi [**Od*ta***], Section II.3, for a summary of the basic notions of the λ-calculus and McEvoy [**McE84**] for a fuller summary of details concerning the graph model.

One problem in finding a model for the (type-free) lambda calculus is that of giving an interpretation of the term $x(x)$. It is not usually possible for a function to be applied to itself as we think of a function as being an object of higher type than its arguments. However, for an e-operator there is no difficulty: an e-operator operates on sets, but an e-operator itself is essentially an r.e. set. The r.e. sets are closed under this definition of application and so the class of computable sets will be the domain of the model. The definition of application is that of evaluating an e-reduction: If x, y are variables in the domain of the model then $x(y) = \{m \mid (\exists D)[\langle m, D\rangle \in x \,\&\, D \subset y]\}$. λ-abstraction describes how we define an e-operator from an r.e. set $\lambda x.\tau = \{\langle m, D\rangle \mid m \in \tau[D/x]\}$ where $\tau[D/x]$ denotes the value of the term τ when every free occurrence of x is replaced by D. We follow Odifreddi [**Od*ta***] in summarising the details of the formulation of the model given by Scott.

The two-way correspondence between e-operators and r.e. sets is given by:

Definition 2.1. If A is an r.e. set then Ψ_A is the enumeration operator defined by it, i.e.

$$x \in \Psi_A(B) \Leftrightarrow (\exists u)(\langle x, u\rangle \in A \,\&\, D_u \subseteq B).$$

If Θ is an enumeration operator then G_Θ is a canonical r.e. set defining it, i.e.

$$\langle x, u\rangle \in G_\Theta \Leftrightarrow x \in \Theta(D_u).$$

It is straightforward to verify that $\Psi_{W_i} = \Psi_i$ and $\Psi_{G_\Theta} = \Theta$ (that is, that the e-operator corresponding to W_i is Ψ_i, and that Θ is the e-operator defined by its corresponding canonical set). In later sections we will follow the usual convention of identifying Θ and G_Θ. Under this identification any standard recursive sequence $\{W_i^s\}_{i,s\geq 0}$ of finite approximations to the r.e. sets immediately yields such a sequence $\{\Psi_{i,s}\}_{i,s\geq 0}$ for the e-operators.

We can now define a model of λ-calculus, by associating to every term t an r.e. set $[t]$. If t has n free variables then $[t]$ will be interpreted as an enumeration operator of n set variables (defined by an r.e. expression positive in the set variables). The idea is that closed terms correspond to elements and will be interpreted as r.e. sets, while terms with free variables describe collection of elements, and receive uniform interpretations by interpreting their variables.

DEFINITION 2.2 (Plotkin [Pl72], Scott [Sc75a]). We associate to every variable x of the language of λ-calculus a variable X intended to range over r.e. sets. To every λ-term t with free variables among $x_1, \ldots, x_n$ we inductively associate an arithmetical expression $[t]$ in the variables $X_1, \ldots, X_n$:

(1) $[x] = X$,
(2) $[t_1 \cdot t_2] = \Phi_{[t_1]}([t_2])$,
(3) $[\lambda x.t] = G_{\lambda X.[t]}$.

Intuitively, $[t]$ is interpreted as denoting the value of an e-operator of its set variables, which range over r.e. sets. Thus $[\lambda x.t]$ is inductively interpreted as the function over r.e. sets induced by the expression $[t]$ with respect to the variable corresponding to x. But formally the expression $[t]$ denotes not a function but a set, and so e-operators must be coded by their coded graphs.

It is again straightforward to check, by an induction on the definition of $[t]$ (cf. Plotkin [Pl72], Scott [Sc75a] or Odifreddi [Od*ta*]), that this definition produces what it is intended to:

PROPOSITION 2.3. *For any term t, $[t]$ is well-defined as an enumeration operator of its set variables. In particular, if t is closed then $[t]$ is an r.e. set.*

Looking more closely at the inductive step corresponding to part (3) of Definition 2.2, we have to show that if Ψ denotes the value of an e-operator of the variables $X, Y_1, \ldots, Y_n$ then $G_{\lambda X.\Psi}$ denotes an e-operator of the variables $Y_1, \ldots, Y_n$. But

$$\langle x, u\rangle \in G_{\lambda X.\Psi} \Leftrightarrow x \in \Psi(D_u).$$

By hypothesis, $\Psi(X)$ is an r.e. expression positive in $X, Y_1, \ldots, Y_n$, and it follows that $\Psi(D_u)$ is uniformly r.e. in u and positive in $Y_1, \ldots, Y_n$, as required. A corollary of this part of the proof is that one can iterate the process of taking the graph of an r.e. expression. Thus *any e-operator Ψ of n set variables can be written as the successive composition of e-operators of one variable*, in accordance with the interpretation of functions of many variables in λ-calculus.

We can now verify (again following Plotkin [Pl72] and Scott [Sc75a]) that the interpretation defined above provides a model of λ-calculus:

PROPOSITION 2.4. *The interpretation preserves β-equality, i.e.*

$$t_1 \stackrel{\beta}{=} t_2 \;\Rightarrow\; [t_1] = [t_2].$$

PROOF: The interpretation preserves applications of the β-rule, since

$$\begin{aligned}
[(\lambda x.t)a] &= \Phi_{[\lambda x.t]}([a]) && \text{by definition of } [t_1 \cdot t_2] \\
&= \Phi_{G_{\lambda X.[t]}}([a]) && \text{by definition of } [\lambda x.t] \\
&= (\lambda X.[t])([a]) && \text{because } \Phi_{G_\Psi} = \Psi \\
&= [t][X/[a]] && \text{by definition of } \lambda X.[t] \\
&= [t[x/a]] && \text{by induction on } t.
\end{aligned}$$

Then, by induction on the number of reduction steps, the interpretation also preserves β-equality. □

We can now use essentially the above construction to describe the graph model for lambda calculus in more generality. This entails identifying which properties of e-operators are actually used in the above construction, and abstracting from this an appropriate notion of *generalised enumeration operator*.

We first define the topological space $\mathcal{P}\omega$. For each finite $D \subset \omega$ we define $U_D \subseteq 2^\omega$ by

$$U_D = \{A \subseteq \omega \mid D \subseteq A\}.$$

The sets U_D are closed under finite intersections, and $\cup\{U_D \mid D \in [\omega]^{<\omega}\} = 2^\omega$, and hence these sets form a countable basis for a topology on 2^ω. Denoting this topological space by $\mathcal{P}\omega$, it follows that the enumeration operators are the computable continuous functions $f : \mathcal{P}\omega \to \mathcal{P}\omega$, and they form the smallest class of functions which is large enough to interpret the lambda calculus.

However, we are now interested in the class of *all* continuous functions $f : \mathcal{P}\omega \to \mathcal{P}\omega$, that is those defined by arbitrary sets $A \in 2^\omega$ with A not necessarily r.e.

DEFINITION 2.5 (Case [Ca71]). A set $A \in 2^\omega$ defines a *generalised enumeration operator* $\Lambda : 2^\omega \to 2^\omega$ iff for each $B \in 2^\omega$, each $m \in \omega$,

$$m \in \Lambda(B) \Leftrightarrow (\exists D)[\langle x, D\rangle \in A \,\&\, D \subseteq B].$$

The definition of the graph model for λ-calculus will now proceed much as before, working with generalised enumeration operators in place of e-operators, given the following:

PROPOSITION 2.6. *An operator $\Psi : 2^\omega \to 2^\omega$ is a generalised enumeration operator if and only if it is a continuous function $\Psi : \mathcal{P}\omega \to \mathcal{P}\omega$.*

PROOF: Let Ψ be a generalised enumeration operator and $O = \cup\{U_D \mid D \in \xi\}$, some $\xi \subseteq [\omega]^{<\omega}$, be an open set in $\mathcal{P}\omega$. Then for any $A \in 2^\omega$ we have

$$\begin{aligned} A \in \Psi^{-1}(O) &\Leftrightarrow \Psi(A) \in O \\ &\Leftrightarrow (\exists D \in \xi)[D \subseteq \Psi(A)] \\ &\Leftrightarrow (\exists D \in \xi)(\exists F \subseteq A)[D \subseteq \Psi(F)]. \end{aligned}$$

So $\Psi^{-1}(O) = \cup\{U_F \mid (\exists D \in \xi)[D \subseteq \Psi(F)]$ and is open.

Conversely, let $\Psi : \mathcal{P}\omega \to \mathcal{P}\omega$ be continuous, and define a set $A \in 2^\omega$ by $\langle m, D\rangle \in A \Leftrightarrow m \in \Psi(D)$. To show that the generalised enumeration operator defined by A is Ψ, it only remains to show that given $m \in \omega$ and $B \in 2^\omega$, $m \in \Psi(B)$ if and only if $(\exists D \subseteq B)[m \in f(D)]$. Let $O = \{X \in 2^\omega \mid m \in X\}$. O is a basis element and so O, and hence $\Psi^{-1}(O)$, is open. It is now easy to use the openness of $\Psi^{-1}(O)$ to verify that $B \in \Psi^{-1}(O)$ if and only if $(\exists D \subseteq B)[D \in \Psi^{-1}(O)]$, from which the result follows. □

Then the definitions corresponding to application and λ-abstraction still apply for arbitrary sets $A \in 2^\omega$, giving the required $\mathcal{P}\omega$ model for the λ-calculus.

Finally, we note that apart from Scott's use of generalised enumeration operators in defining the graph model for the λ-calculus, more general types of e-operators have been studied as interesting objects in themselves. For instance, taking the defining set A in Definition 2.5 to be a member of some standard listing $\{A_i\}_{i\in\omega}$ of the arithmetical sets, we get:

DEFINITION 2.7 (Case [**Ca71**]). A is ***arithmetically enumerable in*** B (written $A \leq_{ae} B$) iff there is some arithmetical set A_i such that

$$(\forall x)[x \in A \Leftrightarrow (\exists D)[\langle x, D\rangle \in A_i \,\&\, D \subseteq B].$$

One can define (as in Case [**Ca71**]) a relation $\equiv_{ae}$ with the usual properties, define a structure $\mathcal{D}_{ae}$ (the ***partial arithmetical degrees*** or ***ae-degrees***) in the usual way, and derive a theory for $\mathcal{D}_{ae}$ which parallels that for $\mathcal{D}_e$ (most of the results of the next section can also be stated for the ae-degrees).

Selman ([**Se71**] and [**Se72**]) examines a sequence of positive reducibilities corresponding to the individual levels of the arithmetical hierarchy.

DEFINITION 2.8 (Selman [**Se71**]). $A \mathfrak{S}_n B \Leftrightarrow (\forall X)[B \in \Sigma_n^X \Rightarrow A \in \Sigma_n^X]$.

For each n, $\mathfrak{S}_n$ is a Σ_n*-reducibility* (that is, a reflexive, transitive subrelation of "Σ_n-in"), and (an alternative characterisation of e-reducibility) $\mathfrak{S}_1 = \leq_e$:

THEOREM 2.9 (Selman [**Se71**]).

$$A \leq_e B \Leftrightarrow \forall X (B \text{ r.e. in } X \Rightarrow A \text{ r.e. in } X).$$

PROOF: The left-to-right implication is immediate. Conversely, assume that $A \not\leq_e B$, and construct $C = \bigcup_{s \geq 0} C_s$ such that B is r.e. in C but A is not r.e. in C. Satisfy "B r.e. in C" by imposing an overall requirement

$$\exists \langle x, y\rangle \in C \Leftrightarrow x \in B \tag{2-1}$$

for each $x \geq 0$. Call a finite $D \supseteq C_s$ *admissible* if it satisfies (2-1) with D in place of C, but with the right-to-left half of (2-1) restricted to $x \leq s$ (so that the admissible D's can be enumerated from an enumeration of B and a *finite* amount of information about $\bar{B}$). Satisfy $A \neq W_s^C$ (at stage $s+1$) by looking for some admissible $D \supseteq C_s$ with $x \in W_s^D - A$. If D exists, choose $C_{s+1} = D$ giving $A \neq W_s^C$. Otherwise, either $x \in A - W_s^D$ for some x, all admissible D (so $A \neq W_s^C$ again), or

$$\forall x \, (x \in A \Leftrightarrow \exists \text{ an admissible } D \text{ such that } x \in W_s^D),$$

giving $A \leq_e B$, a contradiction. □

(The above simplification of Selman's proof is essentially due to Copestake [**Copeta3**]).

Selman also proves a hierarchy theorem for $\{\mathfrak{S}_n\}_{n\geq 1}$ and shows that $\mathfrak{S}_1$ ($= \leq_e$) is a *maximal* subrelation of "Σ_1-in" (that is, of "r.e. in"). In fact, Case [**Ca74**] extends these results to show that *every* $\mathfrak{S}_n$, $n \geq 1$, is a maximal Σ_n-reducibility, that there are continuously many such maximal Σ_n-reducibilities for each $n \geq 1$, and that for each $n \geq 1$ $E \in \mathfrak{S}_n$ if and only if

$$AEB \Leftrightarrow (\exists C \in \Sigma_n)(\forall x)[x \in A \Leftrightarrow (\exists D)(\langle x, D\rangle \in C)].$$

Case also extends the proof that $\mathcal{D}_e$ is not a lattice to the degree structures for $\mathfrak{S}_n$, $n \geq 2$ (where the appropriate zero degree is Σ_n in each case) and derives results on *quasi-minimal* $\mathfrak{S}_n$-degrees (that is, degrees with no non-Σ_n total predecessors).

Another generalisation of e-reducibility occurs in Sanchis ([**San78**] and [**San79**]) in the form of *hyperenumeration reducibility*, which relates to enumeration reducibility as hyperarithmetic reducibility relates to (total) Turing reducibility.

§3. The total degrees, sets and their complements and the jump operator within the enumeration degrees.

We start by looking at the relationship of the enumeration degrees to $\mathcal{D}$ via the embedding of $\mathcal{D}$ onto the total e-degrees. Since any countable partial ordering is embeddable in $\mathcal{D}$ [**KP54**] we immediately get all such embeddings in $\mathcal{D}_e$. The following simple characterisation of the total e-degrees is due to Case ([**Ca71**]).

THEOREM 3.1. *If* **a** *is an e-degree then the following are equivalent:*

(1) **a** *is total.*
(2) **a** *contains an infinite retraceable set.*
(3) **a** *contains an infinite regressive set.*

PROOF: (2)⇒(3) is immediate.

For (3)⇒(1) let A be an infinite regressive set in **a**. Using the regressing function of A, an enumeration of A in some fixed order $f(0), f(1), \ldots$, given by some function f, can be uniformly effectively obtained from any enumeration of A. Since $f \equiv_e A$ we immediately get **a** total.

(1)⇒(2): Given a function $f \in \mathbf{a}$ let A be the set of sequence numbers of the form $\langle f(0), \ldots, f(n) \rangle$, $n \in \omega$. Then $A \equiv_e f$ and is retraceable via the function that removes the final entry from any sequence number. □

Many familiar sets of recursion theory turn out to be of total e-degree. Case **[Ca74]** notes that for each set A we have $\chi_A \equiv_e A \oplus \bar{A} \equiv_e A'$ (since $A, \bar{A} \leq_1 A'$ and A' is r.e. in A), so that A' is always of total e-degree. Moreover, if $\bar{A} \leq_e A$ we get $\chi_A \equiv_e A \oplus \bar{A} \equiv_e A$, giving $\deg_e(A)$ total ($= \deg_e(A')$). Hence (McEvoy **[McE84]**) if $\mathbf{a} \geq \mathbf{0}^{(n)}$, $n \geq 0$, and **a** contains a Π^0_{n+1} set then **a** is total. In particular (Gutteridge **[Gu71]**) the e-degrees of Π^0_1 sets are total, and the Rogers **[Rog67]** isomorphism between the r.e. Turing degrees and the Π^0_1 e-degrees (depending on the fact that $\chi_A \leq_e \chi_B \Leftrightarrow \chi_A \leq_T \chi_B$) is in fact the isomorphism induced by the natural isomorphism from $\mathcal{D}$ onto the total e-degrees. More generally, we observe that if $D = A - B$ is d-r.e. ($A, B = \cup_{s \geq 0} A^s, \cup_{s \geq 0} B^s$ respectively being r.e.) then if C is the r.e. set

$$C = \{\langle s, x \rangle \mid x \in A^s \,\&\, x \in B^t, \text{ some } t \geq s\},$$

we have $D \leq_e \bar{C}$ and $\bar{C} \leq_e D$, so $D \equiv_e C \oplus \bar{C} \equiv_e \chi_C$. So each d-r.e. e-degree is Π_1 and hence total. A similar argument shows that if D is $(n+1)$-r.e. then $D \equiv_e$ some co-n-r.e. C.

The next result shows that $\mathcal{D}_e$ properly extends the total e-degrees. We first need some notation and terminology for strings. Let $\omega^* = \omega \cup \{\uparrow\}$, where the intended interpretation of "$\varphi(m) = \uparrow$" is "φ is undefined on argument m". $(\omega^*)^\omega$ is then the set $\mathcal{P}$ of partial functions. We use α, β, γ etc. for *binary strings*, mapping from a finite initial segment of ω into $\{0,1\}$, and τ, ρ, σ etc. for ω^**-valued strings*, mapping from a finite initial segment of ω into ω^*. The *length* of string τ is $\ell h(\tau) = \mu x[\tau(x)\uparrow]$. We say τ *strongly extends* σ ($\tau \tilde{\supset} \sigma$) iff $\forall x < \ell h(\sigma)[\tau(x) = \sigma(x)]$. We say τ *extends* σ ($\tau \supset \sigma$) iff $\forall x < \ell h(\sigma)[\tau(x) = \sigma(x)$ or $\sigma(x) = \uparrow]$. σ is a *beginning* of φ iff $\sigma \tilde{\subset} \varphi$, and $\varphi \restriction x$ is the beginning of φ of length x. Similarly, α is a *beginning* of A means $\alpha \subset \chi_A$, and $A \restriction x$ is the beginning of A of length x. τ and σ are *compatible* iff one of them strongly extends the other. We write $\tau\hat{\ }\sigma$ for the standard *concatenation* of the two strings τ and σ.

DEFINITION 3.2. A non-r.e. e-degree (or partial degree) **a** is *quasi-minimal* iff there are no total predecessors $\mathbf{b} \leq \mathbf{a}$ other than $\mathbf{0}_e$.

THEOREM 3.3 (Medvedev [**Me55**]). *There exists a quasi-minimal enumeration degree*

PROOF: We construct a partial function φ to satisfy the following requirements

$$R_{2i}: \quad \varphi \neq \{i\} \text{ (the } i^{th} \text{ p.r. function)}$$
$$R_{2i+1}: \quad \Psi_i(\varphi) \text{ total } \Rightarrow \Psi_i(\varphi) \text{ p.r.}$$

We build φ by finite initial segments, starting with $\sigma_0 = \phi$ (the empty function). At stage $s+1$, let σ_s be given.

* If $s = 2i$ then we satisfy R_{2i} by choosing the least x such that $\sigma_s(x)$ is not yet defined, and extending σ_s to σ_{s+1} by defining

$$\sigma_{s+1}(x) \neq \{i\}(x).$$

* If $s = 2i+1$ then we try to trivially satisfy R_{2i+1} by making $\Psi_i(\varphi)$ not single-valued. To do this we look for an ω^*-valued string $\sigma \tilde{\supseteq} \sigma_s$ such that, for some x, y and z:

$$y \neq z \ \& \ \langle x, y\rangle, \langle x, z\rangle \in \Psi_i(\sigma).$$

If such a σ exists (subcase (a)), choose one and define $\sigma_{s+1} = \sigma$. Otherwise, we further look for $\pi, \rho \tilde{\supseteq} \sigma_s$ such that for some w $\exists \langle w, u\rangle \in \Psi_i(\pi), \langle w, v\rangle \in \Psi_i(\rho)$, with $u \neq v$. If π, ρ exist (subcase (b)), define

$$\sigma_{s+1} = \sigma_s \cup \{\langle z, 0\rangle\},$$

where $z \geq \max\{z' \mid z' \in \mathrm{dom}(\pi) \cup \mathrm{dom}(\rho)\}$, and otherwise (subcase (c)) let $\sigma_{s+1} = \sigma_s$.

To see that this strategy satisfies R_{2i+1}, we first notice that if subcase (b) applies then we have R_{2i+1} satisfied through $\Psi_i(\varphi)$ not being total (since if $\Psi_i(\varphi)(w)\downarrow$ some beginning of $\varphi \cup \pi$ or $\varphi \cup \rho$ would put us in subcase (a)). We need now only consider subcase (c) in which no such σ exists satisfying (a) or (b) and $\Psi_i(\varphi)$ is total. It is then straightforward to verify that (for $s = 2i+1$)

$$\langle x, y\rangle \in \Psi_i(\varphi) \Leftrightarrow (\exists \sigma \supseteq \sigma_s)[\langle x, y\rangle \in \Psi_i(\sigma)],$$

so that $\Psi_i(\varphi)$ is p.r. as required. It is important to notice the use of $\Psi_i(\varphi)$ total in proving the right-to-left implication - if $\Psi_i(\varphi)$ is not total, we merely get that $\Psi_i(\varphi)$ has a partial recursive extension. □

If we replace φ with a set A which is not necessarily single valued in the above proof we can dispense with subcase (b) (see Odifreddi [**Od89**]). The slightly stronger construction is needed for Theorem 4.3 below.

In [**My61**] Myhill extended Theorem 3.3 by showing that 'most' e-degrees are quasi-minimal in that the set of quasi-minimal degrees is co-meager in $\mathcal{D}_e$. Lagemann [**Lag72**] showed that the measure of the sets of quasi-minimal degree is one. In the next section we look at how finite extension arguments such as the one above can be presented using a forcing framework.

If we recursively approximate the construction of Theorem 3.3, we can show that there exists a 3-r.e. quasi-minimal e-degree (so that not all 3-r.e. degrees are total). We can construct via a finite injury priority argument a co-d-r.e. set A satisfying

$$R_{2i}: \quad A \neq W_i$$
$$R_{2i+1}: \quad \Psi_i(A) \text{ total } \Rightarrow \Psi_i(A) \text{ p.r.}$$

The R_{2i} requirements are satisfied by choosing some follower x to be extracted from A if x is enumerated into W_i. For R_{2i+1} we wait for some finite D, and some x, y, z, $y \neq z$, with $\langle x,y\rangle, \langle x,z\rangle \in \Psi_i^D$ at stage $s+1$ (say), and if such a D appears, seek at later stages to maintain $D \subset A$. If we avoid ever choosing a follower x for a requirement R_{2i} which has already appeared in such a D, we obtain A co-d-r.e., as required. Incidentally, we notice that we can adapt the construction (see[**CLW89**]) of a properly n-r.e. Turing degree to show that the hierarchy of sets of n-r.e. e-degrees is proper.. Nothing seems to be known about the structure of the n-r.e. e-degrees for $n \geq 3$.

Case [**Ca71**] gives a number of incomparability results concerning the possible e-degrees of $A, \overline{A}$ (see also [**Se71**], Theorem 2.9). For instance:

THEOREM 3.4 (Jockusch). *Every non-recursive Turing degree contains a set A such that $A \mid_e \overline{A}$.*

PROOF: In [**Jo68**] Jockusch showed that every non-recursive Turing degree contains a semi-recursive set A such that $A, \overline{A}$ are not r.e. The theorem follows by showing that

$$(\forall A)[A \text{ is semi-recursive } \& \ \overline{A} \leq_e A] \Rightarrow \overline{A} \text{ is r.e.}]. \square$$

Case notes that the theorem implies the non-existence of total e-degrees which are minimal (cf. section 5 below). This follows from the fact that if f is total then there is a set A such that $A, \overline{A} \leq_e \chi_A \equiv_e f$ and $A \mid_e \overline{A}$. Case also shows that among such sets A with $A \mid_e \overline{A}$, one can have both $A, \overline{A}$ of total e-degree, or both $A, \overline{A}$ of non-total e-degree (either possibilities with $A \leq_T \phi'$). One can also obtain the possibility with $\deg_e(A)$ total and $\deg_e(A)$ non-total (in fact quasi-minimal - see [**Gu71**], [**Mo74**] and [**Sor88**]).

An obvious consequence of Theorem 3.4 is that A and χ_A often have distinct e-degrees. For this reason we will avoid the usual convention in Turing degree theory of identifying a set A with its characteristic function. This will enable us to unambiguously identify Turing degrees with corresponding total e-degrees so long as we write $\deg_T(\chi_A)$ for $\deg_T(A)$.

Selman ([**Se71**]) showed that for each e-degree there is a larger one by proving that for each set A $\deg_e(A) <_e \deg_e(A'')$, while remarking that $\deg_e(A')$ is not always greater than $\deg_e(A)$ (as pointed out previously, $A' \equiv_e \chi_A$). Given that $A'' \equiv_e \chi_{A'}$, one might hope to extend the jump for the Turing degrees via the Turing double-jump. Unfortunately, the fact that A'' uses negative information about A, so (cf. Theorem 3.1) is dependent on a particular enumeration of A, means that A'' is not invariant with respect to e-equivalence (not even over $\mathbf{0}_e$). Instead:

DEFINITION 3.5 (Cooper [Co84], McEvoy [McE1985]). Let $K_A = \{x \mid x \in \Psi_x(A)\}$. We define the*e-jump* of A by $J(A) = \chi_{K_A}$, and the*jump* of an e-degree $\mathbf{a}$ by $\mathbf{a}' = \deg_e J(A)$, any $A \in \mathbf{a}$.

It is straightforward to verify (see [McE85]) that this jump is well-defined on the e-degrees, that $\mathbf{a} < \mathbf{a}'$ for all $\mathbf{a}$, and that $\chi_{A'} \equiv_e J(\chi_A)$, so that this jump on the e-degrees agrees with the jump on the Turing degrees (under the natural embedding). Inductively defining $J^{(n)}(X)$ by $J^{(n+1)}(X) = J(J^{(n)}(X))$ in the usual way, we can also get an analogue of Post's Theorem [Po44] for the Turing degrees:

THEOREM 3.6 (COOPER [Co84], McEvoy [McE85]). *For any sets A, B $A \in \Sigma^B_{n+1}$ if and only if $A \leq_e J^{(n)}(\chi_B)$.*

PROOF: By induction on n. We just give the case $n = 1$ and leave the inductive step to the reader.

Assuming $A \leq_e J(\chi_B)$, so that $x \in A \Leftrightarrow x \in \Psi_i(J(\chi_B))$ for some i, we get $A \in \Sigma^B_2$ by direct calculation of the quantifier form of $x \in \Psi_i(J(\chi_B))$.

Conversely, if $A \in \Sigma^B_2$ then A is r.e. in $B' = (\chi_B)'$. So $A \leq_e \chi_{(\chi_B)'} \equiv_e J(\chi_B)$ since the jumps agree via the natural embedding. □

COROLLARY 3.7. *For any set A $A \in \Sigma_{n+1}$ if and only if $A \leq_e J^{(n)}(\phi)$. In particular, $A \in \Sigma_2$ if and only if $A \leq_e J(\phi)$.*

We identify $\mathbf{0}_e^{(n)} = \deg_e(J^{(n)}(\phi))$ in the natural way with $\mathbf{0}^{(n)}$, $n \geq 1$. Then as a corollary of the Friedberg Jump Inversion Theorem it follows that the range of the jump over the e-degrees is exactly the set of total degrees $\geq \mathbf{0}'$. By analogy with the Cooper jump inversion theorem [Co73]:

THEOREM 3.8 (McEvoy [McE85]). *Given any total e-degree $\mathbf{b} \geq \mathbf{0}'$ there is a quasi-minimal degree $\mathbf{a}$ such that $\mathbf{a}' = \mathbf{b}$.*

PROOF: We add extra stages (analagous to those in Rogers' proof [Ro67] of the Friedberg jump inversion theorem) to the finite extension construction for Theorem 3.3. At these extra stages, we make $J(A) \equiv_e B$ by:

(1) Coding information about $B \in \mathbf{b}$ into A in such a way that we can retrieve the construction from ϕ' and A, giving $B \leq_e A \oplus \phi' \leq_e \chi_{K_A} = J(A)$, and
(2) Trying to ensure $\alpha \subset \chi_A$ for some finite extension $\alpha \supset \alpha_{4s+3}$ (say) with $s \in \Psi_s(\alpha^+)$, $\alpha^+ = \{x \mid \alpha(x) = 1\}$ (that is, we try to make A *force its jump* in an analagous sense to that of [Le83]). Hence, by making the construction recursive in B and ϕ' we get $J(A) = \chi_{K_A} \leq_e B \oplus \phi' \leq_e B$.

□

In addition, as a corollary of the Sacks Jump Inversion Theorem [Sa63] for the r.e. Turing degrees, McEvoy [McE85] verifies that the range of the jump over the Π_1 e-degrees is exactly the set of total Π_2 e-degrees $\geq \mathbf{0}'$.

Finally, we can define (generalised) high/low hierarchies of e-degrees exactly as for the Turing degrees (see [So87] or [Le83]). We will examine the lower levels of these hierarchies in more detail below.

§4. Genericity, basic structure and immunity properties.

The most basic structural results for $\mathcal{D}_e$, like those for $\mathcal{D}$ (see for example Lerman [**Le83**]), are best presented in the context of generic functions and sets, or in some roughly equivalent framework using category, measure or games. At this level (as can be seen from the original such treatments by Case in [**Ca71**] or by Moore in [**Mo74**]) the differences in technique needed in the two structures are interesting but undramatic.

Genericity for partial functions can be defined as in Case [**Ca71**] in terms of Cohen forcing, relative to the language of first order arithmetic augmented in such a way as to allow statements about partial functions to be directly expressed. An equivalent formulation (compare Jockusch [**Jo80**]) is:

DEFINITION 4.1. φ is *generic* iff for all arithmetical sets S of ω^*-valued strings either

(1) $\exists\sigma\tilde{\subset}\varphi$ such that $\sigma \in S$, or
(2) $\exists\sigma\tilde{\subset}\varphi$ such that $\forall\tau\tilde{\supset}\sigma(\tau \notin S)$.

An e-degree (or partial degree) **a** is *generic* iff it contains a generic function φ (that is, contains $graph(\varphi)$ for some generic φ).

We also define in the usual way

DEFINITION 4.2. A is *set generic* iff for all arithmetical sets S of binary strings either

(1) $\exists\alpha \subset \chi_A$ such that $\alpha \in S$, or
(2) $\exists\alpha \subset \chi_A$ such that $\forall\beta \supset \alpha(\beta \notin S)$.

An e-degree **a** is *setgeneric* iff it contains a generic set A.

Case's emphasis on generic functions rather than sets has the advantage that the results below for generic and n-generic e-degrees also hold in $\mathcal{P}$ for generic and n-generic partial degrees (we will see later that the notions do not always agree).

The existence of generic and set generic e-degrees follows in the usual way (as in [**Fe65**]). We summarise below the more interesting properties of the generic e-degrees.

THEOREM 4.3 (Case [**Ca71**]). *If φ is a generic function then:*

(1) *graph(φ) is infinite and contains no infinite arithmetical subset (so is immune).*
(2) *φ has no partial recursive extension.*
(3) *$\deg_e(\varphi)$ is quasi-minimal.*
(4) *If $\varphi = \varphi_0 \oplus \varphi_1$ then φ_0, φ_1 are generic and $\varphi_0 \mid_e \varphi_1$.*
(5) *$\deg_e(\varphi_0), \deg_e(\varphi_1)$ form aminimal pair (that is, $\deg_e(\varphi_0), \deg_e(\varphi_1)$ are non-zero and share no predecessors other than* **0** *) below $\deg_e(\varphi)$.*
(6) *There is a countably infinite e-independent set $\{\varphi_i\}_{i\in\omega} \leq_e \varphi$ (that is, each $\varphi_i \leq_e \varphi$ and no φ_i is $\leq_e$ the recursive join of a finite subset of $\{\varphi_i\}_{i\in\omega}$).*

(The theorem also holds with a generic set A in place of φ and graph(φ)).

PROOF: With the proof of Theorem 3.3 as background, the proofs of (1) to (4) are fairly straightforward forcing arguments.

To prove (5) let φ be generic. By (4) $\deg_e(\varphi_0), \deg_e(\varphi_1) > \mathbf{0}$. It remains to show that for each $i, j \geq 0$ either (a) $\Psi_i(\varphi_0) \neq \Psi_j(\varphi_1)$, or (b) $\Psi_j(\varphi_1)$ is r.e. Let

$$R = \{\sigma \mid \sigma \Vdash \Psi_i^{\varphi_0} \neq \Psi_j^{\varphi_1}\}$$

(that is, $R =$ the set of all σ such that there is some x for which $\Psi_i^{\tau_0}(x) \neq \Psi_j^{\tau_1}(x)$ for each $\tau \tilde{\supseteq} \sigma$). Then R is arithmetical (in fact $\in \Sigma_2$). So since φ is generic either there is some $\sigma \tilde{\subset} \varphi$ in R (in which case (a) holds), or there is some $\sigma \tilde{\subset} \varphi$ such that no $\tau \tilde{\supseteq} \sigma$ is in R. In the latter case we verify that

$$x \in \Psi_j(\varphi_1) \Leftrightarrow (\exists \tau \tilde{\supseteq} \sigma)[x \in \Psi_j^{\tau}]. \tag{4-1}$$

The left-to-right implication is immediate. Assume $x \in \Psi_j^{\tau}$, $\tau \tilde{\supset} \sigma$. Define

$$S_x = \{\pi \tilde{\supseteq} \tau \mid x \in \Psi_i^{\pi_0}\} \in \Sigma_1.$$

Then since φ_0 is generic (by (4)), and since case (a) does not hold, $x \in \Psi_i^{\varphi_0}$ giving $x \in \Psi_j(\varphi_1)$. It immediately follows from (4-1) that $\Psi_j(\varphi_1)$ is r.e., and (5) is proved.

(6) is obtained by extending the recursive decomposition of φ in (4) to a countably infinite such decomposition. □

We notice that (6) implies, using a similar embedding technique to that of Kleene and Post [**KP54**] (see also section 5 below), that any countable partial ordering $\leq_e \varphi$ is embeddable in the e-degrees below $\deg_e(\varphi)$. Case also obtains, by constructing a binary branching tree of e-independent generic functions, a continuum of independent quasi-minimal degrees. Also, all the above results are extended to $\leq_{ae}$ and the ae-degrees. Copestake [**Cope87**] observes that if $\mathbf{a}, \mathbf{b}$ are set generic then (arguing as in [**Jo80**] for the Turing degrees) the structures $\mathcal{D}_e(\leq_e \mathbf{a})$ and $\mathcal{D}_e(\leq_e \mathbf{b})$ are elementarily equivalent.

Unfortunately, by part (1) of the above theorem generic sets and functions have the disadvantage of being extremely non-constructive. Following Hinman [**Hi69**] (introducing the notion of n*-genericity*), Posner [**Pos77**] (formulating 1-genericity in terms of sets of strings) and Jockusch [**Jo80**] (defining n-generic in this way), Copestake [**Cope88**] uses the arithmetical hierarchy to refine the notion of genericity for e-degrees:

DEFINITION 4.4. φ is n*-generic* iff for all Σ_n sets S of ω^*-valued strings either (i) $\exists \sigma \tilde{\subset} \varphi$ such that $\sigma \in S$, or (ii) $\exists \sigma \tilde{\subset} \varphi$ such that $\forall \tau \tilde{\supset} \sigma (\tau \notin S)$. An e-degree is n*-generic* iff it contains an n-generic function.

Corresponding to Definition 4.2 we have the parallel definitions of n*-generic set* (as in [**Jo80**] or [**Le83**]) and *set* n*-generic e-degree.*

THEOREM 4.5 (Copestake [**Cope88**]).

(1) *There exists an* n*-generic e-degree below* $\mathbf{0}^{(n)}$ *for each* $n \geq 1$.
(2) *There is no* $(n+1)$*-generic e-degree below* $\mathbf{0}^{(n)}$, *any* $n \geq 1$.
(3) *Every* 1*-generic e-degree is quasi-minimal.*
(4) *Every* 2*-generic e-degree bounds a minimal pair of e-degrees.*
(5) *If* $\mathbf{a}$ *is a* 1*-generic e-degree then every r.e. partial ordering can be embedded below* $\mathbf{a}$ *(in the* 1*-generic degrees below* $\mathbf{a}$*).*

(All results holding with 'set n*-generic' in place of '*n*-generic').*

PROOF: (1) follows from an examination of the standard existence proof for generic functions.

For (2), follow the proof of Theorem 4.3, part (1), in showing that if φ is n-generic then graph(φ) contains no infinite Σ_n subset, and graph(φ) contains no infinite single-valued Σ_n subset. Then the result follows since $\varphi \leq_e \phi^{(n)} \Rightarrow \varphi \in \Sigma_{n+1}$.

An examination of the proof of Theorem 3.3 shows that a 1-generic set suffices to replicate the finite extension argument required, which gives (3).

For (4), we need only notice that the set R of strings in the proof of Theorem 4.3, part (5), is Σ_2.

Finally, considering Theorem 4.3, part (6), we see that the proof can be adapted to show that if φ is n-generic then the recursive decomposition $\{\varphi_i\}_{i\in\omega}$ gives an e-independent set of n-generic functions. Then (5) follows using the case $n = 1$ with the Kleene-Post embedding argument mentioned previously. □

In contrast to part (4) of Theorem 4.5, Copestake [**Cope*ta*2**] shows, using an infinite injury priority construction, that there is a 1-generic e-degree below $\mathbf{0}'$ which bounds no minimal pair of e-degrees. Jockusch has observed that the direct construction of a minimal pair $\deg_e(A)$, $\deg_e(B)$ is actually more easily achieved when one of the sets A or B is given (so every non-zero e-degree is cappable, despite, as we shall see below, the non-existence of minimal e-degrees). The Jockusch technique occurs in the proof of the Exact Pair Theorem below.

Structural results obtained by the methods of Theorem 4.5 fairly easily relativise to $\mathcal{D}_e(\geq \mathbf{d})$, any given $\mathbf{d}$. This is not possible in general (unlike for the Turing degrees), as the e-reducibility of $D \in \mathbf{d}$ to the sets constructed must be independent of the particular construction used. See Rozinas [**Roz78a**] for details of the relativisation of the minimal pair construction, and [**MC85**] for further results using these techniques, such as $(\forall \mathbf{b})(\forall \mathbf{d})[\mathbf{d} < \mathbf{b} \Rightarrow (\exists \mathbf{a} > \mathbf{d})\, \mathbf{a} \cap \mathbf{b} = \mathbf{d}]$ and a correct proof of the:

EXACT PAIR THEOREM 4.6 (Case [**Ca71**]). *Given a countable ideal* $\{\mathbf{b}_0, \mathbf{b}_1, \ldots\}$ *of e-degrees, there exist degrees* $\mathbf{a}$ *and* $\mathbf{b}$ *such that:*

(i) For every n, $\mathbf{b}_n < \mathbf{a}, \mathbf{b}$, *and*

(ii) If $\mathbf{d} \leq \mathbf{a}, \mathbf{b}$ *then for some* n, $\mathbf{d} \leq \mathbf{b}_n$.

PROOF: Similar to Spector's proof [**Sp56**] for the Turing degrees. Let $B_n \in \mathbf{b}_n$ for each n and define B by $\langle k, n\rangle \in B \Leftrightarrow k \in B_n$, each n, k. In the same way we ensure

$B_n \leq_e A$ by coding B_n into the nth row of A at stage $n+1$ of the construction, modulo a finite set of previously defined special values of A concerned with making $\Psi_i(A) = \Psi_j(B) \Rightarrow \Psi_i(A) \leq_e \oplus_{i \leq n} B_i$, each $\langle i, j \rangle \leq n$. The special values defined at stage $n+1$ are intended to make some $x \in \Psi_i(A) - \Psi_j(B)$, $\langle i, j \rangle = n$. If no such special values can be found, we are able to argue that if $\Psi_i(A) = \Psi_j(B)$ then $\Psi_i(A) \leq_e \oplus_{i \leq n} B_i$ as required. □

This shows that infinite ascending sequences of degrees do not have least upper-bounds (little is known about minimal upper-bounds for ascending sequences). We also get:

COROLLARY 4.7 (Case [**Ca71**], Selman [**Se71**]). *$\mathcal{D}_e$ is not a lattice.*

PROOF: Apply Theorem 4.7 to any infinite ascending sequence of e-degrees. □

Rozinas [**Roz78a**] shows that $\mathcal{D}_e$ is not distributive by adapting the minimal pair construction to get sets A, B for which $A \cap B \not\leq_e B$ and $\deg_e(A) \cap \deg_e(A \cap B) = \mathbf{0}$.

The 1-generic sets and functions are of special interest, as they produce structural results below $\mathbf{0}'$. An examination of the relationship between the (function) 1-generic and the set 1-generic degrees yields some unexpected consequences. We first define:

DEFINITION 4.8 (Copestake [**Cope88**]). $\mathbf{a} > \mathbf{0}$ is *minimal-like* iff there exists $\varphi \in \mathbf{a}$ such that each $\psi <_e \varphi$ has a partial recursive extension. $\mathbf{a} > \mathbf{0}$ is ***strongly minimal-like*** iff there exists $A \in \mathbf{a}$ such that each $\psi \leq_e A$ has a partial recursive extension.

This terminology is motivated by a consideration of the usual minimal (Turing) degree construction in the context of $\mathcal{P}$. We say that a binary string α ***i-splits*** if it has extensions β_1, β_2 such that, for some x, y and z with $y \neq z$,

$$\langle x, y \rangle \in \Psi_i(\beta_1) \text{ and } \langle x, z \rangle \in \Psi_i(\beta_2).$$

Making the obvious modifications in the proof of the Computation Lemma (see for instance [**Le83**], p.105), we see that if A is on the recursive tree T then: (i) If T is i-splitting we have $\chi_A \leq_e \Psi_i(\chi_A)$, and (ii) If T has no i-splittings then $\Psi_i(\chi_A)$ has a partial recursive extension whenever it is single-valued. Doing the modified minimal degree construction below $\mathbf{0}'$ ([**Sa61**]), Copestake [**Cope87**] proves that there is a total minimal-like e-degree $< \mathbf{0}'$. Since it is easy to extend Theorem 4.3, part (2) to show that no 1-generic φ has a p.r. extension, and we know that no 1-generic φ has a non-recursive total predecessor, Copestake immediately gets the corollary that there is an e-degree ($< \mathbf{0}'$) incomparable with all the 1-generic e-degrees.

By Theorem 4.5, part (5), no 1-generic e-degree is minimal-like. On the other hand:

THEOREM 4.9 (Copestake [**Cope88**]). *(i) Every set 1-generic e-degree is strongly minimal-like, and hence no set 1-generic degree has a 1-generic predecessor.*
(ii) A set A is n-generic if and only if it is the domain of some n-generic function φ. Hence there is a set n-generic $\mathbf{b}$ below any given n-generic $\mathbf{a}$, and there is a n-generic $\mathbf{a}$ above an arbitrary set n-generic $\mathbf{b}$.

PROOF: (i) Suppose A is 1-generic and $\varphi \leq_e A$, so $\varphi = \Psi_i^A$, some i. Show that some $\alpha \subset \chi_A$ forces each Ψ_i^X, $\chi_X \supset \alpha$, to be single-valued (that is, show that $\exists \alpha \subset \chi_A$

such that no $\beta \supset \alpha$ is in $S = \{\beta \mid \exists x, y, z\,(\langle x, y\rangle \in \Psi_i^{\beta^+} \;\&\; \langle x, z\rangle \in \Psi_i^{\beta^+} \;\& y \neq z)\}\}$. Then ψ defined by

$$\psi(a) = b \Leftrightarrow \exists\beta \supset \alpha[\langle a, b\rangle \in \Psi_i^{\beta^+}]$$

is a partial recursive function extending φ.

(ii) ($\Leftarrow$) Associate with a given 2-valued set $S \in \Sigma_n$ an ω^*-valued Σ_n set $T = \{\tau \mid \mathrm{d}(\tau) \in S\}$, where $\mathrm{d}(\tau)(x) = 0$ if $\tau(x) = \uparrow$, $= 1$ otherwise, each $x \leq \ell h(\tau)$. If φ is n-generic, either some $\sigma \tilde{\subset} \varphi$ is in T or for some $\sigma \tilde{\subset} \varphi$ no $\tau \tilde{\supset} \sigma$ is in T. Hence, respectively, either $\mathrm{d}(\sigma) \subset \chi_{\mathrm{dom}(\varphi)}$ is in S or $\mathrm{d}(\sigma) \subset \chi_{\mathrm{dom}(\varphi)}$ and no $\alpha \supset \mathrm{d}(\sigma)$ is in S.

($\Rightarrow$) Conversely, let A be n-generic, and define an n-generic $\varphi = \cup_{s\geq 0}\varphi_s$ where $\forall s\,(\mathrm{d}(\varphi_s) \subset \chi_A)$ gives $\mathrm{dom}(\varphi) = A$. At stage $s+1$, assume σ_s constructed, and take care of the n-genericity requirement for the sth Σ_n set S_s of ω^*-valued strings by using the n-genericity of A to find either (i) a $\sigma_{s+1} \tilde{\supset} \sigma_s$ for which $\mathrm{d}(\sigma_{s+1}) \subset \chi_A$ and $\sigma_{s+1} \in S_s$ or (ii) a $\sigma_{s+1} \tilde{\supset} \sigma_s$ for which $\mathrm{d}(\tau) \not\subset \chi_A$, each $\tau \tilde{\supset} \sigma_{s+1}$ with $\tau \in S_s$. The rest of part (ii) follows from the fact that $\mathrm{dom}(\varphi) \leq_e \varphi$, each φ. □

We therefore get the surprising result that not only are the notions of 1-generic and set 1-generic degree distinct, they are mutually exclusive. This reinforces our intuition that, for instance, it is *harder* constructing a quasi-minimal function than a quasi-minimal set (as remarked following the proof of Theorem 3.3), namely because our constructions produce different degrees.

A number of questions arise by analogy with the situation for the 1-generic Turing degrees, such as those concerned with jump and downward closure.

As for the Turing degrees, an e-degree $\mathbf{a}$ is said to be *low*$_n$ (with *low*=low$_1$) if $\mathbf{a}^{(n)} = (\mathbf{a} \cup \mathbf{0}')^{(n-1)}$, and *high*$_n$ (with *high*=high$_1$) if $\mathbf{a}^{(n)} = (\mathbf{a} \cup \mathbf{0}')^{(n)}$. We can characterise the low e-degrees [**MC85**] as those consisting entirely of Δ_2 sets. Since every 1-generic Turing degree $\leq \mathbf{0}'$ is low (see for instance [**Le83**], p.80), we have for each 1-generic $A \in \Delta_2$ that $J(A) \leq_e J(\chi_A) \equiv_e \chi'_A \in \mathbf{0}'$, so (Copestake [**Cope*tal***]) every Δ_2 set 1-generic $\mathbf{a}$ is low. On the other hand, there are 1-generic sets of degree $< \mathbf{0}'$ that are not low:

DEFINITION 4.10. An e-degree $\mathbf{a} < \mathbf{0}'$ is *properly*-Σ_2 iff it contains no Δ_2 sets.

THEOREM 4.11 (Copestake [**Cope*tal***]). *There exists a set 1-generic e-degree $< \mathbf{0}'$ which is properly Σ_2, and hence is not low.*

PROOF: Construct a Σ_2 set A by means of a Σ_2 *approximation* $\{A^s\}_{s\geq 0}$ (that is, $\{A^s\}_{s\geq 0}$ is a recursive sequence of finite sets with

$$x \in A \Leftrightarrow \exists t \forall s > t\,(x \in A^s)).$$

We need a standard listing $\{B_i\}_{i\geq 0}$ of the Σ_2 sets in which the Δ_2 sets appear with Δ_2 approximations $\{B_i^s\}_{s\geq 0}$, and take $W_i = \cup_{s\geq 0} W_i^s$, $i \geq 0$, to be a standard listing of all r.e. sets of binary strings. The requirements to be satisfied are

$$S_i: \;\; \exists\alpha \subset \chi_A (\alpha \in W_i) \text{ or } \exists\alpha \subset \chi_A \forall\beta \supset \alpha (\beta \notin W_i),$$
$$R_i: \;\; [A = \Psi_i(B_i) \;\&\; B_i = \Theta_i(A)] \Rightarrow \exists x \lim_s B_i^s(x) \uparrow,$$

where $\{(\Psi_i, \Theta_i)\}_{i\geq 0}$ is a standard listing of all pairs of e-operators.

We satisfy the R_i requirements by looking for a follower (x, D, E) (D, E finite) satisfying the 'set-up' $x \in \Psi_i(D)$ & $D \subseteq \Theta_i(E)$. If some such triple is found we fix $E - \{x\}$ in A and also put x in A. We then wait until $D \subseteq B_i^s$. If this happens we extract x from A and insert it again only if $D \not\subseteq B_i^t$ at a later stage t. We repeat the extraction and insertion of x so that if the set-up returns infinitely often some element of D is forced in and out of B_i, thus preventing it from being Δ_2. (We call this the *'properly Σ_2 strategy'* - note the parallel with the d-r.e. strategy in, for instance, [**CLW89**]).

We satisfy S_i by looking for a string in W_i^s that is compatible with action taken on higher priority requirements, and making it a beginning of A. Because of the infinite outcome to R_i in which $\lim_s B_i(x)$ does not exist, a tree of strategies is required to provide requirements with sufficient information concerning higher priority outcomes and to help with the coordination of actions on the different requirements. □

By Theorem 4.9, part (ii), we have the corollary that not every 1-generic e-degree below $\mathbf{0}'$ is low. We are left with:

PROBLEM **4.12**. Characterise the jumps of the (set) 1-generic e-degrees below $\mathbf{0}'$.

We can also ask (cf. Haught [**Ha86**]):

QUESTION **4.13**. Are the e-degrees of 1-generic sets (below $\mathbf{0}'$) closed downwards?

We have seen that 1-generic sets and the graphs of 1-generic functions are immune. In fact, 1-generic sets ([**Jo80**]) and the graphs of 1-generic functions ([**Cope87**]) are hyperimmune (but not cohesive). It follows from the following result that all $\mathbf{b} \geq$ a set 1-generic $\mathbf{a}$ are hyperimmune.

THEOREM **4.14** (Rozinas [**Roz78b**]). *(i) The immune e-degrees are closed upwards, and (ii) so are the hyperimmune e-degrees.*

PROOF: (i) Say $B \geq_e$ an immune set A. Define an infinite set $C \equiv_e B$ by $C = \{\langle m, n\rangle \mid m \in A \ \&\ n \in B \ \&\ m \geq n\}$. Verify that if W is r.e. and $\subseteq C$, then $X = \{x \mid \exists n [\langle x, n\rangle \in W]\}$ is $\subseteq A$ and so finite, giving W finite.
(ii) If A is now hyperimmune, define $C \equiv_e B$ as before. Assume that C is not hyperimmune and let f be a recursive function for which:

(4-2) $\qquad \forall x, y\, [x \neq y \Rightarrow D_{f(x)} \cap D_{f(y)} = \phi]$, and

(4-3) $\qquad \forall x\, [D_{f(x)} \cap C \neq \phi]$.

Define g recursive by $g(0) = f(0)$ and for all x, $g(x+1) = f(n_x)$ where

$$n_x = \mu n \geq x + 1\, [\forall z \leq x\, \forall i, j, k\, [\langle i, j\rangle \in D_{g(z)} \Rightarrow \langle i, k\rangle \notin D_{f(n)}]].$$

By (4-3) f is total. Define a recursive h by

$$x \in D_{h(n)} \Leftrightarrow \exists y\, [\langle x, y\rangle \in D_{g(n)}],$$

each x, n. Verify that h witnesses that A is not hyperimmune, a contradiction. □

However, in contrast to the situation in the Turing degrees:

THEOREM 4.15 (Rozinas [Roz78b]). *There exist immune-free e-degrees (that is, degrees containing no immune sets).*

PROOF: Inductively define approximations A^s, B^s to sets A, B (A^s r.e., B^s finite) satisfying for each s

$$R_s: \quad A^s \cap B^s = \phi,\ A^s, B^s \subset A^{s+1}, B^{s+1} \text{ respectively, and } \bar{A}^s \text{ contains an infinite r.e. subset } C_s, \text{ and}$$

$$S_s: \quad \Psi_s(\omega - B^s) \text{ is finite, or } \Psi_s(A^s) \text{ contains an infinite r.e. set.}$$

Assume also some steps to make $A \neq W_s$. At stage $s+1$, we look for a finite $D \supset B^s$ with $A^s \cap D = \phi$ and $\Psi_s(\bar{D})$ finite. If such a D exists, defining $B^{s+1} = D$ satisfies S_s.

Otherwise, we are able to inductively define at substages $t \geq 0$ finite extensions $P^t \supset P^t \supset \ldots \supset A^s$, $C^t_{s+1} \supset C^{t-1}_{s+1} \supset \ldots \supset B^s$ (using $C^t_s \supset C^{t-1}_s \supset \ldots \supset B^s$), satisfying

(1) $P^t \cap C^t_{s+1} = \phi$,
(2) $\Psi_s(P^t) \supset \Psi_s(P^{t-1})$ (using the fact that no such D exists), and
(3) $P = \cup_{t \geq 0} P^t$ is r.e.

In this case, defining $A^{s+1} = P$ satisfies S_s. □

It is clear from the above proof that there exist immune-free e-degrees below $\mathbf{0}''$ (while in section 6 below we see that this result is best possible in that below $\mathbf{0}'$ the situation is simpler). See Solon [Sol78] for properties of e-degrees containing hyperimmune retraceable sets. The main gap in our knowledge concerns possible analogues of the Jockusch [Jo73] upward-closure results for the cohesive Turing degrees.

QUESTION 4.16. Are the e-degrees of cohesive sets (and of hyperhyperimmune sets) closed upwards in $\mathcal{D}_e$?

It is not clear yet what role immunity properties play in the theory of the e-degrees, and to what extent such properties naturally relate to structure and jump.

§5. Density in the enumeration degrees.

As we saw above, in the context of the e-degrees the Spector minimal degree construction gives a degree which is minimal-like rather than minimal. In fact, Gutteridge [Gu71] was able to show that no minimal e-degrees exist. The proof falls into two parts. The first part, which relativises to show that any e-degree has at most countably many minimal covers, is a demonstration that any set of minimal e-degree must be Δ^0_2. The second is a very different proof, which only relativises above total degrees, of the fact that no Δ^0_2 set is of minimal e-degree.

THEOREM 5.1 (Gutteridge [Gu71]). *If* $\deg_e(B)$ *is a minimal cover for* $\deg_e(A)$ *then* $B \in \Delta^A_2$.

PROOF: Enumerate an e-operator Θ such that for *any* sets A, B

$$[\Theta^B \leq_e A \text{ or } B \equiv_e A \oplus \Theta^B] \Rightarrow B \in \Delta^A_2.$$

Then given $A <_e B$ we will have $B \in \Delta_2^A$ or $A <_e A \oplus \Theta^B <_e B$, which will prove the theorem. The requirements to satisfy are

$$N_i: \quad B = \Psi_i(A \oplus \Theta^B) \Rightarrow B \leq_e A,$$
$$P_i: \quad \Theta^B = \Psi_i^A \Rightarrow B \in \Delta_2^A,$$

each $i \geq 0$. We recursively enumerate Θ at stages $s \geq 0$ by a combination of ticking and crossing, where to *tick* $\langle n, x\rangle$ (via Θ) means to enumerate $\langle\langle n, x\rangle, \{n\}\rangle$ into Θ, and to *cross* $\langle n, x\rangle$ (via Θ) means to enumerate $\langle\langle n, x\rangle, \phi\rangle$ into Θ. The strategy for P_i is to tick an initial segment of the xth column up to some $\langle n_x, x\rangle$ (each x) in such a way that $\langle n_x, x\rangle$ never gets crossed. Hence if $\Theta^B = \Psi_i^A$ we have

$$x \in B \Leftrightarrow \langle n_x, x\rangle \in \Theta^B \Leftrightarrow \langle n_x, x\rangle \in \Psi_i^A$$

where $\lambda x\,[n_x] \leq_T \phi'$, so $B \in \Delta_2^A$. The strategy for N_i is to ensure that if $x \in \Psi_i(A \oplus \Theta^B)$ then $x \in \Psi_i(A \oplus \Theta^{B\restriction x})$ modulo a finite number of pairs $\langle n, y\rangle$ ticked through $y < i$. This is achieved by crossing numbers $\langle n, y\rangle \in D - [0, x-1]$, $y \geq i$, for some $D \subseteq \Theta^B$ with $x \in \Psi_i(A \oplus D)$, if we get $x \in \Psi_i(A \oplus \Theta^B)$ at some stage. This means that, if we are given an enumeration of A and have $B = \Psi_i(A \oplus \Theta^B)$, then we can inductively get $x \in B$ enumerated in B from an enumeration of $B \restriction x$. The P_i-strategy is allowed to succeed because we only allow crossing of $\langle n, x\rangle$ through N_j if $j \leq x$, and only finitely many numbers are crossed through any given N_j. □

COROLLARY 5.2 (Gutteridge [**Gu71**]). *If* **b** *is a minimal cover for* **a** *then* $\mathbf{b} \leq \mathbf{a}'$. *Hence any e-degree has at most countably many minimal covers.* □

THEOREM 5.3 (Gutteridge [**Gu71**]). *If $B \in \Delta_2^0$ is not r.e. then there is an e-operator Θ such that $\phi <_e \Theta^B <_e B$. Hence there is no minimal e-degree.*

PROOF: Let $\{B^s\}_{s\in\omega}$ be a recursive sequence of finite approximations to $B \in \Delta_2^0 - \Sigma_1^0$. We again enumerate Θ by ticking and crossing at stages $s \geq 0$. We satisfy the requirements

$$N_i: \quad B \neq \Psi_i(\Theta^B),$$
$$P_i: \quad \Theta^B \neq W_i.$$

Attend to P_i by monitoring the initial segment of agreement $L(i,s)$ between W_i and Θ^B at stage $s+1$, and as agreement grows tick as yet unmarked numbers $\langle j, i\rangle < L(i,s)$ at stages $s+1 > i$. Attend to N_i by monitoring the initial segment of agreement $\ell(i,s)$ between B and $\Psi_i(\Theta^B)$ at stage $s+1$, and try to make the segment of agreement independent of the approximation B^s by crossing numbers $\langle j, i'\rangle \in \Theta^B$ used by $\Psi_i(\Theta^B) \restriction \ell(i,s)$ for which $i' > i$. Inductively verify that the construction is finite injury and each requirement is satisfied. P_i is satisfied, since otherwise $\langle j, i\rangle \in W_i = \Theta^B \Leftrightarrow j \in B$ for all but a finite number of j's, giving B r.e., a contradiction. N_i is satisfied, since otherwise all but a finite set of numbers $\langle j, i'\rangle$ (with $i' \leq i$) are crossed, giving Θ^B r.e. Then use $\langle x, D\rangle \in \Theta \Rightarrow |D| \leq 1$ to deduce $B = \Psi_i(\Theta^B)$ from the unboundedness of $\ell(i,s)$, $s \geq 0$, again contradicting B not r.e. □

COROLLARY 5.4 (Gutteridge [**Gu71**],Lagemann [**Lag71**]). *If either* **a** *or* **b** *is total, then* **b** *is not a minimal cover for* **a**.

PROOF: By Corollary 5.2, we need only show that if $A <_e B$ and either (1) A is total with $B \leq_T A'$, or (2) B is total, then there is an e-operator Θ with $A <_e \Theta^B \oplus A <_e B$.

For (1), just relativise the above proof to get Θ r.e. in A, so $\Theta^B = \Psi_{\Phi(A)}(B)$ for some e-operator Φ, $= \tilde{\Theta}^B$, say, for some e-operator $\tilde{\Theta}$, since $A \leq_e B$.

(2) Let Ψ be an e-operator with $A = \Psi^B$. If $B = \{\langle x, b(x)\rangle \mid x \in \omega\}$ where b is total, let $B^s = \{\langle x, b(x)\rangle \mid x < s\}$. Take as requirements: $W_0^{\Psi(B)} \neq \Theta^B$, $B \neq \Psi_0(\Theta^B \oplus \Psi^B), \ldots$, etc., and proceed as in the theorem to get Θ r.e. in B. Then, as in part (1), $\Theta^B = \tilde{\Theta}^B$ for some e-operator $\tilde{\Theta}$. □

Having eliminated the possibility of minimal e-degrees (so that $\mathcal{D}_e$ is not elementarily equivalent to most of the standard degree structures, including the Sasso T-degrees), we can now examine the extent to which the above techniques can be extended to characterise the possible initial segments of $\mathcal{D}_e$.

Let $B \in \Delta_2^0$. By combining the ticking and crossing techniques for the P_i- and N_i-strategies (respectively) in Theorem 5.3, Lagemann showed that e-operators Θ_j, Θ_k can be enumerated satisfying incomparability requirements of the form $\Theta_j^B \neq \Psi_i(\Theta_k^B)$. Hence (cf. Theorem 4.5, part (5)):

COROLLARY 5.5 (Lagemann [**Lag71**]). *If* **b** *is a non-zero* Δ_2^0 *e-degree, then any r.e. partial ordering can be embedded below* **b**. □

This result relativises (McEvoy [**McE84**]) along the lines of Corollary 5.4. For further details of the above proofs see [**Co82**], [**MC85**] and [**Od89**].

The total-ness requirements in Corollary 5.4 and the Δ_2^0 restriction in Corollary 5.5 are necessary, as we will see in the following sections. However, below $\mathbf{0}'$ we we can get:

THEOREM 5.6 (Cooper [84]). *The structure of the e-degrees below* $\mathbf{0}'$ *is dense*

PROOF: Given Σ_2^0 sets A, B where $B \not\leq_e A$, build an e-operator Θ with $A <_e \Theta^{B\oplus A} \oplus A <_e B \oplus A$. Choose Σ_2 approximations $\{A^s\}_{s\in\omega}, \{B^s\}_{s\in\omega}$ to A, B so that $\{A^s \oplus B^s\}_{s\in\omega}$ is a *thin* Σ_2 approximation to $A \oplus B$ (that is, there are infinitely many *stages* s - *called thin* stages - at which $A^s \oplus B^s \subseteq A \oplus B$). The requirements to satisfy are:

$$N_i: \quad B \neq \Psi_i(\Theta^{B\oplus A} \oplus A),$$
$$P_i: \quad \Theta^{B\oplus A} \neq \Psi_i^A,$$

each $i \geq 0$. Monitor N_i, P_i with respective length (of agreement) functions $\ell(i,s)$ and $L(i,s)$. The thin stages ensure that despite our approximations to A, B not being Δ_2, we still have infinitely many stages at which $\ell(i,s)$ and $L(i,s)$ approximate a true eventual outcome for N_i and P_i respectively, and at which our N_i- and P_i-strategies are effectively directed towards satisfying their corresponding requirements (at thin stages we are on the 'true path' of an implicit tree of outcomes).

Attend to P_i at stage $s+1$ by enumerating $\langle\langle x,i\rangle, B^s \oplus A^s\rangle$ into Θ for each $x \in B^s$ with $x \leq L(i,s)$. Attend to N_i at stage $s+1$ by enumerating $\langle z, B^s \restriction \delta \oplus A^s\rangle$ into Θ for each $z = \langle x,j\rangle$ (say) $\in \Theta(B^s \oplus A^s)$, with $j > i$, which is selected to be used in enumerating $\Psi_i(\Theta(B^s \oplus A^s) \oplus A^s) \restriction \ell(i,s)$ at stage $s+1$, where δ is chosen to be the least number we can take without interfering with axioms enumerated in Θ via higher priority requirements N_j, $j < i$. The verification is similar to that of Theorem 5.3 but restricted to the thin stages. Inductively check that we attend to an N_i or P_i at at most finitely many thin stages, and that each requirement is satisfied.

If we attend to P_i infinitely often we get $\Psi_i^A = \Theta^{B\oplus A}$ through the numbers enumerated in Θ at thin stages. Then P_i is satisfied, since otherwise $\langle j,i\rangle \in \Psi_i^A = \Theta^{B\oplus A} \Leftrightarrow j \in B$ for all but a finite number of j's, giving $B \leq_e A$, a contradiction. By looking at thin stages again, if we attend infinitely often to N_i we get $B = \Psi_i(\Theta^{B\oplus A} \oplus A)$. Then N_i is satisfied, since otherwise all but a finite set of numbers $z \in \Theta(B \oplus A)$ selected to be used by $\Psi_i(\Theta^{B\oplus A} \oplus A)$ are made independent of all but a finite part of B, giving $B = \Psi_i(\Theta^{B\oplus A} \oplus A) \leq_e A$, a contradiction. Even though requirements may receive attention at infinitely many non-thin stages, the axioms for Θ so defined will not injure lower priority requirements. □

We look at the structure of the Σ_2^0 e-degrees in more detail in section 7.

§6. Techniques for proving non-density.

Whereas A-partial recursive trees (due to Shoenfield [**Sh66**]) provide the framework for relativisations of the Spector minimal degree construction, for the construction of a minimal cover in the e-degrees we need the following sequence of definitions (see [**Co87**] and [**Co*ta*2**]). Let S denote the set $\{0,1\}^{<\omega}$ of all binary strings.

Definition 6.1. A *celling* $\mathcal{C}$ is a partial function $\lambda\alpha C_\alpha$ from binary strings to sets of numbers (where the C_α with $C_\alpha\downarrow$ are the *cells* of $\mathcal{C}$), satisfying for all $\alpha, \beta \in S$ with $C_\alpha, C_\beta \downarrow$:
(a) $\alpha \subseteq \beta$ implies $C_\alpha \subseteq C_\beta$, and (b) If η is a set of strings and $C_\alpha \subseteq \bigcup_{\beta\in\eta} C_\beta$ then $\alpha \subseteq \beta$ some $\beta \in \eta$.

Definition 6.2. The *α-increment* in $\mathcal{C}$ is defined by $I_\phi = C_\phi$ and $I_\alpha = C_\alpha \setminus C_{\alpha^-}$ if $\alpha = (\alpha^-)\hat{\ } i$ for some $i \leq 1$. We write $C_\alpha \in \mathcal{C}$ iff C_α is a cell of $\mathcal{C}$ and $I_\alpha \in \mathcal{I}$ iff I_α is an increment of $\mathcal{C}$. Sometimes we identify $\mathcal{C}$ ($\mathcal{I}$) with $\{\langle x,\alpha\rangle \mid x \in C_\alpha,\ \alpha \in S\}$ ($\{\langle x,\alpha\rangle \mid x \in I_\alpha,\ \alpha \in S\}$).

Definition 6.3. If $C_\alpha \in \mathcal{C}$, we say that C_α *is in* $\mathcal{C}$. If $\alpha \subseteq \beta$, we say that C_β *includes* C_α (in $\mathcal{C}$). $\mathcal{C}$ is *closed* iff for each C_α in $\mathcal{C}$ and each $\beta \subset \alpha$ we have a cell $C_\beta\downarrow$ in $\mathcal{C}$.

Definition 6.4. A set A *weakly respects* the celling $\mathcal{C}$ iff $A = \bigcup_{\alpha\in\xi} C_\alpha$ for some set $\xi \subseteq S$. A *respects* $\mathcal{C}$ iff ξ is a chain in S.

DEFINITION 6.5. We say that $\{C_{\alpha_0}, C_{\alpha_1}, \ldots\}$ is a *skeleton* for a set X of cells of $\mathcal{C}$ iff
(a) $X \subseteq \{C_\beta \mid \beta, \alpha_i$ are comparable for some $i = 0, 1, \ldots\}$,
(b) If $i \neq j$ then $\alpha_i \mid \alpha_j$, and
(c) For each $i = 0, 1, \ldots$ we have $X \cap \{C_\beta \mid \beta \supseteq \alpha_i\} \neq \phi$.
We say that the skeleton $\{C_{\alpha_0}, C_{\alpha_1}, \ldots\}$ *omits* A iff $A \mid C_{\alpha_i}$ for each $i = 0, 1, \ldots$. If $\{C_{\alpha_0}\}$ is a skeleton, we say that $\{C_{\alpha_0}\}$ is a *principal skeleton*.

DEFINITION 6.6. We say that C_α is an *outside cell* iff $C_\alpha \downarrow$ but $C_\beta \uparrow$ for each $\beta \supset \alpha$. We say that C_α is a *terminal cell* iff $C_\alpha \downarrow$ and C_α is omitted by a skeleton for $\{C_\beta \mid lh(\beta) \geq n\}$ for some $n \geq 0$. (That is, C_α is terminal iff $C_\alpha \downarrow$ and there is some $n \geq 0$ such that for each C_β which includes C_α we have $lh(\beta) < n$.)

DEFINITION 6.7. The *restriction* $\mathcal{C}[\zeta]$ of the celling $\mathcal{C}$ to a set of strings ζ is defined by

$$\mathcal{C}[\zeta]_\beta = \begin{cases} C_\beta & \text{if } C_\beta \downarrow \text{ and } \beta \in \zeta \\ \text{undefined} & \text{otherwise.} \end{cases}$$

We write $\mathcal{C}[\beta]$ for $\mathcal{C}[\{\alpha \mid \beta \subseteq \alpha\}]$.

DEFINITION 6.8. The *$\mathcal{C}$-enumerating set of D for i following ζ, W* (or, just the *enumerating set of D following ζ, W*), written $ES^{\mathcal{C},i,\zeta,W}(D)$ (or just $ES^{\zeta,W}(D)$), is defined by

$$ES^{\mathcal{C},i,\zeta,W}(D) = \{C_\beta \mid \beta \in \zeta \;\&\; D \subseteq \Psi_i^{W \cup C_\beta}\}.$$

We write $ES^{\alpha,W}(D)$ for $ES^{\{\beta \mid \alpha \subseteq \beta\},W}(D)$, $ES^{\alpha,W}(x)$ for $ES^{\alpha,W}(\{x\})$, $ES^{\alpha}(D)$ for $ES^{\alpha,\phi}(D)$ and $ES(D)$ for $ES^{\phi,\phi}(D)$.
More generally, we define the *$\mathcal{C}$-weak enumerating set of D for i following ζ, W* by

$$WES^{\mathcal{C},i,\zeta,W}(D) = \{\{C_{\beta_0}, C_{\beta_1}, \ldots, C_{\beta_m}\} \in [\mathcal{C}]^{<\omega} \mid \{\beta_0, \beta_1, \ldots, \beta_m\} \subseteq \zeta \;\&\; D \subseteq \Psi_i(W \cup C_{\beta_0} \cup C_{\beta_1} \cup \ldots \cup C_{\beta_m})\}.$$

DEFINITION 6.9. We say that $\{C_{\alpha_0}, C_{\alpha_1}, \ldots\}$ is an *i-skeleton* for D (beyond ζ, W) iff $\{C_{\alpha_0}, C_{\alpha_1}, \ldots\}$ is a skeleton for $ES(D)$ ($ES^{\zeta,W}(D)$ respectively).

DEFINITION 6.10. We say that C_α is *i, D-distinguished* (beyond ζ, W) iff $D \subseteq \Psi_i^{C_\alpha}$ ($D \subseteq \Psi_i^{W \cup C_\alpha}$, respectively) and $\{C_\alpha\}$ is a principal i-skeleton for D (beyond ζ, W).

DEFINITION 6.11. We say that C_α is *strongly i, D-distinguished (beyond ζ, W)* iff C_α is i, D-distinguished (beyond ζ, W respectively) and for each A weakly respecting $\mathcal{C}$ ($\mathcal{C}[\zeta]$) we have $D \subseteq \Psi_i^A \Leftrightarrow C_\alpha \subseteq A$ ($D \subseteq \Psi_i^{W \cup A} \Leftrightarrow C_\alpha \subseteq A$).

Definition 6.12. $\mathcal{C}$ is (*strongly*) *i-distinguished with* A (weakly, respectively) respecting $\mathcal{C}$ iff there is a function $f : dom\mathcal{C} \to [\omega]^{<\omega}$, called a (*strong*, respectively) *i-labelling* of $\mathcal{C}$ *with* A, and a cofinite set $\zeta \subseteq S$, such that for each $\alpha \in \zeta$ we have that

$$C_\alpha \subseteq A \Leftrightarrow \left[(\forall\beta \subset \alpha)C_\beta \subseteq A \& f(\alpha) \subseteq \Psi_i\left(C_\alpha \cup \left(A \cap \left(\bigcup_{\beta \notin \zeta} C_\beta\right)\right)\right)\right].$$

Definition 6.13. Let A be a set respecting $\mathcal{C}$. Then A is *i-undistinguished* in $\mathcal{C}$ iff A is omitted by no nonempty i-skeleton in $\mathcal{C}$. (This corrects Definition 3.16 of [Co87].)

Definition 6.14. A *maximally i-enumerates over* $\mathcal{C}$ ($\mathcal{C}[\zeta]$, some $\zeta \subseteq S$) iff A weakly respects $\mathcal{C}$ and $\Psi_i\left(\bigcup_{\alpha \in S} C_\alpha\right) = \Psi_i^A$ ($\Psi_i\left(A \cup \bigcup_{\alpha \in \zeta} C_\alpha\right) \subseteq \Psi_i^A$, respectively).

Definition 6.15. We say $f : \omega \times dom\mathcal{C} \to [\omega]^{<\omega}$ is a *strong labelling* (or just *labelling*) *of* $\mathcal{C}$ *with* A (weakly respecting $\mathcal{C}$) iff for each $i \geq 0$ there is some cofinite $\zeta \subseteq S$ such that either:

(i) A maximally i-enumerates over $\mathcal{C}[\zeta]$, or

(ii) $\lambda\alpha f(i,\alpha)$ is a strong i-labelling of $\mathcal{C}$ (with corresponding set ζ in Definition 6.12).

We note that when discussing the *degree* of $\mathcal{C}$, or placing $\mathcal{C}$ in the arithmetical hierarchy, we refer to the enumeration degree or quantifier form of the set $\{\langle x,\alpha\rangle \mid x \in C_\alpha\}$. We identify a labelling f with the set $\{\langle D,\alpha,i\rangle \mid D = f_i(\alpha)\}$.

The basic lemmas corresponding to the two parts of the usual Computation Lemma are:

Lemma 6.16. *Let $\mathcal{C}$ be a strongly i-distinguished celling with A with strong i-labelling f. Let A be a set weakly respecting the celling $\mathcal{C}$. Then $A \leq_e f \times \mathcal{C} \times \Psi_i^A$.*

Proof: It is straightforward to verify that the following describes an algorithm for enumerating A from enumerations of Ψ_i^A, f and $\mathcal{C}$: Start by setting up an enumeration of Ψ_i^A, $\{\langle x,\alpha\rangle \mid x \in C_\alpha, \alpha \in S\}$ and the graph of f. Say $(\alpha, f(\alpha))$ is enumerated in the graph of f and the members of $f(\alpha)$ are enumerated in Ψ_i^A. Then enumerate the members of C_α into A. □

Lemma 6.17. *Let A be a set weakly respecting the celling $\mathcal{C}$, and let A maximally i-enumerate over $\mathcal{C}$. Then $\Psi_i^A \leq_e \mathcal{C}$.*

Proof: If A satisfies the conditions of the Lemma then $\Psi_i^A = \Psi_i(\cup_{\alpha\in S} C_\alpha)$, giving $\Psi_i^A \leq_e \cup_{\alpha \in S} C_\alpha \leq_e \mathcal{C}$. □

We include in this section a definition from [Co87] which will be useful in the discussion below:

DEFINITION 6.18. For each $i, x \geq 0$

$$\varepsilon_x^i = \{D \in [\omega]^{<\omega} \mid \langle x, D\rangle \in \Psi_i\},$$
$$\uparrow \varepsilon_x^i = \{K \in [\omega]^{<\omega} \mid (\forall D \in \varepsilon_x^i)(K \cap D \neq \phi)\},$$
$$\downarrow \varepsilon_x^i = \{W \in [\omega]^{<\omega} \mid (\forall K \in \uparrow \varepsilon_x^i)(W \cap K \neq \phi)\}.$$

We often write ε_x for ε_x^i. More generally, given a pair W, K of finite sets of numbers, we define

$$\varepsilon_x^i(W,K) = \{D \in [\omega]^{<\omega} \mid (D \cap K = \phi) \& (x \in \Psi_i^{D \cup W})\},$$
$$\uparrow \varepsilon_x^i(W,K) = \{K' \in [\omega]^{<\omega} \mid K' \cap W = \phi \& K' \cup K \in \uparrow \varepsilon_x^i\},$$
$$\downarrow \varepsilon_x^i(W,K) = \{W' \in [\omega]^{<\omega} \mid W' \cap K = \phi \& W' \cup W \in \downarrow \varepsilon_x^i\}.$$

As expected, the existence of minimal covers can now be reduced to the problem of finding appropriate cellings satisfying Lemmas 6.16 or 6.17. But unlike the analagous use of splitting and anti-splitting trees for getting minimal total degrees it is not possible to nest distinct cellings corresponding to individual requirements.

THEOREM 6.19 (Cooper [Co*ta*2]). *There is a celling $\mathcal{C}$ with labelling f, and a set A weakly respecting $\mathcal{C}$, such that*

(i) $f, \mathcal{C} \leq_e A$,

(ii) $A \not\leq_e f \oplus \mathcal{C}$, *and*

(iii) *For each $i \geq 0$, either $\Psi_i^A \leq_e \mathcal{C}$ or $A \leq_e \Psi_i^A \oplus f \oplus \mathcal{C}$.*

COROLLARY 6.20. *The enumeration degrees are not dense.*

PROOF: Take $\mathbf{a} = deg_e(A)$, $\mathbf{b} = deg_e(f \oplus \mathcal{C})$ in Theorem 6.19. □

PROOF OF THEOREM 6.19: Construct a triple $\mathcal{C}, f, P$, where $\mathcal{C}$ is a celling with labelling f, and a set A weakly respecting $\mathcal{C}$ satisfying the requirements

$\mathcal{R}_i$: $A \neq \Psi_i^{f \oplus \mathcal{C}}$

$\mathcal{S}_i$: $\exists \alpha^* \in dom(\mathcal{C})$ such that either A maximally i-enumerates over $\mathcal{C}[\alpha^*]$ or $\mathcal{C}[\alpha^*]$ is a strongly i-distinguished celling with A with strong i-labelling $\lambda\alpha\, f(i,\alpha)$ (written f_i),

where P is the *prohibiting function* $dom(\mathcal{C}) \to [\omega]^{<\omega} \cup \{\omega\}$ concerned with policing the construction of $\mathcal{C}$ on behalf of a successful outcome to the definition of the labelling f.

DEFINITION 6.21. We say $\mathcal{D}, g, Q$ is a *finite triple* iff $\mathcal{D}$ is a closed finite celling, $g : \omega \times dom(\mathcal{D}) \to [\omega]^{<\omega}$ and $Q : dom(\mathcal{D}) \to [\omega]^{<\omega} \cup \{\omega\}$, satisfying:

(a) $dom(\mathcal{D}) = dom(Q)$ and $(\forall \alpha, \beta \in dom(\mathcal{D}))(\alpha \subset \beta \Rightarrow D_\beta \cap Q(\alpha) = \phi)$, and

(b) $(\forall i \geq 0)(dom(\mathcal{D}) \supseteq dom(g_i))$ and $(\forall \alpha, \beta \in dom(\mathcal{D}))(\alpha \subseteq \beta \Rightarrow (g_i(\alpha)\downarrow \Rightarrow g_i(\beta)\downarrow))$, and

(c) $(\forall \alpha, \beta \in dom(\mathcal{D}))(\alpha \subseteq \beta \Rightarrow Q(\alpha) \subseteq Q(\beta))$.

Definition 6.22. If $\mathcal{D}, g, Q$ is a finite triple and D_β (say) $\subseteq$ some boundary cell $D_\alpha \in \mathcal{D}$ with $Q(\alpha) \neq \omega$, we say that D_β is *potentially nonterminal*, or just *nonterminal* when there is no ambiguity.

We say that $\mathcal{D}, g, Q$ is *potentially nonterminal*, or just *nonterminal*, (*beyond* ζ) iff (for each $\alpha \in \zeta$, respectively) there is some potentially nonterminal cell $D_\beta \in \mathcal{D}$ ($\in \mathcal{D}[\alpha]$, respectively).

We say that $\mathcal{D}, g, Q$ is *uniquely nonterminal* (*beyond* ζ) iff (for each $\alpha \in \zeta$, respectively) there is exactly one such boundary cell $D_\beta \in \mathcal{D}$.

Definition 6.23. We say C', f', P' is a *finite extension* of $\mathcal{D}, g, Q$, and write $C', f', P' \succ \mathcal{D}, g, Q$ iff C', f', P' is a finite triple with C', f' and P' extensions of $\mathcal{D}$, g, Q respectively.

During the construction construct an infinite nest (with respect to $\succ$) of finite triples whose union will be the required triple C, f, P. In order to get $f, C \leq_e A$, progressively code f and C into every infinite A respecting C using a uniformly recursive family $\{\xi(x, \alpha, E, i) \mid i, x \geq 0, \alpha \in S, E \in [\omega]^{<\omega}\}$ of disjoint infinite sets $\xi(x, \alpha, E, i) \subset \omega$. The aim will be to have for each infinite A weakly respecting C, and each (x, α, E, i) that

$$A \cap \xi(x, \alpha, E, i) \neq \phi \Leftrightarrow x \in C_\alpha \& E = f_i(\alpha).$$

Assume as usual that each $z \in \xi(x, \alpha, E, i)$ is greater than $x, i, max\{y \in E\}$ and $lh(\alpha)$.

Definition 6.24. We say that C', f', P' is *alright codewise* iff for each A respecting C' we have

$$A \cap \xi(x, \alpha, E, i) \neq \phi \Rightarrow x \in C'_\alpha \& E = f'_i(\alpha).$$

Since the $\mathcal{R}_i$-strategy (a straightforward diagonalisation) requires the building into the construction of a certain amount of choice in defining A weakly respecting C, the coding process presents difficulties for the $\mathcal{S}_i$-strategy. When constructing at stage $s+1$, say, certain extensions $C_{\beta^\frown\gamma_0}, C_{\beta^\frown\gamma_1}$ for C_β, we will want to do this as far as possible so as to have $C_{\beta^\frown\gamma_j}$ strongly $i, \tilde{f}_i(\beta^\frown\gamma_j)$-distinguished with A for each $j \leq 1$, each $i \leq s$. But numbers in an appropriate $I_{\beta^\frown\gamma_j}$ may be codes z for quadruples (x, α, E, i) with C_α as yet undefined. As long as $x \notin P(\beta)$, this will not in general prevent the definition of $I_{\beta^\frown\gamma_j}$ with $z \in I_{\beta^\frown\gamma_j}$. Merely extend the (so far defined part of) C, f, P (call it $\widehat{C}, \widehat{f}, \widehat{P}$) to an alright codewise finite extension C', f', P' with, in particular, $z \in I'_{\beta^\frown\gamma_j}$, $x \in C'_\alpha$ and $E = f'_i$.

The problem then is that no precautions can be taken to make the whole of C', f', P' suitably strongly i-distinguished. This means we only want C', f', P' to give us, in effect, the extensions $C_{\beta^\frown\gamma_0}, C_{\beta^\frown\gamma_1}$. To do this ensure during the rest of the construction that A is only allowed a nontrivial choice (in satisfying the $\mathcal{R}_i$ requirements) between designated *unique* nonterminal boundary cells C_{α_j} beyond $C_{\beta^\frown\gamma_j}$, $j \leq 1$. This means that the strong $i, \tilde{f}_i(\beta^\frown\gamma_j)$-distinguishing of each $C_{\beta^\frown\gamma_j}$, $j \leq 1$, will be sufficient to get us to the next pair of designated extensions beyond C_β, proceeding within the A which weakly respects C, given an enumeration of C, f_i and Ψ_i^A.

However, in order to obtain a detailed enumeration of exactly how much of $\cup\{C'_\alpha \mid \alpha \in dom(C') - dom(\widehat{C})\}$ is in A, some care needs to be taken in defining the labelling over the parts of $C' - \widehat{C}$ which we are forced to include because they are coded by some $z \in C_{\beta^\frown\gamma_j}$, some $j \leq 1$. Care also needs to be taken in defining $A \cap [\cup\{C'_\alpha \mid \alpha \in dom(C') - dom(\widehat{C})\}]$.

Achieve the former by asking that the labelling f satisfies (roughly speaking) $f_i(\alpha) \subseteq \Psi_i^{C_\alpha}$ for each $\alpha \in dom(f_i)$. Then satisfy the latter need by defining some $C_{\beta^\frown\gamma_j} \subset A$, and then taking (again roughly speaking)

$$A \cap \{\cup_{\alpha \in dom(C') - dom(\widehat{C})} C'_\alpha\} = \cup_{\alpha \notin C'[\beta^\frown\gamma_{(1-j)}]} C'_\alpha.$$

There is another problem, arising from the way in which the S_i-requirements need to be satisfied when there are not sufficient strong i, D-distinguishings available. In this case one cannot use i-undistinguished cellings (as for the compact reducibilities discussed in section 8 below), but must construct A to be maximally i-enumerating.

In this situation there are many α's with $\cup_{\alpha' \supseteq \alpha} C_{\alpha'}$ maximally i-enumerating as far as those numbers x for which $\uparrow \varepsilon^i_x(C_\beta, P_\beta) \neq \phi$ goes. Deal with the other numbers by including in the construction some simple actions to ensure that (again roughly speaking) modulo the coding constraints if $\uparrow \varepsilon^i_x(C_\beta, P_\beta) = \phi$ then for some $B \subseteq A$ weakly respecting C we have $x \in \Psi_i^B$. Then, choosing such an α which does not interfere with the steps to satisfy the $\mathcal{R}_i$-requirements, obtain A maximally i-enumerating by making $\cup_{\alpha' \supseteq \alpha} C_{\alpha'} \subset A$. To prevent such interference, it is necessary to include precautions to ensure that if $C_{\beta^\frown\gamma_j} \subset A$ is chosen, as above, then no B respecting $C[\beta^\frown\gamma_{(1-j)}]$ can be maximally i-enumerating over C. Having done this, take $\cup_{\alpha' \supseteq \alpha} C_{\alpha'} \subset A$ for *all* such α's (in a trivial way).

A problem then is that the weak respecting by A of C becomes nontrivial. And that means that in defining strongly i'-distinguished extensions for C_β, say, in C ($i' \neq i$) it is necessary to know about $A \cap \{\cup_{\alpha \not\subseteq \alpha'} C_{\alpha'}\}$. This requires that A is constructed at the same time as C, f, P is constructed, using the knowledge of that part of $A \cap \{\cup_{\alpha \not\subseteq \alpha'} C_{\alpha'}\}$ which has already been defined in defining extensions for C_β, and then using the prohibiting function P to protect the strong i'-distinguished extensions from injury in defining the rest of $A \cap \{\cup_{\alpha \not\subseteq \alpha'} C_{\alpha'}\}$. (We notice that that the interdependence of C and A during the construction is the main impediment to constructing a continuum of minimal covers of the form $deg_e(A)$ for $deg_e(C \oplus f)$.)

To summarise: Assume that by stage $s+1$ of the construction a finite triple C^s, f^s, P^s and a finite set A^s weakly respecting C^s have been defined . Also assume given a boundary cell $C^s_{\beta^s} \subseteq A^s$ for C^s which has been designated A^{s+1}-*burgeoning* at the end of stage s. Then:

(1) First examine what freedom the actions on requirements S_i, $i < s$, have given for satisfying S_i through the first part of S_i.
(2) If there is no such freedom, decide best how to make A maximally s-enumerate while still providing room to satisfy lower priority requirements.
(3) Mop up numbers x we want to be in Ψ_i^A, $i \leq s$.
(4) Look for appropriately distinguished extensions of the A^{s+1}-burgeoning cell C^s_β within the context of a finite triple $C^{s+1}, f^{s+1}, P^{s+1} \succ C^s, f^s, P^s$. The

main concern here is the satisfaction of the $\mathcal{S}_i$-requirements for $i \leq s$ in such a way as to allow the satisfying of the requirement $\mathcal{R}_s$ when A^{s+1} is defined.

(5) Take advantage of the choice of distinguished cellings to satisfy $\mathcal{R}_s$.

(6) Code up most of what has been defined in $\mathcal{C}^{s+1}, f^{s+1}$ within A^{s+1} while choosing A^{s+1} to weakly respect $\mathcal{C}^{s+1}$.

To complete the proof verify that: (a) A is an infinite set weakly respecting the celling $\mathcal{C}$, (b) $f \oplus \mathcal{C} \leq_e A$ (so the coding works), (c) for each $i \geq 0$ there is a $\mathcal{C}[\alpha(i)]$ such that either A maximally i-enumerates over $\mathcal{C}[\alpha(i)]$ or $\mathcal{C}[\alpha(i)]$ is a strongly i-distinguished celling with A with a strong i-labelling f_i (=that part of f relating to Ψ_i), so that each $\mathcal{S}_i$ is satisfied, and (d) $\mathcal{R}_i$ is satisfied for all $i \geq 0$. □

A direct calculation of the quantifier form of A in Theorem 6.19 yields a Σ^0_7 relation, so that A is r.e. in $\phi^{(6)}$, giving $A \leq_e \phi^{(6)}$. Hence:

COROLLARY 6.24. *The enumeration degrees below* $0^{(6)}$ *are not dense.* □

The gap between Theorem 5.6 and Corollary 6.24 leads to some obvious but important questions:

QUESTION 6.25. What is the smallest n, $2 \leq n \leq 6$, for which we can prove nondensity of $\mathcal{D}_e(\leq 0^{(n)})$?

QUESTION 6.26. Are the enumeration degrees below $0''_e$ dense? (We conjecture that at least the e-degrees of the Π_2 sets are dense).

QUESTION 6.27. Are all finite distributive lattices embeddable as (non-initial) segments of the e-degrees?

QUESTION 6.28. Can the techniques of Theorem 6.19 be adapted to provide minimal upper bounds in the e-degrees?

Little seems to be known (along the lines of the classic Spector result **[Sp56]** for ascending sequences of Turing degrees) concerning the existence of minimal upper bounds in the e-degrees, although every degree $\leq 0'$ *is* a minimal upper bound for some ascending sequence (using the density result).

Corollary 6.20 provides the only result so far concerning the global theory of the e-degrees: $\mathcal{D}_e$ is not homogeneous. Slaman and Woodin **[SW*ta*]** have used their coding technique to obtain results concerning definability in the e-degrees. Otherwise everything is still open. For example:

QUESTION 6.29. Are the total degrees definable in $\mathcal{D}_e$? Is the jump definable in $\mathcal{D}_e$?

QUESTION 6.30. Classify the automorphisms of $\mathcal{D}_e$.

Rogers **[Rog67]** asks:

QUESTION 6.31. Is the collection of total degrees order-theoretic, i.e., invariant under all automorphisms $\mathcal{D}_e$?

§7. The structure of the enumeration degrees below 0'.

The e-degrees below $\mathbf{0}'$ are of special interest, being the degrees of the Σ_2^0 sets. In this section we examine structure below $\mathbf{0}'$, in particular in relation to the high/low hierarchy and the classes Δ_2^0, Π_1^0. We assume below that all standard listings $\{B_i\}_{i\geq 0}$ of the Σ_2^0 sets come with thin Σ_2 approximations $\{B_i^s\}_{s\geq 0}$ (see Jockusch [Jo68] for a proof that this can be done), where every Δ_2^0 set appears somewhere in the list with a Δ_2 approximation. We sometimes use the notation $A[s]$ to denote expression A evaluated at stage s.

DEFINITION 7.1 (McEvoy [McE84], McEvoy and Cooper [MC85]). If $\{A^s\}_{s\geq 0}$ is a Σ_2 approximation to a set A, define a *computation function* C_A for A by

$$C_A(x) = \mu s > x\, [A^s \restriction x \subset A],$$

each $x \geq 0$. A Σ_2*-high approximation* to A is a Σ_2 approximation to A for which C_A is total and dominates every recursive function. A set is Σ_2*-high* iff it has a Σ_2 approximation, and a degree is Σ_2*-high* iff it contains a Σ_2-high set.

It is easy to see that on the total degrees the notion of Σ_2-high coincides with the usual notion of high for the Turing degrees (under the natural embedding). To see that the Σ_2-high degrees contain the high Turing degrees (cf. [McE85]), we just notice that if $\chi_A \in \mathbf{a}$ with A high, then $A \oplus \bar{A} \in \mathbf{a}_e$ is Σ_2-high. For the converse, observe that $A \leq_e C_A$, and if A is a total set (so $\bar{A} \leq_e A$) then $A \equiv_e C_A$. The result follows from the fact (cf. Robinson [Rob68]) that any $A \in$ a Σ_2-high degree $\mathbf{a}$ has a Σ_2-high approximation, so $\mathbf{a}$ contains a C_A dominating every recursive function. McEvoy [McE85] showed that the Σ_2-high degrees properly extend the high Turing degrees by constructing a quasi-minimal Σ_2-high degree. In fact:

THEOREM 7.2 (Cooper and Copestake [CC88]). *There exists a Σ_2-high properly Σ_2 degree.*

PROOF: Construct A with Σ_2 approximation $\{A^s\}_{s\geq 0}$. Get C_A to dominate $\{i\}$ total by delaying extraction from A of members of the ith column. Reconcile this with the properly Σ_2 strategy (see Theorem 4.11) by means of a tree of outcomes. □

However:

QUESTION 7.3. Characterise the jumps of the Σ_2-high e-degrees.

From Shoenfield's construction [Sh59] of a non-r.e. Turing degree below $\mathbf{0}'$, we know that the Δ_2^0 e-degrees properly extend the class of Π_1^0 e-degrees. Yates' construction [**Ya65**] of a Turing degree below $\mathbf{0}'$ which is incomparable with all the r.e. Turing degrees other than $\mathbf{0}$ and $\mathbf{0}'$ has an immediate corollary in the e-degrees:

PROPOSITION 7.4. *There is a Δ_2^0 e-degree incomparable with all Π_1^0 e-degrees other than $\mathbf{0}$ and $\mathbf{0}'$.* □

Unlike the total degrees below $\mathbf{0}'$, the e-degrees below $\mathbf{0}'$ properly extend the degrees of Δ_2^0 sets (by Cooper and Copestake [**CC88**], there exist properly Σ_2 e-degrees). In fact there is the following analogue of the Yates result:

THEOREM 7.5 (Cooper and Copestake [**CC88**]). *Let* $\mathbf{h}$ *be a* Σ_2*-high degree. Then there exists an e-degree* $\mathbf{a}$ *below* $\mathbf{h}$ *incomparable with all the* Δ_2^0 *e-degrees below* $\mathbf{h}$ *(other than* $\mathbf{0}$ *and* $\mathbf{h}$*).*

PROOF: Build $A \in \Sigma_2^0$ by recursive approximation to satisfy the requirements:

E_i: If $A = \Psi_i(B_i)$, then either (a) there exists some x such that $\lim_s B_i^s(x)$ does not exist, or (b) $H \leq_e B_i$,

F_i: If $B_i = \Psi_i(A)$, then either (a) there exists some x such that $\lim_s B_i^s(x)$ does not exist, or (b) B_i is r.e.

The basic strategy for F_i is to try and make B_i r.e. by fixing D in A at stages $t > s$ whenever we get some $x \in B_i \cap \Psi_i(D)[s]$. But to leave room for the E_i-strategy in the presence of this infinitary outcome, split F_i into subrequirements $F_{i,x}$, $x \geq 0$, where $F_{i,x}$ will be satisfied if either (a*) $\lim_s B_i^s(x)$ fails to exist, or (b*) if $x \in B_i \cap \Psi_i(D)$ for some D, then there exists such a D which does not interfere with any E_j, $j \leq \max\{i, x\}$, and we fix D in A. Seek to produce such a D for x by working with the equation $B_i = \Psi_i(A)$ as in the properly Σ_2 strategy, using 'temporary' negative restraints for $F_{i,x}$ which are taken in turn via a 'line' which will serve the purpose of isolating the true path in the tree of outcomes from the 'Σ_2 noise' produced by the temporary restraints.

The E_i requirements conflict with the F_i requirements in that they will require extraction of numbers from A. In satisfying E_i define values of a one-to-one coding $\alpha_i : \omega \to \omega$. For each x aim to have

(7-1) $$\alpha_i(x) \in \Psi_i(D) \text{ with } D \subset B_i \Leftrightarrow (x \in H[\max\{y \in D\}] \Rightarrow x \in H),$$

resulting in $H \leq_e B_i$. Work with the equation $A = \Psi_i(B_i)$ as in the properly Σ_2 strategy using the highness of H, either getting a y for which $\lim_s B_i^s(y)$ does not exist, or getting all the B_i-extractions needed for stage-by-stage rectification of the coding in (7-1), giving $H \leq_e B_i$. □

Little is known about jump restricted interpolation in the Σ_2^0 e-degrees, although we can extend the proof (Theorem 7 of [**MC85**]) that every Δ_2^0 $\mathbf{b} > \mathbf{0}$ has a low predecessor $\mathbf{a} > \mathbf{0}$ to get:

THEOREM 7.6. *If* **h** *is* Σ_2*-high then there is a non-zero low* $\mathbf{a} < \mathbf{h}$.

PROOF: Construct $A = \Theta(H)$, H Σ_2-high, satisfying

$$N_i: \quad \Theta(H) \neq W_i,$$
$$P_i: \quad \lim_s \Psi_i^s(\Theta^s(H^s))(i) \text{ exists},$$

each $i \geq 0$. Satisfy N_i in the usual way, but use the Σ_2-highness of H to get a follower $x \notin \Theta(H)[s]$ if $x \in W_i^s$. Satisfy P_i by enumerating $\langle y, H^t \restriction y \rangle$ into Θ at all stages $t > s$ when we get $i \in \Psi_i^s(D)$ and $D \subset \Theta^s(H^s)$, unless in conflict with a higher priority N_j. □

We now look at minimal pairs below $\mathbf{0}'$. Using the previously mentioned observation of Jockusch that capping in the e-degrees is often easier than constructing minimal pairs, we show that we can always cap low degrees below Σ_2-high degrees (and, in particular, below $\mathbf{0}'$).

THEOREM 7.7 (McEvoy and Cooper [**MC85**]). *If* $\mathbf{0} < \mathbf{a} < \mathbf{h}$ *with* **a** *low and* **h** Σ_2*-high, then there is a degree* $\mathbf{c} < \mathbf{h}$ *such that* $\mathbf{a} \cap \mathbf{c} = \mathbf{0}$.

PROOF: Carry out a Jockusch-style construction of a minimal pair **a**, **c** by recursive approximation. Use the lowness of **a** to make the construction finite injury, and use Σ_2-highness to get it permitted below **h**. □

Theorems 7.6 and 7.7 combine to give (compare [**Co74**]) the following extension of Corollary 7.1 of [**MC85**]:

COROLLARY 7.8. *If* **h** *is* Σ_2*-high then there is a minimal pair of degrees below* **h**. □

Every low minimal pair of r.e. Turing degrees is a minimal pair in the e-degrees (under the natural embedding):

THEOREM 7.9 (McEvoy and Cooper [**MC85**]). *If* **a**, **b**, *with* **a** *low, is a minimal pair in the* Π_1^0 *e-degrees, then* **a**, **b** *is a minimal pair in the e-degrees.*

This follows immediately from:

PROPOSITION 7.10. *If* $\mathbf{g} \leq \Pi_1^0$ *e-degrees* **a**, **b**, **a** *low, then there is a* Π_1^0 **e** *with* $\mathbf{g} \leq \mathbf{e} \leq \mathbf{a}, \mathbf{b}$.

PROOF OF THE PROPOSITION: Let $\{A^s\}_{s \geq 0}$ be a low Π_1 approximation to $A \in \mathbf{a}$ and let $\{B^s\}_{s \geq 0}$ be a Π_1 approximation to $B \in \mathbf{b}$. Let G, i and j be such that $G \in \mathbf{g}$ and $G = \Psi_i(A) = \Psi_j(B)$. Define $E \in \mathbf{e}$ by:

$$\langle x, t \rangle \in E \Leftrightarrow (\forall s \geq t)[x \in \Psi_i^s(A^s) \vee x \in \Psi_j^s(B^s)]. \square$$

This means that any lattice embedding in the low r.e. degrees is a lattice embedding in the e-degrees. McEvoy and Cooper give some immediate consequences of the Lachlan/Yates construction of a minimal pair of (low) r.e. degrees, and of Lachlan's embedding results [**La72**] for the r.e. degrees:

COROLLARY 7.11. *(i) (Gutteridge [**Gu71**]) There exists a minimal pair of Π_1^0 e-degrees.*

(ii) Any countable distributive lattice can be embedded in the e-degrees, and the two five-element nondistributive lattices can be embedded in the e-degrees. Hence the upper semi-lattice of the e-degrees below $\mathbf{0}'$ is not distributive. □

McEvoy and Cooper show that the lowness condition in Theorem 7.9 is necessary.

THEOREM 7.12. *There is a minimal pair in the Π_1^0 degrees which is not a minimal pair in the e-degrees.*

PROOF: Adapt the Lachlan [**La66**] construction of a high minimal pair of r.e. degrees. Construct co-r.e. $A, B \in \mathbf{a}, \mathbf{b}$ respectively, with $\mathbf{a}, \mathbf{b}$ a minimal pair in $\boldsymbol{\mathcal{D}}$. Introduce enough changes in membership of A, B to enable us to get e-operators Γ, Λ for which $C = \Gamma^A = \Lambda^B$ with $C \neq W_i$ each i. Permit extractions of followers from C using non-simultaneous extractions of elements of columns from A and B which do not (through the timing of the extractions) injure the minimal pair strategy. This works because we do not need, as in the Turing case, Δ_2 'use functions' for Γ and Λ. □

The lowness condition in Theorem 7.7 is also necessary:

DEFINITION 7.13. $\mathbf{a} < \mathbf{c}$ is *noncappable below* $\mathbf{c}$ iff for no nonzero $\mathbf{b} < \mathbf{c}$ do we have $\mathbf{a} \cap \mathbf{b} = \mathbf{0}$, and $\mathbf{a}$ is *noncappable* iff $\mathbf{a}$ is noncappable below0'.

We say there is a *breakdown in capping below* $\mathbf{c}$ iff there is an $\mathbf{a}$ noncappable below $\mathbf{c}$.

McEvoy and Cooper show that for each low $\mathbf{b} > \mathbf{0}$ there is a breakdown in capping below some nonzero $\mathbf{a} \leq \mathbf{c}$. Further development of these techniques yields:

THEOREM 7.14. *There exists a noncappable e-degree* $\mathbf{a}$.

PROOF: Build $A \in \Sigma_2^0$ by recursive approximation satisfying:

$$P_i: \quad \overline{K} = \Psi_i^A \Rightarrow \Psi_i^A \text{ r.e.},$$
$$N_{\langle i,j\rangle}: \quad B_i \text{ r.e.} \vee (\exists G_i)[G_i \leq_e A \,\&\, G_i \leq_e B_i \,\&\, G_i \neq W_j],$$

each $i, j \geq 0$. Satisfy P_i by an analogue of Sacks restraints. To get $G_i \leq_e A$ and B_i build e-operators Γ_i and Λ_i with $G_i = \Gamma_i^A = \Lambda_i^{B_i}$. Appoint followers x of $N_{\langle i,j\rangle}$ which are held in G_i via Γ_i and Λ_i as long as $x \notin W_j$. If eventually $x \in W_j$, apply the following easily proved lemma:

If $A \in \Sigma_2^0$ and the computation function C_A is dominated by a recursive function then A is r.e.

to get B_i-permission (along with A-permission) to extract (in the limit) some x from G_i if B_i is not r.e. Use a tree of outcomes to enable the P_i-strategy to live with (for instance) the infinitary outcome when B_i is r.e. □

We have already mentioned the result of Copestake [**Cope*ta*2**] that there is a 1-generic e-degree below $\mathbf{0}'$ which bounds no minimal pair. The existence of nonbounding e-degrees (cf. [**La79**]) also follows (independently) from:

THEOREM 7.15 (Cooper and Sorbi [**CS***ta*]). *There is a linearly ordered initial segment of the Σ_2^0 e-degrees.*

PROOF: Let $\{\Psi_i, \Theta_i\}_{i\geq 0}$ be a standard listing of all pairs of e-operators. Construct $A \in \Sigma_2^0$ and e-operators Γ_i, Λ_i, $i \geq 0$, satisfying:

$$N_i: \quad A \neq W_i,$$

$$P_i: \quad \forall i\,[\Theta_i^A = \Gamma_i(\Psi_i^A) \vee \Psi_i^A = \Lambda_i(\Theta_i^A),$$

each $i \geq 0$. Pursue the usual N_i-strategy involving extraction of followers x when $x \searrow W_i$ ('x enters W_i'). To satisfy P_i we need to rectify (in the limit) Γ_i or Λ_i. Very roughly speaking, we play off a Γ_i-strategy of fixing certain $D \subset A$ to keep Θ_i^A against a Λ_i-strategy of extracting certain a z from Θ_i^A via some $y \in E \subset A$ to rectify Λ_i. These strategies contrast in that the former requires many D's, the latter few E's. They allow room for the N_j-strategies by the rectification of Γ_i, Λ_i for x 'favouring' requirements N_j, $j > x$. Uncertainty as to the true path on the tree of outcomes results in N_i possibly requiring attention infinitely often, giving $A \notin \Delta_2^0$. □

By Lagemann's embedding results (Corollary 5.5 above) all the degrees in the constructed linear ordering must be properly Σ_2, and by the density of the Σ_2^0 degrees it must have order type that of the rationals between 0 and 1. A corollary of Theorem 7.15 is that $\mathcal{D}_e(\leq \mathbf{0}')$ is not elementarily equivalent to the class of Δ_2^0 e-degrees.

QUESTION 7.16. Characterise the possible order-types of the initial segments of $\mathcal{D}_e$ (or of $\mathcal{D}_e(\leq \mathbf{0}')$.

It also follows from Theorem 7.15 that there are degrees below $\mathbf{0}'$ which cannot be nontrivially split. By the Sacks Splitting Theorem (p.124 of [**So87**]) each Π_1^0 e-degree can be split, but Ahmad [**Ah89**] has independently announced that there is a Δ_2^0 (in fact, low) e-degree which cannot be split.

Cooper and Sorbi [**CS***ta*] have the following positive result:

THEOREM 7.17. *Every Σ_2-high degree can be split.*

PROOF: Given H Σ_2-high, define low sets $A = \Gamma^H$, $B = \Lambda^H$ with $H = \Omega^{A\oplus B}$ by using the highness of H to ensure that we can always get H-permission via Γ or Λ to rectify Ω via A or B respectively with regard to the higher priority lowness requirements. □

Other interesting results are announced by Ahmad in [**Ah89**], including her analagous result to the diamond theorem ([**Co72**]) for the Δ_2^0 degrees:

THEOREM 7.18 (Ahmad [**Ah***ta*]). *There exist incomparable e-degrees* $\mathbf{a}, \mathbf{b}$ *with* $\mathbf{a} \cup \mathbf{b} = \mathbf{0}'$ *and* $\mathbf{a} \cap \mathbf{b} = \mathbf{0}$.

PROOF: Construct low sets A and B such that $\overline{K} \leq_e A \oplus B$ and $\forall i, j\,[\Psi_i^A = \Psi_j^B \Rightarrow \Psi_i^A$ is r.e.]. Get $\overline{K} \leq_e A \oplus B$ by ensuring $\forall x\,[x \in A \cap B \Leftrightarrow x \in \overline{K}]$. Carry out the minimal pair construction by recursive approximation (analagous to the minimal r.e. Turing pair construction in its alternation between A- and B-restraints), while favouring higher priority lowness and minimal pair restraints in deciding on the extraction of an A- or B-trace for $x \in K$. As usual, use a tree to identify 'windows' correctly, using the non-r.e.-ness of A and B to recover from actions taken off the true path. □

By the Lachlan Nondiamond Theorem ([**So87**], p.162), Theorem 7.18 shows that the e-degrees below $\mathbf{0}'$ are not elementarily equivalent to the structure of the r.e. degrees (as does Theorem 7.15). Many questions concerning the e-degrees below $\mathbf{0}'$, including a number suggested by the Turing case, remain to be answered, including:

QUESTION 7.19. Characterise the degree of the first-order theory of $\mathcal{D}_e(\leq \mathbf{0}')$.

It would also be interesting to see results exploring the context of the total e-degrees within $\mathcal{D}_e(\leq \mathbf{0}')$.

§8. Strong enumeration reducibilities.

The need for a theoretical counterpart to real computational situations in which only restricted information is available has motivated various areas of recursion theory, including that of strong Turing reducibilities (see [**Rog67**] or [**Od89**]) such as many-one, truth-table, weak truth-table and bounded truth-table reducibilities. We now look at strong reducibilities which use only positive information (in providing positive information), that is those which can be considered as restrictions of $\leq_e$. Most of the material in this section can be found in a fuller form in [**Co87**].

A number of truth-table reducibilities can be defined as restrictions of $\leq_e$. Recall (dropping the i in Definition 6.18) that $\varepsilon_x = \{D \in [\omega]^\omega \mid \langle x, D\rangle \in \Psi\}$, and define:

DEFINITION 8.1. The *norm* $\|\varepsilon_x\|$ of ε_x is defined by $\|\varepsilon_x\| = \sup\{\, |D| \mid D \in \varepsilon_x\}$.

Then:

DEFINITION 8.2.

(1) (*Many-one reducibility*, Post [**Po44**]). Ψ is a *many-one operator* iff $(\forall x)\ |\bigcup \varepsilon_x| = 1$. A is *many-one reducible to* B ($A \leq_m B$) iff $A \leq_e B$ via a many-one Ψ. The *one-one reducibility* $\leq_1$ is the special case where the ε_x's are mutually disjoint.

(2) (*Conjunctive reducibility*, Jockusch [**Jo66**]). Ψ is a *c-operator* iff $(\forall x)\ |\varepsilon_x| = 1$. $A \leq_c B$ iff $A \leq_e B$ via a c-operator.

(3) (*Bounded conjunctive reducibility*, Jockusch [**Jo66**]). Ψ is a *bc-operator* iff it is a c-operator and $(\exists n)(\forall x)\, \|\varepsilon_x\| \leq n$. $A \leq_{bc} B$ iff $A \leq_e B$ via a bc-operator.

(4) (*q-reducibility*, Friedberg and Rogers [**FR59**]). Ψ is a *q-operator* iff $\lambda x\, \varepsilon_x$ is recursive and $(\forall x)\, \|\varepsilon_x\| = 1$. $A \leq_q B$ iff $A \leq_e B$ via a q-operator. ($\leq_q$ is also called *disjunctive reducibility* and written $\leq_d$.)

(5) (*Bounded q-reducibility*, Jockusch [**Jo66**] and Lachlan [**La65**]). Ψ is a *bq-operator* iff $(\exists n)(\forall x)(\|\varepsilon_x\| = 1\ \&\ |\varepsilon_x| = n)$. $A \leq_{bq} B$ iff $A \leq_e B$ via a bq-operator.

(6) (*Positive reducibility*, Jockusch [**Jo66**]). Ψ is a *p-operator* iff $\lambda x\, \varepsilon_x$ is recursive. $A \leq_p B$ iff $A \leq_e B$ via a p-operator.

(7) (*Bounded p-reducibility*, Jockusch [**Jo66**] and Lachlan [**La65**]). Ψ is a *bp-operator* iff Ψ is a p-operator and $(\exists n)(\forall x)\ |\bigcup \varepsilon_x| \leq n$.

We say that a reducibility E is a *truth-table e-reducibility* iff E is a subreducibility of $\leq_p$: that is, iff $\lambda x\, \varepsilon_x$ is a recursive function for each E-operator Ψ.

Remembering the close relationship between e-reducibility and nondeterministic relative computability of partial functions, p-reducibility and q-reducibility can be thought of as nondeterministic versions of c-reducibility and m-reducibility respectively (see section 9 on polynomial time-bounded e-reducibilities).

In **[PR79]** Polyakov and Rozinas consider a generalisation of the notion of a truth-table e-reducibility: if $\leq_E$ is some e-reducibility, denote by rE the restriction of E (=the set of E-operators) to operators $\Psi \in E$ which are recursive. Unfortunately, if $\leq_E$ is not already a truth-table reducibility, $\leq_{rE}$ will, in general be non-transitive. Moreover, for most natural E we have $A \leq_E B \Leftrightarrow A \leq_{rE} B \oplus \omega$, where $B \equiv_{rE} B \oplus \omega$.

Polyakov and Rozinas **[PR77]** make the useful distinction between *decision reducibilities* (such as $\leq_T$ and $\leq_{tt}$) and *enumeration reducibilities* (such as $\leq_e$ and $\leq_s$). Each of the above truth-table reducibilities has a corresponding degree structure which can be looked at *either* as that for a decision reducibility *or* as that for an enumeration reducibility. Since for each reducibility $\leq_R$ above $\mathbf{0}_R$ consists of recursive sets only, the former view seems more appropriate.

DEFINITION 8.3. We say that an e-reducibility $\leq_R$ is a *proper enumeration reducibility* iff $\mathbf{0}_R$ includes all r.e. sets.

If we allow *partial* versions of the truth-table e-reducibilities, where we only ask for $\lambda x\, \varepsilon_x$ to be partial recursive, we do obtain proper e-reducibilities. In doing this, we will of course want the resulting reducibility $\leq_{pR}$ to be transitive. Satisfactory existing notions are:

DEFINITION 8.4.

(1) (*Partial many-one reducibility*, Rogers **[Rog67]**). Ψ is a *pm-operator* iff $(\forall x)\ |\bigcup \varepsilon_x| \leq 1$. $A \leq_{pm} B$ iff $A \leq_e B$ via a pm-operator. (Rogers also defines *partial one-one reducibility* $\leq_i$ in the natural way.)

(2) (*pc-reducibility*). Ψ is a *pc-operator* iff $|\varepsilon_x| \leq 1$. $A \leq_{pc} B$ iff $A \leq_{pc} B$ via a pc-operator. (The notation $\leq_{pc}$ is due to Polyakov and Rozinas **[PR77]**, although the notion appears earlier in Skordev **[Sk72]** and in Sasso **[Sas73]** via his $\leq_T$ restricted to the semi-characteristic functions.)

Polyakov and Rozinas **[PR77]** note that transitivity fails for the partial counterparts $\leq_{pp}$, $\leq_{pq}$ for $\leq_p$ and $\leq_q$. There is no problem in giving partial counterparts $\leq_{pbp}$, $\leq_{pbc}$ and $\leq_{pbq}$ for the bounded reducibilities $\leq_{bp}$, $\leq_{bc}$ and $\leq_{bq}$.

Referring again to the trees of nondeterministic computations which underly a given enumeration operator, we can identify three natural classifications of strong e-reducibilities.

DEFINITION 8.5 (Cooper [Co87]).

(1) (*Finite e-reducibility*). Ψ is ***branch finite*** iff $(\forall x)(\varepsilon_x$ is finite$)$. $A \leq_{fe} B$ iff $A \leq_e B$ via a branch finite Ψ. If $\leq_r$ is an e-reducibility, we sometimes denote by $\hat{r}$ the restriction of r to branch finite e-operators.

(2) (*Bounded e-reducibility*). Ψ is ***norm bounded*** iff $(\exists n)(\forall x)(\|\varepsilon_x\| \leq n)$. $A \leq_{be} B$ iff $A \leq_e B$ via a norm bounded Ψ.

(3) (*btt-like e-reducibility*). Ψ is ***btt-like*** iff $(\exists n)(\forall x)(\mid \cup \varepsilon_x \mid \leq n$. $A \leq_{btte} B$ iff $A \leq_e B$ via a btt-like Ψ.

Apart from $\leq_{be}$, all the reducibilities considered so far have been branch finite. A particularly important subreducibility of $\leq_{be}$ is the branch infinite counterpart for $\leq_{pm}$:

DEFINITION 8.6 (*Singleton e-reducibility*, Friedberg and Rogers [**FR59**]). Ψ is an ***s-operator*** iff $(\forall x)(\|\varepsilon_x\| \leq 1)$. $A \leq_s B$ iff $A \leq_e B$ via an s-operator. (The branch finite version $\leq_{\hat{s}}$ of $\leq_s$ appears in Odifreddi [**Od81**] as $\leq_Q$.)

The definition given here is actually that of McEvoy's $\leq_{se}$ in [**McE84**], Friedberg and Rogers originally defining: $A \leq_s B$ if and only if there is a recursive f such that

$$(\forall x)(x \in A \Leftrightarrow W_{f(x)} \cap B \neq \phi).$$

Polyakov and Rozinas [**PR77**] showed that the requirement for f to be total could be dropped. In fact (see [**Co87**]) all three definitions are equivalent. s-operators occur commonly in constructions in the e-degrees (cf. section §5 above). In fact, Watson [**Wa***ta*] notices that the following transitive subreducibilities of $\leq_{\hat{s}}$ are those most often used:

(*Nearly m-reducibility/nearly pm-reducibility*). $A \leq_{nm} B \,/\, A \leq_{npm} B$ iff $A = \Psi^B \cup W$ for some r.e. W, some m-operator/pm-operator (respectively) Ψ.

We notice that even though $\leq_{be}$ and $\leq_s$ are branch infinite, combinatorially the be- and s-operators are not that far removed from the fe-operators (Cooper [**Co87**]): If Ψ is norm bounded then Ψ is $\downarrow$branch finite, where we say Ψ is ***$\downarrow$branch finite*** iff

$$(\forall x)(\exists D_{i_x})(\forall D)[D \in \downarrow \varepsilon_x \Rightarrow D \cap D_{i_x} \in \downarrow \varepsilon_x).$$

All the reducibilities so far defined are transitive, so that we can refer to their corresponding degree structures. We can summarise the known implications between the various proper e-reducibilities in the following diagram:

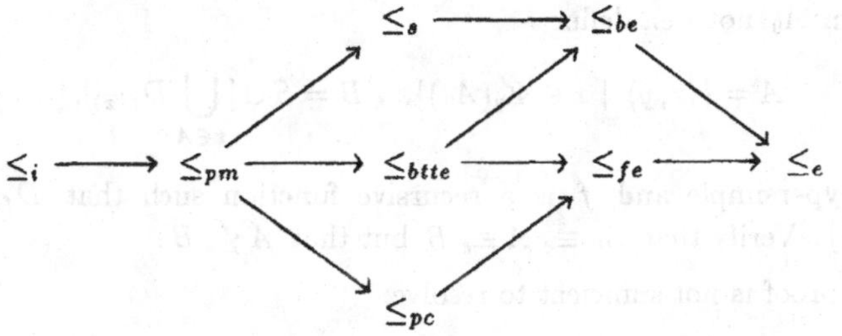

Most of the non-implications are easy to prove (see [**PR77**], [**PR79**] and [**Co87**]). See [**Co87**] for a summary how the truth-table e-reducibilities fit into this picture.

We now examine the degree structures for the proper e-reducibilities, with special reference to the s-degrees and their close relationship to the e-degrees. Following Vuckovic [**Vu74**] we define:

DEFINITION 8.7. A is *strong* iff $A \equiv_{pm} [A]^{<\omega}$. If $\leq_r$ is some transitive reflexive e-reducibility, then we say the r-degree $\mathbf{a}_r$ is *strong* iff there is some strong $A \in \mathbf{a}_r$.

Since $A \equiv_e [A]^{<\omega} \equiv_{pm} [[A]^{<\omega}]^{<\omega}$, every e-degree $\mathbf{a}$ contains a strong r-degree $\mathbf{a}_r$, say, and if $\leq_r$ includes $\leq_s$ then this strong $\mathbf{a}_r$ is of largest r-degree in $\mathbf{a}$ (cf. Vuckovic [**Vu74**]). In particular:

Every e-degree contains a largest s-degree.

Since $[\overline{K}]^{<\omega}$ is Π^0_1, so $\equiv_m \overline{K}$, $\overline{K}$ is an example of a non-r.e. strong set. If $A (\neq \omega)$ is r.e., then $\chi_{\overline{A}} \equiv_{pm} \overline{A}$, so if $\leq_r$ includes $\leq_{pm}$, then $\overline{K}$ is of strong total r-degree.

We easily adapt Definition 3.5 to get a jump on the r-degrees:

DEFINITION 8.8. $\mathbf{a}'_r = \deg_r J(A)$.

Hence, for any $\leq_r$ including $\leq_{pm}$ we have that $\mathbf{0}'_r = \deg_r(\overline{K})$ is a strong total Π^0_1 r-degree, and if $\leq_r$ includes $\leq_s$, then $\mathbf{0}'_r$ is of largest r-degree in $\mathbf{0}'$.

DEFINITION 8.9. Given two reducibilities $\leq_y$ and $\leq_z$ with $\leq_y$ included in $\leq_z$, we say $\mathbf{a}_z \neq \mathbf{0}_z$ is a *y-contiguous z-degree* iff $\forall A, B \in \mathbf{a}_z\, [A \equiv_y B]$: that is, $\mathbf{a}_z$ consists of a single y-degree.

The non-existence of s-contiguous e-degrees follows from:

THEOREM 8.10 (Zakharov [**Za84**]). *Every non-zero e-degree contains at least two distinct s-degrees.*

PROOF: Given A_0 not r.e., define

$$A = \{\langle x, y\rangle \mid x \in \Psi_y(A_0)\}, \quad B = S \cup [\bigcup_{x \in A} D_{f(x)}],$$

where S is hypersimple and f is a recursive function such that $D_{f(x)} = \{2^x - 1, \ldots, 2^{x+1} - 2\}$. Verify that $A_0 \equiv_e A \equiv_e B$ but that $A \not\leq_s B$. □

The above proof is not sufficient to resolve:

QUESTION 8.11. Does every non-zero e-degree contain infinitely many distinct s-degrees?

Zakharov [**Za84**] showed that if B is as in Theorem 8.10, then $B <_m B \times B <_m \ldots <_m [B]^n <_m \ldots$, so every e-degree contains an infinite ascending sequence of m-degrees. Using an analogue of Beigel, Gasarch and Owings' [**BGO87**] notion of *terse set*, Copestake [**Cope*tal***] showed that every 1-generic e-degree **a** contains an infinite ascending chain of s-degrees, namely the s-degrees got from the sequence

$$A <_s [A]^2 <_s \ldots <_s [A]^n <_s \ldots,$$

A a terse set in **a**. Watson [**Wa*ta***] gave a constructive version of Zakharov's theorem for the Σ_2^0 degrees, and proved:

THEOREM 8.12 (Watson [**Wa*ta***]). *If* $\mathbf{a} > \mathbf{0}$ *is* Δ_2^0 *or* Σ_2*-high, then* **a** *contains no s-degree minimal in* **a**. *Hence every* $\mathbf{a} > 0$ *which is* Δ_2^0 *or* Σ_2*-high contains an infinite descending sequence of s-degrees.*

PROOF: Let $\widehat{\Psi}_i$ be a standard listing of all s-operators. Given a non-r.e. set $A \in \Delta_2^0$ construct a $B \in \Delta_2^0$ satisfying $R_i : A \neq \widehat{\Psi}_i^B$, each $i \geq 0$, $P : A = \Gamma^B$ (Γ an e-operator to be constructed) and $Q : B = \Lambda^A$ (Λ an s-operator to be constructed). Monitor the initial segment of agreement $\ell(i, s)$ for R_i, and as this grows seek to preserve it (working against the pseudo-outcome $A = \widehat{\Psi}_i^B$ r.e.) by restraining a corresponding singleton y in B for each $x \in A$, $x < \ell(i, s)$.

Ensure the rectification of Γ in P eventually leaves room for the R_i-strategy by using large sets D in $\langle x, D\rangle \in \Gamma$, so that if $x \nearrow A$ ('x leaves A') we can get $x \nearrow \Gamma^B$ by extracting some $y \in D$ from B while respecting the R_i-restraints corresponding to numbers $y \leq x < \ell(i, s)$. Achieve Q by making $\langle y, \{x\}\rangle \in \Lambda$ if $\langle x, D\rangle \in \Gamma$ with $y \in D$, and using crossing for rectifying Γ when $x \nearrow A^t$ and $y \in B^t$. In the A Σ_2-high part of the theorem the strategies are similar. □

It is not known whether there exist e-degrees containing minimal or least s-degrees, or which are closed under formation of meets of pairs of s-degrees. Watson [**Wa88**] conjectures that there exists a minimal pair of s-degrees contained in a single e-degree. More generally Watson [**Wa*ta***] asks:

QUESTION 8.13. Do there exist (distinct) proper e-reducibilities $\leq_y$ and $\leq_z$, with $\leq_y$ contained in $\leq_z$, for which there exists a y-contiguous z-degree?

On the positive side, Watson [**Wata**] is able to show: There exists a Π_1^0 pc-degree $\mathbf{a}$, $\mathbf{0} < \mathbf{a} < \mathbf{0}'$, such that if A and B are Π_1^0 sets in $\mathbf{a}$ then $A \equiv_s B$.

Turning to general structural questions, [**Co87**] gives an in-depth analysis of density and non-density in a wide range of strong e-degree structures. We are unable to do more here than summarise the main results, and hint at the techniques involved. Fundamental to this analysis are notions of *compactness*, trivial for the Turing case but of crucial importance in deriving existence results for suitably i-distinguished cellings.

DEFINITION 8.14. (1) Let ζ be a set of numbers. We say that ζ is a *Ψ-tower* iff there is a pair W, K such that

$$(\forall x)[x \in \zeta \Leftrightarrow W \notin \downarrow \varepsilon_x \& K \notin \uparrow \varepsilon_x] \tag{8-1}$$

and

$$(\forall x, y \in \zeta)[\downarrow \varepsilon_x(W,K) \subseteq \downarrow \varepsilon_y(W,K) \vee \downarrow \varepsilon_y(W,K) \subseteq \downarrow \varepsilon_x(W,K)]. \tag{8-2}$$

(2) We say that W^*, K^* *decides* the Ψ-tower ζ iff $W^* \supseteq_{\text{fin}} W$ and $K^* \supseteq K$ and for all x, $W^* \in \downarrow \varepsilon_x$ or $K^* \in \uparrow \varepsilon_x$.
(3) We say that Ψ is *compact* iff every Ψ-tower can be decided by some pair W^*, K^*. We say that $\leq_E$ is *compact* iff each $\Psi \in E$ is compact.

PROPOSITION 8.15. *Each of the reducibilities* $\leq_m$, $\leq_1$, $\leq_c$, $\leq_{bc}$, $\leq_q$, $\leq_{bq}$, $\leq_{bp}$, $\leq_i$, $\leq_{pm}$, $\leq_{pc}$, $\leq_s$ $\leq_{\hat{s}}$ $\leq_{btte}$ $\leq_{be}$ *or* $\leq_T$ *(via its corresponding e-reducibility) is compact.*

PROOF: Observe that each of these reducibilities is norm bounded or $\uparrow$ *norm bounded* (that is, for each Ψ $(\exists n)(\forall x)(\exists K \in \uparrow \varepsilon_x)[\mid K \mid \leq n]$), and show that all such reducibilities are compact. □

It will follow that for most known proper e-reducibilities, compactness characterises those reducibilities $\leq_r$ for which we can find degrees with uncountably many minimal covers in the r-degrees. As remarked in [**Co87**], the truth-table reducibilities are approached differently, within the context of the decision reducibilities (see for example Odifreddi [**Od81**] or Lachlan [**La70**]). For compact operators one can derive i-distinguished and i-undistinguished cellings (instead of just i-maximally enumerating cellings as in the proof of Theorem 6.19) in such a way that they can be nested without too much loss of effectiveness. Then Lemmas 6.16 and 6.17 can be adapted (in 6.17 replacing i-maximally enumerating with i-undistinguished), to provide the satisfaction of a wide range of minimality and minimal cover requirements. The nesting depends technically on the notion (analagous to that of subtree in the Turing case) of *a continuation* C' of a celling C.

THEOREM 8.16 (Cooper [**Co87**]). *Let $\leq_r$ be a partial truth-table reducibility. Then, if $\leq_r$ is a proper e-reducibility, there exists a continuum of minimal r-degrees.*

In particular, there exists a continuum of minimal pm-degrees, pc-degrees and of T-degrees.

PROOF: The construction is formally the same as that for a continuum of minimal Turing degrees (see [**Le83**]). For each $\alpha \in 2^\omega$ construct a distinct nest $\{C_i(\alpha)\}_{i\in\omega}$ of appropriate cellings where for each $i \in \omega$ $C_i(\alpha) \in \Sigma_1^0$, $C_{i+1}(\alpha)$ is a continuation of $C_i(\alpha)$, and (roughly speaking) either $C_{i+1}(\alpha)$ is i-undistinguished or $C_{i+1}(\alpha)$ is i-distinguished with Σ_1^0 i-labelling. Define A_α to be the unique set respecting all the cellings $C_i(\alpha)$, $i \in \omega$, in the usual way, and get the minimality of $\deg_r(A_\alpha)$ by applying suitable variants of Lemmas 6.16 and 6.17.

The result for the T-degrees follows from the earlier observation (derived from Sasso's construction [**Sas73**] of a minimal T-degree $< \mathbf{0}'$) that sets of minimal pc-degree yield unique semicharacteristic functions of minimal T-degree. □

It is clear from the proof of Gutteridge's Theorem (5.3) that there do not exist minimal r-degrees for any e-reducibility $\leq_r$ which includes $\hat{s}$-reducibility. We now see that for most compact proper e-reducibilities relativisation of the Gutteridge result breaks down completely.

DEFINITION 8.17. (1) If f is a (possibly partial) function, then the f-contraction $f\Psi$ of Ψ is defined by $\langle x, D\rangle \in f\Psi \Leftrightarrow \langle x, f^{-1}D\rangle \in \Psi$.
(2) We say a reducibility $\leq_r$ is *combinatorially compact* iff whenever $\Psi \in r$ we have that each f-contraction $f\Psi$ is compact.

Since the proof of Proposition 8.15 only depends upon norm boundedness or $\uparrow$norm boundedness, all the reducibilities is actually seen to be combinatorially compact.

THEOREM 8.18 (Cooper [**Co87**]). *Let $\leq_r$ be a combinatorially compact e-reducibility. Then, if $\leq_r$ includes $\leq_s$, there exists a continuum of minimal covers for $\mathbf{0}'_r$.*

In particular, there exists a continuum of minimal covers for $\mathbf{0}'_s$ in the s-degrees, and of $\mathbf{0}'_{be}$ in the be-degrees.

PROOF: The proof is similar to that for Theorem 8.16, except that we now get $C_i(\alpha)$ and its corresponding increment $\mathcal{I} \in \Sigma_2^0$ for each $i \in \omega$, and similarly any i-labelling $f^{\mathcal{I}} \in \Sigma_2^0$. Again applying suitable variants of Lemmas 6.16 and 6.17, we get that for each i either $A_\alpha \leq_r f^{\mathcal{I}} \times \mathcal{I} \times \Psi_i^{A_\alpha}$ or $\Psi_i^{A_\alpha} \leq_r C$. Since $\overline{K}$ is r-complete in Σ_2^0, we get $A_\alpha \leq_r \overline{K} \times \Psi_i^{A_\alpha}$ or $\Psi_i^{A_\alpha} \leq_r \overline{K}$. To obtain $\deg_r(A_r)$ minimal over $\mathbf{0}'_r$ make $\overline{K} \leq_r A_\alpha$ by coding $\overline{K}$ into $C_0(\alpha)$ in a trivial way. □

Further detailed consideration confirms the existence of continuously many minimal covers for $\mathbf{0}'_{\hat{s}}$ in the ŝ-degrees, but the problem for the T-degrees and the btte-degrees remains open.

We briefly return to the non-compact case.

DEFINITION 8.19. We say that a branch finite Ψ is ***anti-compact*** iff there is a Ψ-tower (with $W, K = \phi$) and for any two sets A, B we have that $\Psi^A = \Psi^B$. (It follows from the definition that no compact Ψ can be anti-compact.)

PROPOSITION 8.20. *There exists an anti-compact p-operator.*

PROOF: Define the *Slaman operator* $\widehat{\Psi}$ as follows: Associate with each D_x a rational number

$$a_x = \tfrac{1}{2} + (-1)^{D_x(0)+1}(\tfrac{1}{2})^2 + \ldots + (-1)^{D_x(k)+1}(\tfrac{1}{2})^{k+2}$$

where $k = \max\{z \in D_x\}$. Let $D_x \in \widehat{\mathcal{E}}_x$ if and only if $a_x < a_y$ and

$$\mu w\,[D_x(w) \neq D_y(w) \& w \leq \max\{z \in D_x\} + 1 \& w \leq \max\{z \in D_y\}]$$

exists $= \max\{z \in D_y\}$. □

The following generalises the theorem of Gutteridge (Corollary 5.2) on minimal covers in the e-degrees.

THEOREM 8.21 (Cooper [Co87]). *(i) If an e-reducibility $\leq_r$ contains an anti-compact Ψ, then there are at most countably many minimal covers for any $\deg_r(B)$, where each such minimal cover $\leq_T B'$. In particular, there are at most countably many minimal covers for a given degree in each of the e-degrees, the fe-degrees and the p-degrees.*

(ii) If a truth-table e-reducibility $\leq_r$ contains an anti-compact Ψ, then any minimal cover for $\deg_r(A)$ in the r-degrees is r.e. in A. In particular, every minimal p-degree is r.e.

PROOF: For (i), show that if Ψ is an anti-compact e-operator then for any sets A, B $\Psi^A \leq_e B \Rightarrow A \leq_T B'$. Define a new anti-compact Θ by relabelling the outputs for Ψ on ω to allow us, given $A = \Psi_i(\Theta^A \oplus B)$, to uniformly enumerate $x+1$ from $\Psi^{A \restriction x} \oplus B)$ just in the case $x + 1 \in A$, giving $A \leq_e \Theta^A \oplus B \Rightarrow A \leq_e B$. Proposition 8.20 gives the particular cases.

(ii) is similar. Show that if Ψ is an anti-compact truth-table operator, then $\Psi^A \leq_p B \Rightarrow A \leq_p B$, and that we can modify Ψ to get an anti-compact truth-table operator Θ for which $A \leq_p \Theta^A \oplus B \Rightarrow A \leq_p B$. □

On the other hand, in contrast to Gutteridge's Theorem for the e-degrees, one can construct a (r.e.) minimal p-degree (see [Co*ta*2]).

There are many open questions (see §5 of [Co87]), including:

QUESTION 8.22. Are there techniques for producing upper cones with required properties in the e- or s-degrees?

QUESTION 8.23. Are the structures of the pm- or pc-degrees above $\mathbf{0}'_{pm}$ or $\mathbf{0}'_{pc}$ respectively dense?

Polyakov and Rozinas [**PR79**] have proved some basic results concerning s-, pm- and pc-degrees. Also:

THEOREM 8.24 (Zakharov [**Za84**]). *The elementary theories of the r-degrees and s-degrees are different for* r∈ {e,pc,p,c}.

PROOF: The proposition

$$(\exists \mathbf{a} \neq 0)(\forall \mathbf{b})(\forall \mathbf{c})[\mathbf{a} \leq \mathbf{b} \cup \mathbf{c} \Rightarrow \mathbf{a} \leq \mathbf{b} \vee \mathbf{a} \leq \mathbf{c}] \tag{8-3}$$

is true in $\mathcal{D}_s$ since if A is a retraceable set we have

$$(\forall B)(\forall C)[A \leq_s B \oplus C \Rightarrow A \leq_s B \vee A \leq_s C].$$

However, direct construction shows that (8-3) is not true in $\mathcal{D}_r$ each r∈ {e, pc, p, c}. (In the e case Zakharov shows that one can join above **a** with quasi-minimal e-degrees **b**, **c**.) □

Using Theorem 7.14 above, we can show that $\mathcal{D}_s(\leq \mathbf{0}'_s)$ is not elementarily equivalent to $\mathcal{D}_s(\leq \mathbf{0}'_s)$:

THEOREM 8.25. *If* **a** *is an s-degree below* $\mathbf{0}'_s$ *then* **a** *is cappable in the s-degrees below* $\mathbf{0}'_s$.

PROOF: Let $\{(\widehat{\Psi}_i, \widehat{\Theta}_i)\}_{i \geq 0}$ be a standard listing of all pairs of s-operators. Given $A \in \Sigma^0_2$, $\overline{K} \not\leq_s A$, construct $B \in \Sigma^0_2$ satisfying:

$$N_i: \quad B \neq W_i,$$
$$P_i: \quad \widehat{\Psi}^A_i = \widehat{\Theta}^B_i \Rightarrow \widehat{\Theta}^B_i \text{ r.e.},$$

each $i \geq 0$. The N_i-strategy is as usual. For P_i monitor the initial segment of agreement, of length $\ell(i,s)$. If almost all $y < \ell(i,s)$ at some stage s which are in $\widehat{\Psi}^A_i[s]$ are enumerated into $\widehat{\Theta}^B_i[s]$ via more than one singleton $\{y\} \subset B[s]$, then we can protect $\ell(i,s)$ via restraints which still allow room for the N_j's. Otherwise, we can use infinitely many members of $\widehat{\Theta}^B_i$ to code $\overline{K}$, when $x \nearrow \overline{K}$ using a B extraction to achieve an extraction of the x-trace from Θ^B_i, forcing $\lim\inf_s \ell(i,s)$ to be finite. □

QUESTION 8.26. Prove density of the s- and be- degrees $\leq \mathbf{0}'$. Characterise the degrees of the first-order theories of these structures.

§9. Polynomial time-bounded enumeration reducibilities.

All the reducibilities introduced above give rise to computation bounded reducibilities. These take various forms, mainly dependent on the type of bound on time or space allowed for the reduction and on the presence or otherwise of nondeterminism in the reduction procedures. Polynomial time bounds are closely related to practical limits on computations, and so far this is the area in which notions are well developed. The connection between relative e-reducibility of sets and relative nondeterministic computability of functions gives polynomial bounded enumeration reducibilities an important role in clarifying the nature of polynomial time reducibility between functions.

Before giving definitions, the bounding process requires us to be more precise in describing computations. For convenience, we take the multi-tape Turing machine as our model of computation, inputs being words in a finite alphabet (we choose $\{0,1\}$). Numbers are identified with their binary presentations, and we write $|x|$ for the length of $x \in \{0,1\}^*$ (so $|x| \simeq \log x$, the binary logarithm of x). A Turing machine M (either deterministic or nondeterministic) runs in polynomial time if there is a polynomial p such that on any input of length n, any computation of M halts in at most $p(n)$ steps. M *accepts* on input x, and *recognises* a set A, have their usual meanings. Turing machines equipped with distinguished output tapes for computing functions are called *transducers*. In general, nondeterministic transducers will be permitted to compute many-valued functions. For relative computations we use oracle Turing machines with a designated oracle tape. An oracle Turing machine M runs in polynomial time if there is a polynomial p such that on any input of length n, any computation of M halts in at most $p(n)$ steps, independently of the oracle used. As usual, $\mathcal{P}$ denotes the class of all subsets of $\{0,1\}^*$ (deterministically) computable in polynomial time and $\mathcal{NP}$ is the class of all such sets nondeterministically computable in polynomial time.

DEFINITION 9.1. (1) (*Polynomial time Turing reducibility*, Cook **[Coo71]**/*Nondeterministic polynomial time Turing reducibility*, Meyer and Stockmeyer **[MS72]**). $A \leq_T^{\mathcal{P}} B$ / $A \leq_T^{\mathcal{NP}} B$ iff there is a deterministic/ nondeterministic (respectively) oracle Turing machine M which runs in polynomial time such that $x \in A$ iff some computation sequence of M on input x with oracle B accepts.

(2) (*Polynomial time many-one reducibility*, Karp **[Ka72]**/*Nondeterministic polynomial time many-one reducibility*, Ladner, Lynch and Selman **[LLS75]**). $A \leq_m^{\mathcal{P}} B$ / $A \leq_m^{\mathcal{NP}} B$ iff there is a deterministic/ nondeterministic transducer M which runs in polynomial time such that $x \in A$ iff M, on input x, produces some output $y \in B$.

Both deterministic reducibilities are reflexive, transitive relations giving rise to degree structures with least degree $= \mathcal{P}$ (ignoring ϕ and $\{0,1\}$ in the former case). $\leq_m^{\mathcal{NP}}$ (but not $\leq_T^{\mathcal{NP}}$) is transitive with least degree $= \mathcal{NP}$. Ladner, Lynch and Selman **[LLS75]** added further polynomial time bounded positive reducibilities to $\leq_m^{\mathcal{P}}$ and $\leq_m^{\mathcal{NP}}$. For technical reasons, these need to be approached via truth table reducibility, where an abstract approach is taken in order to make the bounds independent of the particular presentation of the Boolean functions giving the truth tables.

Let Δ be a fixed finite alphabet (for example $\{\wedge, \vee, \neg\}$) for encoding Boolean functions and let $c \notin \Delta \cup \{0,1\}^*$.

DEFINITION 9.2 (Ladner, Lynch and Selman [**LLS75**]). (i) A *tt-condition* is a member of $\Delta^* c(c\{0,1\}^*)^*$.

(ii) A *tt-condition generator / nondeterministic tt-condition generator* is a deterministic/ nondeterministic (respectively) transducer M such that each computation sequence of M produces a tt-condition.

(iii) A *tt-condition evaluator* is a recursive mapping of $\Delta^* c\{0,1\}^*$ into $\{0,1\}$.

(iv) Let e be a tt-condition evaluator. A tt-condition $\alpha c c y_1 \ldots c y_k$ is e-satisfied by B iff $e(\alpha c \chi_B(y_1) \ldots \chi_B(y_k)) = 1$.

(v) (*Polynomial time truth table reducibility / Nondeterministic polynomial time truth table reducibility*). $A \leq_{tt}^{\mathcal{P}} B$ / $A \leq_{tt}^{\mathcal{NP}} B$ iff there is a deterministic/ nondeterministic polynomial time computable generator M and a polynomial time computable evaluator e such that $x \in A$ iff M, on input x, computes some tt-condition which is e-satisfied by B.

Ladner, Lynch and Selman look at various alternative formulations of these definitions (such as allowing nondeterminism in the evaluator in $\leq_{tt}^{\mathcal{NP}}$) and show them to lead to the same notions.

DEFINITION 9.3 (Ladner, Lynch and Selman [**LLS75**]). (i) (*Polynomial time positive reducibility / Nondeterministic polynomial time positive reducibility*). $A \leq_p^{\mathcal{P}} B$ / $A \leq_p^{\mathcal{NP}} B$ iff $A \leq_{tt}^{\mathcal{P}} B$ / $A \leq_{tt}^{\mathcal{NP}} B$ (respectively) with evaluator e having the property that if $e(\alpha c \sigma_1 \ldots \sigma_k) = 1$ and $\sigma_i = 1$ implies $\tau_i = 1$ for $1 \leq i \leq k$, then $e(\alpha c \tau_1 \ldots \tau_k) = 1$.

(ii) (*Polynomial time conjunctive reducibility / Nondeterministic polynomial time conjunctive reducibility*). $A \leq_c^{\mathcal{P}} B$ / $A \leq_c^{\mathcal{NP}} B$ iff $A \leq_{tt}^{\mathcal{P}} B$ / $A \leq_{tt}^{\mathcal{NP}} B$ with evaluator e having the property that $e(\alpha c \sigma_1 \ldots \sigma_k) = 1$ iff $\sigma_i = 1$ for $1 \leq i \leq k$.

(iii) (*Polynomial time disjunctive reducibility / Nondeterministic polynomial time disjunctive reducibility*). $A \leq_d^{\mathcal{P}} B$ / $A \leq_d^{\mathcal{NP}} B$ iff $A \leq_{tt}^{\mathcal{P}} B$ / $A \leq_{tt}^{\mathcal{NP}} B$ with evaluator e having the property that $e(\alpha c \sigma_1 \ldots \sigma_k) = 0$ iff $\sigma_i = 0$ for $1 \leq i \leq k$.

(Bounded versions of these reducibilities can be defined by adding suitable restrictions on the generator.)

In general, following Baker, Gill and Solovay [**BGS75**] for the Turing case, the nondeterministic reducibilities (in contrast to the situation without oracles) can be shown to properly include the deterministic notions. Ladner, Lynch and Selman [**LLS75**] summarise the implications and non-implications (these often, but not always, proved on the sets computable in 2^n time) between the various reducibilities in the following diagram (showing a collapse of $\leq_d^{\mathcal{NP}}$ and $\leq_p^{\mathcal{NP}}$):

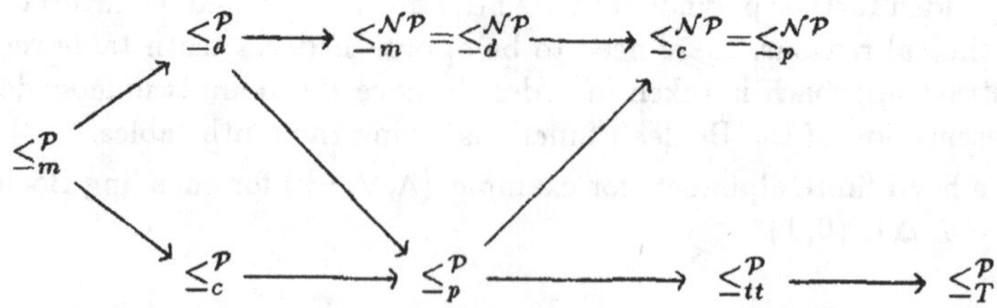

The transitivity of $\leq_c^{\mathcal{NP}}$ with corresponding least degree $= \mathcal{NP}$ leads Ladner, Lynch and Selman to suggest $\leq_c^{\mathcal{NP}}$ as a polynomial time bounded analogue of enumeration reducibility. There are many questions concerning the polynomial time bounded degree structures, often intimately connected with the question of $\mathcal{P} =?\mathcal{NP}$.

Following from Adleman and Manders **[AM77]** (for the $\leq_m$ case), Long **[Lo82]** introduces the notion of *strong* nondeterministic polynomial time reducibility in which there is a consistency condition imposed on different branches of the nondeterministic reduction procedure corresponding to a given input x, and shows the strong reducibilities to lie between the deterministic and unrestricted nondeterministic notions. Other related reducibilities include the log space reducibilities of Meyer and Stockmeyer **[MS73]** and Jones **[Jon73]**.

We now consider notions of polynomial time bounded e-reducibility (referring the reader to Selman **[Se78]** and Copestake **[Cope*ta*3]** for a fuller discussion). Arguing by analogy with Theorem 2.9 above, Selman **[Se78]** defines:

DEFINITION 9.4. A is polynomial time enumeration reducible to B ($A \leq_{pe} B$) iff

$$\forall x\,[B \leq_T^{\mathcal{NP}} X \Rightarrow A \leq_T^{\mathcal{NP}} X].$$

This is intended to capture the notion that A is polynomial time enumeration reducible to B just in the case that every 'polynomial-enumeration' of B (relative to any set X) yields some polynomial-enumeration of A (relative to X). Then, following Selman's proof **[Se71]** that $\leq_e$ is a maximal transitive subrelation of 'r.e. in', we have:

THEOREM 9.5 (Selman **[Se78]**). *$\leq_{pe}$ is a maximal transitive subrelation of $\leq_T^{\mathcal{NP}}$.*

The definition of $\leq_{pe}$ can be presented so as to look more like the usual definition of enumeration reducibility.

THEOREM 9.6 (Selman **[Se78]**). *$A \leq_{pe} B$ iff for every polynomial q, there is a set W in $\mathcal{NP}$, a polynomial p, and a constant k such that*

$$\begin{aligned}\forall x\,[x \in A \Leftrightarrow \exists \alpha\,[\,|\,\alpha\,| \leq p(\,|\,x\,|\,)\;\&\; xc\alpha \in W \\ \&\; \alpha = cy_1\ldots cy_n ccz_1\ldots cz_m \\ \&\; y_1,\ldots,y_n,z_1,\ldots,z_m \in \{0,1\}^* \\ \&\; \forall j \leq m\;\; q(\,|\,z_j\,|\,) \leq k \log |\,x\,| \\ \&\; \{y_1,\ldots,y_n\} \subseteq B \;\&\; \{z_1,\ldots,z_m\} \subseteq \overline{B}]].\end{aligned}$$

The unexpected negative information about B appearing in this technically more useful definition of $\leq_{pe}$ is due to the fact that a polynomial-enumeration of B will give us usable information about $\overline{B}$. Selman shows that $\leq_{pe}$ lies strictly between $\leq_c^{\mathcal{NP}}$ and $\leq_T^{\mathcal{NP}}$, but that on TIME($2^{\mathcal{P}}$) (=the class of sets recognised by deterministic Turing machines which run in time $2^{p(n)}$, $p(n)$ a polynomial) $\leq_{pe}$ and $\leq_c^{\mathcal{NP}}$ coincide. Partially justifying the definition of $\leq_{pe}$ Selman verifies that for polynomial bounded functions we have:

$$f \leq_T^{\mathcal{NP}} g \text{ if and only if } \operatorname{graph}(f) \leq_{pe} \operatorname{graph}(g).$$

In contrast to the situation with the deterministic polynomial time reducibilities (see for example [**Am***ta*] or [**SS***ta*]), little is known about the degree structures for the nondeterministic positive reducibilities. Silver [**Si***ta*] examines various polynomial time analogues of singleton reducibility ($\leq_s$), including $\leq_m^{\mathcal{NP}}$, Long's $\leq_m^{SN}$ (originally appearing in Adleman and Manders [**AM77**] as γ-reducibility) and a singleton version of $\leq_{pe}$.

There is a problem with $\leq_{pe}$ arising from the nonconstructiveness of its definition (there are no accessible pe-operators). Copestake [**Cope***ta***3**] convincingly argues that what is needed is a polynomial bounded analogue of the *full* description of $\leq_e$, namely: $A \leq_e B$ iff *there is an algorithm* which given any enumeration of B will provide us with an enumeration of A. See [**Cope***ta***3**] for an alternative definition of polynomial time bounded e-reducibility.

§10. The Medvedev lattice and its extension to partial functions.

Enumeration operators underly Medvedev's [**Me55**] definition of the lattice of *degrees of difficulty*, derived from a natural way of comparing relative computability of classes of functions (or *mass problems*). Of special interest is the *Dyment lattice* [**Dy76**] which generalises the concept of degree of difficulty to the classification of classes of partial functions. We briefly review the basic definitions (see [**Rog67**] or [**Sor***ta***l**]).

DEFINITION 10.1. $\mathcal{A}$ is a *mass problem* iff $\mathcal{A}$ is a collection of total functions. $\mathcal{A}$ is a *mass problem of partial functions* iff $\mathcal{A}$ is a collection of partial functions.

Rogers gives a number of examples of natural mass problems (consisting of sets of functions with something in common). In order to formalise the idea that $\mathcal{A}$ (say) is reducible to $\mathcal{B}$ iff from any 'solution to' (that is member of) $\mathcal{B}$ we can effectively get a solution to $\mathcal{A}$, we recall the definition (1.12 above) of recursive operator.

Then every e-operator Ψ_i defines a partial recursive operator $\Theta : \mathcal{P} \to \mathcal{P}$. The Fundamental Operator Theorem (Rogers [**Rog67**], Theorem XXIII) says that although not every Ψ_i^f, $f \in \mathcal{P}$ is necessarily a function (so Θ is not necessarily a recursive operator), we can uniformly define $\Psi_{\sigma(i)}$ so that it defines a recursive operator which agrees with Θ on the total functions. This provides us with an effective enumeration $\{\Theta_i\}_{i\in\omega}$ of the recursive operators on $\mathcal{F}$ (the set of total functions), where Θ_i is the recursive operator defined by $\Psi_{\sigma(i)}$.

DEFINITION 10.2 (Medvedev [**Me55**]). Let $\mathcal{A}$ and $\mathcal{B}$ be mass problems.

(i) $\mathcal{A}$ is *reducible* to $\mathcal{B}$ ($\mathcal{A} \leq \mathcal{B}$) iff $\exists i\,[\Theta_i(\mathcal{B}) \subseteq \mathcal{A}]$.

(ii) $\mathcal{A}$ is *equivalent* to $\mathcal{B}$ ($\mathcal{A} \equiv \mathcal{B}$) iff $\mathcal{A} \leq \mathcal{B}$ and $\mathcal{B} \leq \mathcal{A}$ (where $\equiv$ is easily seen to be an equivalence relation).

(iii) The resulting equivalence classes are called the *degrees of difficulty* (denoted by M) and have a *reducibility ordering* $\leq$ induced by the reducibility ordering. We call $\mathcal{M} = \langle M, \leq \rangle$ the *Medvedev lattice*.

If $\mathcal{A}$ and $\mathcal{B}$ are mass problems of partial functions, define $\mathcal{A} \oplus \mathcal{B} = \{\varphi \oplus \psi \mid \varphi \in \mathcal{A} \,\&\, \psi \in \mathcal{B}\}$ and, for each $x \in \omega$, $x \circledast \mathcal{A} = \{x \circledast \varphi \mid \varphi \in \mathcal{A}\}$ where $x \circledast \varphi$ is the partial function θ such that $\theta(0) = x$, and $\theta(y) = \varphi(y-1)$ if $y > 0$. Then, denoting the degree of difficulty of $\mathcal{A}$ by $[\mathcal{A}]$, we can define a *meet* $\bigvee$ and *join* $\bigwedge$ on $\boldsymbol{\mathcal{M}}$ by

$$[\mathcal{A}] \bigwedge [\mathcal{B}] = [\mathcal{A} \oplus \mathcal{B}] \quad \text{and} \quad [\mathcal{A}] \bigvee [\mathcal{B}] = [0 \circledast \mathcal{A} \cup 1 \circledast \mathcal{B}].$$

Medvedev [**Me55**] showed that $\boldsymbol{\mathcal{M}}$ is in fact a distributive lattice with these operations (see [**Rog67**]), with a least degree$\mathbf{0}$ consisting of those mass problems containing a recursive function, and a greatest degree$\mathbf{1}$ (the degree of the empty mass problem). Also, a special source of interest in $\boldsymbol{\mathcal{M}}$, there is a natural monomorphism, preserving least element, of $\boldsymbol{\mathcal{D}}$ into $\boldsymbol{\mathcal{M}}$ (and onto the *degrees of solvability*) got by mapping $\deg_T(A)$ to $[\{\chi_A\}]$. Similarly, $\boldsymbol{\mathcal{D}}_e$ is isomorphically embeddable in $\boldsymbol{\mathcal{M}}$ (onto the *degrees of enumerability*) via the mapping taking $\deg_e(A)$ to $[\{f \mid \text{range } f = A\}]$.

We can extend the above definitions to obtain a classification of mass problems of partial functions:

Definition 10.3 (Dyment [**Dy76**]). Let $\mathcal{A}$ and $\mathcal{B}$ be mass problems of partial functions. $\mathcal{A} \leq_e \mathcal{B}$ iff

$$(\exists \text{ partial recursive operator } \Theta)(\forall \varphi \in \mathcal{B})[\varphi \in \text{dom}(\Theta) \,\&\, \Theta(\varphi) \in \mathcal{A}].$$

Let $\mathcal{A} \equiv_e \mathcal{B}$ iff $\mathcal{A} \leq_e \mathcal{B}$ and $\mathcal{B} \leq_e \mathcal{A}$.

As before $\equiv_e$ is an equivalence relation and, defining meet and join operations on M_e (the set of equivalence classes under $\equiv_e$, or *degrees of difficulty*) in a similar way as before, we obtain the *Dyment lattice* $\boldsymbol{\mathcal{M}}_e = \langle M_e, \leq_e \rangle$ with least element $\mathbf{0}_e = [\{\varphi \mid \varphi \text{ is p.r.}\}]_e$ and greatest element$\mathbf{1}_e = [\phi]_e$. There a natural embedding of $\boldsymbol{\mathcal{D}}_e$ onto the *degrees of enumerability in* $\boldsymbol{\mathcal{M}}_e$ (those degrees having the form $[\{\varphi\}]$ for some partial function φ) given by $V_e(\deg_e(\varphi)) = [\{\varphi\}]_e$.

Both structures have many attractive and interesting features, and relate closely to $\boldsymbol{\mathcal{D}}$ and $\boldsymbol{\mathcal{D}}_e$. We can only refer briefly to some of the more recent work in this area. Dyment [**Dy76**] examines basic questions of incomparability and interpolation. Answering a question of Rogers [**Rog67**], it is shown that the property of being a degree of solvability in $\boldsymbol{\mathcal{M}}$ is lattice-theoretic: a degree is a degree of solvability if and only if there is a smallest degree among all the degrees greater than it. In [**Dy80**], Dyment examines analogues of Spector's work on upper bounds for countable ideals in $\boldsymbol{\mathcal{D}}$. One of the original motivations for the study of $\boldsymbol{\mathcal{M}}$ was a relationship with certain nonstandard logics ($\boldsymbol{\mathcal{M}}$ is a Brouwer algebra), which is followed through in a sequence of three papers [**Sor*ta*1**], [**Sor*ta*2**] and [**Sor*ta*3**], in which Sorbi carries through an intensive study of the algebraic properties of $\boldsymbol{\mathcal{M}}$ and $\boldsymbol{\mathcal{M}}_e$. It is shown in [**Sor*ta*1**] that $\boldsymbol{\mathcal{M}}$ is not a Heyting algebra. In [**Sor*ta*2**] the author investigates some non-principal filters and ideals of $\boldsymbol{\mathcal{M}}$, as well as the corresponding quotient lattices, and in [**Sor*ta*3**] studies in more detail the structure of $\boldsymbol{\mathcal{M}}$ as a Brouwer algebra.

There are still many questions concerning the Medvedev and Dyment lattices, including two originally asked by Rogers [**Rog67**]:

QUESTION 10.4. Is the property of being a degree of enumerability lattice-theoretic in $\mathcal{M}$?

QUESTION 10.5. Is the jump operation on degrees of solvability lattice-theoretic or lattice-theoretic with respect to the degrees of solvability (that is, invariant under all automorphisms of the lattice that carry the degrees of solvability onto the degrees of solvability)?

REFERENCES

Ah89. S. Ahmad, *Some results on the structure of the Σ_2 enumeration degrees*, Recursive Function Theory Newsletter **38** (July, 1989), item 373.

Ah*ta*. S. Ahmad, *Embedding the diamond in the Σ_2 enumeration degrees*, to appear.

Am*ta*. K. Ambos-Spies, *Minimal pairs for polynomial time reducibilities*, to appear.

AM77. L. Adleman and K. Manders, *Reducibility, randomness, and intractibility*, Proc. 9th ACM STOC (1977), 151-163.

BGS75. T. Baker, J. Gill and R. Solovay, *Relativizations of the P=NP? question*, SIAM J. Comput. **4** (1975), 431-442.

BGO87. R. Beigel, W. Gasarch and J. Owings, *Terse sets and verbose sets*, Recursive Function Theory Newsletter **36** (1987), item 367.

Cas85. P. Casalegno, *On the T-degrees of partial functions*, J. Symbolic Logic **50** (1985), 580-588.

Ca71. J. Case, *Enumeration reducibility and partial degrees*, Ann. Math. Log. **2** (1971), 419-439.

Ca74. J. Case, *Maximal arithmetical reducibilities*, Z. Math. Logik Grundlag. Math. **20** (1974), 261-270.

Coo71. S. Cook, *The complexity of theorem proving procedures*, Proc. 3rd ACM STOC (1971), 151-158.

Co72. S.B. Cooper, *Degrees of unsolvability complementary between recursively enumerable degrees, Part I*, Ann. Math. Logic **4** (1972), 31-73.

Co73. S.B. Cooper, *Minimal degrees and the jump operator*, J. Symbolic Logic **38** (1973), 249-271.

Co74. S.B. Cooper, *Minimal pairs and high recursively enumerable degrees*, J. Symbolic Logic **39** (1974), 655-660.

Co82. S.B. Cooper, *Partial degrees and the density problem*, J. Symbolic Logic **47** (1982), 854-859.

Co84. S.B. Cooper, *Partial degrees and the density problem. Part 2: The enumeration degrees of the Σ_2 sets are dense*, J. Symbolic Logic **49** (1984), 503-513.

Co87. S.B. Cooper, *Enumeration reducibility using bounded information: Counting minimal covers*, Z. Math. Logik Grundlag. Math. **33** (1987), 537-560.

Co*ta*1. S.B. Cooper, *The jump is definable in the structure of the degrees of unsolvability*, to appear Bull. Amer. Math. Soc. (1990).

Co*ta*2. S.B. Cooper, *The enumeration degrees are not dense*, to appear.

CC88. S.B. Cooper and C.S. Copestake, *Properly Σ_2 enumeration degrees*, Z. Math. Logik Grundlag. Math. **34** (1988), 491-522.

CLW89. S.B. Cooper, S. Lempp and P. Watson, *On the degrees of d-r.e. sets*, Israel J. Math. **67** (1989), 137-152.

CS*ta*. S.B. Cooper and A. Sorbi, *Initial segments of the enumeration degrees*, to appear.

Cope87. C.S. Copestake, *The Enumeration Degrees of Σ_2 Sets*, Ph.D. Thesis, Leeds University, 1987.

Cope88. C.S. Copestake, *1-genericity in the enumeration degrees*, J. Symbolic Logic **53** (1988), 878-887.

Cope*ta*1. C.S. Copestake, *1-generic enumeration degrees below* $\mathbf{0}_e'$, to appear in the proceedings of Heyting '88, Bulgaria, 1988 (Plenum Press).

Cope*ta*2. C.S. Copestake, *A 1-generic enumeration degree which bounds no minimal pair*, to appear.

Cope*ta*3. C.S. Copestake, *Nondeterminacy, enumeration reducibility and polynomial bounds*, to appear.

Cr53. W. Craig, *On axiomatizability within a system*, J. Symbolic Logic **18** (1953), 30-32.

Da58. M. Davis, "Computability and Unsolvability," McGraw-Hill, New York, 1958.

Dy76. E.Z. Dyment, *Certain properties of the Medvedev lattice*, Mat. Sb. (new series) **101 (143)** (1976), 360-379 (Russian).

Dy80. E.Z. Dyment, *Exact bounds of denumerable collections of degrees of difficulty*, Mat. Zametki **28** (1980), 899-910 (Russian).

Fe57. S. Feferman, *Degrees of unsolvability associated with classes of formalized theories*, J. Symbolic Logic **22** (1957), 161-175.

Fe65. S. Feferman, *Some applications of the notions of forcing and generic sets*, Fund. Math. **56** (1965), 325-345.

FR59. R.M. Friedberg and H. Rogers, Jr., *Reducibility and completeness for sets of integers*, Z. Math. Logik Grundlag. Math. **5** (1959), 117-125.

Gu71. L. Gutteridge, *Some Results on Enumeration Reducibility*, Ph.D. Dissertation, Simon Frazer University, 1971..

Ha86. C.A. Haught, *The degrees below a 1-generic degree and less than* $\mathbf{0}'$, J. Symbolic Logic **51** (1986), 770-777.

He79a. F. Hebeisen, *Masstheoretische Ergebnisse fur WT-Grade*, Z. Math. Logik Grundlag. Math. **25** (1979), 33-36.

He79b. F. Hebeisen, *Uber Halbordnungen von WT-Graden in e-Graden*, Z. Math. Logik Grundlag. Math. **25** (1979), 209-212.

Hi69. P.G. Hinman, *Some applications of forcing to hierarchy problems in arithmetic*, Z. Math. Logik Grundlag. Math. **15** (1969), 341-352.

Jo66. C.G. Jockusch Jr., *Reducibilities in Recursive Function Theory*, Ph.D. Dissertation, MIT, Cambridge, Mass.,1966.

Jo68. C.G. Jockusch Jr., *Semirecursive sets and positive reducibility*, Trans. Amer. Math. Soc. **131** (1968), 420-436.

Jo73. C.G. Jockusch Jr., *Upward closure and cohesive degrees*, Israel J. Math. **15** (1973), 332-335.

Jo80. C.G. Jockusch Jr., *Degrees of generic sets*, in F.R. Drake and S.S. Wainer, eds., Recursion Theory: its Generalisations and Applications, Proc. of Logic Colloquium, Leeds, 1979 (Cambridge University Press, 1980), 110-139.

Jon73. N.D. Jones, *Reducibility among combinatorial problems in log n space*, in Proc. of Seventh Annual Princeton Conference on Information Sciences and Systems, 1973.

Ka72. R. Karp, *Reducibility among combinatorial problems*, in Miller and Thatcher, eds., Complexity of Computer Computations (Plenum Press, New York, 1972), 85-103.

Kl52. S.C. Kleene, "Introduction to Metamathematics," Van Nostrand, New York, 1952.

KP54. S.C. Kleene and E.L. Post, *The upper semi-lattice of degrees of recursive unsolvability*, Ann. of Math. (2) **59** (1954), 379-407.

La65. A.H. Lachlan, *Some notions of reducibility and productiveness*, Z. Math. Logik Grundlag. Math. **11** (1965), 17-44.

La66. A.H. Lachlan, *Lower bounds for pairs of recursively enumerable degrees*, Proc. London Math. Soc. **16** (1966), 537-569.

La70. A.H. Lachlan, *Initial segments of many-one degrees*, Canad. J. Math. **22** (1970), 75-85.

La72. A.H. Lachlan, *Embedding nondistributive lattices in the recursively enumerable degrees*, in W. Hodges, ed., Conference in Mathematical Logic, London, 1970 (Springer-Verlag, Berlin, 1972), 149-177.

La79. A.H. Lachlan, *Bounding minimal pairs*, J. Symbolic Logic **44** (1979), 626-642.

LLS75. R. Ladner, N. Lynch and A. Selman, *A comparison of polynomial time reducibilities*, Theoret. Comput. Sci. **1** (1975), 103-123.

Lag72. J. Lagemann, *Embedding Theorems in the Reducibility Ordering of the Partial Degrees*, Ph.D. Dissertation, MIT, 1972.

Le83. M. Lerman, "Degrees of Unsolvability," Perspectives in Mathematical Logic, Omega Series, Springer-Verlag, Berlin, Heidelberg, New York, Tokyo, 1983.

Lo82. T.J. Long, *Strong nondeterministic polynomial-time reducibilities*, Theoret. Comput. Sci. **21** (1982), 1-25.

McE84. K. McEvoy, *The Structure of the Enumeration Degrees*, Ph.D. Thesis, Leeds University, 1984.

McE85. K. McEvoy, *Jumps of quasi-minimal enumeration degrees*, J. Symbolic Logic **50** (1985), 839-848.

MC85. K. McEvoy and S.B. Cooper, *On minimal pairs of enumeration degrees*, J. Symbolic Logic **50** (1985), 983-1001.

Me55. Yu.T. Medvedev, *Degrees of difficulty of the mass problem*, Dokl. Akad. Nauk SSSR, N.S. **104** (1955), 501-504 (Russian).

MS72. A.R. Meyer and L.J. Stockmeyer, *The equivalence of regular expressions with squaring requires exponential space*, in Proc. Thirteenth Annual IEEE Symposium on Switching and Automata Theory (1972), 125-129.

MS73. A.R. Meyer and L.J. Stockmeyer, *Word problems requiring exponential time*, in Proc. Fifth Annual Symposium on Theory of Computing (1973).

Mo74. B.B. Moore, *Structure of the Degrees of Enumeration Reducibility*, Ph.D. Dissertation, Syracuse University, 1974.

My61. J. Myhill, *A note on degrees of partial functions*, Proc. Amer. Math. Soc. **12** (1961), 519-521.

MSh55. J. Myhill and J.C. Shepherdson, *Effective operations on partial recursive functions*, Z. Math. Logik Grundlag. Math. **1** (1955), 310-317.

NS80. A. Nerode and R.A. Shore, *Second order logic and first order theories of reducibility orderings*, in J. Barwise et al, eds., The Kleene symposium (North-Holland, Amsterdam, 1980), 181-200.

Od81. P. Odifreddi, *Strong reducibilities*, Bull. Amer. Math. Soc. (N.S.) **4** (1981), 37-86.

Od89. P. Odifreddi, "Classical Recursion Theory," North-Holland, Amsterdam, New York, Oxford, 1989.

Od*ta*. P. Odifreddi, "Classical Recursion Theory, Vol. II," North-Holland, Amsterdam, New York, Oxford, to appear.

Pl72. G.D. Plotkin, *A set-theoretical definition of application*, Memo. MIP-R-95, School of Artificial Intelligence, University of Edinburgh, 1972.

PR77. E.A. Polyakov and M.G. Rozinas, *Enumeration reducibilities*, Sib. Math. J. **18** (1977), 594-599.

PR79. E.A. Polyakov and M.G. Rozinas, *Relationships between different forms of relative computability*, Math. USSR-Sbornik **35** (1979), 425-436.

Pos77. D. Posner, *High Degrees*, Ph.D. Dissertation, University of California, Berkeley, 1977.

Po44. E.L. Post, *Recursively enumerable sets of positive integers and their decision problems*, Bull. Amer. Math. Soc. **50** (1944), 304-337.

Rob68. R.W. Robinson, *A dichotomy of the recursively enumerable sets*, Z. Math. Logik Grundlag. Math. **14** (1968), 339-356.

Rog67. H. Rogers Jr., "Theory of Recursive Functions and Effective Computability," McGraw-Hill, New York, 1967.

Roz74. M.G. Rozinas, *Partial degrees and r-degrees*, Siberian Math. J. **15** (1974), 935-941.

Roz78a. M.G. Rozinas, *The semilattice of e-degrees*, Ivanov. Gos. Univ., Ivanovo (1978), 71-84 (Russian).

Roz78b. M.G. Rozinas, *Partial degrees of immune and hyperimmune sets*, Siberian Math. J. **19** (1978), 613-616.

Sa61. G.E. Sacks, *A minimal degree less than* $\mathbf{0}'$, Bull. Amer. Math. Soc. **67** (1961), 416-419.

Sa63. G.E. Sacks, *Recursive enumerability and the jump operator*, Trans. Amer. Math. Soc. **108** (1963), 223-239.

San78. L.E. Sanchis, *Hyperenumeration reducibility*, Notre Dame J. Formal Logic **19** (1978), 405-415.

San79. L.E. Sanchis, *Reducibilities in two models for combinatory logic*, J. Symbolic Logic **44** (1979), 221-233.

Sas71. L.P. Sasso Jr., *Degrees of Unsolvability of Partial Functions*, Ph.D. Dissertation, University of California, Berkeley, 1971.

Sas73. L.P. Sasso Jr., *A minimal partial degree* $\leq \mathbf{0}'$, Proc. Amer. Math. Soc. **38** (1973), 388-392.

Sas75. L.P. Sasso Jr., *A survey of partial degrees*, J. Symbolic Logic **40** (1975), 130-140.

Sc75a. D. Scott, *Lambda calculus and recursion theory*, in Kanger, ed., Proc. Third Scandinavian Logic Sympos. (North-Holland, Amsterdam, 1975), 154-193.

Sc75b. D. Scott, *Data types as lattices*, in Proc. Logic Conf., Kiel, SVLNM **499** (1975), 579-651.

Se71. A.L. Selman, *Arithmetical reducibilities I*, Z. Math. Logik Grundlag. Math. **17** (1971), 335-350.

Se72. A.L. Selman, *Arithmetical reducibilities II*, Z. Math. Logik Grundlag. Math. **18** (1972), 83-92.

Se78. A.L. Selman, *Polynomial time enumeration reducibility*, SIAM J. Comput. **7** (1978), 440-457.

SS*ta*. J. Shinoda and T.A. Slaman, *On the theory of the polynomial degrees of the recursive sets*, to appear.

Sh59. J.R. Shoenfield, *On degrees of unsolvability*, Ann. of Math. (2) **69** (1959), 644-653.

Sh66. J.R. Shoenfield, *A theorem on minimal degrees*, J. Symbolic Logic **31** (1966), 539-544.

Sita. A. Silver, *Polynomial time singleton enumeration reducibilities*, to appear.

Sk72. D.G. Skordev, *On partial conjunctive reducibility*, in Second All-Union Conference on Math. Logic, Inst. Prikl. Mat. (1972), 43-44.

SWta. T.A. Slaman and W.H. Woodin, *Definability in the degrees*, to appear.

Sol78. B.Ya. Solon, *e-powers of hyperimmune retraceable sets*, Siberian Math. J. **19** (1978), 122-127.

So87. R.I. Soare, "Recursively Enumerable Sets and Degrees," Springer-Verlag, Berlin, New York, London, Tokyo, 1987.

Sor88. A. Sorbi, *On quasi-minimal e-degrees and total e-degrees*, Proc. Amer. Math. Soc. **102** (1988), 1005-1008.

Sorta1. A. Sorbi, *Some remarks on the algebraic structure of the Medvedev lattice*, to appear.

Sorta2. A. Sorbi, *On some filters and ideals of the Medvedev lattice*, to appear.

Sorta3. A. Sorbi, *Embedding Brouwer algebras in the Medvedev lattice*, to appear.

Sp56. C. Spector, *On degrees of recursive unsolvability*, Ann. of Math. (2) **64** (1956), 581-592.

Vu74. V. Vuckovic, *Almost recursivity and partial degrees*, Z. Math. Logik Grundlag. Math. **20** (1974), 419-426.

Wata. P.Watson, *On restricted forms of enumeration reducibility*, to appear.

Ya65. C.E.M. Yates, *Three theorems on the degree of recursively enumerable sets*, Duke Math. J. **32** (1965), 461-468.

Za84. S.D. Zakharov, *e- and s-degrees*, Algebra and Logic **23** (1984), 273-281.

NOTES ON THE 0''' PRIORITY METHOD WITH SPECIAL ATTENTION TO DENSITY RESULTS

Rod Downey[1]
Mathematics Department, Victoria University of Wellington
PO Box 600, Wellington, New Zealand

(*e-mail* DOWNEY @ RS1. VUW. AC. NZ)

§1 *INTRODUCTION*

The use of trees control strategies in priority arguments has become ubiquitous in modern recursion theory. One of the crucial reasons for this seems to be that the use of trees gives us the ability to lay out the outcomes of each atomic strategy and reduces our verification to a 'coherence lemma'. That is, our strategies are so devised that Harrington's 'golden rule' is satisfied: each requirement has a version that can live with any particular sequence of outcomes of the other relevant requirements.

The power of this technique is that it often reveals the 'real reasons' that earlier results employing technical devices, such as the 'hat trick' of Soare/Lachlan (cf [16]) or Sacks density theorem (cf (2.3) of the present paper) work. In many ways the early infinite injury arguments were being described originally in a finite injury (linear) way, where in reality as **0''** arguments they would be best understood via a nonlinear tree of strategies. We remark that combinatorially, such a tree argument may be more difficult then other models(witness the use of pinball arguments for embedding nondistributive lattices into **R**) but the intuition seems more readily apparent.

As yet, we do not feel the situation for **0''** and **0'''** arguments has been similarly clarified. In many ways we feel that current **0'''** arguments are really being written in a **0''** framework, and hence we decided to investigate the power of a general **0'''** framework. That is trees specifically designed for **0'''** arguments. In particular, our ideas were inspired by Shore's **0'''** argument [15] which uses an $\omega + 1$ branching tree where the true path of the construction requires a **0'''** oracle. It seemed to us that this idea ought to be general enough to encompass all **0'''** arguments and furthermore gives the potential for $\mathbf{0}^{(n)}$ arguments in a similar framework.

1. The author's research partially supported by a US/NZ binational grant.

The idea is that if we have $\omega + 1$ branches representing the outcomes of a single π_3 requirement (not spread out on the tree), then the ω outcomes $\{i : i \in \omega\}$ will be π_2 outcomes collectively giving the Σ_3 outcome and the "1" will be the π_3 outcome. Now, although one can be left of the π_3 outcome infinitely often, the π_3 outcome might be the on left most path *visited* infinitely often. In Shore's argument, this caused no problems but for general $\mathbf{0}'''$ arguments it becomes necessary to develop some machinery to allow the construction to live with this feature. Primarily one develops a 'local priority ordering' that is different from the normal priority ordering and we can allow injury from left to right and right to left.

This idea could then allow extensions to a $\mathbf{0}^{(4)}$ argument. For example, we could then use linking on the $\mathbf{0}'''$ tree, or we could use a non-uniform $\mathbf{0}'''$ argument (e.g Downey - Slaman [6]) or we could change the order type of the outcomes to be a $1 + \omega^*$sequence of $\omega + 1$ outcomes. In the last option the true path would be $\mathbf{0}^{(4)}$.

In §2 we develop the machinery in the guise of a detailed sketch of new proof of Slaman's density theorem : [14] that if $\mathbf{e} < \mathbf{f}$ then there exist $\mathbf{a} \mid \mathbf{b}$ with $\mathbf{e} < \mathbf{a}, \mathbf{b} < \mathbf{f}$ and with $\mathbf{a} \cap \mathbf{b}$ existing. It is fair to say that this is a difficult $\mathbf{0}'''$ argument and hence provides a good testing ground for the technique. Much of the discussion to (2.15) is devoted to this specific result. We remark that much of the delicacy and length of this discussion stems from the problem of keeping the construction $\leq_T F$. One of the reasons we choose Slaman's density theorem rather than an easier $\mathbf{0}'''$ argument is the fact that we feel our technique will be very good for other density questions (see §3).

We develop some general machinery in (2.15) and (2.16), and in (2.17) discuss how one could use the framework for other theorems. In particular in (2.18) we look at the Lachlan nonbounding theorem under this framework.

Finally in §3 we examine some extensions of our results. First we show that the superbranching degrees (of [4]) are dense (by observing that the strategies are compatible with §2). The other result is to use the machinery to solve a question of J.B. Remmel : We show that there exists an r.e degree $\mathbf{a} \neq \mathbf{0}, \mathbf{0}'$ such that if B is r.e and deg (B) $\geq \mathbf{a}$ then there exists an r.e splitting $B_1 \cup B_2 = B$ with deg $(B_1) = \mathbf{a}$.

Notation is standard and follows Soare [16]. All uses are monotone in stage and argument where defined, and bounded by s at stage s. We do assume the reader already familiar with tree arguments at least on the $\mathbf{0}''$ level and hope that he or she will be familiar with Soare [15, 16].

§2 A NEW PROOF OF SLAMAN'S DENSITY THEOREM

2.1 The Components

The goal of this section is to (eventually) give a new proof of Slaman's result that if **e** < **f** then there exist **a**, **b** with **a** < **b**, **a** < **f**, **a** ∩ **b** existing and **a** | **b** . In view of the fact that we shall later extend the ideas of this construction to a general approach to **0'''** arguments, we shall proceed very slowly breaking the argument into small components : streaming (and the infimum requirements below a degree), the density requirements, the coding and coherence.

Let E and F be given r.e sets with $E <_T F$. We build re sets A, B and Q to meet the requirements A, B, $Q \leq_T F$ with $Q \leq_T$ A, B and

$$P_{2e} : \Phi_e(\hat{A} \oplus E) \neq B$$
$$P_{2e+1} : \Phi_e(\hat{B} \oplus E) \neq A \text{ and}$$
$$R_e : \Phi_e(\hat{A} \oplus E) = \Phi_e(\hat{B} \oplus E) = f \text{ total} \Rightarrow f \leq_T Q \oplus E$$

where $\hat{A} = A \oplus Q$, $\hat{B} = B \oplus Q$ and $\{\Phi_e : e \in \omega\}$ is an enumeration of all procedures.

2.2 Streaming and Fejer's Result

The first component of our construction technique that we examine is the one we call *streaming* and should be thought of as the R_e requirements first duty : to *process* numbers into a will behaved *stream.* This is the way we meet the P_i for E = Ø, that is, prove the result from Fejer's thesis [7] that each nonzero r.e degree bounds a diamond lattice (c.f. also Lachlan [11] and Downey[3]).

Thus let E = Ø and suppose then we wish to meet the requirements above. In this case it would suffice to meet A, B, $Q \leq_T F$ and

$$\hat{P}_{2e} : \Phi_e(\hat{A}) \neq B,$$
$$\hat{P}_{2e+1} : \Phi_e(\hat{B}) \neq A \text{ and}$$
$$\hat{R}_e : \Phi_e(\hat{A}) = \Phi_e(\hat{B}) = f \text{ total} \Rightarrow f \leq_T Q.$$

By Lachlan's nonbounding theorem [10] know that Q = Ø is not always possible for all given F. Technically this means, roughly speaking, that the idea of "preserving one side of the computation between expansionary stages" fails when combined with permitting. The idea then is to meet the $\hat{R}_e$ by enumeration into Q whenever " both sides" of a computation fail between expansionary stages.

Define $\hat{l}(e, s) = \max\{x : (\forall y < x)[\Phi_{e,s}(\hat{A}_s, y) = \Phi_{e,s}(\hat{B}_s ; y)]\}$ (similarly $l(e, s) \max\{x : (\forall y < x)[\Phi_{e,s}(\hat{A}_s \oplus E_s ; y) = \Phi_{e,s}(\hat{B}_s \oplus E_s ; y)]\}$) $m\hat{l}(e, s) = \max\{0, \hat{l}(e, t) : t < s\}$ (similarly $ml(e, s)$), $m\hat{u}(e, s) = \max\{u(\Phi_{e,s}(\hat{A}_s ; x)), u(\Phi_{e,s}(\hat{B}_s ; x))\}$.

The basic idea to meet the R_e is to allow the R_e' to create a well behaved *stream* of numbers as follows. R_e will be given a set (ω for the basic module, the numbers processed by higher priority R_j (in the α - module) that it will *process* as follows. Initially it is given x_0 (say 0). At the first stage s_0 where $l(e, s_0) > x_0$, R_e will process x_0 and pick a new number x_1 with $x_1 > m\hat{u}(e, x_0, s_0)$ (eg $x_1 = s$), and from the point of view of cooperation with other requirements, cancel (or restrain) all numbers z with $x_0 < z < x_1$ and so stop them from entering A.

Remark Variations are possible here. For example, we can use a "dump" construction and ask that these numbers enter A iff x_0 enters A at the same stage.

In general we continue this process, and, assuming nothing has yet been enumerated, we will have at any stage a stream of numbers $x_0 < x_1 < x_2 < \_\_ < x_n$ such that each $x_i > m\hat{u}(e, x_j, s)$ for all $j < i$ (recall we assume uses monotone where defined).

What is the point of this procedure? The first thing to notice is that x_{i+1} is "good" for x_j for $j \leq i$ in the sense that if x_{i+1} enters $\hat{A}$, $\hat{B}$ or both, *it will not affect the computations for* x_j *for* $j \leq i$. The idea then is for those versions of $\hat{P}_j$ (guessing that $\hat{R}_e$'s effect is infinitary) of lower priority than $\hat{R}_e$ will use only numbers from the $\hat{R}_e$ - stream for followers, and themselves will also process this stream in essentially the same way.

Thus to meet $\hat{P}_{2j}$ (for example) $j > e$, $\hat{P}_{2j}$ will take $x_0 = x(2j, 0, s)$ and in a similar way, wait till $\hat{L}(2j, s) > x(2j, 0, s)$ where

$$\hat{L}(2j, s) = \max\{x : \forall y \leq x\,(\Phi_{j,s}(\hat{A}_s ; y) = B_s(y))\}$$
$$\hat{L}(2j+1, s) = \max\{x : \forall y \leq x\,(\Phi_{j,s}(\hat{B}_s ; y) = A_s(y))\}.$$

At this stage (when we see $\hat{L}(2j, s) > x(2j, 0, s)$) $\hat{P}_{2j}$ will request the next member of $\hat{R}_e$'s stream and cancel/restrain all current members of $\hat{R}_e$'s stream $> x_0$. At the stage that we see a new member of $\hat{R}_e$'s stream appear, say x_n, we will assign $x(2j, 1, s) = x_n$ etc. Thus the first action of $\hat{P}_{2j}$ is to *refine* $\hat{R}_e$'s stream to look like

$$x(2j, 0, s) = \underbrace{x_0, x_1, x_2 \_\_, x_n}_{\text{don't really exist}} = x(2j, 1, s), \underbrace{x_{n+1} \_\_ x_m}_{\text{nor do these.}} = x(2j, 2, s), \_$$

Now $\hat{P}_{2j}$'s final act is to wait till we see F permit i at some stage s where x(j, i + 1, s) is defined. When this occurs, we wish to enumerate x(2j, i, s) into B_i so winning $\hat{P}_{2j}$ (a Friedberg - Muchnik procedure).

The problems stem from the fact that other $\hat{P}_k$ with k > e *also* must select their x(k, n, t) from R_e's stream. Arguing by priorities, at any stage the R_e - stream will have been refined to look like

$$x(2e, 0, s), \_, x(2e, n_{2e}, s), x(2e+1, 0, s), \_, x(2e+1, n_{2e+1}, s), \_\_\_,$$
$$x(m, 0, s), \_\_, x(m, n_m, s) \_\_$$

where the x (k, d, s) are devoted to R_k awaiting F to permit d. The bad scenario is that at some stages for some k we see F permit k and so we might enumerate (say) x(2r + 1, k, s) into A_{s+1} to win R_{2r+1} for some r > j. Since we did not win (e.g.) R_{2j} it must have been that x(2j, k, s) was not defined. But now, at some later stage we see F permit i and we wish to enumerate x(2j, i, t) =(2j, i, s) into B_{t+1} to win R_{2j}. The problem is that our enumeration of x(2r + 1, k, s) into the A-side earlier might have destroyed some A-computation and if we now enumerate x(2j, i, t) we might *also destroy the B-side of the same computation.*

The *solution* is to *also* enumerate x(2r + 1, k, s) into B_{t+1} and Q_{t+1} at the same stage as we enumerated x(2j, i, t) = x(2j, i, s) into the A-side. In this way Q can comprehend the fact that both sides may have changed. Notice that at least one side remains valid for x_q < x(2r + 1, k, s). For our purposes, it is worthwhile to think of this as follows. If the R_e - stream looks like x(e, 0, s), x(e, 1,s), ___ at any stage, then we regard x(e,n + 1, s) as the Q-use for x(e, n, s). Hence if x(e, n + 1, s) is enumerated into Q (for example, x(e, n + 1, s) = x(2r + 1, k, s) as above) then at the next e-expansionary stage t we could redefine x(e, n + 1, t).

It is not hard to see that - in this construction - we can ensure that if $l(e, s) \to \infty$ then $\lim_s x(e, n, s) = x(e, n)$ exists. Then suppose $\Phi_e(\hat{A}) = \Phi_e(\hat{B}) = f$ with f total. Let z be given. Note that x(e, z, s) > x for all z, s. Now find the least e-expansionary stage where x(e, z + 1, s) is defined and $x(e, z+1, s) \notin Q$. Then $\Phi_{e,s}(\hat{A}_s; z) = \Phi_e(\hat{A}; z)$.

The remaining details are to implement the above for all $\hat{R}_e$ and $\hat{P}_e$ via a tree of strategies. Note that all of the actions are compatible as each $\hat{R}_e$ and $\hat{P}_e$ essentially wishes to do the same thing : refine a set of numbers they are given into a well behaved stream. The only conflict ocurs due to the fact that we wish to make sure that $(\forall i)(\lim x(e, i, s)$ exists), but in the present argument this causes no problems since such injury is only caused *by us via the finitary requirements* $\hat{P}_i$. Similar arguments that involve processing numbers can be combined with the strategy above. For example, it is not difficult to extend the ideas above to embed the countable atomless boolean algebra below any nonzero r.e degree ([3]), or to construct a *superbranching* degree (i.e. $\mathbf{a} \neq \mathbf{0}'$ such that $\forall\ \mathbf{b} > \mathbf{a}\ \exists\ \mathbf{c}, \mathbf{d}\ (\mathbf{a} < \mathbf{c}, \mathbf{d} < \mathbf{b}\ \&\ \mathbf{c} \cap \mathbf{d} = \mathbf{a})$). Downey - Mourad [4]) or finally to construct a contiguous nonbranching degree (Downey [2]). (This last paper and [5] were the places where the terminology and ideas were developed).

We remark that the above streaming is a little too simple minded for Slaman's density theorem due to the interaction of E-coding and P_j action as we shall see, but the ideas really underpin the construction.

2.3 *The Density Requirements*

We return to the Sacks requirements

$$P_{2e} : \Phi_e(A \oplus E) \neq B,\ P_{2e+1} : \Phi_e(B \oplus E) \neq A.$$

When Sacks density theorem was first proved, it was apparently accomplished by a series of clever tricks one of which is the famous Sacks coding strategy. Today, using tree of strategies it is possible to expose the underlying intuition behind this strategy, and to see that it is really no more than the original Friedberg - Muchnik method and the *delayed permitting method.*

In this section we let $Q = \emptyset$ so that $\hat{A} = A$ and $\hat{B} = B$.

Suppose that also $E \equiv_T \emptyset$ and so we needed to only build $A, B \leq_T F$ so that $\Phi_e(A) \neq B$ (and $\Phi_e(B) \neq A$). In this case we`would the a familiar process of the last section : define a stream of followers $x(2e, i+1, s)$ for $i \leq n = n(s)$ so that $x(2e, i+1, s)$ exceeded the use of $x(2e, i, s)$, we would wait till $i \in F_{\text{at } s}$. If this occurs, we'd enumerate $x(2e, i, s)$ into B winning R_{2e}.

If $E \not\equiv_T \emptyset$ the problem is that E can later code numbers below the use of $x(2e, i, s)$ to upset this win. For a single requirement P_{2e} our solution is to define our stream to essentially obey the following rules (2.4) - (2.6).

2.4 ***Cancellation*** If $x = x(2e, i, s)$ is currently *active* (that is, waiting for an F-permission and $x(2e, i+1, s)$ is defined) and we see that $\Phi_{e,s}(A_s \oplus E_s; x)$ is E-incorrect cancel $x(2e, j, s)$ for $j > i$. If $x(2e, i, s) \in B_s$ also cancel $x(2e, i, s)$ and declare $x(2e, i, s)$ asinactive.

2.5 ***Activation (Appointment)*** If $x(2e, i, s)$ is currently defined but not active (and hence $x(2e, i+1, s)$ is not defined) and $L(2e, s) > x(2e, i, s)$ set $x(2e, i+1, s) > s$, $r(2e, s+1) = s$ and declare $x(2e, i, s)$ as active.

2.6 ***Permission*** If $x(2e, i, s)$ is active and $i \in F_{at\, s}$ enumerate $x(2e, i, s)$ into B_{s+1} cancel $x(2e, j, s)$ for $j > i$ but regard $x(2e, k, s)$ for $k \leq i$ as active.

The reader should note that, because of the last clause of (2.6), P_{2e} can still receive attention (via some $x(2e, k, s)$ for $k < i$) whilst it appears satisfied, as this attack (via k) is more likely to succeed. The crucial point is that whilst P_{2e} appears temporarily satisfied (via $x(2e, i, s)$) it *cannot get any new followers appointed to it.*

The rules above suffice for a single P_{2e}.

2.7 ***Lemma*** *Suppose* $(\forall x)[\Phi_e(A \oplus E; x)\downarrow]$. *Then* $(\exists y)[\Phi_e(A \oplus E; y) \neq B(y)]$ *and* P_{2e} *acts only finitely often.*

Proof Suppose not. We show $F \leq_T E$. It suffices to show that

(a) $(\forall s)(\lim_s x(2e, i, s) = x(2e, i)$ exists and $\notin B)$

(b) $(\forall i)(\exists s)\big(x(2e, i, s) = x(2e, i)$ and $l(2e, s) > x(2e, i)$ and $u(\Phi_{e,s}(A_s \oplus E_s; x(2e, i)) = u(\Phi_e(A \oplus E; x(2e, i))\big)$.

(c) $(\forall s)\big(x(2e, i+1, s) > u(\Phi_{e,s}(A_s \oplus E_s; x(2e, i, s))\big)$.

(d) E can recognise when (a) and (b) occur.

Once we have (a) - (d) we can E-recursively compute F as follows. Let $z \in \omega$. E-recursively find a stage s where $x(2e, z+1)$ is defined. Then $x(2e, z)$ is active and for all $j \leq z$ the Φ_e - computation on $x(2e, j)$ are E-correct. By restraints, the computations are final. Hence $z \in F$ iff $z \in F_{at\, s}$ since otherwise z's entry into F would cause us to win P_{2e} at or below $x(2e, z)$.

To verify (a) - (d), suppose we have already computed $x(2e, 0)$, _ _, $x(2e, k)$ and a stage s_0 where $\forall s \geq s_0\, (x(2e,j) = x(2e, j, s))$ for $j \leq k$. By hypothesis (a) - (d) hold for

$j < k$: Note $x(2e, k) \notin B_{so}$ otherwise the $\Phi_{e, so}(A_{so} \oplus E_{so} ; x(2e, k))$ computations are E-incorrect (so that x(2e, k) not final a contradiction). Now E-recursively find a stage s_1 where $L(2e, s_1) > x(2e, k, s_1)$ via E-correct computation. Then $x(2e, k + 1, s_1 + 1) = x(2e, k + 1)$.

Thus the method above solves the problem for a single P_{2e}. A problem is caused by the interaction of (2.5) and (2.6) for the coherence of several P_j. Our concern is as follows.

Suppose that for some least k we have $L(2e, s) \to \infty$ but $\Phi_e(A \oplus E ; x(2e,k)) \uparrow$ so that the use $\to \infty$. Now the x(2e, j) for $j < k$ have finite effect, but infinitely often the x(2e, j, s) - list is chopped back to x(2e, 0), _, x(2e, k) (a π_2 - event). Each time this is chopped back, at the next e-expansionary stage, we reset r(2e, t) to $u(\Phi_{e,t}(A \oplus E ; x(2e,k)))$. We thus meet P_{2e} by divergence.

Consider some P_j for $j > 2e$ desiring to put some numer into A. Such P_j can only put unrestrained numbers into A, and if P_{2e} is met by divergence then potentially its restraint is infinite. The usual way dilemmas like the above are overcome is to allow the restraint to "drop back" at E-nondeficiency stages a'la Soare [16]. In our case we need F-permitting too (and this is the heart of Sacks trick). The bad scenario is as follows. We have some $\hat{x} = x(j, p, s)$ targeted for B. We see F permit $\hat{x}$ at stage s but at this stage $r(2e, s) > \hat{x}$. It may be the case that at some stage $t > s$ it is found that r(2e, t) is E-incorrect and the restraint drops. But we have no longer E-permission to allow us to put $\hat{x}$ in.

The key modification is to replace (2.6) by *delayed permission* The crucial point is that E knows if r(2e, s) is E-correct or not (remember r(2e, s) only drops due to E-incorrect computations (essentially).). Further as $E \leq_T F$, whatever E knows, F knows too. This when we see $\hat{x}$ F-permitted we *declare it so*. Now, whilst $\hat{x}$ is still active and F-permitted, should we discover r(2e, s) to be E-incorrect and so drop back we then allow $\hat{x}$ to enter A. The whole point is that F can still decide the fate of $\hat{x}$ and so the construction remains $\leq_T F$.

The implementation of the above on a tree of strategies is as follows. A node σ devoted to P_{2e} has $\omega + 1$ many outcomes labelled from left to right
(0,u),(0,d),(1,u),(1,d), ___, ω where

(i,u) denotes unbounded use as x(2e, i)

(i,d) denotes disagreement preserved at x(2e, i)

w denotes wait (i.e $L(2e, s) \not\to \infty$).

A version of P_j guessing (0,u) will be guessing that all the $\Phi_{e,s}(A_s \oplus E_s; x(2e, 0))$ computations are E-incorrect (whose $\lim_s x(2e, 0, s) = x(2e, 0)$ exists). Thus a follower of this version of P_j in some sense "doesn't believe" an F-permission until it sees the Φ_e restraint drop to zero. Note that although F *cannot decide* which is the correct *outcome,* for any follower x, F can decide if the conditions that will allow x to be enumerated will ocur.

For example if a follow x of P_j had guess $\sigma = (0,u)\wedge(1,d)\wedge(5,u)$, then we would permit x to enter the relevant set provided that we saw σ appear correct (again), which would happen by the next E-nondeficiency stage, if it will happen at all.

The ideas embodied in Sacks delayed permitting are really at the heart of our actions in the full construction.

2.8 *Coding and Infimum*

In this section we shall see what happens when we combine (2.2),(2.3) and the coding of E into the infimum. It is here the argument gets complicated and becomes at **0'''** one. We will essentially meet the P_j as we did in (2.3) but must re-examine the infimum requirements.

$$R_e : \Phi_e(\hat{A} \oplus E) = \Phi_e(\hat{B} \oplus E) = f \text{ total} \Rightarrow f \leq_T Q \oplus E.$$

In (2.2) with $E \equiv_T \emptyset$ we showed how to meet Friedberg type requirements in the presence of such R_e whilst keeping the sets $\leq_T F$. The idea was to use Φ_e to process numbers into a well behaved stream, x_0, x_1 --- so that $x_{i+1,s} > mu(e, x_i, s)$ when appointed, and hence $x_{i+1,s} >$ one of the uses whilst $x_{i+1,s}$ is not enumerated. The P_j predicated on infinitary R_e beavior then choose numbers from the R_e - stream and this allows us to enumerate x_{i+1} into both sides should come x_j for $j \leq i$ be enumerated into one side.

In fact it is easy to combine this strategy with the $\hat{P}_j$ and hence show that for all $\mathbf{e} < \mathbf{f}$ there exist **a**, **b** such that $\mathbf{a} \cup \mathbf{e} \mid \mathbf{b} \cup \mathbf{e}$, $\mathbf{a} \cap \mathbf{b}$ exists and $\mathbf{a}, \mathbf{b} < \mathbf{f}$. All of our problems stem from E-coding in the infimum requirements.

Note that E-coding in the R_e might cause infinitary behavior for two reasons. First the use on some side for some y might be unbounded (as with a P_j). The second reason is that the use might be bounded for all y but $l(e,s) \to \infty$ (the π_3 reason).

Our hope is to develop some reasonably general machinery to deal with such π_3 situations.

2.9 The Basic (R_e -) Module

The main idea is to, along the lines of Shore [15], represent R_e by a single outcome on the tree. Specifically σ devoted to R_e has $\omega + 2$ (primary) outcomes labelled (o, u),(1, u), _ _ _, b, w where.

(i, u) denotes *unbounded* use at x_i (on either A - or B- side)

b represents the π_3 outcome *bounded* use for all x_i and $l(e, s) \rightarrow \infty$

w represents the *waiting* "non-recovery of computation" outcome.

Note that the collection of π_2 outcomes $\{(i, u) : i \in \omega\}$ forms the "Σ_3 outcome". The idea is that the leftmost math *visited* infinitely often will be on the tree path (TP). The problems will stem from the combinatorics. It is possible that σ^b $\subset$ TP and yet in the construction we are left of σ^b infinitely often.

For the *basic module* (for R_e) the idea is to play outcome (i, u) whenever s is a σ - stage and one of the computations $\Phi_{e,s}(\hat{A}_s \oplus E_s ; x_{i,s})$ or $\Phi_{e,s}(\hat{B}_s \oplus E_s ; x_{i,s})$ prove to be E-incorrect since the last σ-stage. It will be the case that we will have enumerated $x_{i+1,s}$ into Q so that Q can comprehend this fact. Note that if has a limit x_i then if σ^(i, u) $\subset$ TP, it must be that $\Phi_e(A \oplus E, x_i) \uparrow$.

We must make sure that σ^(i, u) produces a stream of good numbers or those $\gamma \supset \sigma$^(i, u) should σ^(i, u) be correct. Thus σ produces many types of streams. It will be given a stream $\{x(\sigma^+, i, s) : s, i \in \omega\}$ where σ^+ ^a = σ and will refine this stream.

The final σ-stream $\{x(\sigma, i, s) : i \in \omega\}$ will depend on the outcomes of σ. In particular, we focus on the outcomes σ^(i, u) and σ^b. To indicate that those are two possible types of stream produced for these outcomes, we will write them as y(σ^a, j, s) where a = (i, u) or b. Each outcome (i, u) has associated with it a number p(i, s) to test it. The idea is that y(σ^b, 0, s) = x(σ, 0, s) =y(σ^(k, u), 0, s) all s and p(0, s) = 0 but that y(σ^b, i, s) = x(σ, p(i, s), s) for some p(i, s) $\geq$ i (with p(i + 1, s) > p(i, s)). Initially we will define p(i, s) = i. When we get to first define y(σ^b, i, s) we will have defined y(σ^b, i, s) =

$x(\sigma, p(i, s), s)$. Suppose this is its limit value and write $x_i = y(\sigma\hat{}b, i, s)$. Now at the next e-expanssionary stage s we will define $x_{i+1,s} = y(\sigma\hat{}b, i+1, s) = x(\sigma, p(j+1, s), s)$ (=s (for the basic module)) which will equal $y(\sigma, p(j, s)+1, s)$ if this is the first stage so that $x_{i+1,s} > mu(e, x_i, s)$. Now this would be a $\sigma\hat{}b$ - stage and playing $\sigma\hat{}b$ we'd be free to assign these numbers to P_j for P_j guessing $\sigma\hat{}b \subset TP$.

If, at a later stage t we see that the current Φ computations on x_i were not both E-correct (this is slightly modified when we consider the interaction with the various P_j), we should have enumerated $x_{i+1,t} = x(\sigma, p(k+1, s), s) = x(\sigma, p(k+1, t), t)$ into Q cancelling any P_j restraint working with $x_{i+1,t}$. At the next e-expansionary stage t_1, we would then play a $\sigma\hat{}(i, u)$ stage.

One notable point here is that such $x_{i+1,t}$ need to be enumerated as soon as they appear E-correct. This is because the construction must remain $\leq_T F$ and hence we cannot wait until t_1. We must however wait until t_1 to *appoint* the next $x_{i+1,t}$ as we now see.

At this stage we would set aside a collection of numbers $C = \{x(\sigma, p(k)+i, t_1), \_\_\_ x(\sigma, p(k)+t, t_1)\}$ (say) and then assign $p(k+1, t_1) = p(k)+t+1$ and $x(\sigma\hat{}b, i+1, t_1) = x(\sigma, p(k)+t+1, t_1)$.

The "back-up" stream is $\{y(\sigma\hat{}(i, u), j, s : j \in \omega\}$. This would have the collection C added to it (in order). Such numbers can only become followers of nodes $\gamma \supset \sigma\hat{}(i, u)$ and only be *assigned* during $\sigma\hat{}(i, u)$ stages. Since we need the sets must be $\leq_T F$, it is necessary that they enter during other than $\sigma\hat{}(i, u)$ stages. These elements will have a finite collection of restraints they must respect. Obviously won't respect any restraint based on $x_{j,s}$ for $j \geq i+1$, but as we shall see, they may respect restraints based on $x_{k,s}$ for $k \leq i$. These restraints nevertheless behave in a friendly fashion (as in (2.3)) and drop down at E-nondeficiency stages. This allows us to decide if such a follower will enter.

Note that, in the limit, if $\sigma\hat{}(i, u) \subset TP$ then all of the σ - stream is of the form $y(\sigma\hat{}(i, u), j, s)$ and all the $y(\sigma\hat{}b, k, s)$ die (for $k > i$).

On the other hand if $\sigma\hat{}b$ is correct then eventually we will stop building $y(\sigma\hat{}(i, u), k, s)$ as the uses of $y(\sigma\hat{}b, i)$ come to a limit.

2.10 Coherence of One P_j of Lower Priority with an R_e

There are two types of P_j we must consider : those associated with $\rho > \sigma$^(i, u) and those associated with $\tau > \sigma$^b. When to play the ρ versions is clear enough (at present) namely when y(σ^b, i + 1, s) appears incorrect, and such a ρ will act during such "gaps". (A delicate part of the construction is to keep the F-recursive, but this is like the delayed permitting part of (2.3), as we see later).

Where we do have some real potential coherence problems is with a P_b associated with some $\tau \supset \sigma$^b.

Suppose that such a P_b targets an element x(τ, n, s) for B and that $x_{k,s}$ = y(σ^b, k, s) = x(τ, n, s).

Now we wish to implement the strategy of (2.3). Recall this was given in the rules (2.4) - (2.6) and basically consisted of a follower x being initially inactive, then active and then eventually perhaps enumerated after an E-correct permission. Note that the activation of x will probably occur after $x_{k+1,s}$ has been appointed.

Here we are also committed to something like the processing of (2.2). Thusx(τ, n + 1, s) will be appointed after $x_{k,s}$ activated, but x(τ, n + 1, s) is probably larger than $x_{k+1,s}$ and indeed perhaps $x_{k+1,s} < u(\Phi_{j,s}(A_s ; x_{k,s}))$ where j = e(τ).

The reader should note that in the discussion to follow, the difference between on situation here and (2.2) is that in (2.2) the only change to such a set-up is due to P_j activity, and in the basic module for R_e, the only change in $x_{k+1,s}$ is due to E-coding making a Φ_e - computation E-incorrect. Here we are concerned with the *combination* of such events. If E-coding destroys such an $x_{k,s}$ = x(τ, n, s) - set up before we add $x_{k,s}$ to B we have no problems.

The bad situation is the following *(which does not occur in* (2.2)*)*. Suppose at some stage s_1 we enumerate $x_{k,s}$ into B_{s_1}. This will cause the e = e(σ) computation concerning $x_{k,s}$ to be destroyed. Consider the situation at the next e-expansionary σ-stage t_1. This is possibly as given in the diagram 1 below.

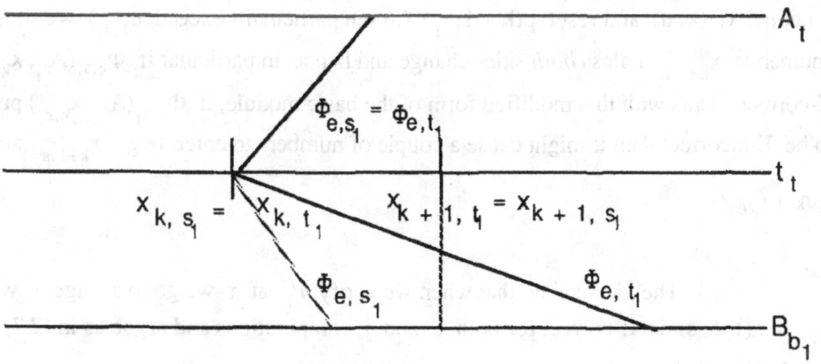

Diagram 1

In particular $x_{k+1,t_1} = x_{k+1,s_1} < u = u(\Phi_{e,s_1}(B_{t_1}; x_{k+t_1}))$. This causes the following potential problem : Remember the verification of (2.7). This consists of arguing that should we not win (e.g.) P_b then $F \leq_T E$. We do so by arguing that once we see an E-correct set up on X_{k,s_1} F cannot permit n lest we win on x_{k,s_1}. Now the reader should note that in the argument of (2.3) this followed as the only way that x_{k+1,t_1} entered into A or B and so $\hat{A}$ would be because of some $\hat{P}$'s activity. Here also E-coding for R_e via σ can cause problems via such enumeration.

Certainly in the definition of "E-correct set up for $x_{k,s}$" we can ask that for all $x_{m,s}$ if $x_{m,s} < u_{1,s_1}(\Phi_{j,s}(A_{s,j}\ x_{k,s}))$ then $x_{m,s}$ has currently E-correct Φ_e-computations too. This is consistent with $\tau > \sigma$^b after all.

Nevertheless, in the diagram above, although Φ_{e,s_1} was E-correct for $x_{k,s_1} = x_{k,t_1}$, by enumerating x_{k,s_1} into B we can cause Φ_{e,t_1} to have a much bigger use at t_1. Suppose that E now decides to cause a change in B below u_{1,s_1} *(but not below* mu(e, x_{k,s_1}, s_1)). If we now implement the basic module for R_e as stated, we must enumerate x_{k+1,s_1} into Q and so code it into $\hat{A}$ causing us to perhaps destroy the $\Phi_{j,s_1}(A_{s_1}; x_{k,s_1})$ computations. But now lemma (2.7) fails. Although $\hat{A}$ was E-correct for all potentially injurious numbers (and so $\hat{B}$ was E-correct on mu(e, x_{k,s_1}, s_1)) B was not E-correct on u_{1,s_1} allowing F to permit n late.

Our solution here is to note that, after all, the common value f can only change if *both sides* of the Φ_e - computation change. This, although we should clearly play on outcome $\sigma\hat{}(k, u)$ and reset $p(k+1, t_1)$ (to - in particular - exceed u_{1,s_1}) we should not enumerate x_{k+1,t_1} unless *both* sides change and hence, in particular if $\Phi_{e,s_1}(A_{s_1}; x_{k,s_1})$ is E-correct. Thus with this modified form of the basic module, if $\Phi_{e,s_1}(A_{s_1}; x_{k,s_1})$ proves to be E-incorrect then it might cause a couple of numbers to enter (e.g. x_{k+1,s_1} and $x_{p(k+1,t_1),t_1}$).

The idea is then that when we verify P_b at τ we go to a stage s where $x_{k,s} = x(\tau, n, s)$ is E-correct for both e and j computations and argue as in (2.7) that $n \in F$ iff $n \in F_{\text{at } s}$.

2.11 Coherence of R_e with Two P_j (of lower priority)

The "α - module" for R_e of (2.10) was okay for one P_j, but the problems become more subtle for two P_j. Let P_a and P_b be two such requirements and suppose that P_a has guess γ and P_b as before has guess τ, P_a targets for A, (P_b for B) and that for simplicity $\sigma\hat{}b < \gamma < \tau$.

Suppose that $e(\gamma) = m$ and that we are concerned with a follower of P_a of the form $x_{r,s} = x(\gamma, d, s)$ and note that (by priorities) $r < k$.

Now if $x_{r,s}$ was active before $x_{k,s}$ was enumerated, then the priority set up will ensure that $x_{k,s} > x(\gamma, d+1, s)$ and so, should we enumerate $x_{r,s}$ into A we can enumerate (safely) all $x_{t,s}$ for $x_{t,s} \geq x(\gamma, d+1, s)$ and so reduce this case to the basic module.

Note that we are really here considering the α - correct version of (2.2) : what can we do so that Q can comprehend the fact that two sides have changed? The method of (2.10) works in (2.2) and works for the case above.

However a case that is seriously different (due to E-coding) is the following sequence:

(2.12) $x_{k,s}$ enumerated at s_1
(2.13) $x_{r,s}$ is activated at the next γ - stage s_2
(2.14) $x_{r,s}$ is enumerated at $s_3 > s_2$.

Now in (2.2) although we know that since (2.13) occurred after (2.12), it did occur *at a σ-expansionary stage*.

Hence, although $x_{k+1,s}$ might be this below $u_2 = u(\Phi_{m,s_2}(B_{s_2}; x_{r,s}))$ this doesn't matter since now the B-side will hold the computation (in (2.2) arguing by priorities) that there are no followers z left alive with $x_{r,s} < z < u_2$).

The situation is given in diagram 2 below.

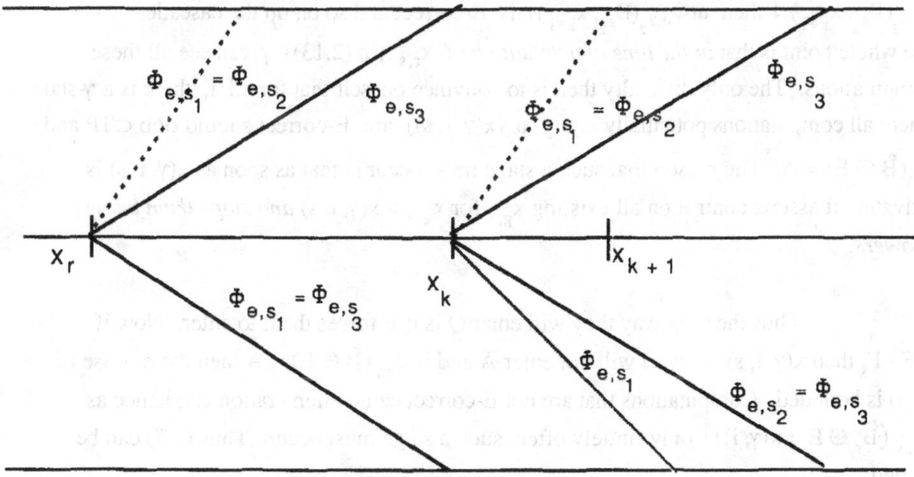

Diagram 2

Note that it is now possible for both sides of the x_k - computation to have changed. E-coding causes the following problem to the (2.7) argument.

We are forced to enumerate x_{k+1} into Q should E permit both sides after s_3. Indeed, note that in the situation above, it is also possible for x_{k+2,s_2} to affect x_{k+1} as perhaps $x_{k+2} < q_1 = u(\Phi_{e,s_3}(A_{s_3}; x_{k+1}))$, $q_2 = u(\Phi_{e,s_3}(B_{s_3}; x_{k+1}))$ for example E might permit q_1 and q_2 causing us to enumerate x_{k+2} which might be below both $u(\Phi_{e,s_2}(A_{s_2}; x_k))$ and $u\Phi_{e,s_2}(A_{s_2}; x_k))$ so that we are forced to enumerate x_{k+1} by a cascade effect (Diagram 3)

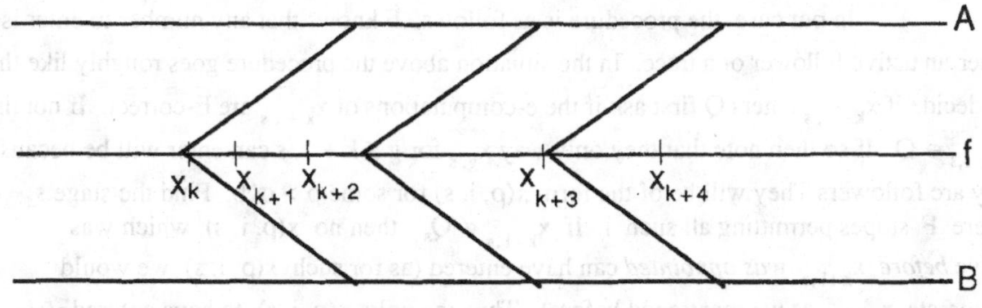

Diagram 3 : A cascade effect with x_{k+4} causing x_{k+1} to be enumerated

Because of the fact that $x_{i+1,s} > mu(e, x_i\ s,)$ we know that the only way this situation can occur is for two numbers like x_k and x_r to enter B and A, in the sequence described above.

Again it is possible for $x_{k+1} < u_2$. We must still be above to rescue (2.7) as with (2.10). Evidently now we must ask that all computations 'potentially injury' u_2 must be E-correct (at activation). This in particular, in the verification, γ would demand that if $\Phi_{e,s_2}(B_{s_2}; x_{k+1})\downarrow$ then $u(\Phi_{e,s_2}(B_{s_2}; x_{k+1}))$ is E-correct and so on up the cascade. The whole point is that *at the time of activitation* of $x_{r,s}$ (in (2.13)) γ can see all these computations. The only difficulty then is to convince oneself that for all i, there is a γ-state where all computations potentially injury $u_2(x(\gamma, i, s))$ are E-correct should $\sigma\hat{}b \subset TP$ and $\Phi_m(\hat{B} \oplus E) = \hat{A}$. The reason that such a stage must occur is that as soon as $x(\gamma, i, s)$ is activated, it asserts control on all existing $x_{g,s}$ for $x_{g,s} > x(\gamma, i, s)$ *and stops them being followers.*

Thus the only way they will enter Q is if E forces them to enter. Now if $i \notin F - F_s$ then $x(\gamma, i, s) = x(\gamma, i)$ will not enter A and if $\Phi_m(\hat{B} \oplus E) = \hat{A}$ then the m - use of $x(\gamma, i)$ is bounded. Computations that are not E-correct cause enumeration and hence as $\Phi_{m,s}(\hat{B}_s \oplus E_s; x(\gamma, i))\uparrow$ only finitely often, such a stage must occur. Thus (2.7) can be rescued.

Another subtle point is that the construction must be kept $\leq_T F$. The situation above has the potential to cause mischief since now perhaps large E-changes can cause small enumeration, i.e. a relatively large change in E can cause a relatively small number like x_{k+1} to enter Q (perhaps $u_3 = u(\Phi_{e,s_2}(B_{s_2}; x_k))$ and $u_4 = u(\Phi_{e,s_3}(B_{s_3}; x_k))$ are very big.

For the basic R_e-module F had no problems as $x_{k+1,s} \in Q$ iff the $x_{k+1,s}$ is appointed to trace $x_{k,s}$ and either computation is E-incorrect. The way traces will be appointed will ensure this is $\leq_T F$.

In our case the procedure is as follows. F knows that any number to enter is either an active follower or a trace. In the situation above the procedure goes roughly like this. To decide if $x_{k+1,s}$ enters Q first ask if the e-computations of $x_{k+1,s}$ are E-correct. If not then $x_{k+1,s} \in Q$. If so then note that they only way $x_{g,s}$ for $g \leq k + 1$, s can enter will be because they are followers.They will be of the form $x(\rho, i, s)$ for some $\rho > \sigma\hat{}b$. Find the stage s_4 where F stopes permitting all such i. If $x_{k+1,s} \notin Q_{s_4}$ then no $x(\rho, i, s)$ which was active *before* $x_{k+1,s}$ *was appointed* can have entered (as for such $x(\rho, i, s)$ we would enumerate $x_{k+1,s}$ as we mentioned before). Thus the only $x(\rho, i, s)$ to have entered after s would be ones which became active *at* σ^b-*stages after s.*

Let $s_5 \leq s_4$ be the stage where the last such $x(\rho, i, s)$ entered. By the argument above (for the α-module) since we won't enumerate any existing x_{g,s_5} unless both sides change (and be E-incorrect) we see that $x_{k+1,s} \in Q$ iff $x_{k+1,s} \in Q_{s_6}$ where $s_6 = \mu s(E[s_5] = E_s[s_5])$. In this way, A, B and $Q \leq_T F$.

2.15 The Priority Ordering and the Problems with $\omega + 2$ Branches : General Machinery

In a normal **0''** priority construction one defines a priority ordering $\leq_p$ via lexicographic ordering on the tree of strategies (eg Soare [15, 16]). Then if σ is played during stage s we would initialise all τ to right of σ. The argument is that if σ is the leftmost math visited infinitely often than we are *left of* σ *only finitely often*. This is no longer true in our construction. Here we have $\omega + 2$ branches, and, should b be the correct outcome we might nevertheless actually be left of b infinitely often. Were we to use the standard **0''** intialization strategy (i.e. initialize nodes right of τ when we vist τ), we would intialise all $\gamma \supset \sigma$^b cofinally in the construction.

Thus we will be guided by the principle that we cannot intialise all of $\gamma \supset \sigma$^b each time we are left of σ^b. On the other hand, $\beta > \sigma$^(k, u) cannot respect all of the$\gamma \supset \sigma$^b when we play σ^(k, u). There are two reasons for this. First - many such γ would be using $x_{\hat{k},s}$ for $\hat{k} > k$ which would 'appear wrong' when we visit σ^(k, u). Second, from more general grounds, σ^(k, u) ought to only respect a finite number of nodes extending σ^b, so as to be above to be met if σ^(k, u) $\subset$ TP.

Our idea is to define a (local) priority ordering $<^*$ so that $\tau <^* \gamma$ implies γ must respect τ's restraints. We delay the exact definition of τ until later, but we ask that $<^*$ is a well ordering.

To motivate the following, consider a simple situation where $\sigma = \lambda$ and so we ask what sort of $\tau > b$ say $\gamma > (k, u)$ should respect. A natural choice would be that (e.g.) $\beta \supset (0, u)$ should not respect anything, and that $\beta \supset (1, u)$ perhaps might only respect b and perhaps $\beta \supset (2, u)$ should respect for example b^(0, u), b^(0, d) and b^(1, u), b^(1, d) and b^w (see diagram 4)

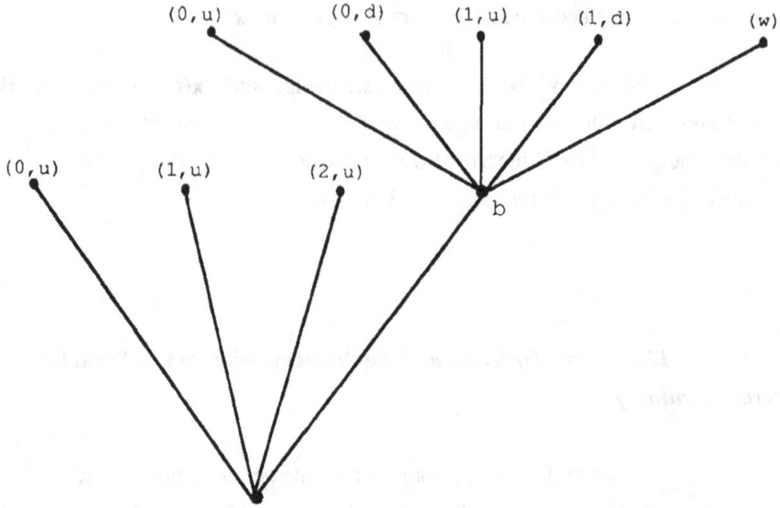

Diagram 4

The idea is that $x_{0,s}$ $x_{1,s}$,and $x_{3,s}$ can only be assigned to such nodes $\tau <^*(2, u)$ and hence if (2, n) appears correct as $x_{3,s}$ appears E-incorrect any restraints associated with it vanish. When we visit (2, u) we only respect this finite set of restraints generated by γ extending (0, u), (1, u) or $\gamma <^*(2, u)$ and $(2, u) \leq_L \gamma$. Note that as a node such as b can have infinitary outcome (e.g. (0, u)) there is a little problem for say $\beta > (2, u)$. In some sense when we visit β it may be that β must respect b's restraint on say $x_{1,s}$.

If we suppose (2, u) is the correct outcome and $x_{1,s} = x(b, 0, s)$ with $0 \notin F$, it may be that the correct b coutcome is (0, u) but when we visit (2, u) b's restraint is up. The point is this. We are - after all - going to have followers apointed for $\gamma \supset (2, u)$. Such followers need to know when it is appropriate for them to enter. This is a familiar enough problem. After all, the potentially bad numbers are only $x_{0,s}$ $x_{1,s}$ and $x_{2,s}$ and we really only need to guess the π_2 behavior of the nodes to which they are assigned.

We can either do this implicitly via the so called "hattrick" approach, or we can be more thematic and expand the tree of outcomes of (2, u). Really we only need to know the *number* of such x_i have infinite activity associated with them, and therefore to guess the π_2 behavior of nodes associated with x_0, x_1, x_2, (2, u) could have a tree of 8 outcomes in diagram 5 below.

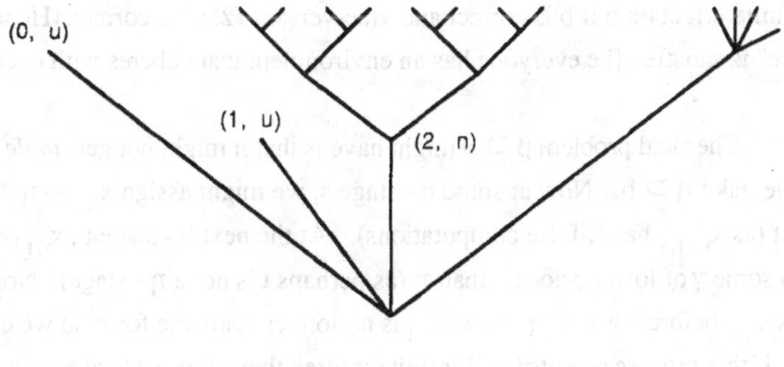

Diagram 5

Doing this expansion would allow us to know exactly the behavior of the nodes associated with x_0, x_1, x_2 if (2, u) $\subset$ TP, and this would be the thematic way of performing the construction.

We shall perform a combinatorially simpler approach rather more along the lines of a pinball machine - by noticing we can have only one version of (2, u) if we allow injury from the right. Imagine y is a follower with guess $\gamma \supset$ (2, u) that is F-permitted. We need to add y when the restraints for $\tau <^* \gamma$ become E-incorrect (or at least don't restrain y). The construction will ensure that if $x_{i, s}$ is associated with ρ and $x_{j, s}$ with η and $i \leq j$ then $\rho \leq^* \eta$. The problem is that F needs to sort out if y will enter.

F essentially goes to a stage where $x_{0, s}$, $x_{1, s}$, $x_{2, s}$ cannot later be F-permitted and asks if the computations corresponding to there x_i restraining y are E-correct. If so then y can't enter. The process it must avoid as if (say) $x_{2, s}$ restrains y then later $x_{1, s}$ acts (cancelling the restraint associated with $x_{2, s}$) but it turns out that $x_{1, s}$ has E-incorrect restraint. We note that this can only happen if the node associated ρ with $x_{1, s}$ *initializes* the associated node η with $x_{2, s}$. When this happens we will initialize (2, u) and ensure that, thereafter $x_{2, s}$ can only be associated with nodes $\leq^* \rho$. This device makes the argument and the priority tree such simpler cominatorially and can be used purely because of the way we appoint numbers. (In some other arguments it does not seem to be applicable.) Note that the construction still remains $\leq_T$ F.

From the other point of view, nodes $\gamma \supset$ (2, u) will set up restraints whilst they appear satisfied.

Collectively $\{(i, u) : i \in \omega\}$ might restrain all those $\beta \supset$ b from being met. Again we must allow $\tau <^*$(2, u) priority over the restraints of such γ. This tradeoff ensures that

(2, u) has finite effect on b if b is correct and vice-versa if (2, u) is correct. Hence Harrington's "golden rule" is satisfied (i.e everyone has an environment that coheres with everyone else).

The final problem $\beta \supset b$ might have is that it might not get *stable* followers. For example, take $\eta \supset b$. Now at some b - stage s, we might assign $x_{i,s}$ to η. Now $x_{1,s}$ may be incorrect (as $x_{i-1,s}$ has E-false computations). At the next b - stage t, $x_{i,t}$ ($\neq x_{i,s}$) may be assigned to some γ of lower priority than η (as perhaps t is not a η - stage). Now perhaps γ enumerates $x_{i,t}$ before we visit η. Now $x_{i,t}$ is no longer available for η so we choose $x_{j,t}$ some $j > i$. If this process repeats itself infinitely often then after we lose at η, and we would say η does not get a stable follower.

To overcome this problem, at each stage s for each node γ we have the task of ensuring that there will be infinitely many numbers set aside for γ should γ be visited, and this task is met if $\gamma \subset$ TP. The idea is to simply use a queue. We look back on the construction at the next b - stage t (as above) and see that η was the hightest priority node previously visited with out an inactive follower and see that $x_{i,s}$ was asigned to η. We then declare $x_{i,t}$ as *only to be available for* η (or higher priority nodes) *thereafter* (this is consistent with our earlier devices). Note that activiation only occurs if η is *visited* so the only time infinitely many η - numbers are so "directed" to η is if some $\hat{\eta} \leq \eta$ has the π_2 outcome.

It is clear that the local ordering must be continued throughout the tree and we will give details in (2.16) below. We remark that before we discuss the full construction we will discuss the in (2.18) general machinery of (2.15) in the context of another construction the *Lachlan nonbounding theorem* [10] which is the best known and probably the easiest of the $\mathbf{0}'''$ arguments. If the reader has not altogether followed the previous discussion he or she might like to read this section.

2.16 The Priority Tree and Local Ordering

We shall define the tree T in the obvious way we want "levels" where R_0 occurs at level 0, P_0 at level 1, R_1 at level 2 etc.

The easiest thing to do is to assign a rank for $\gamma \in$ T and define T inductively. Thus λ has rank 0. The outcomes of λ are $\{(i, u) : \tau \in \omega\}$ and b and w. We call λ the *rank 0 node* (written rk(0) = λ).

Now on each of λ's outcomes we will place a copy of P_0 i.e the outcomes

$\{(i, u), (i, d)\ i \in \omega\}$ and w giving each of the outcomes of λ rank 1. Now put a copy of λ's outcomes on each version of P_0's outcomes and these have rank 2 etc.

Similarly we define the local ordering $<^*$ on T inductively in any reasonable way as we did for λ.

This we start λ and for each outcome (j, n) define $\leq^*$ via.

All (i, u) and b are $<^*$ w.

(0, u) : $(0, u) <^* \gamma$ all $\gamma \in T$ and $\gamma \neq (0, u)$ (& $\gamma \neq \lambda$)
(1, u) : $\tau <^* (1, u)$ for all $\tau \in T$ with $\tau \supset (0, u)$
$b <^* (1, u)$ and $(1, n) <^* \tau$ all $\tau \in$ T otherwise.

(2, u) : $\tau <^*$ (2, u) for all $\tau \in$ T with $\tau \supset (1, u)$
$\tau <^*$ (2, u) for all $\tau \in$ T with $\tau <^*$ (1, u) and $\tau <^*$ 02, u)
$\tau <^*$ (2, u) if τ = b^(0, u), b^(0, d), b^(1, u) ,b^(1, d)
or τ = b^w. We have (2, n) $<^*$ τ otherwise.

In general one extends the ordering above in any reasonable fashion. I am using the idea that (n, u) should be $<^*$ all $\tau <^*$ (n - 1, u) and should respect an increasing finite number of rank $m \leq n$ nodes in such a way as the whole tree to the right of b is eventually enumerated. Also consistency principle is that if σ is a node devoted to P_j then σ^j $<^*$ σ^i iff σ^j $\leq_L$ σ^i and so that the $<^*$ differs only because of even rank nodes.

One can now extend $<^*$ to all of T by an inductive procedure. The only real consistency proviso is that if $\sigma \supset$ b and σ is devoted to R_e then $\tau \supset$ σ^(i, u) with $\tau \supset \sigma$^b will agree with the set of $\tau <^*$ (j, u) with $\tau \supset$ σ^b for some j.

2.17 Remark

The reader should think of $<^*$ as the sort of ordering one gets if we were to write the R_e requirements and then constructed the tree with no b - outcome and the (i, u) outcomes (of λ) scattered over the tree. The outcomes (i, u) correspond roughly to a "linking" procedure and the region above (i, u) as a pruned tree constructed via lists as in Slaman/Soare account [16] of there Lachan Nonbounding theorem. We expand on the signifigance of this below (in 2.18).

Theorem (Lachlan [10) $\exists\, \mathbf{a} \neq \mathbf{0}\, \forall\, (\mathbf{b}, \mathbf{c} \leq \mathbf{a}\, (\mathbf{b} \cap \mathbf{c} = 0 \rightarrow (\mathbf{b} = 0 \vee \mathbf{c} = \mathbf{0}))$

Here one meets the requirements.

$P_e : \bar{A} \neq W_e$

$R_e : \Phi_e (A) = V_e$ and $\Gamma_e (A) = U_e \rightarrow (V_e$ recursive $\vee\, U_e$ recursive or $((\forall\, i)\, (R_{e,\, i}))$
where
$R_{e,\, 1} : Q_e \neq \overline{W}_i$.

Here we build $Q_e \leq_T V_e, U_e$ and A and $< \Phi_e, V_e \Gamma_e U_e >_{e \in \omega}$ is an enumeration of all 4-tuples consisting of 2 r.e sets and two functionals. The strategies associated with the above have been discussed in great detail in [10, 15, 16] and we will only give a very brief account (for the sake of completeness).

The basic module for R_e is to have two restraints r_1 (e, s) and r_2 (e, s) we attempt to meet R_e by followers. If we fail one of $V_e \equiv_T \emptyset$ or $U_e \equiv_T \emptyset$ will hold.

The basic module consists of the steps below:-

Step 1 Pick a follower x = s at stage s and now wait till l(e, t) > x and $x \in W_{i,\,t}$ for some $t \geq s$, where *for this section* $l(e, s) = \max \left\{ x : (\forall y < x)\, (\Phi_{e,\, s} (A_s\, ; y) = V_{e,\, s} (y) \text{ and } \Gamma_{e,\, s} (A_s\, ; y) = U_{e,\, s} (y)) \right\}$.

Step 2 When t occurs, open a V_e - gap by setting r_1 (e, t) = 0, potentially allowing V_e to change.

Step 3 Wait till the least stage $t_1 > t$ such that $l(e, t_1) > l(e, t)$. Adopt the first case below to pertain.

Case 3a (Successful closure) $V_{e,t_1}[x] \neq V_{e,t}[x]$

Action Go to 4, setting $r_2(e, t_1) = 0$

Case 3b (Unsuccessful closure) $V_{e,t_1}[x] = V_{e,t}[x]$

Action Set r_1 (e, t_1) = t_1 choose a new follower x = t, and go to step 1.

Step 4 Wait till $t_2 > t_1$ occurs with $l(e, t_2) > l(e, t_1)$ adopt the first case to pertain

<u>Case 4a</u> (Successful closure) $U_{e,t_2}[x] \neq U_{e,t}[x]$

<u>Action</u> Put x into Q_{e,t_2} meeting $R_{e,i}$. Stop.

<u>Case 4b</u> (Unsuccessful closure) $U_{e,t_2}[x] = U_{e,t}[x]$.

<u>Action</u> Set $r_1(e, t_2) = r_2(e, t_2) = t_2$, pick up a new x and go to step 1.

The outcomes of the module above are, in order of priority f (we reach 4a), g_2 (infintely many case 4b), g_1(infintely many 3b, but only finitely many 4b) and w (stay waiting). To use our set up, a node σ on the tree devoted to R_e would have $\omega + 2$ outcomes labelled

$$(0, g_2), 0, g_1), (1, g_2), (1, g_1), \_\_\_, f, w.$$

where (i, g_i) denotes the outcome that $R_{e,\,i}$ has outcome g_j, f is the π_3 outcome : that all the $R_{e,\,i}$ have finite behaviour and w is $l(e, w) \not\to \infty$ outcome.

Again we would define a local ordering $<^*$ on the tree made up from such primary outcomes where outcome (i, g_1) would be refined by having a finite tree of suboutcomes corresponding to the π_2 behaviour of those $\tau <^* (i, g_i)$ with $\tau \supset s$^ f(this is easier the most general such theorem as we don't need the whole tree only the number of nodes that exhibit π_2 behaviour).

We play outcome f whenever it looks correct. here that would mean that we would have a monotone maker m(e, s) and, whenever we see that $R_{e,\,0}, \_\_, R_{e,\,m\,(e,\,s)}$ all exhibit finite behaviour (i.e. either waiting for $R_{e,\,i}$'s current x to enter $W_{i,\,s}$ or we get the 4a for $R_{e,\,i}$) we play f and set m(e, s + 1) = m(e, s) + 1.

Note that if (i, g_j) is the correct primary outcome of R_e at σ then those $\gamma \leq_L \sigma$^(i, g_j) only have finite effect, and there are only finitely many $\beta <^* \sigma$ with $\beta \leq_L \sigma$. These exhibit at worst π_2 beaviour and this beaviour can be guessed as the correct sub -outcome μ of σ^(k, g_j). This means that the restraint $r_j(\sigma$^(i, g_j)^ μ, s) holdsV_e (j = 1) or U_e (j = 2) during the co-gaps and there is no change in the gaps. It follows that if e.g. j = 1, then V_e is recursive.

We remark that we feel in some sense that this technique is to **0'''** - arguments is what the tree method was to **0''** - arguments, as in some sense, it is the 'natural' method of doing such arguments. We shall expand on these comments later in §3.

2.18 Notes on the Nonbounding Theorem, the General Machinery and the 0''' Method

As we remarked in (2.17) the $<^*$ priority ordering can be thought of as the alternative method for the construction. (We also remark that one can use a more dynamic notion of local ordering. This device was used by Shore[13] and a forth coming paper of Downey and Shore (see the note at end of paper).) The module for Slaman's density theorem is, itself, quite complicated and the fact that the construction remains $\leq_T$ F quite delicate. To aide the reader will take time out to discuss the machinery of (2.15) and (2.16) in the context of a much better know (and easier construction). Here we recall the nonbounding theorem of Lachan.

2.19 The Construction and Verification

The inductive strategies in this result are the same as those for the α-module described earlier. The basic rules are that if y is associated with $\sigma \in T$ then $y < |\sigma|$. If $\gamma \supset \sigma$ and $x_{i,s}$ is a member of σ's stream at s and associated with γ then for all stages $t \geq s$, $x_{i,s}$ is associated with only η such that $\eta \leq^* \gamma$, and if $x_{i,s}$ is associated with γ and some $\eta <^* \gamma$ asserts control of $x_{i,s}$ at stage s (and declares $x_{i,s}$ to no longer follow η (e.g. $x_{i,s} < r(n, s)$) then for all $t \geq s$, $x_{i,t}$ can only be associated with $\rho \leq^* \eta$.

If σ is associated with R_e and we see some $x_{i,s}$ e-computation have E-incorrect use, we enumerate $x_{i+1,s}$ as described in the α-module. Namely $x_{i+1,s}$ enters if nobody controls $x_{i+1,s}$ or if both sides change, and $x_{i+1,s}$ does not enter if only one side changes but the other side has higher priority control asserted on it. We will play a σ^ (i, u) stage the next time we visit σ. If r (e, s) is preserving or measuring a computation at ρ and this is E-incorrect at s + 1 we cancel the restraint. If $x_{i,s} = x(\gamma, j, s)$ ($|\gamma|$ odd) is active and j occurs in F at s, we declare $x_{i,s}$ as F-permitted at γ. This remains so for all stages $t \geq s$ unless some higher priority $x_{j,s}$ acts, the region over which γ assert control is E-incorrect.

The construction then proceeds in the obvious way in substages. Starting at λ we see find the highest priority option amongs σ^(i, u) appears correct, σ^b appears correct, σ^w appears correct, or some τ requires attention (that is some $x(\tau, j, s)$ which is F-permitted is, (now) unrestrained).

One then chooses the relevant option and this gives $\sigma_0 = \lambda$, and $\sigma_1 \supset \sigma_0$. We proceed until $|\sigma_s| = s$ inductively.

The verification of the construction is virtually the same as the earlier discussion and we will this omit it, and let this conclude our notes on Slaman's density theorem.

§3 ***EXTENSIONS, VARIATIONS AND OTHER RESULTS USING THE TECHNIQUE***

The proof of Slaman's density thereom, as well as the general machinery of §2 admits several further extensions and applications. First, a miner modification to the P_j will give an embedding of the countable atomless boolean algebra into any **[e, f]** with **e < f**. This follows by virtually the same argument (see e.g. [3]).

Actually we feel that by varying especially the Friedberg type requirements P_j one can obtain many other density results. One example is the density of the *superbranching* degrees. Here a degree $\mathbf{a} \neq \mathbf{0'}$ is superbranching (Downey - Mourad [4]) if for all $\mathbf{b} > \mathbf{a}$ there exist $\mathbf{c}, \mathbf{d}$ such that $\mathbf{a} < \mathbf{c}, \mathbf{d} < \mathbf{b}$ and $\mathbf{c} \cap \mathbf{d} = \mathbf{a}$.

To see this we need to consider how one might make such an **a**. We would build $A = \bigcup_s A_s$,

$C_e = \bigcup_s C_{e,s}$ and $D_e = \bigcup_s D_{e,s}$ to meet

$P_{2<e,i>} : (W_e \leq_T A) \vee (\Phi_i(A) \neq C_e)$

$P_{2<e,i>+1} : (W_e \leq_T A) \vee (\Phi_i(A) \neq D_e)$ with $C_e, D_e \leq_T W_e \oplus A$, and

$N_{e,i} : \Phi_e(\hat{C}_e) = \Phi_e(\hat{D}_e) = t$ total $\Rightarrow t \leq_T A$ where $\hat{C}_e = C_e \oplus A$ and $\hat{D}_e = D_e \oplus A$.

Let $l(e, i, s) = mx\{x : (\forall y < x)(\Phi_{e,s}(C_{i,s} ; y) = \Phi_{e,s}(D_{i,s} ; y))\}$ and $ml(e, i, s) = mx\{l(e, i, t) : t < s\}$.

To meet the $P_{2(e,i)}$ we will define a stream of followers $\{x(e, i, j, s) : j \in \omega\}$ - we shall wait until $L(e, i, s) > x(e, i, j, s)$ where

$L(e, i, s) = \max\{x : (\forall y < x)[\Phi_{i,s}(A_s ; y) = C_{e,s}(y)]\}$, and then appoint $x(e, i, j + 1, s) > u(\Phi_{i,s}(A_s ; x(e, i, j, s)))$ in the same way as (2.3). Again we shall wait until we see W_e permit some j and then put $x(e, i, j, s)$ into C_e.

This wins $P_{2(e,i)}$ by a Friedberg argument if we restrain A on (e.g.) $x(e, i, j + 1, s)$. As we don't know whether $W_e \leq_T A$, we may get all $x(e, i, j, s)$ defined for $j \in \omega$ (perhaps $W_e = \emptyset$). Subsequent P_k must this take numbers from $P_{2<e,i>}$'s stream but this causes no new problems save to ensure that $\lim_s x(e, i, j, s)$ exists, this being simply done by controlling which P_k have access to which $x(e, i, j, s)$ (see [4] for more details).

The $N_{e,i}$ are met in exactly the same way as we did in (2.2). The only problems with coherence with density are those that have occurred in the last section and virtually the same reasoning shows that:

3.1 ***Corollary.*** *The Superbranching Degrees are Dense*

It seems to me that the same technique will also show that many other degree classes are dense. For example the same technique ought to show that the contiguous degrees are nowhere dense in **R**, that is, if $\mathbf{e} < \mathbf{f}$ then there exists $\mathbf{a} < \mathbf{b}$ with $\mathbf{a} < \mathbf{a} < \mathbf{b} < \mathbf{f}$ such that for all $\mathbf{c} \in [\mathbf{a}, \mathbf{b}]$, $\mathbf{c}$ is not contiguous.

As a final example of the use of this technique, we will sketch the proof of the following result which negatively answers a conjective of Remmel.

3.2 ***Theorem*** *There exists an re degree* $\mathbf{a}$ *with* $\mathbf{0} < \mathbf{a} < \mathbf{0}'$ *such that if* B *is any* r.e *set with* $\deg(B) > \mathbf{a}$ *then there exists an r.e splitting* $B_1 \sqcup B_2 = B$ *of* B *with* $\deg(B_1) = \mathbf{a}$. .

Proof Let $E \not\equiv_T \emptyset$ be given. We construct $A = \bigcup_s A_s$ and auxiliary re. set E_e, C_e and D_e meet

$$P_e : \bar{A} \neq W_e$$
$$N_e : \Phi_e(A) \neq E, \text{ and}$$
$$R_e : \Gamma_e(V_e) = A \rightarrow (C_e \sqcup D_e = V_e \text{ and } C_e \equiv_T A.).$$

Here (Γ_e, V_e) is an enumeration of all pairs consisting of an r.e. set and a functional. We meet the P_e by followers as usual. We shall meet the N_e by a Sacks restraint and shall only have problems with the R_e. For the sake of there we shall define reductions (dropping the "e") $\Delta(C) = A$ and $\Psi(A) = C$.

The basic idea for R is to wait till
$l(e, s) > x$ (where $l(e, s) = \max\{x : (\forall y < x)(\Gamma_{e,s}(V_{e,s}; y) = A_s(y)\}$ and define
$\delta(x, s) = \gamma(x, s) = u(\Gamma_{e,s}(V_{e,s}; x))$ and $\psi(\delta(x, s)) = \langle e + 1, x, s \rangle$.

Were there no N_e requirements around, the idea is then quite simple: Whenever

$V_{e,s}[\delta(x,s)] \neq V_{e,t}[\delta(x,s)]$ at some least e-expansionary stage $t > s$ enumerate the change into $C_{t+1} - C_t$ and $\psi(\delta(x,s))$ into $A_{t+1} - A_t$ and then choose a new $\psi(\delta(x, t+1))$ as, say, $<e+1, x, t+1>$ and again set $\delta(x, t+1) = \gamma(x, t+1)$. Note that if we use $y = <0, x, s>$ to follow P_e then if we ever put such y into A it will cause a change in $\gamma(y, s)$ and hence in $\delta(x, s)$. Note also that if $\Gamma(V) = A$ then $\lim_s \delta(x, s) = \delta(x)$ and $\lim_s \psi(\delta(x, s)) = \psi(\delta(x))$ exists, and further as the use functions are, by convention monotone and $\delta(x) \geq x$, we will have a reduction $\Psi(A) = C$.

Unfortunately the procedure above makes A = C and complete. However, the reader should note that R_e has already $\omega + 2$ outcomes, that is outcome (i, u) indicating i is least with $\Gamma(V, i)$ unbounded (for $i \in \omega$) plus b and w as in §2. Note that outcome (i, u) essentially codes a recursive set into A provided we ensure that $\psi(\delta(x)) > \psi(\delta(y))$ when $x > y$. This follows because we get to reset all traces $\psi(\delta(x, s))$ if $\delta(x, s)$ changes.

N_e wishes to stop A from changing. It asserts control of $\psi(\delta(x, s))$ (say) when it sees $L(e, s) > y$ $\left(\text{where } L(i, s) = \text{mx}\left\{x : (\forall y < x)\,(\Phi_{e,s}(A_s; y) = E_s(y))\right\}\right)$ and $\psi = \psi(\delta(x, s)) < u(i, y, s) = u(\Phi_{i,s}(A_s; y))$. This control asks us to keep ψ out of A.

This causes problems. As V is not under our control $V[\delta(x, s)]$ might change after N_i asserts control. If we do not enumerate ψ our only option is to enumerate the relevant changes into D (and not C). But note now that x is finished as a possible follower in the sense that as $\delta(x, s)$ is now no longer equal to $\gamma(x, s)$, if x enters A it may not cause a change in V below $\delta(x, s)$. Hence we may not be able to get C to comprehend that such an x enters.

The machinery of (2.15) and (2.16) handles this situation very nicely. We put the R_e, P_e and N_e on a tree in the same way as in §2 defining a local priority ordering $<^*$. Then if τ corresponds to an N_e and σ to R_e with $\tau \supset \sigma\hat{\ }b$ then $\sigma\hat{\ }(i, u)$ respect τ's control precisely if $\tau <^* \sigma\hat{\ }(i, u)$. This restraint again has finite lim inf (and so has finite\limit on the true path on the expanded tree for $\sigma\hat{\ }(i, u)$) and so if $\Gamma_e(V_e; i)\uparrow$ then almost all of the $\psi_e(\delta_c(x, s))$ enter A (i.e. all those $\geq$ j some $j \geq i$). Those N_e guessing $\sigma\hat{\ }(i, u)$ will not believe computations until this recursive set enters. On the other hand, for $\tau \supset \sigma\hat{\ }b$ then if some $\gamma \supset \sigma\hat{\ }b$ higher priority than τ is devoted to D_e, it will request an x to follow it. It will wait until it sees a fresh x provided by $\sigma\hat{\ }b$. It asserts control of x by asking that $\psi(x)$ and $\delta(x)$ move everytime $\gamma(x)$ changes.

The remaining details fit together in exactly the same way as the other arguments and this concludes the sketch of the proof of this result.

3.3 Final Notes on the Framework

Finally, we point out a couple of remarks on the framework. We feel that the beauty of the approach is that for a general **0'''** argument are simply writes down a π_3 requirement and then give a module for *the whole requirement exactly as in a* **0''** *argument*. The $<^*$ machinery then ought to take care of the coherence problems (in the same way as the $\leq_L$ tree machinery takes care of **0''** arguments).

The disadvantage of our approach is that the combinatorics seem more difficult, than, for example, the linking of Slaman/Soare (at least for some arguments).

This would give a simultaneous extension of Slaman's density theorem and Lachlan's decomposition theorem. It appears to be quite difficult.

Richard Shore has pointed out that one can also construct the local priority ordering during the construction. For example, if the ω π_2 outcomes are labelled i and the π_3 outcome is labelled b then the number of times i is accessed can control the number nodes $\sigma \supset b$ that $i + 1$ must respect. Note that if b is correct then this number will be finite. Similar comments apply from σ to i. Shore used this device in his construction, and it can make the combinatorics easier. Downey and Shore have employed this technique in a forthcoming paper entitled "Decomposition and infima in the r.e. degrees". There Downey and Shore obtain the following generalization of Slaman's density theorem: $\forall$ **a**, **b** (**a** | **b** $\rightarrow$ ($\exists$ **c**) (**a** $\cup$ **c** | **b** $\cup$ **c** and **a** $\cup$ **c**, **b** $\cup$ **c** < **a** $\cup$**b** and (**a** $\cup$ **c**) $\cap$ (**b** $\cup$ **c**) = **c**)). Finally Lerman has pointed out that the use of π_3 trees also occurs in his paper on degrees that do not bound minimal degrees.

REFERENCES

1. Ambos - Spies K., On pairs of recursively enumerable degrees Trans. Amer. Math. Soc. *283* (1984) 507-531.

2. Downey, R. G., A contiguous nonbranching degree, Z. Math. Logik. Grundlagen Math *35* (1989) 375-383.

3. _________, Lattice nonembeddings and initial segments of the r.e. degrees. Annals Pure and Aplied Logic (to appear)

4. _________, and J Mourad, Superbranching degrees, these proceedings

5. _________, and T Slaman, Completely mitotic r.e. degrees, Annals Pure and Appl. Logic *41* (1989) 119-152.

6. _________, and T Slaman, On co-simple isols and their intersection types, in preparation.

7. Fejer, P., *The Structure of Definable Subclasses of the Recursively Enumerable Degrees,* Ph. D. Diss., Univ. of Chicago, 1980.

8. _________, The density of the nonbranching degrees, Annals Pure and Appl. Logic *24* (1983) 113-130.

9. Lachlan, A. H., A recursively enumerable degree which will not split over all lesser ones, Ann. Math. Logic *9* (1975) 307-365.

10. _________, Bounding minimal pairs, J. Symb. Logic *44* (1979) 626-642.

11. _________, Decomposition of recursively enumerable degrees, Proc. Amer. Math. Soc. *79* (1980) 629-634.

12. Sacks, G. E., The recursively enumerable degrees are dense, Ann. of Math *80* (1964) 300-312.

13. Shore, R. A., A non-inversion theorem for the jump operator, Ann. Pure and Appl. Logic *40* (1988) 277-303.

14. Slaman, T. A., The recursively enumerable branching degrees are dense in the recursively enumerable degrees, handwritten notes, University of Chicago, 1981.

15. Soare, R. I., Tree arguments in recursion theory and the $\mathbf{0}'''$ priority method, in *Recursion Theory* (ed. A. Nerode and R. Shore) A.M.S. publ. Providence, Rhode Island (1985) 53-106.

16. ________, *Recursively Enumerable Sets and Degrees,* Springer-Verlag, New York (1987).

Array nonrecursive sets and multiple permitting arguments *

Rod Downey, Victoria University
Carl Jockusch, University of Illinois
Michael Stob, Calvin College

April 2, 1990

Abstract

We study a class of permitting arguments in which each positive requirement needs multiple permissions to succeed. Three natural examples of such constructions are given. We introduce a class of r. e. sets, the array nonrecursive sets, which consists of precisely those sets which allow enough permission for these constructions be performed. We classify the degrees of array nonrecursive sets and so classify the degrees in which each of these constructions can be performed.

1 Introduction

Permitting is the name given to a class of techniques for constructing an r. e. set B which is recursive is some fixed r. e. set A. In a permitting argument, enumeration into B is allowed or "permitted" only if some event related to the enumeration of A occurs. For example, in Yates permitting (often called permitting or simple permitting), we allow x to be enumerated in B at stage $s+1$ only if some integer $y \leq x$ is enumerated in A at stage $s+1$. It is obvious that this ensures that $B \leq_T A$. Various notions of permitting can be found in the literature corresponding to various classes of sets A and various types of requirements which appear in the specification of B. Obviously, permitting functions as a negative requirement on B and a notion of permitting may or may not cohere with a positive requirement desired for the enumeration of B. For example, Yates permitting described above and the standard positive requirements for constructing a simple set are compatible, producing the theorem that every r. e. degree bounds an r. e. degree containing a simple set. The most common notions of permitting are Yates permitting, Martin (or high) permitting, and prompt permitting. Each corresponds to a natural class of r. e. degrees and for each there is a large class of constructions that can be done precisely in those degrees.

*This work was partially supported by National Science Foundation Grants DMS88-00030 to Stob and DMS86-01242 to Jockusch and Downey and also a New Zealand–United States Cooperative Science Program Grant from the National Science Foundation to all three authors. This paper is in final form and no similar paper has been or will be submitted elsewhere.

In this paper we analyze a class of permitting arguments which are characterized by the fact that each positive requirement requires multiple permissions to succeed. To see how our notion of multiple permitting differs from other standard notions of permitting, and to place all these methods in a common framework, we now review the basic (Yates) permitting method and make some remarks about it. We do this by means of the following theorem.

Theorem 1.1 *If A is r. e. and nonrecursive, then there is a simple set B such that $B \leq_T A$.*

Proof. We construct B to be coinfinite and to meet for every $e \in N$ the requirement

$$\mathbf{P}_e: \quad W_e \text{ infinite } \Rightarrow W_e \cap B \neq \emptyset.$$

Given an enumeration $\{A_s\}_{s \in N}$ of A, we enumerate B in stages. The requirement that $B \leq_T A$ is met as follows. We say that a number x is permitted by A at stage $s+1$ if $(\exists y \leq x)[y \in A_{s+1} - A_s]$. We enumerate x into B at stage $s+1$ only if x is permitted by A at stage $s+1$. This guarantees that $B \leq_T A$ for if s is a stage such that $(\forall y \leq x)[y \in A_s \leftrightarrow y \in A]$, then $x \in B_s \leftrightarrow x \in B$.

CONSTRUCTION.

Stage $s+1$

For every $e < s$, if $W_{e,s} \cap B_s = \emptyset$, and $(\exists x)[x \in W_{e,s}, x > 2e$, and x is permitted by A at $s+1]$, enumerate the least such x in B.

By permitting, $B \leq_T A$. The clause $x > 2e$ guarantees that B is coinfinite. To see that $\mathbf{P}_e$ is satisfied, suppose that W_e is infinite but that $W_e \cap B = \emptyset$. We argue that A is recursive, contrary to hypothesis. To determine if $y \in A$, enumerate W_e until a stage s and integer x are discovered such that $x > y$, $x > 2e$, and $x \in W_{e,s}$. Since $\mathbf{P}_e$ is not satisfied, x is never enumerated in B. This implies that x is never permitted by A after stage s. In particular, $y \in A \leftrightarrow y \in A_s$. ■

We notice the following key features of $\mathbf{P}_e$ which allows the above permitting argument to succeed.

(1) If W_e is infinite, there are infinitely many potential witnesses for $\mathbf{P}_e$. (Any $x \in W_e$ such that $x > 2e$ will do.)

(2) The construction requires that only one witness for $\mathbf{P}_e$ needs to be permitted only once for $\mathbf{P}_e$ to succeed.

(3) Witnesses, once discovered, do not disappear and are available at any later stage. (In this case, if $x \in W_{e,s}$, then $x \in W_{e,t}$ for all $t \geq s$.)

Any set of positive requirements satisfying these three properties can be combined with simple permitting in the manner of Theorem 1.1. Different notions of permitting arise from positive requirements which do not have one or the other of the features above. The two most important examples are high permitting and prompt permitting.

The high permitting method of Martin [M] results from replacing (2) above by

$(2)_{high}$ The construction requires that cofinitely many of the witnesses for $\mathbf{P}_e$ be permitted each once.

As the name suggests, the method of permitting which results from (1), $(2)_{high}$, and (3) can only be used with sets of high r. e. degree. Martin used it to show that maximal r. e. sets exist in all high r. e. degrees.

The prompt permitting method of Maass, [MSS,AJSS] results from keeping (1) and (2) but replacing (3) by

$(3)_{prompt}$ Witnesses, once discovered, need to be permitted immediately (promptly) if they are to be used in satisfying $\mathbf{P}_e$.

The method of permitting which results from (1), (2), and $(3)_{prompt}$ can only be used with r. e. sets of promptly simple degree. The class of promptly simple degrees is a filter in the upper semilattice of all r. e. degrees which contains low degrees but not all high degrees. Thus prompt permitting is up to degree a different notion than high permitting or standard Yates permitting.

The notion of permitting that we study here arises from positive requirements that satisfy (1) and (3) but in which we modify (2) to

$(2)_{mp}$ At least one witness x needs to be permitted $f(x)$ times; f is some fixed recursive function.

Note that $(2)_{mp}$ is a stronger requirement than

$(2)_n$ At least one witness x needs to be permitted n times; n a fixed positive integer.

It is easily seen that $(2)_n$ is no harder to guarantee than (2).

We study arguments which have positive requirements with the characteristics (1), $(2)_{mp}$, and (3) in a somewhat indirect manner. We first introduce a class of r. e. sets, the array nonrecursive sets.

The array nonrecursive sets are defined as follows. Recall that a sequence of finite sets $\{F_n\}_{n\in N}$ is called a *strong array* if there is a recursive function f such that $F_n = D_{f(n)}$ for every $n \in N$ where D_y denotes the finite set with canonical index y.

Definition 1.2 A strong array $\{F_n\}_{n\in N}$ is a *very strong array* (v. s. a.) if

(4) $$\bigcup_{n\in N} F_n = N,$$

(5) $$F_n \cap F_m = \emptyset \text{ if } n \neq m, \text{ and}$$

(6) $$0 < |F_n| < |F_{n+1}| \text{ for all } n \in N.$$

Definition 1.3 An r. e. set A is *array nonrecursive with respect to* $\{F_n\}_{n\in N}$ (F-a. n. r.) if

(7) $$(\forall e)(\exists n)[W_e \cap F_n = A \cap F_n].$$

Definition 1.4 An r. e. set A is *array nonrecursive* (a. n. r.) if there is a v. s. a. $\{F_n\}_{n\in N}$ such that A is F-a. n. r.

Definition 1.5 An r. e. degree **a** is *array nonrecursive* if there is an r. e. set $A \in \mathbf{a}$ such that A is array nonrecursive.

We note the following facts about these definitions. First, if A is a. n. r., then A is nonrecursive. Second, F-a. n. r. sets exist for any v. s. a. $\{F_n\}_{n \in N}$, since $A = \bigcup_{e \in N} W_e \cap F_e$ is F-a. n. r. Finally, (7) is equivalent to

$$(8) \qquad (\forall e)(\overset{\infty}{\exists}\, n)[W_e \cap F_n = A \cap F_n].$$

The condition (7) translates to a notion of multiple permitting in roughly the following way. Suppose that A is F-a. n. r. and that we are constructing an r. e. set $B \leq_T A$ using Yates permitting. If we enumerate an r. e. set V we are entitled to assume that $(\exists n)[V \cap F_n = A \cap F_n]$. Since we enumerate V, for this n equality implies that we can force up to $|F_n|$ many integers all less than $\max(F_n)$ to enter A. This gives several Yates permissions for a large enough number. (If all that is assumed is that A is nonrecursive, Yates permitting guarantees a single permission on a large enough number.) The simplest example of such a multiple permitting argument is Theorem 2.5 below.

We show in Section 4 that such multiple permitting arguments arise naturally in recursion theory by showing that three constructions from elsewhere in recursion theory can be carried out precisely below those r. e. degrees which are array nonrecursive. These theorems are as follows.

Theorem 1.6 *Let f be a strictly increasing recursive function. Then an r. e. degree* **a** *is a. n. r. iff there is a degree* $\mathbf{b} \leq \mathbf{a}$ *(not necessarily r. e.) such that some set B of degree* **b** *is not f-r. e. (A Δ^0_2 set is f-r. e. if it has a recursive approximation $\{B_s\}_{s \in N}$ as a Δ^0_2 set such that $|\{s | B_s(x) \neq B_{s+1}(x)\}| \leq f(x)$ for all x.)*

The next theorem arises from a construction performed by Jockusch and Soare [JS, Theorem 1] to show that every degree which contains a consistent extension of Peano arithmetic bounds an incomparable pair of degrees. In that proof, sets B_0, C_0, B_1, C_1 were constructed satisfying the conditions in part (c) of Theorem 1.7.

Theorem 1.7 *For r.e. sets A, the following are equivalent:*

(a) A has a. n. r. degree,

(b) there are disjoint r.e. sets B and C each recursive in A such that $B \cup C$ is coinfinite and no set of degree $\mathbf{0}'$ *separates B and C,*

(c) there exist two disjoint pairs of r.e. sets B_0, C_0 and B_1, C_1 such that $B_i \cup C_i$ is coinfinite for $i = 0, 1$, each set B_i, C_i is recursive in A, and each set which separates (B_0, C_0) is Turing incomparable with each which separates (B_1, C_1).

The third major theorem concerns a class of r. e. theories called the Martin Pour-El theories. To define this class let Q be the free countable, atomless Boolean algebra and let $\{p_n | n \in N\}$ be a set of generators for it. Then a theory T can be identified with a filter of Q. We call such a theory *well-generated* if there are sets B and C such that T is generated by a set of the form $\{p_n | n \in B\} \cup \{\neg p_n | n \in C\}$. An r. e. theory T is *Martin-Pour-El* if it is well-generated, essentially undecidable, and every r. e. theory $W \supseteq T$ is principal over T. The existence of such theories is due to Martin and Pour-El [MP, Theorem I]. They have been extensively studied by Downey [D1].

Theorem 1.8 *An r. e. degree* **a** *is a. n. r. iff there is a theory T of degree* **a** *which is Martin Pour-El.*

In sections 2 and 3 we initiate an investigation into the properties of a. n. r. sets and degrees. Of particular interest because of Theorems 1.6, 1.7, and 1.8 is the classification of a. n. r. degrees. Our principal results are as follows.

- The array nonrecursive degrees are closed upwards in **R**, the class of all r. e. degrees (Corollary 2.8).
- There are low a. n. r. degrees (Theorem 2.1).
- All r. e. degrees **a** such that $\mathbf{a}'' > \mathbf{0}''$ are a. n. r. (Corollary 4.3).
- There exist promptly simple degrees which are not a. n. r. Thus, since promptly simple degrees are noncappable and the non-a. n. r. degrees are closed upwards, every nonzero r. e. degree bounds a nonzero r. e. degree which is not a. n. r. (Corollary 2.11).
- Every a. n. r. degree bounds a low a. n. r. degree (Corollary 3.8, due to Cameron Smith).
- The r. e. weak-truth-table degrees containing no a. n. r. set form an ideal in the upper-semilattice of r. e. wtt-degrees (Corollary 3.14).

Our notation is standard; a reference is Soare [S]. All sets and degrees are r. e. unless otherwise noted. The principal exceptions to this convention are in Theorems 1.6 and 1.7.

2 Basic Existence Theorems

Given a very strong array $\{F_n\}_{n\in N}$, the F-a. n. r. set $A = \bigcup_{n\in N} W_n \cap F_n$ is clearly Turing-complete and, in fact, is creative. The next theorem shows that low F-a. n. r. sets exist. It also clearly exhibits the construction of an a. n. r. set as a finite injury priority argument.

Theorem 2.1 *Let $\{F_n\}_{n\in N}$ be a very strong array. Then there is an r. e. set A of low degree such that A is F-a. n. r.*

Proof. To make A F-a. n. r., it suffices to meet for every $e \in N$ the requirement

$$\mathbf{R}_e : \qquad (\exists n)[W_e \cap F_n = A \cap F_n].$$

The requirements to make A of low degree are

$$\mathbf{N}_e : \qquad (\overset{\infty}{\exists}\, s)[\{e\}_s^{A_s}(e)\downarrow] \Rightarrow \{e\}^A(e)\downarrow .$$

Recall that the requirement $\mathbf{N}_e$ is met by preserving the restraint function $r(e,s) = u(A_s, e, e, s)$ at all but finitely many stages s. Let $q(e,s) = \max\{r(i,s) | i \leq e\}$. To meet $\mathbf{R}_e$, we reserve the sets $F_{\langle e,0\rangle}, F_{\langle e,1\rangle}, \ldots$. The construction assigns priority to the requirements in the order $\mathbf{N}_0$, $\mathbf{R}_0$, $\mathbf{N}_1$, $\mathbf{R}_1$, ...

CONSTRUCTION.

Stage $s+1$

Requirement $\mathbf{R}_e$ *requires attention* at stage $s+1$ if there is $i \in N$ such that

(9) $$\min(F_{\langle e,i\rangle}) > q(e,s), \text{ and}$$

(10) $$(\forall j \leq i)[W_{e,s+1} \cap F_{\langle e,j\rangle} \neq A_s \cap F_{\langle e,j\rangle}].$$

Let e be least such that $\mathbf{R}_e$ requires attention and let i be least such that (9) and (10) hold. Enumerate all of $W_{e,s+1} \cap F_{\langle e,i\rangle}$ into A. This ends the construction.

Note that the construction ensures that $A_s \cap F_{\langle e,i\rangle} \subseteq W_{e,s} \cap F_{\langle e,i\rangle}$ for every e, i, and s, so that if $\mathbf{R}_e$ receives attention at stage $s+1$ and i is the least integer satisfying (9) and (10), then $W_{e,s+1} \cap F_{\langle e,i\rangle} = A_{s+1} \cap F_{\langle e,i\rangle}$. It is now easy to show by simultaneous induction on e that

(a) $\mathbf{N}_e$ is satisfied,

(b) $\lim_s q(e,s) < \infty$,

(c) $\mathbf{R}_e$ is satisfied, and

(d) $\mathbf{R}_e$ receives attention only finitely often. ■

A. Kučera has pointed out that the following extension of Theorem 2.1 holds: For any very strong array $\{F_n\}_{n\in N}$, there is a complete extension T of Peano arithmetic of low degree such that there is an F-a. n. r. set A recursive in T.

The next two results, Theorems 2.2 and 2.5, clarify the role of the very strong array $\{F_n\}_{n\in N}$ in the definition of array nonrecursive sets. In particular, Theorem 2.5 shows that up to degree, the notion of array nonrecursveness is independent of the choice of very strong array. It will also be used in the proof of many subsequent results.

Theorem 2.2 *For every r. e. set A there is a very strong array $\{F_n\}_{n\in N}$ such that A is not F-a. n. r.*

Proof. If A is recursive, then A is not F-a. n. r. for any F. If A is not recursive, let R be an infinite recursive subset of A. Choose a v. s. a. $\{F_n\}_{n\in N}$ such that $F_n \cap R \neq \emptyset$ for every $n \in N$. Let $W = \overline{R}$. Then for every $n \in N$, $W \cap F_n \neq A \cap F_n$ witnessing that A is not F-a. n. r. ■

The following definition and lemma will be used in the proof of Theorem 2.5 and elsewhere.

Definition 2.3 Suppose that A is r. e. with a given enumeration $\{A_s\}_{s\in N}$ and $\{F_n\}_{n\in N}$ is a strong array. *A F-permits y at stage $s+1$* if

$$(\exists z \leq y)(\exists x \leq \max(F_z))[x \in A_{s+1} - A_s].$$

Lemma 2.4 *Suppose that A is r. e. with a given enumeration $\{A_s\}_{s\in N}$ and $\{F_n\}_{n\in N}$ is a strong array. Suppose that f is a recursive function. Suppose that B is an r. e. set with enumeration $\{B_s\}_{s\in N}$ such that for every x, $x \in B_{s+1} - B_s$ only if A F-permits $f(x)$ at stage $s+1$. Then $B \leq_T A$. In fact $B \leq_{wtt} A$.*

Theorem 2.5 *Suppose that $\{F_n\}_{n\in N}$ and $\{E_n\}_{n\in N}$ are very strong arrays, that A is F-a. n. r., and that* **b** *is an r. e. degree such that $deg(A) \leq$* **b**. *Then there is $B \in$* **b** *such that B is E-a. n. r.*

Proof. Fix a set $\hat{B} \in \mathbf{b}$. To ensure that $\hat{B} \leq_T B$, we will reserve the sets $E_{\langle 0,i\rangle}$, $i \in N$, for coding $\hat{B}$. Namely, we will enumerate all of $E_{\langle 0,i\rangle}$ in B if and only if $i \in \hat{B}$. The requirements $\mathbf{R}_e$ to make B E-a. n. r. are similar to those of Theorem 2.1:

$$\mathbf{R}_e : \quad (\exists n)[W_e \cap E_n = B \cap E_n].$$

We will reserve the sets $E_{\langle e+1,0\rangle}, E_{\langle e+1,1\rangle}, \ldots$ for meeting $\mathbf{R}_e$. To aid in meeting $\mathbf{R}_e$ we shall also enumerate an r. e. set V_e and since A is F-a. n. r., (8) guarantees that

$$(\overset{\infty}{\exists} n)[V_e \cap F_n = A \cap F_n]. \tag{11}$$

For each e, let $n(e)$ be the least integer n such that $|F_n| > |E_{\langle e+1,0\rangle}|$. For each $n \geq n(e)$ let $g(e,n)$ be the greatest pair of the form $\langle e+1, i\rangle$ such that

$$|F_n| > |E_{g(e,n)}|. \tag{12}$$

Note that if $e \neq f$, then $g(e,n) \neq g(f,m)$ for all n, m. However it is possible that $g(e,n) = g(e,m)$ for some $n \neq m$. However, for every x, the set $\{n \mid g(e,n) = x\}$ is finite (uniformly in x).

We will replace the requirement $\mathbf{R}_e$ with the following requirements $\mathbf{R}_{e,n}$ for $n \geq n(e)$:

$$\mathbf{R}_{e,n} : \quad V_e \cap F_n = A \cap F_n \Rightarrow W_e \cap E_{g(e,n)} = B \cap E_{g(e,n)}.$$

To ensure that action taken for $\mathbf{R}_{e,n}$ does not interfere with the requirement to make $B \leq_T \hat{B}$ we will allow $x \in E_{g(e,n)}$ to enter B at stage $s+1$ only if A F-permits n at stage $s+1$. By Lemma 2.4, we have that $\bigcup_{e,i\in N} E_{\langle e+1,i\rangle} \cap B \leq_T A \leq_T \hat{B}$.

Before giving the construction, which is quite simple, we describe the strategy for one requirement. This strategy is the same one that is used throughout the paper when it is necessary to construct a set recursive in some given a. n. r. set A. It essentially captures the notion of multiple permitting allowed by an a. n. r. set.

Fix e and $n \geq n(e)$. $\mathbf{R}_{e,n}$ is met if either $V_e \cap F_n \neq A \cap F_n$ or $W_e \cap E_{g(e,n)} = B \cap E_{g(e,n)}$. We view our attempts to establish this disjunction as a two-state finite automaton. At any stage s of the construction, we say that requirement $\mathbf{R}_{e,n}$ is in state S_1 if $W_{e,s} \cap E_{g(e,n)} = B_s \cap E_{g(e,n)}$ and in state S_2 otherwise. The construction is intended to ensure that if $\mathbf{R}_{e,n}$ is in state S_2 at stage s then $V_{e,s} \cap F_n \neq A_s \cap F_n$ as indicated in Figure 1.

Suppose that stage $s+1$ is such that $\mathbf{R}_{e,n}$ is in state S_1 at stage s but not at stage $s+1$. Since we enumerate B, this is because an element of $E_{g(e,n)}$ is enumerated in W_e at stage $s+1$ (and thus there are at most $|E_{g(e,n)}|$ such stages). To guarantee that the

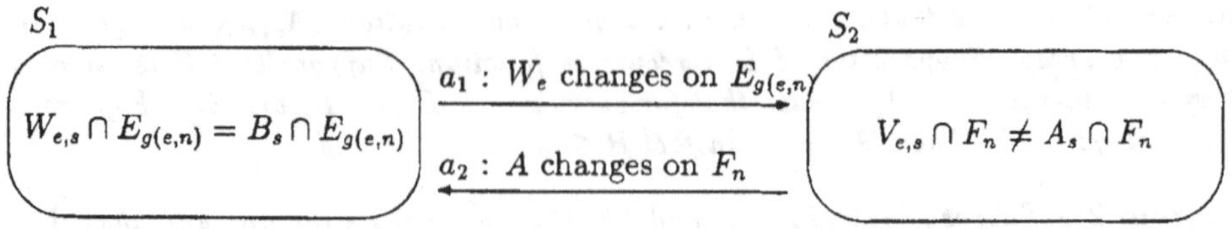

Figure 1: State diagram of the construction

condition of state S_2 holds at stage $s+1$, we enumerate, if necessary, one element of F_n into V_e to cause $V_{e,s+1} \cap F_n \neq A_{s+1} \cap F_n$. This constitutes the action of arrow a_1. Since this is the only action which causes us to enumerate elements of F_n into V_e and since $|F_n| > |E_{g(e,n)}|$, it is always possible to perform this action at such a stage $s+1$.

We also need to guarantee that while $\mathbf{R}_{e,n}$ remains in state S_2, the condition $V_{e,s} \cap F_n \neq A_s \cap F_n$ continues to hold. Thus, let s be a stage such that $V_{e,s} \cap F_n \neq A_s \cap F_n$ but $V_{e,s+1} \cap F_n = A_{s+1} \cap F_n$. It must be the case that an element of F_n is enumerated in A at stage $s+1$. This is just the condition that A F-permits n at stage $s+1$. Thus at stage $s+1$ we may enumerate all of $W_{e,s+1} \cap E_{g(e,n)}$ into B, thereby guaranteeing that $\mathbf{R}_{e,n}$ is in state S_1 at stage $s+1$. This constitutes the action of arrow a_2. Note that to perform such action, we must require that $B_s \cap E_{g(e,n)} \subseteq W_{e,s} \cap E_{g(e,n)}$ for all e, n, s. This is guaranteed by the construction described above and by the fact that $E_{g(e,n)}$ is disjoint from $E_{g(f,m)}$ if $e \neq f$. Because of this stipulation, there are no conflicts between the various requirements. We now give the formal details of the construction.

CONSTRUCTION.

Stage $s+1$

Step 1. (Coding.) If $i \in \hat{B}_{s+1} - \hat{B}_s$, enumerate all of $E_{\langle 0,i \rangle}$ into B.

Step 2. (Arrow a_2.) For every e and $n \geq n(e)$, if A F-permits n, enumerate all of $W_{e,s+1} \cap E_{g(e,n)}$ into B.

Step 3. (Arrow a_1.) For every e and $n \geq n(e)$ if

$$B_s \cap E_{g(e,n)} = W_{e,s} \cap E_{g(e,n)} \quad \text{and} \quad B_{s+1} \cap E_{g(e,n)} \neq W_{e,s+1} \cap E_{g(e,n)} \tag{13}$$

then enumerate one element of $F_n - V_{e,s}$ into V_e, if necessary, to cause $A_{s+1} \cap F_n \neq V_{e,s+1} \cap F_n$. There is such an element since $|F_n| > |E_{g(e,n)}|$ and an element of F_n is enumerated in V_e only if (13) holds; i. e., if arrow a_1 is traversed.

Lemma 2.6 $B \equiv_T \hat{B}$.

Proof. $B \leq_T \hat{B}$ since B restricted to $\bigcup_{i \in N} E_{\langle 0,i \rangle}$ is recursive in $\hat{B}$ by step 1 of the construction and B restricted to $\bigcup_{e,i \in N} E_{\langle e+1,i \rangle}$ is recursive in A by step 2 and Lemma 2.4 (applied to the array $\{F_n\}_{n \in N}$ and the function f where $f(x)$ is the greatest n such that $x \in E_{g(e,n)}$). □

Lemma 2.7 *For each* $e \in N$, $\mathbf{R}_e$ *is satisfied.*

Proof. It is enough to show that for every e and $n \geq n(e)$, that $\mathbf{R}_{e,n}$ is satisfied. At every stage of the construction, $\mathbf{R}_{e,n}$ is either in state S_1 or state S_2. Thus, since $\mathbf{R}_{e,n}$ changes state finitely often, $\mathbf{R}_{e,n}$ is in state S_1 at cofinitely many stages of the construction or $\mathbf{R}_{e,n}$ is in state S_2 at cofinitely many stages of the construction. If the former holds, $W_e \cap E_{g(e,n)} = B \cap E_{g(e,n)}$. If the latter holds $V_e \cap F_n \neq A \cap F_n$ since the construction guarantees that if $\mathbf{R}_{e,n}$ is in state S_2 at stage s, then $V_{e,s} \cap F_n \neq A_s \cap F_n$. ∎

The following are easy corollaries of Theorem 2.5.

Corollary 2.8 *Suppose that* $\mathbf{a}$ *is a. n. r. and that* $\mathbf{b} \geq \mathbf{a}$. *Then* $\mathbf{b}$ *is a. n. r. That is, the a. n. r. degrees form a filter in the upper semilattice of the r. e. degrees. Since the Turing reductions employed in Theorem 2.5 are weak-truth-table reductions, this result holds also for the weak-truth-table degrees.*

Corollary 2.9 *Suppose that* $\{F_n\}_{n\in N}$ *is a very strong array and that* A *is a. n. r. Then there is a set* B *of the same weak-truth-table degree as* A *such that* B *is* F*-a. n. r. That is, up to (weak-truth-table) degree, the notion of array nonrecursiveness is independent of array.*

We turn now to existence theorems for array recursive sets and degrees; a set (degree) is *array recursive* just in case it is not array nonrecursive. The following result shows that our notion of multiple permitting is strictly stronger than ordinary (Yates) permitting.

Theorem 2.10 *There is an r. e. degree* $\mathbf{a} > \mathbf{0}$ *such that* $\mathbf{a}$ *is array recursive.*

Proof. Fix a very strong array $\{F_n\}_{n\in N}$. By Corollary 2.9, it suffices to prove that no set of degree $\mathbf{a}$ is F-a. n. r. (We will actually prove that no set of degree less than or equal to that of $\mathbf{a}$ is F-a. n. r., which is equivalent by Corollary 2.9.)

Let $(\Phi_e, B_e)_{e\in N}$ be an effective listing of all pairs (Φ, B) of recursive functionals Φ and r. e. sets B. We will enumerate r. e. sets V_e, $e \in N$, satisfying the following requirement for every $e \in N$ and for every $n > e$:

$$\mathbf{R}_{e,n}: \quad \Phi_e(A) = B_e \Rightarrow V_e \cap F_n \neq B_e \cap F_n.$$

Requirements $\mathbf{R}_{e,n}$ for $n > e$ suffice to make B not F-a. n. r. by (8). To make A nonrecursive we have for every $e \in N$ the requirement

$$\mathbf{P}_e: \quad A \neq \overline{W}_e.$$

We use the following priority ordering of the requirements: $\mathbf{R}_{01}$, $\mathbf{P}_0$, $\mathbf{R}_{02}$, $\mathbf{R}_{12}$, $\mathbf{P}_1$, $\mathbf{R}_{03}$, $\mathbf{R}_{13}$, $\mathbf{R}_{23}$, $\mathbf{P}_2$, ... The key fact about this priority ordering is that $\mathbf{R}_{e,n}$ can only be injured by $\mathbf{P}_i$ for $i \leq n-2$ or at most $n-1$ times.

The strategy for meeting $\mathbf{R}_{e,n}$ is as follows. Wait until $l(e,s) > \max(F_n)$ (where $l(e,s)$ measures the length of agreement between the computation $\Phi_{e,s}(A_s)$ and the set $B_{e,s}$). Cause $V_{e,s+1} \cap F_n$ to be unequal to $B_{e,s} \cap F_n$ (by enumerating at most one element of F_n into V_e). Restrain A on the use of the computations involved in establishing that length of agreement. Thus, if $\mathbf{R}_{e,n}$ is not injured by a higher priority requirement, either $V_e \cap F_n \neq B_e \cap F_n$ or $\Phi_e(A) \neq B_e$ and, in either case, $\mathbf{R}_{e,n}$ imposes only a fixed finite restraint on A for the rest of the construction. Since $\mathbf{R}_{e,n}$ can be injured at most $n-1$ times, at most $n-1$ attempts of the above form need be made and this can be done since $|F_n| \geq n$. Note that the requirements $\mathbf{R}_{e,n}$ are purely negative requirements on A (although positive on V_e) and so do not conflict with each other. We omit further details of the construction and its verification. ∎

The next two corollaries follow by making the obvious modifications to the construction suggested in the proof above. Alternatively, the second follows from the first, Corollary 2.8, and the fact that no promptly simple degree is half of a minimal pair [MSS, Theorem 1.11].

Corollary 2.11 *There is a promptly simple degree* **a** *which is array recursive.*

Corollary 2.12 *For every r. e. degree* $\mathbf{b} > \mathbf{0}$, *there is an r. e. degree* **a** *such that* $\mathbf{0} < \mathbf{a} < \mathbf{b}$ *and* **a** *is array recursive.*

To state the final theorem of this section, we need the following definition.

Definition 2.13 An r. e. set is *semirecursive* if there is a recursive function $f : N^2 \to N$ such that

$$f(x,y) \in \{x,y\} \tag{14}$$

$$f(x,y) \in A \Rightarrow \{x,y\} \subseteq A \tag{15}$$

Thus, the function f of Definition 2.13 chooses of x and y the one "least likely" to be an element of A.

Theorem 2.14 *If r. e. set A is semirecursive, then A is not a. n. r.*

Proof. Let $\{F_n\}_{n \in N}$ be a very strong array and let A be semirecursive with f the recursive function satisfying (14) and (15). We enumerate V so that if $|F_n| \geq 2$, $V \cap F_n \neq A \cap F_n$. To do this, for each n such that $|F_n| \geq 2$, we wait for a stage such that for some pair $\{x_n, y_n\} \subseteq F_n$, we have that $x_n \neq y_n$ and $f(x_n, y_n)$ converges. We then enumerate $f(x_n, y_n)$ and no other element of F_n into V. Thus $V \cap F_n = \{f(x_n, y_n)\}$ but if $f(x_n, y_n) \in A$, $A \cap F_n \supseteq \{x_n, y_n\} \neq V \cap F_n$. ∎

Corollary 2.15 *Every r. e. truth-table degree contains an array recursive set.*

Proof. This is immediate from Theorem 2.14 since every r. e. truth-table degree contains a semirecursive r. e. set [J1, Corollary 3.7(ii)]. ∎

3 Properties of a. n. r. sets and degrees

The first two theorems in this section locate the array nonrecursive sets in the hierarchy of simplicity properties.

Theorem 3.1 *If A is a. n. r., then*

(a) A is not dense simple, and

(b) A is not strongly hypersimple.

Proof. (a). An r. e. set A is dense simple if $p_{\bar{A}}$, the principal function of the complement of A, dominates every recursive function. We use an alternate characterization of dense simplicity due to Robinson [R, Theorem 3]. Namely, A is dense simple if and only if for every strong array $\{F_n\}_{n\in N}$ of disjoint sets,

$$(\exists m)(\forall n \geq m)[|F_n \cap \bar{A}| < n]. \tag{16}$$

Now suppose that A is F-a. n. r. Using $W_e = \emptyset$ and the characterization of F-a. n. r. in (8), we have

$$(\overset{\infty}{\exists} n)[A \cap F_n = \emptyset]. \tag{17}$$

But for any such n, $|F_n \cap \bar{A}| \geq n$, and thus by (16) A is not dense simple. □

(b). A is strongly hypersimple if for every weak array, $\{W_{f(n)}\}_{n\in N}$, of disjoint sets such that $\bigcup_{n\in N} W_{f(n)} = N$ there is an n such that $W_{f(n)} \subseteq A$. Now suppose again that A is F-a. n. r. Define $W_{f(n)}$ for all $n \in N$ as follows. Given F_m, enumerate the least element of F_m in $W_{f(0)}$, the next least in $W_{f(1)}$, and so forth. Obviously because $\{F_n\}_{n\in N}$ is a very strong array, $\bigcup_{n\in N} W_{f(n)} = N$ and the sets $W_{f(n)}$, $n \in N$, are disjoint. By (17), $W_{f(n)} \cap \bar{A} \neq \emptyset$ for every n. (In fact, $W_{f(n)} \cap \bar{A}$ is infinite). Thus A is not strongly hypersimple. ∎

Corollary 3.2 *No array nonrecursive set is maximal, hyperhypersimple, or r-maximal.*

The following theorem shows that Theorem 3.1 is the best possible as far as the standard list of simplicity properties is concerned.

Theorem 3.3 *There is an r. e. set A such that A is array nonrecursive and finitely strongly hypersimple.*

Proof. Fix a v. s. a. $\{F_n\}_{n\in N}$. As usual, the requirements to make A F-a. n. r. are

$$\mathbf{R}_e \qquad (\exists n)[W_e \cap F_n = A \cap F_n].$$

The requirements to make A finitely strongly hypersimple are

Q$_e$: the sets $W_{\{e\}(n)}, n \in N$ are not disjoint or
$\bigcup_{n\in N} W_{\{e\}(n)} \neq N$ or
$(\exists n)[W_{\{e\}(n)}$ is infinite] or
$(\exists n)[W_{\{e\}(n)} \subseteq A]$.

For ease of notation, we will write V_n^e for $W_{\{e\}(n)}$ and $V_{n,s}^e$ for $W_{\{e\}_s(n),s}$ (where we understand that $V_{n,s}^e = \emptyset$ if $\{e\}_s(n)$ does not converge). The strategy for meeting $\mathbf{Q}_e$ while respecting $\mathbf{R}_0, \mathbf{R}_1, \ldots \mathbf{R}_{e-1}$ is as follows. Wait for a stage s so that $\bigcup_{n \in N} V_{n,s}^e$ contains all the elements of each set F_m assigned to any requirement $\mathbf{R}_i$, $i < e$. If such a stage does not exist then $\bigcup_{n \in N} W_{\{e\}(n)} \neq N$ and $\mathbf{Q}_e$ is satisfied. At stage s, choose n such that $V_{n,s}^e$ does not contain any element of any such F_m. Assign V_n^e to $\mathbf{Q}_e$. We then attempt to meet $\mathbf{Q}_e$ by enumerating all of V_n^e into A. However this threatens to interfere with requirements $\mathbf{R}_i$, $i \geq e$, as this V_n^e may contain elements from almost every F_m. To avoid this conflict, we enumerate an element of V_n^e into A only if we discover a new F_m which we can certify is disjoint from V_n^e (by virtue of being entirely contained in the union of other sets V_i^e for $i \neq n$). Thus if $V_n^e \cap A$ is infinite the requirement is met since V_n^e is infinite but also we are assured of having infinitely many sets F_m not interfered with by $\mathbf{Q}_e$ and so available for use by requirements $\mathbf{R}_i$, $i \geq e$.

The details of combining the strategies for various $\mathbf{Q}_e$ are straightforward and are omitted. ■

The next four theorems and their corollaries concern degree-theoretic and set-theoretic splitting properties of array nonrecursive sets.

Theorem 3.4 *For every array nonrecursive set A there are disjoint array nonrecursive sets A_0 and A_1 such that $A = A_0 \cup A_1$.*

Proof. Suppose that A is array nonrecursive with respect to the very strong array $\{F_n\}_{n \in N}$. For each $e \in N$ and $i \in \{0, 1\}$ we have the requirement

$$\mathbf{R}_{e,i}: \qquad (\exists n)[W_e \cap F_n = A_i \cap F_n].$$

To meet $\mathbf{R}_{e,i}$ we will enumerate a certain set $V_{e,i}$ and use the fact that

$$(18) \qquad (\overset{\infty}{\exists}\, n)[V_{e,i} \cap F_n = A \cap F_n].$$

During the course of the construction, we will reserve certain n for $\mathbf{R}_{e,i}$. Each n may be reserved for at most one requirement $\mathbf{R}_{e,i}$ at any one stage, but the reservation may be cancelled at a later stage for the purpose of reserving n for a requirement of higher priority. (The intention of these reservations is that there will be some n which is reserved for $\mathbf{R}_{e,i}$ and for which $W_e \cap F_n = A \cap F_n$.) The priority order of the requirements $\mathbf{R}_{e,i}$ is in order of increasing $\langle e, i \rangle$.

CONSTRUCTION.

Stage $s+1$

Step 1. For each $x \in A_{s+1} - A_s$ let n be the integer such that $x \in F_n$. If n is reserved for the requirement $\mathbf{R}_{e,i}$, then enumerate x in A_i. If n is not reserved for any requirement, enumerate x in A_0.

Step 2. For each x and e, if $x \in W_{e,s+1} - W_{e,s}$, $x \in F_n$, and n is reserved for a requirement $\mathbf{R}_{e,i}$, then enumerate x in $V_{e,i}$.

Step 3. $\mathbf{R}_{e,i}$ *requires attention* at stage $s+1$ if

(19) $(\forall n)[n$ is reserved for $\mathbf{R}_{e,i} \Rightarrow W_{e,s} \cap F_n \neq A_{i,s} \cap F_n]$, and

(20) $(\exists n)[A_s \cap F_n = \emptyset$ and n is not reserved for any $\mathbf{R}_{f,j}$ such that $\langle f,j\rangle \leq \langle e,i\rangle]$

If such a pair e,i exists, choose the pair such that $\langle e,i\rangle$ is least and let n be the least integer satisfying (20) for e,i. Perform the following actions for these fixed e,i,n. Reserve n for $\mathbf{R}_{e,i}$. Cancel any other reservation of n. Enumerate all of $W_{e,s+1} \cap F_n$ into $V_{e,i}$. This ends the construction.

Lemma 3.5 *If n is reserved for $\mathbf{R}_{e,i}$, and that reservation is never cancelled, then $W_e \cap F_n = V_{e,i} \cap F_n$ and $A_i \cap F_n = A \cap F_n$.*

Proof. The first clause of the conclusion is by steps (2) and (3) of the construction. To see that $A_i \cap F_n = A \cap F_n$, notice that at the stage that n is first reserved for $\mathbf{R}_{e,i}$, $A_{i,s} \cap F_n = A_s \cap F_n$ $(= \emptyset)$ by (20). Step (1) guarantees that this equality is maintained for all later stages. □

Lemma 3.6 *If $V_{e,i} \cap F_n \neq \emptyset$, then n is reserved for $\mathbf{R}_{e,i}$ or some requirement of higher priority at cofinitely many stages.*

Lemma 3.7 *Each requirement $\mathbf{R}_{e,i}$ receives attention only finitely often and is satisfied.*

Proof. Given e,i, let s_0 be such that if $\langle f,j\rangle < \langle e,i\rangle$, $\mathbf{R}_{f,j}$ does not receive attention after s_0. By (18), there are infinitely many n such that $V_{e,i} \cap F_n = A \cap F_n$. Let n be any such n which is not reserved for $\mathbf{R}_{f,j}$ for any $\langle f,j\rangle < \langle e,i\rangle$. There are two cases.

Case (i): n is reserved for $\mathbf{R}_{e,i}$ at some stage of the construction. Then by Lemma 3.5, $W_e \cap F_n = V_{e,i} \cap F_n = A \cap F_n = A_i \cap F_n$. Thus $\mathbf{R}_{e,i}$ is satisfied. Let s_1 be a stage such that $W_{e,s_1} \cap F_n = W_e \cap F_n$ and $A_{i,s} \cap F_n = A_i \cap F_n$. Then by (19), $\mathbf{R}_{e,i}$ never receives attention after stage s_1.

Case (ii): n is never reserved for $\mathbf{R}_{e,i}$. Then by Lemma 3.6, $V_{e,i} \cap F_n = \emptyset$. Thus $A \cap F_n = \emptyset$. Thus (20) applies to n at cofinitely many stages of the construction. Since n is never reserved for $\mathbf{R}_{e,i}$, it must be that $\mathbf{R}_{e,i}$ receives attention only finitely often and that at cofinitely many stages of the construction (19) fails. This implies the existence of m such that $W_e \cap F_m = A_i \cap F_m$ and hence that the requirement is satisfied. ∎

It is clear that the requirements to make each set A_0 and A_1 of low r. e. degree can be combined with the construction of Theorem 3.4. Thus we have the following corollary which was first proved (directly) by Cameron Smith.

Corollary 3.8 *For every array nonrecursive degree* **a** *there is an array nonrecursive degree* **b** $<$ **a** *such that* **b** *is low.*

It is not true that if A is a. n. r. and A is the disjoint union of sets A_0 and A_1, then at least one of A_0 or A_1 is anr. However this result is true up to degree. In fact we have the stronger result of the next theorem.

Theorem 3.9 *Suppose that $A \leq_{wtt} A_0 \oplus A_1$ and that A is array nonrecursive. Then there are r. e. sets B_0 and B_1 such that $B_i \leq_{wtt} A_i$ and one of B_0 or B_1 is array nonrecursive.*

Proof. Let $\{F_n\}_{n\in N}$ and $\{E_n\}_{n\in N}$ be very strong arrays such that $|E_n| > 2|F_{\langle i,n\rangle}|$ for every i and $n > i$. We first show that we may assume that A is E-a. n. r. and $A = A_0 \cup A_1$. To see this we first notice that since A is array non-recursive, the wtt-degree of A contains an array-nonrecursive set $\hat{A}$. This follows from Corollary 2.9. We next rely on the following lemma of Lachlan [L].

Lemma 3.10 *Suppose that B, B_0, and B_1 are r. e. sets such that $B \leq_{wtt} B_0 \oplus B_1$. Then there are r. e. sets C_0 and C_1 such that $C_0 \leq_{wtt} B_0$, $C_1 \leq_{wtt} B_1$, and $B = C_0 \cup C_1$.*

Applying the lemma with $B = \hat{A}$ gives sets $\hat{A}_0$ and $\hat{A}_1$ such that $\hat{A} = \hat{A}_0 \cup \hat{A}_1$ and $\hat{A}_i \leq_{wtt} A_i$. The sets B_i which result from the proof of the theorem satisfy $B_i \leq_{wtt} \hat{A}_i$ and thus $B_i \leq_{wtt} A_i$. We shall also assume that A, A_0, and A_1 are enumerated so that

(21) $$A_s = A_{0,s} \cup A_{1,s}.$$

We will meet the following requirements for every $e, j \in N$:

$$\mathbf{R}_{e,j}: \quad (\exists n)[W_e \cap F_n = B_0 \cap F_n \text{ or } W_j \cap F_n = B_1 \cap F_n].$$

(These requirements suffice to make one of B_0 or B_1 F-a. n. r. since if e is such that there is no n with $W_e \cap F_n = B_0 \cap F_n$ then the satisfaction of $\mathbf{R}_{e,j}$ for all $j \in N$ implies that B_1 is F-a. n. r.) As in Theorem 2.5, we will reserve the sets $F_{\langle i,0\rangle}, F_{\langle i,1\rangle}, \ldots$ for requirement $\mathbf{R}_{e,j}$ where $i = \langle e,j\rangle$. We will use the fact that A is a. n. r. by enumerating r. e. sets V_i and assuming that

$$(\overset{\infty}{\exists}\, n)[V_i \cap E_n = A \cap E_n].$$

To insure that $B_i \leq_{wtt} A_i$ we will use permitting as follows. We allow $y \in F_{\langle i,n\rangle}$ to enter B_0 (B_1) at stage $s+1$ only A_0 (A_1) E-permits n at stage $s+1$.

Fix e and j and let $i = \langle e,j\rangle$. Requirement $\mathbf{R}_{e,j}$ is split into the following subrequirements for all $n > \langle e,j\rangle$.

$$\mathbf{R}_{e,j,n}: \quad V_i \cap E_n = A \cap E_n \Rightarrow [W_e \cap F_{\langle i,n\rangle} = B_0 \cap F_{\langle i,n\rangle} \text{ or } W_j \cap F_{\langle i,n\rangle} = B_1 \cap F_{\langle i,n\rangle}].$$

We describe the construction for $\mathbf{R}_{e,j,n}$ as a two-state automaton as in Theorem 2.5. As in Theorem 2.5, we say that $\mathbf{R}_{e,j,n}$ is in state S_1 at stage s if the condition for state S_1 in Figure 2 holds. Otherwise $\mathbf{R}_{e,j,n}$ is in state S_2 at stage s and the construction guarantees that if this happens, the condition in the diagram for state S_2 holds. In order to accomplish this, the action corresponding to arrow a_1 is the same as that of Theorem 2.5. That is, if $\mathbf{R}_{e,j,n}$ is in state S_1 at stage s but not at stage $s+1$, we enumerate an element of E_n into V_i if necessary to cause the condition of state S_2 to hold. Since this happens only if an element of $F_{\langle i,n\rangle}$ is enumerated in W_e or W_j at stage $s+1$, this action need only be performed at most $2|F_{\langle i,n\rangle}|$ many times. Since $|E_n| > 2|F_{\langle i,n\rangle}|$ if $n > i$, we will be able to perform this action. Similarly, if s is such that the condition of state S_2 holds at s but fails at $s+1$, we must be able to ensure that the condition

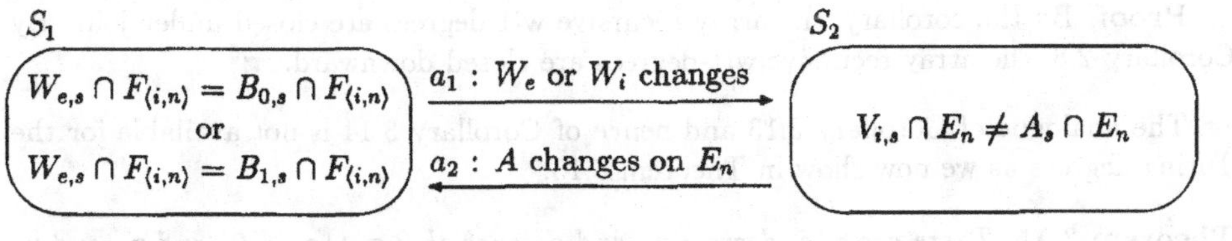

Figure 2: State diagram of the construction.

of state S_1 holds at stage $s+1$. For such an s, it must be the case that an element of E_n is enumerated into A at stage $s+1$, and hence by (21), that element is enumerated in either A_0 or A_1 at stage $s+1$. By our condition on permitting, this allows us to enumerate elements of $F_{\langle i,n\rangle}$ into either B_0 or B_1 at stage $s+1$, thereby guaranteeing that $\mathbf{R}_{e,j,n}$ is in state S_1 at stage $s+1$.

CONSTRUCTION.

Stage $s+1$

Step 1. (Arrow a_2.) For every triple e, j, n such that $\langle e,j\rangle < n$, if $W_{e,s} \cap F_{\langle\langle e,j\rangle,n\rangle} \neq B_{0,s} \cap F_{\langle\langle e,j\rangle,n\rangle}$ and A_0 E-permits n at stage $s+1$, enumerate all of $W_{e,s+1} \cap F_{\langle\langle e,j\rangle,n\rangle}$ into B_0 and similarly for W_j, A_1 and B_1 in place of W_e, A_0 and B_0.

Step 2. (Arrow a_1.) For each triple e, j, n, if

(a) $W_{e,s+1} \cap F_{\langle\langle e,j\rangle,n\rangle} \neq B_{0,s+1} \cap F_{\langle\langle e,j\rangle,n\rangle}$, and

(b) $W_{j,s+1} \cap F_{\langle\langle e,j\rangle,n\rangle} \neq B_{1,s+1} \cap F_{\langle\langle e,j\rangle,n\rangle}$, but

(c) $W_{e,s} \cap F_{\langle\langle e,j\rangle,n\rangle} = B_{0,s} \cap F_{\langle\langle e,j\rangle,n\rangle}$ or $W_{j,s} \cap F_{\langle\langle e,j\rangle,n\rangle} = B_{1,s} \cap F_{\langle\langle e,j\rangle,n\rangle}$,

then enumerate one element of $E_n - V_{\langle e,j\rangle,s}$,if necessary, into $V_{\langle e,j\rangle}$ so that $V_{\langle e,j\rangle,s+1} \cap E_n \neq A_{s+1} \cap E_n$. (Such an element will exist by the construction.) This ends the construction.

The relevant lemmas, parallel in statement and proof (which is omitted), to those of Theorem 2.5 are

Lemma 3.11 $B_0 \leq_{\rm wtt} A_0$; $B_1 \leq_{\rm wtt} A_1$.

Lemma 3.12 *For every* e, j, $\mathbf{R}_{e,j}$ *is satisfied.* ■

The following corollary follows directly from the Theorem and Corollary 2.8.

Corollary 3.13 *Suppose that* $A \leq_{\rm wtt} A_0 \oplus A_1$ *and that* A *is array nonrecursive. Then the weak-truth-table degree of either* A_0 *or* A_1 *contains an array nonrecursive set.*

An immediate consequence of the preceding corollary is the following.

Corollary 3.14 *The array recursive wtt-degrees form an ideal in the uppersemilattice of r. e. wtt-degrees.*

Proof. By the corollary, the array recursive wtt-degrees are closed under join. By Corollary 2.8, the array recursive wtt-degrees are closed downward. ∎

The analogue of Corollary 3.13 and hence of Corollary 3.14 is not available for the Turing degrees as we now show in Theorem 3.15.

Theorem 3.15 *There are r. e. degrees* $\mathbf{a}_0$ *and* $\mathbf{a}_1$ *such that* $\mathbf{a}_0 \cup \mathbf{a}_1 = \mathbf{0}'$ *and* $\mathbf{a}_0$ *and* $\mathbf{a}_1$ *are array recursive.*

Proof. Fix a v. s. a. $\{F_n\}_{n\in N}$ such that $|F_n| > 2^{n^2}$ for all $n \in N$. We construct sets A_0 and A_1 of array recursive degree by showing that every set recursive in either is not F-a. n. r. To do this, as in the proof of Theorem 2.10, we enumerate sets V_e and U_e so that for every e and $n > e$ the following requirements are satisfied.

$$\mathbf{R}_{e,n}: \quad \Phi_e(A_0) = B_e \Rightarrow V_e \cap F_n \neq B_e \cap F_n$$

$$\mathbf{Q}_{e,n}: \quad \Phi_e(A_1) = B_e \Rightarrow U_e \cap F_n \neq B_e \cap F_n$$

Here $(\Phi_e, B_e)_{e\in N}$ enumerates all pairs (Φ, B) of reductions Φ and r. e. sets B. To guarantee that $K \leq_T A_0 \oplus A_1$, we will define a recursive function $\gamma : N^2 \to N$ such that

(22) $$\lim_s \gamma(x,s) \text{ exists;}$$

(23) $\gamma(x, s+1) \neq \gamma(x,s)$ only if $(\exists y \leq \gamma(x,s))[y \in A_{0,s+1} - A_{0,s}$ or $y \in A_{1,s+1} - A_{1,s}]$

(24) if $x \in K_{s+1} - K_s$ then $(\exists y \leq \gamma(x,s))[y \in A_{0,s+1} - A_{0,s}$ or $y \in A_{1,s+1} - A_{1,s}]$.

The existence of such a function γ implies that $K \leq_T A_0 \oplus A_1$; the fact that γ depends on s makes this a Turing reduction rather than a weak-truth-table reduction which is prohibited by Theorem 3.9. We define $\gamma(x, 0) = x$ for all $x \in N$.

The two-state automaton corresponding to requirement $\mathbf{R}_{e,n}$ is in Figure 3.

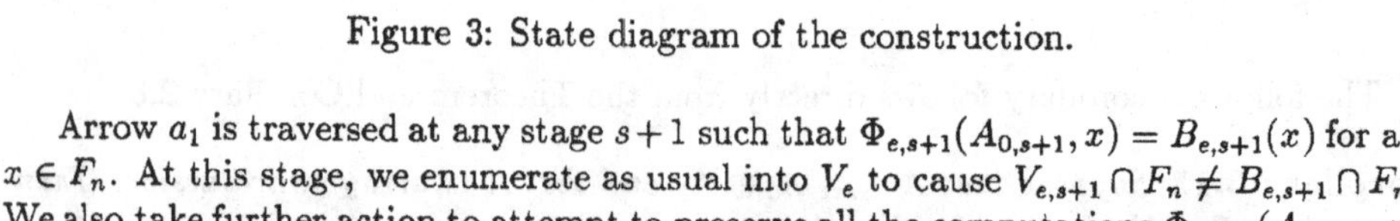

Figure 3: State diagram of the construction.

Arrow a_1 is traversed at any stage $s+1$ such that $\Phi_{e,s+1}(A_{0,s+1}, x) = B_{e,s+1}(x)$ for all $x \in F_n$. At this stage, we enumerate as usual into V_e to cause $V_{e,s+1} \cap F_n \neq B_{e,s+1} \cap F_n$. We also take further action to attempt to preserve all the computations $\Phi_{e,s+1}(A_{0,s+1}, x)$ for $x \in F_n$. Suppose that it is possible to preserve these computations forever and suppose there is a stage $t+1 > s+1$ at which the condition of state S_2 fails. This implies that an integer $x \in F_n$ is enumerated in B_e at stage $t+1$. But then we have that $\Phi_{e,t+1}(A_{0,t+1}) = \Phi_{e,s+1}(A_{0,s+1}) = B_{e,s+1} \neq B_{e,t+1}$ and this disagreement is preserved forever. Thus requirement $\mathbf{R}_{e,n}$ remains in state S_1 forever and is satisfied. The bound on $|F_n|$ above reflects the fact that in taking action a_1 we will not always be able to preserve all computations because of the requirements for coding K. We will ensure that the action a_1 is injured fewer that 2^{n^2} times and thus that arrow a_1 requires traversal at most 2^{n^2} times.

CONSTRUCTION.

Stage $s+1$

Step 1. Let n be the least element of $K_{s+1} - K_s$. Enumerate $\gamma(n, s)$ into A_0. Define $\gamma(y, s+1) = \gamma(y+s, s)$ for all $y \geq n$.

Step 2. (Arrow a_1.) Requirement $\mathbf{R}_{e,n}$ ($\mathbf{Q}_{e,n}$) *requires attention* at stage $s+1$ if

(25) $\Phi_{e,s+1}(A_{0,s+1}, x) = B_{e,s+1}(x)$ $(\Phi_{e,s+1}(A_{1,s+1}, x) = B_{e,s+1}(x))$ for all $x \in F_n$, and

(26) $V_{e,s} \cap F_n = B_{e,s+1} \cap F_n$ $(U_{e,s} \cap F_n = B_{e,s+1} \cap F_n)$.

Let n be least and e least for n such that either $\mathbf{R}_{e,n}$ or $\mathbf{Q}_{e,n}$ requires attention. If $\mathbf{R}_{e,n}$ requires attention do the following. Let u be the maximum element of A_0 used in the computations mentioned in (25). If $\gamma(n, s) \leq u$, enumerate $\gamma(n, s)$ into A_1 and define $\gamma(y, s+1) = \gamma(y+s, s)$ for all $y \geq n$. (By the usual conventions on the use function of a computation, $\gamma(y, s+1) > u$ for all $y \geq n$. Thus this step has the effect of clearing the computations of (25) of lower priority markers.) Also, choose $z \in F_n - V_{e,s}$ (such will exist) and enumerate $z \in V_e$. If instead $\mathbf{Q}_{e,n}$ requires attention but $\mathbf{R}_{e,n}$ does not, attend to $\mathbf{Q}_{e,n}$ just as $\mathbf{R}_{e,n}$ but with U_e, A_0, and A_1 in place of V_e, A_1, and A_0 respectively. This ends the construction.

Lemma 3.16 *For every $e, n \in N$ such that $n > e$, requirements $\mathbf{R}_{e,n}$ and $\mathbf{Q}_{e,n}$ receive attention at most 2^{n^2} times and are satisfied.*

Proof. We assume the lemma is true for all pairs e', n' such that $n' < n$ or $n' = n, e' < e$ and give the proof for $\mathbf{R}_{e,n}$. The proof for $\mathbf{Q}_{e,n}$ is identical. Suppose that $\mathbf{R}_{e,n}$ receives attention at stage $s+1$ and there is $z \in F_n - V_{e,s}$. Then $V_{e,s+1} \cap F_n \neq B_{e,s+1} \cap F_n$. Furthermore, by (25) $\Phi_{e,s+1}(A_{0,s+1}, x) = B_{e,s+1}(x)$ for all $x \in F_n$ so that if these computations are never injured, either $V_e \cap F_n \neq B_e \cap F_n$ or $\Phi_e(A_0) \neq B_e$ and $\mathbf{R}_{e,n}$ never requires attention after stage $s+1$. Now by the definition of $\gamma(y, s+1)$ for $y \geq n$, the computation in (25) can be injured at a later stage $t+1$ only if $\gamma(y, t+1) = \gamma(y, s)$ enters A_0 for some $y < n$. This happens only if such a number y enters K at stage $t+1$ or because a requirement $\mathbf{R}_{e',y}$ or $\mathbf{Q}_{e',y}$ for some e' such that $e' < y < n$ receives attention at stage $t+1$. Therefore there can be at most $n + \sum_{0<y<n} 2^{y^2}$ many stages $s+1$ at which $\mathbf{R}_{e,n}$ receives attention and is later injured. Thus $\mathbf{R}_{e,n}$ receives attention at most $1 + n + \sum_{0<y<n} 2^{y^2} \leq 2^{n^2}$ times. Since $|F_n| > 2^{n^2}$, $F_n - V_e \neq \emptyset$. Thus, if $\Phi_e(A_0) = B_e$, $\mathbf{R}_{e,n}$ will receive attention enough times to enumerate V_e to make $V_e \cap F_n \neq B_e \cap F_n$. $\square$

Lemma 3.17 $K \leq_T A_0 \oplus A_1$.

Proof. The definition of γ satisfies (24) by step (1) of the construction. (23) is satisfied since $\gamma(y,s) \neq \gamma(y,s+1)$ only if some $\gamma(n,s)$ for $n \leq y$ is enumerated in either A_0 or A_1 at stage $s+1$, and γ is increasing in its first argument. To see that (22) is satisfied, note that $\gamma(y,s+1) \neq \gamma(y,s)$ only if some $n \leq y$ enters K at stage $s+1$ or some requirement $\mathbf{R}_{e,n}$ or $\mathbf{Q}_{e,n}$ receives attention for some $n < y$. Because of Lemma 3.16, there are only finitely many such stages and thus (22) is satisfied. ■

4 Natural multiple permitting arguments

In this section we prove the three main theorems, Theorems 1.6, 1.7, and 1.8 promised in Section 1. In each, we show that a certain construction from elsewhere in recursion theory can be done below precisely the a. n. r. degrees. Thus, besides characterizing the degrees which admit these constructions by a simple recursion theoretic property, these constructions show that the notion of multiple permitting considered here is quite natural.

Theorem 4.1 *Let f be a strictly increasing recursive function. Then the r. e. degree* $\mathbf{a}$ *is a. n. r. iff there is a degree* $\mathbf{b} \leq \mathbf{a}$ *(not necessarily r. e.) such that some set B of degree* $\mathbf{b}$ *is not f-r. e.*

Proof. (only if). Let $\{F_n\}_{n\in N}$ be a very strong array such that $|F_n| > f(\langle e,n\rangle)$ for every n and $e \leq n$. Let $A \in \mathbf{a}$ such that A is F-a. n. r. We shall define B by giving a recursive approximation $\{B_s\}_{s\in N}$ of B so that for every pair $\langle e,n\rangle$, $B_{s+1}(\langle e,n\rangle) \neq B_s(\langle e,n\rangle)$ only if A F-permits n. Then, by a suitable analogue to Lemma 2.4, $B \leq_{\mathbf{T}} A$. The requirements to make B not f-r. e. are as follows. Let $\{\phi_e\}_{e\in N}$ be a recursive enumeration of the partial recursive binary functions.

$$\mathbf{R}_e: \quad \text{if } \lim_y \phi_e(x,y) = B(x) \text{ then } (\exists x)[|\{y : \phi_e(x,y) \neq \phi_e(x,y+1)\}| > f(x)].$$

To meet $\mathbf{R}_e$, as usual we enumerate sets V_e and use (8):

$$(27) \qquad (\overset{\infty}{\exists}\, n)[V_e \cap F_n = A \cap F_n].$$

For the witness x mentioned in requirement $\mathbf{R}_e$, we use the numbers $\langle e,0\rangle, \langle e,1\rangle, \ldots$. We recast $\mathbf{R}_e$ as the following sequence of requirements $\mathbf{R}_{e,n}$ for $n \geq e$.

$$\mathbf{R}_{e,n}: \quad V_e \cap F_n = A \cap F_n \Rightarrow \langle e,n\rangle \text{ witnesses } \mathbf{R}_e.$$

In light of (27), the requirements $\mathbf{R}_{e,n}$ for $n \geq e$ are enough.

The strategy for meeting $\mathbf{R}_{e,n}$ is represented by a two-state automaton. For the purpose of describing the machine, we make the following definition. We say that ϕ_e is *correct on* $\langle e,n\rangle$ *at stage* s if $\phi_{e,s}(\langle e,n\rangle, y_s) = B_s(\langle e,n\rangle)$ where y_s is the greatest integer, if such exists, such that $\phi_{e,s}(\langle e,n\rangle, y_s)$ converges. The machine is given by Figure 4. Arrow a_1 is implemented in the usual way. That is, when ϕ_e is not correct on $\langle e,n\rangle$ at s but is correct at $s+1$, we enumerate an element of F_n into V_e if necessary to enter state S_2. For arrow a_2, we define B_{s+1} so that ϕ_e is not correct on $\langle e,n\rangle$ at $s+1$. Since

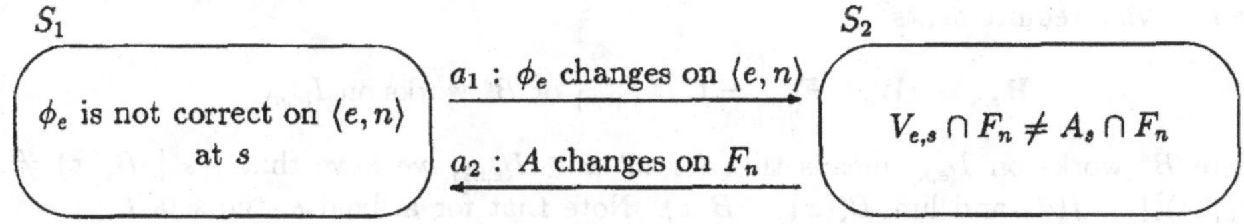

Figure 4: State diagram for construction.

arrow a_2 is traversed only if A F-permits n, we are allowed by our permitting condition to do this. The important thing to notice is that for each complete traversal of the machine from state S_1 to S_2 and back to S_1 again, there must exist a new y such that $\phi_e(\langle e,n\rangle, y) \neq \phi_e(\langle e,n\rangle, y+1)$. Since $|F_n| > f(\langle e,n\rangle)$, we will be able to force that there are more than $f(\langle e,n\rangle)$ such y if $V_e \cap F_n = A \cap F_n$ and $\lim_y \phi_e(\langle e,n\rangle, y) = B(\langle e,n\rangle)$.

CONSTRUCTION.

Stage $s+1$

Step 1. (Arrow a_2) For every e and $n \geq e$ such that A F-permits n at $s+1$ do the following. Let y_{s+1} be maximal, if such exists, such that $\phi_{e,s+1}(\langle e,n\rangle, y_{s+1})$ converges. Define $B_{s+1}(\langle e,n\rangle)$ so that $B_{s+1}(\langle e,n\rangle) \neq \phi_{e,s+1}(\langle e,n\rangle, y_{s+1})$. For all x such that $B_{s+1}(x)$ has not otherwise been defined in this step, define $B_{s+1}(x) = B_s(x)$.

Step 2. (Arrow a_1) For each e and $n \geq e$, if ϕ_e is correct on $\langle e,n\rangle$ at $s+1$ and $V_{e,s} \cap F_n = A_{s+1} \cap F_n$, enumerate the least element of $F_n - V_{e,s}$ into V_e.

To see that $\mathbf{R}_{e,n}$ is satisfied, suppose that $V_e \cap F_n = A \cap F_n$ and that $\lim_y \phi_e(\langle e,n\rangle, y) = B(\langle e,n\rangle)$. Step (2) then implies that arrow a_1 is traversed $|F_n|$ many times. Let $s_1 < s_3$ be stages such that consecutive traversals of arrow a_1 are made at s_1+1 and s_3+1. Let s_2+1 be the intervening stage at which arrow a_2 is traversed. Thus $\phi_{e,s_1+1}(\langle e,n\rangle, y_{s_1+1}) = B_{s_1+1}(\langle e,n\rangle)$, $\phi_{e,s_3+1}(\langle e,n\rangle, y_{s_3+1}) = B_{s_3+1}(\langle e,n\rangle)$, and $\phi_{e,s_2+1}(\langle e,n\rangle, y_{s_2+1}) \neq B_{s_2+1}(\langle e,n\rangle)$ This implies the existence of y such that $y_{s_1+1} \leq y < y_{s_3+1}$ such that $\phi_e(\langle e,n\rangle, y) \neq \phi_e(\langle e,n\rangle, y+1)$. The existence of $|F_n| > f(\langle e,n\rangle)$ such y implies that $\mathbf{R}_{e,n}$ is satisfied. □

(if) Suppose that $B \in \mathbf{b}$ is not f-r. e., and $B \leq_T A$. Let Γ and γ be such that $B = \Gamma(A)$ with use function γ; i.e., $\gamma(x,s)$ is the use of the computation $\Gamma_s(A_s, x)$ at stage s if the computation converges. We may assume that $\gamma(x,s)$ is increasing in x. By speeding up the enumeration of Γ and A, we may also assume that $\Gamma_s(A_s, x)$ is defined for all $s > x$. We construct $C \leq_T A$ such that C is array nonrecursive. Fix a v. s. a. $\{F_n\}_{n \in N}$. The requirements are

$$\mathbf{R}_e : \quad (\exists n)[W_e \cap F_n = C \cap F_n].$$

We devote the sets $F_{\langle e,0\rangle}, F_{\langle e,1\rangle}, \ldots$ to $\mathbf{R}_e$.

To meet $\mathbf{R}_e$, we will construct a recursive approximation $\{B_s^e\}_{s \in N}$ which threatens to witness that B is f-r. e. Define $I_{\langle e,n\rangle} = \{z \mid |F_{\langle e,n\rangle}| < x \leq |F_{\langle e,n+1\rangle}|\}$. We split $\mathbf{R}_e$ into the following requirements

$$\mathbf{R}_{e,n}: \quad W_e \cap F_{\langle e,n\rangle} = C \cap F_{\langle e,n\rangle} \text{ or } B^e \text{ works on } I_{\langle e,n\rangle}$$

where B^e works on $I_{\langle e,n\rangle}$ means that for all $x \in I_{\langle e,n\rangle}$, we have that $|\{s \mid B^e_s(x) \neq B^e_{s+1}(x)\}| \leq f(x)$ and $\lim_s B^e_s(x) = B(x)$. Note that for a fixed e, the sets $I_{\langle e,n\rangle}$ are finite, disjoint, and have cofinite union. It is clear that the requirements $\mathbf{R}_{e,n}$ are sufficient to meet $\mathbf{R}_e$.

The strategy for meeting $\mathbf{R}_{e,n}$ is given by the two-state automaton of Figure 5.

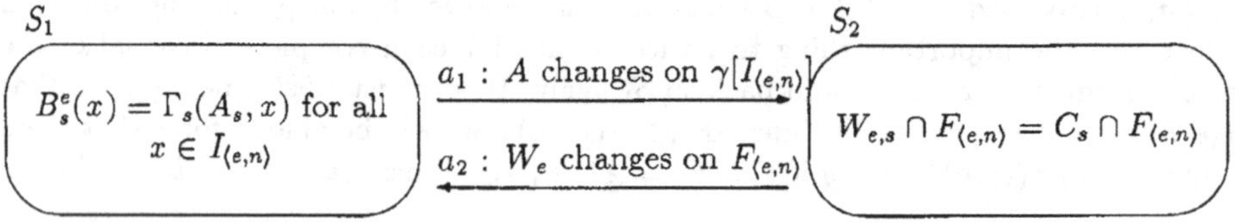

Figure 5: State diagram for the construction.

In Figure 5, $\gamma[I_{\langle e,n\rangle}]$ denotes $\max\{y \mid y \leq \gamma(x,s)$ for some $x \in I_{\langle e,n\rangle}\}$. We begin the strategy for $\mathbf{R}_{e,n}$ at stage s_0 such that $s_0 = \max I_{\langle e,n\rangle} + 1$. At this stage, $\Gamma_{s_0}(A_{s_0}, x)$ converges for all $x \in I_{\langle e,n\rangle}$ and we set $B^e_t(x) = \Gamma_{s_0}(A_{s_0}, x)$ for all $t \leq s_0$. Thus, at stage s_0, we are in state S_1 of figure 5. While $\Gamma_s(A_s, x) = \Gamma_{s+1}(A_{s+1}, x)$ for all $x \in I_{\langle e,n\rangle}$, we set $B_{s+1}(x) = B_s(x)$ and remain in state S_1 (without any changes in our approximation to B). Suppose that $s \geq s_0$ is such that $\Gamma_{s+1}(A_{s+1}, x) \neq \Gamma_s(A_s, x)$ for some $x \in I_{\langle e,n\rangle}$. Then it must be the case that $(\exists y \leq \gamma(x,s))[y \in A_{s+1} - A_s]$. We take this to be our permitting condition and thus are allowed to follow arrow a_1 at stage $s+1$ and set $C_{s+1} \cap F_{\langle e,n\rangle} = W_{e,s+1} \cap F_{\langle e,n\rangle}$. While in state S_2, we also cause $B_{s+1}(x) = B_s(x)$ for all $x \in I_{\langle e,n\rangle}$. We remain in state S_2 unless there is s such that W_e changes on $F_{\langle e,n\rangle}$ at stage $s+1$. For such an s, we pass to state S_1 by defining $B_{s+1}(x) = \Gamma_{s+1}(A_{s+1}, x)$ for all $x \in I_{\langle e,n\rangle}$. Notice that the above construction requires us to change the approximation to B at a stage $s+1$ on $x \in I_{\langle e,n\rangle}$ only if arrow a_2 is traversed at stage $s+1$. This can occur only $F_{\langle e,n\rangle}$ times. Since $f(x) > |F_{\langle e,n\rangle}|$ for all $x \in I_{\langle e,n\rangle}$, our approximation B_e to B does not change too often. Furthermore, if the construction for $\mathbf{R}_{e,n}$ ends in state S_1, we have that $\lim_s B^e_s(x) = B(x)$ for all $x \in I_{\langle e,n\rangle}$ and $\mathbf{R}_{e,n}$ is satisfied. On the other hand, if the construction for $\mathbf{R}_{e,n}$ ends in state S_2, the requirement $\mathbf{R}_e$ is satisfied (on $F_{\langle e,n\rangle}$). Note also that our permitting condition guarantees that $C \leq_{\mathrm{T}} A$. There are no conflicts among the various requirements $\mathbf{R}_{e,n}$. We omit the description of the whole construction as it is now quite routine. ■

We now use Theorem 4.1 to improve the classification begun in Section 3 of the a. n. r. degrees in terms of the jump operator. We use the following result of Jockusch [J2, Theorem 1].

Lemma 4.2 *If* $\mathbf{a}$ *is any r. e. degree then* $\mathbf{a}' \geq \mathbf{0}''$ *iff the recursive sets are uniformly of degree* $\leq \mathbf{a}$.

We now have the following Corollary of Theorem 4.1.

Corollary 4.3 *If* **a** *is an r. e. degree such that* $\mathbf{a}'' > \mathbf{0}''$, *then* **a** *is array nonrecursive.*

Proof. Suppose that **a** is array recursive. By the theorem, this implies that if f is a strictly increasing function, then every set B such that $deg(B) \leq \mathbf{a}$ is f-r. e. It is easy to see that this implies that the sets recursive in **a** are uniformly $\leq_{tt} K$ and hence of degree $\leq \mathbf{0}'$. By the relativization of Lemma 4.2 to **a**, we have that $\mathbf{0}'$ is high over **a**; i. e., $\mathbf{0}'' = \mathbf{a}''$. ∎

The result of Corollary 4.3 is best possible since Downey [D2, Theorem 1.3] has shown that there are array recursive degrees that are low$_2$ but not low. The easiest way that we know to construct such a degree is indirect. First, it is possible to construct an r. e. 1-topped degree which is array recursive. This is done by modifying a construction of Downey and Jockusch [DJ, Theorem 2.1]. Then we use the fact, also proved in [DJ, Theorems 3.1, 3.2], that all nonzero 1-topped r. e. degrees are complete or low$_2$ but not low.

A. Kučera and the authors have observed that Theorem 4.1 can be used in conjunction with other results to give a new proof of Theorem 2.10. First, observe that a straightforward modification of the proof of the low basis theorem [JS, Theorem 2.1] shows that every nonempty, recursively bounded Π^0_1 class has an element A such that, for some recursive function f, every set B r. e. in A is f-r. e. Applying this to such a Π^0_1 class which contains only sets of fixed-point-free degree (see [K, Remark 1]), there is a set A of fixed-point-free degree and a function f such that every set B r. e. in A is f-r. e. Then by Kučera's result that every fixed-point-free degee below $\mathbf{0}'$ bounds a nonzero r. e. degree [K, Theorem 1], there is a nonrecursive r. e. set C recursive in A. The degree of C is array recursive by Theorem 4.1. Corollaries 2.11 and 2.12 of 2.10 follow by the same argument, since every fixed-point-free degree below $\mathbf{0}'$ bounds a promptly simple degree by [K, Remark 2].

In [JS, Theorem 1], Jockusch and Soare show that every degree which contains a consistent extension of Peano arithmetic bounds an incomparable pair of degrees. Used in the proof of that theorem is a construction of two pairs B_0, C_0 and B_1, C_1 of r. e. sets such that $B_0 \cap C_0 = B_1 \cap C_1 = \emptyset$ and whenever S separates B_0 and C_0 and T separates S_1 and C_1, then S and T are Turing incomparable. We now show that this construction can be done below precisely the array nonrecursive degrees.

Theorem 4.4 *For r.e. sets A, the following are equivalent:*

(a) A has a. n. r. degree,

(b) there are disjoint r.e. sets B and C each recursive in A such that $B \cup C$ is coinfinite and no set of degree $\mathbf{0}'$ separates B and C,

(c) there exist two disjoint pairs of r.e. sets B_0, C_0 and B_1, C_1 such that $B_i \cup C_i$ is coinfinite for $i = 0, 1$, each set B_i, C_i is recursive in A, and each set which separates (B_0, C_0) is incomparable with each which separates (B_1, C_1).

Proof. (c)⇒(b) is easy; let $B = B_0$, $C = C_0$.

(a)⇒(c). We first review the construction of B_0, C_0, B_1, and C_1 when there is no requirement that any of these sets be recursive in A. Let $\mathbf{R}_{2e+j}$ ($j = 0$ or $j = 1$) be the requirement that if S is any separating set for B_j, C_j, and T is any separating set for B_{1-j}, C_{1-j}, then $S \neq \{e\}^T$. The basic strategy for $\mathbf{R}_{2e+j}$ is to choose a witness w and wait for a stage s such that for some set T with $\max(T) < s$, T separates $B_{1-j,s}$, $C_{1-j,s}$, and $\{e\}^T_s(w) = 0$ or 1. Let u be the use in the computation $\{e\}^T_s(w)$. At stage $s+1$, we enumerate all elements of T into B_{1-j} and all elements of $\overline{T}$ which are less than u into C_{1-j}. This insures that $T[u] = T'[u]$ for any set T' which separates $B_{1-j,s+1}$ and $C_{1-j,s+1}$ and hence for any set T' which separates B_{1-j} and C_{1-j}. Now if $\{e\}^T_s(w) = 0$ we enumerate w into B_j and if $\{e\}^T_s(w) = 1$ we enumerate w into C_j. This meets the requirement forever.

To combine the requirements, we use many witnesses for each requirement. Specifically, if k witnesses are assigned to requirements $\mathbf{R}_m$ for $m < n$, we use 2^k witnesses for $\mathbf{R}_n$; i.e., one for every subset D of the set W of witnesses for the requirements R_m, for $m < n$. If witness w_D corresponds to $D \subseteq W$, it is handled as above except that one considers only separating sets T with $T \cap W = D$, and $\mathbf{R}_{2e+j}$ does not cause any elements of W to be enumerated into B_{1-j} or C_{1-j}. Whenever a witness w for $\mathbf{R}_{2e+j}$ is enumerated into B_j or C_j by a requirement $\mathbf{R}_{2i+(1-j)}$ of higher priority, then w is replaced by a new witness w' for $\mathbf{R}_{2e+j}$ which is not yet in B_j or C_j. Thus requirements $\mathbf{R}_{2e+j}$ of lower priority than $\mathbf{R}_{2i+j}$ must consider new possibilites for sets D contained in the witnesses assigned to requirements (such as $\mathbf{R}_{2i+j}$) of higher priority than $\mathbf{R}_{2e+j}$. Nonetheless, it is easy to compute a recursive upper bound $w(n)$ for the number of witnesses ever assigned to $\mathbf{R}_n$. This is not important for the basic existence result we have been discussing but it is crucial to carrying it out below a given array nonrecursive degree.

To make the sets we construct recursive in a given array nonrecursive set A, we first choose a very strong array $\{F_n\}_{n\in N}$ such that each F_n has sufficiently large cardinality (to be specified later). We shall assume that A is F-a. n. r. since by Theorem 2.5 there is a set $\hat{A}$ of the same degree as A which is F-a. n. r. If X is any one of the sets B_0, C_0, B_1, C_1, we shall guarantee that $X \leq_{\mathrm{T}} A$ by the condition: $x \in X_{s+1} - X_s$ only if A F-permits x at stage $s+1$. The requirements $\mathbf{R}_k$ above, are replaced by infinitely many requirements

$$\mathbf{R}_{k,i}: \quad A \cap F_i = V_k \cap F_i \Rightarrow \mathbf{R}_k \text{ holds}$$

where V_k is an auxiliary r. e. set enumerated during the construction. We meet $\mathbf{R}_{k,i}$ only for $i \geq k$ which suffices by (8). We assign priorities to the requirements $\mathbf{R}_{k,i}$ in increasing order of $\langle k, i\rangle$.

We now describe the strategy for meeting $\mathbf{R}_{k,i}$ which is similar but not identical to that for $\mathbf{R}_k$. The basic idea is to make $V_{k,s} \cap F_i \neq A_s \cap F_i$ by enumerating an element of F_i whenever permission is needed to enumerate an element $\geq i$ into one of B_0, C_0, B_1, or C_1. If $A \cap F_i = V_k \cap F_i$, then the desired permission must occur. The set F_i will be of sufficiently large cardinality so that an element of F_i will always be available. Let W be the set of numbers which are less than i or are witnesses for requirements of higher priority than $\mathbf{R}_{k,i}$ (and thus either cannot be forced to be permitted by the above method or cannot be enumerated by $\mathbf{R}_{k,i}$). Actually, W depends on the stage. For each set $D \subseteq W$, assign a witness $w_D \geq i$ to $\mathbf{R}_{k,i}$. Suppose that $k = 2e + j$. Wait for a stage s such that there is a set T with $\max(T) < s$, T separates $B_{1-j,s}$ and $C_{1-j,s}$, and $\{e\}^T_s(w_D) = 0$ or 1. Now restrain B_{1-j} and C_{1-j} through the use of this computation. (Notice that this restraint was not involved in the original strategy for $\mathbf{R}_k$. It is needed now to ensure that T is still a separating set when, if ever, we get permission for the desired enumerations.) If $V_{k,s} \cap F_i = A_s \cap F_i$, enumerate the least element of $F_i - V_{k,s}$ into V_k. As usual, this element will exist by the construction and the choice of the F_i. Assign new witnesses to all requirements of lower priority than $\mathbf{R}_{k,i}$. If there is a stage $t > s$ such that A F-permits i at stage t, and no requirement of higher priority than $\mathbf{R}_{k,i}$ has acted between s and t, then the obvious enumerations should be made to meet $\mathbf{R}_{k,i}$ for all separating sets T' with $T' \cap W = D$. Specifically, let T be, as before, a set separating $B_{1-j,s}$ and $C_{1-j,s}$ with $\max T < s$ and $\{e\}^T_s(w_D) = 0$ or 1 with use u and $T \cap W = D$. Enumerate all elements of $T - W$ into B_{1-j} and all elements of $\overline{T} \cap \overline{W}$ into C_{1-j}. If $\{e\}^T_s(w_D) = 0$, put w_D into B_j, and otherwise put w_D into C_j. Assign new witnesses to all requirements of lower priority than $\mathbf{R}_{k,i}$.

It is a standard finite injury argument to see that the above procedure works. The main point is that we can in advance choose the sets F_i to be of sufficiently large cardinality. We first define, by recursion on $\langle k, i\rangle$, a recursive bound $w(k,i)$ on the number of witnesses ever assigned to $\mathbf{R}_{k,i}$. Let $c = \sum\{w(k',i') | \langle k',i'\rangle < \langle k,i\rangle\}$, so that c bounds the total number of witnesses ever assigned to requirements of higher priority than $\mathbf{R}_{k,i}$. Since any requirement acts at most twice using any given witness, there are at most $2c$ stages at which requirements of higher priority than $\mathbf{R}_{k,i}$ act. At any such stage, at most 2^{i+c} witnesses are assigned to $\mathbf{R}_{k,i}$, so we may set $w(k,i) = 2c2^{i+c}$. Finally, it suffices for the cardinality of F_i to be at least the number of witnesses for $R_{k,i}$ for each $k \le i$, since each witness can causes at most one element of F_i to enter V_k. We thus require that $|F_i| = \max\{w(k,i) | k \le i\}$. □

(b)⇒(a). Let $\{F_n\}_{n\in N}$ be a very strong array. We construct $A \le_T B \oplus C$ to meet

$$\mathbf{R}_e: \quad (\exists n)[W_e \cap F_n = A \cap F_n].$$

As usual, we reserve the sets $F_{\langle e,i\rangle}$, $i \in N$ for meeting $\mathbf{R}_e$. We first give the construction of A. We show that if the construction fails to meet $\mathbf{R}_e$ for some e, then there is a Δ^0_2 set X such that $K \le_T X$ and X separates the pair B, C.

Define a recursive function g by the two conditions

$$\begin{aligned} g(e,0) &= 1 + |F_{\langle e,0\rangle}| \quad \text{for every } e \in N \\ g(e,i+1) &= g(e,i) + 1 + |F_{\langle e,i+1\rangle}| \quad \text{for every } e,i \in N. \end{aligned}$$

At any stage in the construction, let $d_{0,s} < d_{1,s} < d_{2,s} \ldots$ be the elements of $\overline{B_s \cup C_s}$. To ensure that $A \le_T B \oplus C$, we require that if $x \in A_{s+1} - A_s$ and $x \in F_{\langle e,i\rangle}$ then $d_{g(e,i+1),s} \ne d_{g(e,i+1),s+1}$. Let d_n denote $\lim_s d_{n,s}$.

CONSTRUCTION.

Stage $s+1$

For every e and i, if $W_{e,s} \cap F_{\langle e,i\rangle} \ne A_s \cap F_{\langle e,i\rangle}$ and $d_{g(e,i+1),s} \ne d_{g(e,i+1),s+1}$, then enumerate all of $W_{e,s+1} \cap F_{\langle e,i\rangle}$ into A at stage $s+1$. This ends the construction of A.

To see that each requirement $\mathbf{R}_e$ is satisfied, suppose otherwise and fix e such that $\mathbf{R}_e$ is not satisfied; i. e., that $W_e \cap F_{\langle e,i\rangle} \neq A \cap F_{\langle e,i\rangle}$ for all $i \in N$. We shall define a Δ^0_2 set X such that $K \leq_T X$, $X \supseteq B$ and $X \cap C = \emptyset$. We specify X by giving a recursive approximation $\{X_s\}_{s \in N}$.

We first describe the idea behind the construction of X. Let $D_0 = \{d_0, d_1, \ldots, d_{g(e,0)}\}$ and let $D_i = \{d_{g(e,i-1)+1}, \ldots, d_{g(e,i)}\}$ for all $i > 0$. Since $d_y \in \overline{B \cup C}$ for every y, we are free to enumerate elements of D_i in or out of X as we wish. Observe that our choice of g guarantees that $|D_i| = 1 + |F_{\langle e,i\rangle}|$ for all i. We will use the set D_i to code into X whether $i \in K$ and also to code into X the set D_{i+1}. (Thus X will be able to compute inductively for each integer n whether $n \in K$.) The coding into D_i will be tied to the attempt to meet R_e via $F_{\langle e,i\rangle}$ so that the fact that $W_e \cap F_{\langle e,i\rangle} \neq A \cap F_{\langle e,i\rangle}$ will allow the coding to be successful. Figure 6 gives the two-state diagram for the coding for i.

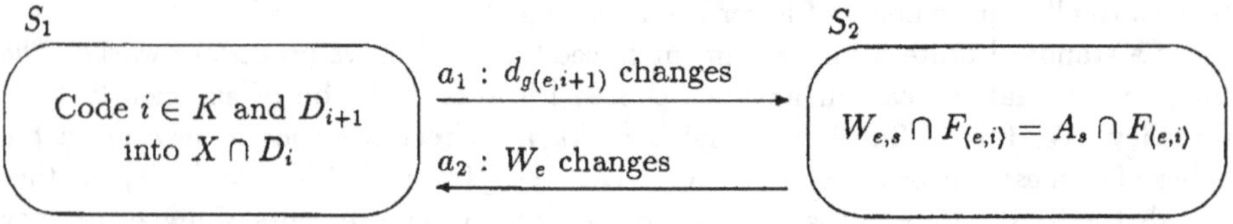

Figure 6: State diagram for the construction of X.

We call Figure 6 the i-module. The action that we take for the i-module at any stage $s+1$ is predicated on the assumption that the current approximation to D_i is correct. Arrow a_1 of the i-module corresponds exactly to the construction of A given above. Also, if a_1 is traversed at stage $s+1$, this indicates that the hypothesis of the j-module for all $j > i$ is false for all stages $\leq s$ so we restart each of these modules at stage $s+1$. This is the intent of Step 1 of the construction of X below. If arrow a_2 is traversed or i is enumerated in K at stage $s+1$, step 2 of the construction below codes this event in X by using an element of (the current approximation to) D_i. This coding is enough for X to recover D_{i+1} from D_i. We now give the formal construction of X and the verification.

CONSTRUCTION.

Define $X_0 = \emptyset$. Having defined X_s, we define X_{s+1} in steps. (For convenience, if we do not specify whether $n \in X_{s+1}$, then $n \in X_{s+1}$ if and only if $n \in X_s$.)

Stage $s+1$

Step 1. Let i be least, if any, such that $d_{g(e,i),s+1} \neq d_{g(e,i),s}$. Remove from X all integers $y > d_{g(e,i-1),s}$ ($y \geq 0$ if $i = 0$).

Step 2. Let i be least, if any, such that $i \in K_{s+1} - K_s$ or such that $W_{e,s+1} \cap F_{\langle e,i\rangle} \neq A_{s+1} \cap F_{\langle e,i\rangle}$ but $W_{e,s} \cap F_{\langle e,i\rangle} = A_s \cap F_{\langle e,i\rangle}$. Let j be least such that $g(e,i-1) < j \leq g(e,i)$ and $d_{j,s+1} \notin X_s$. (We will argue in Lemma 4.7 that such a j exists.) Enumerate $d_{j,s+1}$ into X.

Step 3. Enumerate each y in B_{s+1} into X.

Lemma 4.5 *$\lim_s X_s$ exists.*

Proof. Only step 1 causes removal of any element of X. It is clear from step 1 that any integer y may be removed finitely often. □

Lemma 4.6 *$X \supseteq B$ and $X \cap C = \emptyset$.*

Proof. $X \supseteq B$ by step 3 of the construction. Suppose that $y \in C$ is enumerated into X at some stage $s+1$. Then by the construction, $y = d_{j,s+1}$ for some j. Since $y \in C$, there is $t \geq s+1$ such that $d_{j,t} \neq d_{j,t+1}$. Thus by step 1, y is removed from X at stage $t+1$ (if not before). □

Lemma 4.7 *$K \leq_T X$.*

Proof. Let s_i be the least stage such that $d_{g(e,i),s_i} = d_{g(e,i)}$. By the construction, all integers in D_i are removed from X at stage s_i. Now for any such integer, d_j, d_j is enumerated into X at a stage $s+1 \geq s_i$ only if step 2 applies to i at stage $s+1$; i. e. if $i \in K_{s+1} - K_s$ or W_e changes on $F_{\langle e,i\rangle}$ at stage $s+1$. This can happen at most $1 + |F_{\langle e,i\rangle}|$ times. This proves the claim in the construction that j exists with $g(e,i-1) < j \leq g(e,i)$ and $d_{j,s+1} \notin X_s$.

Now to compute K from X, we assume by induction that we know D_i and show how to compute from X whether $i \in K$ and D_{i+1}. We can assume that we know D_0 since it is a finite set. Given D_i, let s be a stage such that $W_{e,s} \cap F_{\langle e,i\rangle} \neq A_s \cap F_{\langle e,i\rangle}$, $d_{g(e,i),s} = d_{g(e,i)}$, and such that for all $y \leq g(e,i)$, $d_y \in X$ iff $d_y \in X_s$. Then by step 2 of the construction, $i \in K$ iff $i \in K_s$. We claim also that $d_{g(e,i+1),s} = d_{g(e,i+1)}$. For otherwise let $t > s$ be such that $d_{g(e,i+1),t+1} \neq d_{g(e,i+1),t}$. Then $W_{e,t+1} \cap F_{\langle e,i\rangle} = A_{t+1} \cap F_{\langle e,i\rangle}$ by the construction of A. Thus there is $u \geq t+1$ such that $W_{e,u+1} \cap F_{\langle e,i\rangle} \neq A_{u+1} \cap F_{\langle e,i\rangle}$ but $W_{e,u} \cap F_{\langle e,i\rangle} = A_u \cap F_{\langle e,i\rangle}$. Then by step 2 of the enumeration of X, an element of D_i is enumerated in X at stage $u+1$ contrary to the assumption on s. Thus, at stage s we also know D_{i+1}. ■

Let Q denote the free Boolean algebra generated by a fixed recursive set $\{p_i | i \in N\}$ of literals. A (propositional) theory can be viewed as a filter of Q. We consider r. e. theories. An r. e. theory is *well-generated* if it is generated by a pair of sets $\{p_i | i \in B\}$ and $\{\neg p_i | i \in C\}$. It is well-known that if an r. e. theory T is well-generated and the r. e. sets B and C are recursively inseparable, then T is essentially undecidable. We examine such theories which in addition have relatively few r. e. extensions.

Definition 4.8 *An r. e. propositional theory is* Martin–Pour-El *if it is well-generated, essentially undecidable, and every r. e. extension of T is a principal extension. (That is, if $T \subseteq W$ and W is a consistent r. e. theory, then there is $q \in Q$ such that W is the theory generated by T and q.)*

Martin and Pour-El [MP, Theorem I] showed the existence of such theories. Downey in his thesis and in [D1], obtained numerous related results, including a number of results on the possible Turing degrees of such theories. The next theorem, together with the results of sections 2 and 3 considerably extend these results.

Theorem 4.9 *An r. e. degree* **a** *is array nonrecursive if and only if there is a Martin-Pour-El theory T of degree* **a**.

Proof. (only if) We first review the construction of a Martin–Pour-El theory T without the requirement that T be of a particular a. n. r. degree **a**. Let $\{S_n\}_{n\in N}$ be an enumeration of the r. e. consistent theories. If $F \subseteq Q$, we write F^* for the theory generated by F. We will construct T in stages so that T_s will denote the theory constructed by stage s. For each s, T_{s+1} will be of the form $(T_s \cup F)^*$ where F is a finite set of literals or their negations. At each stage s, we will denote by $d_{0,s} < d_{1,s} < \ldots$ the set $\{p_i | p_i, \neg p_i \notin T_s\}$; the ordering of the literals is that given by $p_0 < p_1 < \ldots$. We will use $\epsilon_i p_i$ to denote either p_i or $\neg p_i$.

For each $e \in N$ we have the requirements:

$$\mathbf{R}_e: \quad 0 \notin (T \cup S_e)^* \Rightarrow (\exists x)[(T \cup \{x\})^* = (T \cup S_e)^*], \text{ and}$$

$$\mathbf{N}_e: \quad \lim_s d_{e,s} \text{ exists.}$$

The requirements $\mathbf{N}_e$ guarantee that T is incomplete and consistent. The requirements $\mathbf{R}_e$ guarantee that every r. e. extension of T is principal over T. Together these requirements and the fact that T is well-generated guarantee that T is essentially undecidable. To meet $\mathbf{R}_e$ we shall construct a finite set Q_e such that $x = \bigwedge Q_e$ is the witness for $\mathbf{R}_e$. Q_e will be constructed in stages; $Q_{e,s}$ is the finite set constructed by stage s and $Q_e = \lim_s Q_{e,s}$.

We say that $\mathbf{R}_e$ *requires attention* at stage $s+1$ if $(\exists y)[y \in S_{e,s}, y \notin (T_s \cup Q_{e,s})^*$ and $0 \notin (T_s \cup S_{e,s})^*]$. If y is least with this property, we say that $\mathbf{R}_e$ *requires attention at* $s+1$ *via* y. We will assume that S_e consists of elements of the form $\bigvee \epsilon_i p_i$.

CONSTRUCTION.

Stage 0
Let $T_0 = Q_{e,0} = \emptyset$ for every $e \in N$.
Stage $s+1$
Find the least e such that $\mathbf{R}_e$ requires attention and let y be such that $\mathbf{R}_e$ requires attention via y. Define $F = \{\neg\epsilon_i d_{i,s} | \epsilon_i d_{i,s}$ occurs in y and $i \geq e\}$. Let $T_{s+1} = (T_s \cup F)^*$ and $Q_{e,s+1} = Q_{e,s} \cup \{y\}$. For $i \neq e$, let $Q_{i,s+1} = Q_{i,s}$. This ends the construction.

We show that the construction succeeds in two lemmas.

Lemma 4.10 *If* $\mathbf{R}_e$ *receives attention at stage* $s+1$ *via* y, *then there exists a Boolean combination of* $\{d_{0,s}, d_{1,s} \ldots d_{e-1,s}\} = \{d_{0,s+1}, d_{1,s+1} \ldots d_{e-1,s+1}\}$ *such that* $T_{s+1} \vdash y \leftrightarrow x$.

Proof. We write y as a disjunction of the form

$$\bigvee_{i<e} \epsilon_i d_{i,s} \vee \bigvee_{i \geq e} \epsilon_i d_{i,s} \vee \bigvee_{\epsilon_i p_i \in T_s} \epsilon_i p_i \vee \bigvee_{\neg\epsilon_i p_i \in T_s} \epsilon_i p_i$$

Thus y has the form $x \vee z \vee m \vee n$. Since $\vdash x \rightarrow y$, it suffices to show that $T_{s+1} \vdash y \rightarrow x$. Now if $m \neq 0$, then $y \in T_s$ since $\vdash m \rightarrow y$ and $m \in T_s$. But then $\mathbf{R}_e$ does not require attention via y. Thus $m = 0$. Now $\neg n \in T_{s+1}$ by definition of n, and $\neg z \in T_{s+1}$ by construction so it follows that $T_{s+1} \vdash y \rightarrow x$ as desired. □

Lemma 4.11 *For every e, Q_e is finite (and hence $\mathbf{R}_e$ is satisfied and $\lim_s d_{e,s}$ exists).*

Proof. Assume by induction that for all $i < e$, $\lim_s d_{i,s}$ exists and let s_0 be such that for all $i < e$, $\lim_s d_{i,s} = d_{i,s_0}$. Suppose that $\mathbf{R}_e$ receives attention via y at a stage $s+1 \geq s_0$. Then by Lemma 4.10, there is x, a disjunction of some of the d_{i,s_0}, $i < e$, and their negations such that $T_{s+1} \vdash y \leftrightarrow x$. Thus $(T_{s+1} \cup Q_{e,s+1})^* \vdash x$ (since $y \in Q_{e,s+1}$). However $(T_s \cup Q_{e,s})^* \not\vdash x$ since otherwise $y \in (T_s \cup Q_{e,s})^*$, contradicting that $\mathbf{R}_e$ requires attention via y. Thus, for each stage $s+1 > s_0$ such that $\mathbf{R}_e$ requires attention, a new disjunction x of the literals d_{i,s_0}, $i < e$, or their negations is used. There are only 2^{2e} such disjunctions. Thus $\mathbf{R}_e$ receives attention only finitely often after s_0, Q_e is finite, and thus $\mathbf{R}_e$ is satisfied. □

Of course the key fact in the above construction is that requirement $\mathbf{R}_e$ requires attention only 2^{2e} times, at most, after requirements $\mathbf{R}_i$, $i < e$, have ceased acting.

We now include the requirements that T be Turing computable from A for a fixed array nonrecursive set A. (We construct $T \leq_T A$. A simple coding strategy similar to that of the previous theorem can be used to make $T \equiv_T A$.) Let f be a recursive function defined by the conditions: $f(0) = 1$ and $f(i+1) = 2^{2i+2} \sum_{j=0}^{i} f(j)$ for all $i \geq 0$. We assume that A is array nonrecursive for a fixed v. s. a. $\{F_n\}_{n \in N}$ such that $|F_i| > f(i)$ for all i. As usual, we conceive of requirement $\mathbf{R}_e$ as consisting of subrequirements

$$\mathbf{R}_{e,i}: \qquad V_e \cap F_i = A \cap F_i \Rightarrow \mathbf{R}_e \text{ holds}$$

where V_e is an auxiliary r. e. set which we enumerate. We need only meet cofinitely many of these requirements for each $e \in N$. (In this construction, unlike previous ones, the confinite set of requirements $\mathbf{R}_{e,i}$ which we meet for a fixed e is specified only by recursive approximation.) Requirement $\mathbf{R}_{e,i}$ follows requirements $\mathbf{N}_0, \mathbf{N}_1, \ldots, \mathbf{N}_{i-1}$ in priority. The permitting condition is

$$d_{i,s} \text{ or } \neg d_{i,s} \in T_{s+1} - T_s \text{ only if } A\ F\text{-permits } i \text{ at stage } s+1.$$

Since requirement $\mathbf{R}_{e,i}$ uses F_i and does not attempt to enumerate $d_{0,s}, \ldots, d_{i-1,s}$ or the negations of these into T, this permitting is appropriate.

The strategy for meeting a single $\mathbf{R}_{e,i}$ in isolation is the natural one. That is, $\mathbf{R}_{e,i}$ waits for a stage $s+1$ such that $\mathbf{R}_e$ requires attention at stage $s+1$ via y in the sense of the above construction. At such a stage, we cause $V_{e,s+1} \cap F_i \neq A_{s+1} \cap F_i$ and enumerate $y \in Q_{e,s}$. At a later stage $t+1$ such that A F-permits i at $t+1$, we enumerate $\neg \epsilon_j d_{j,s}$ into T for every $j \geq i$ such that $\epsilon_j d_{j,s}$ occurs in y. Were there no other requirements, $\mathbf{R}_{e,i}$ would be satisfied since $|F_i| > 2^{2i}$ by exactly the proofs of Lemmas 4.10 and 4.11.

The strategy for $\mathbf{R}_{e,i}$ conflicts with that for $\mathbf{R}_{e,j}$, $j \neq i$, and for other requirements $\mathbf{R}_f$. It may be the case, for instance, that by the stage $t+1$ such that A F-permits i at stage $t+1$, $d_{j,s} \neq d_{j,t}$ and indeed that $\epsilon_j d_{j,s} \in T_t$. Then the action specified for $\mathbf{R}_{e,i}$ results in making T_{t+1} inconsistent.

Two devices serve to relieve these conflicts. First, to minimize the interference of $\mathbf{R}_e$ with $\mathbf{R}_f$ for $e < f$, we do the following. We will insure that $\mathbf{R}_e$ acts only finitely often and at each stage $s+1$ such that $\mathbf{R}_e$ acts at stage $s+1$ we will restart $\mathbf{R}_f$. We will assume at stage $s+1$ the literals $d_{j,s}$ that $\mathbf{R}_e$ is concerned with satisfy $j \leq s$. Thus, we specify that after stage $s+1$, we shall only attempt to meet requirements $\mathbf{R}_{f,j}$ for

$j > s$. Since these requirements do not disturb $d_{j,s}$ for $j \leq s$, this insures that $\mathbf{R}_{f,j}$ will not injure whatever action was wanted for $\mathbf{R}_e$ at stage $s+1$. We may also assume for such stages $s+1$, that $V_{f,s+1} = \emptyset$ (by starting a "new" V_f.) In the construction, $m(f,s)$ will denote the least i such that we are attempting to meet $\mathbf{R}_{f,i}$ at stage s.

The device to insure the cooperation of the requirements $\mathbf{R}_{e,i}$ and $\mathbf{R}_{e,j}$ for $j \neq i$ is the following. Requirement $\mathbf{R}_{e,i}$ works under the assumption that for $j < i$ there will be no further permissions for $\mathbf{R}_{e,j}$. By the above, $\mathbf{R}_{e,i}$ may also work under the assumption that $\mathbf{R}_f$ for $f < e$ has ceased acting. Thus it is the assumption of $\mathbf{R}_{e,i}$ at stage s that $d_{0,s}, \ldots d_{i-1,s}$ have attained their final values. Suppose then that at stage $s+1$, $\mathbf{R}_{e,i}$ requires attention via y. Then at stage $s+1$ we activate not only $\mathbf{R}_{e,i}$, but all requirements $\mathbf{R}_{e,j}$ such that $i \leq j \leq s$ (by causing $V_{e,s+1} \cap F_j \neq A_{s+1} \cap F_j$ for all such j). Let $t+1$ be the least stage beyond $s+1$ such that A F-permits some such j at stage $t+1$. Then $\mathbf{R}_{e,j}$ causes us to enumerate $\neg\epsilon_k d_{k,s}$ into T for all $k \geq j$ such that $\epsilon_k d_{k,s}$ occurs in y. Notice first that $d_{k,t} = d_{k,s}$ for all such k because this is the first stage at which such permission occurs. Thus, this action at stage $t+1$ is permissible. Notice also that this action does not injure $\mathbf{R}_{e,i}$ for this action only enumerates in T terms that requirement $\mathbf{R}_{e,i}$ would enumerate in T, given permission. Of course $\mathbf{R}_{e,j}$ must now be allowed to act again after stage $t+1$ (for a different y) since its assumption is that there will be no further permissions for $\mathbf{R}_{e,k}$, $k < j$. However any later action for $\mathbf{R}_{e,j}$ involves only literals $d_{k,t+1}$ for $k \geq j$. By the construction at t, no such literal occurs in y and so such actions do not interfere with $\mathbf{R}_{e,i}$. What this argument shows is that $\mathbf{R}_{e,j}$ does not interfere with $\mathbf{R}_{e,i}$ if $i < j$ and that $\mathbf{R}_{e,i}$ is satisfied (in 2^{2i} attempts) if its hypothesis is correct.

In the construction below, $n(e,s)$ denotes the greatest integer such that $\mathbf{R}_{e,j}$ is waiting for a permission for all j such that $m(e,s) \leq j \leq n(e,s)$. (For convenience, $n(e,s) = m(e,s) - 1$ denotes that there is no such j.) If defined, $z(e,j,s)$ denotes the term which $\mathbf{R}_{e,j}$ wishes to enumerate in T if permitted (either $d_{j,s}$ or $\neg d_{j,s}$).

CONSTRUCTION.

Stage 0
Define $m(e,0) = e$, $n(e,0) = m(e,0) - 1$ and let $z(e,i,0)$ be undefined for all $e, i \in N$.
Stage $s+1$
Requirement $\mathbf{R}_e$ *requires attention* at stage $s+1$ if

(a) A F-permits $n(e,s)$ and $m(e,s) \leq n(e,s)$ or

(b) there is $y \in S_{e,s}$ such that $y \notin (T_s \cup Q_{e,s})^*$ and $0 \notin (T_s \cup S_{e,s})^*$.

Let e be least such that $\mathbf{R}_e$ requires attention at stage $s+1$. For all $f > e$, let $m(f,s+1) = \max\{f, s+1\}$, let $V_{f,s} = \emptyset$, let $n(f,s+1) = m(f,s+1) - 1$, and let $z(f,i,s+1)$ be undefined for all i. For $f < e$ let $m(f,s+1) = m(f,s)$ and let $n(f,s+1) = n(f,s)$.

If $\mathbf{R}_e$ requires attention because of clause (a) above, let i be least such that A F-permits i. For all $j \geq i$ such that $z(e,j,s)$ is defined, enumerate $z(e,j,s)$ into T_{s+1} and let $z(e,j,s+1)$ be undefined. Let $n(e,s+1) = i - 1$. In this case we say that requirements $\mathbf{R}_{e,j}$ for j such that $i \leq j \leq n(e,s)$ *receive permission* at stage $s+1$.

If $\mathbf{R}_e$ requires attention because of (b) above but not (a), let y be least satisfying (b). Enumerate y in $Q_{e,s+1}$. Let y be written as a disjunction

$$y = \bigvee_{i \leq n(e,s)} \epsilon_i d_{i,s} \vee \bigvee_{i > n(e,s)} \epsilon_i d_{i,s} \vee \bigvee_{\epsilon_i p_i \in T_s} \epsilon_i p_i \vee \bigvee_{\neg \epsilon_i p_i \in T_s} \epsilon_i p_i$$

(As we argued in Lemma 4.10, the third term of the disjunction is vacuous.) For each $i > n(e,s)$ such that $\epsilon_i d_{i,s}$ occurs in y, set $z(e,i,s+1) = \neg \epsilon_i d_{i,s}$. Let $n(e,s+1) = s$. For each i such that $n(e,s) < i \leq n(e,s+1)$, enumerate one integer, if necessary, into $V_e \cap F_i$ so as to cause $V_{e,s+1} \cap F_i \neq A_{s+1} \cap F_i$. In this case, we say that requirements $\mathbf{R}_{e,i}$ such that $n(e,s) < i \leq n(e,s+1)$ *receive attention* at stage $s+1$.

If $m(e,s+1)$ is not otherwise specified by this construction, then $m(e,s+1) = m(e,s)$. Similarly, for $n(e,s+1)$ and $z(e,i,s+1)$. This ends the construction.

It suffices to prove the following lemma.

Lemma 4.12 *For every $e \in N$, the requirement $\mathbf{R}_e$ is satisfied and receives attention finitely often.*

Proof. Fix $e \in N$. By induction, let s_0 be the least stage such that for all $f < e$ and $s > s_0$, requirement $\mathbf{R}_f$ does not receive attention at s. Then $m(e,s_0)$ is the final value of $\lambda s m(e,s)$, $n(e,s_0) = m(e,s_0) - 1$, and $V_{e,s_0} = \emptyset$.

By induction on $i \geq m(e,s_0)$, we show that $\mathbf{R}_{e,i}$ receives attention after s_0 fewer than $f(i)$ times and is satisfied. If this is the case, then $\mathbf{R}_e$ is satisfied (via the least $i \geq m(e,s_0)$ such that $V_e \cap F_i = A \cap F_i$) and it is easy to see that this implies that $\mathbf{R}_e$ receives attention only finitely often. Fix $i \geq m(e,s_0)$. We establish the following claim.

Claim Suppose that $t_0 < t_1$ are stages $\geq s_0$ such that no requirement $\mathbf{R}_{e,j}$ for $j < i$ receives attention or permission at any stage s such that $t_0 \leq s \leq t_1$. Then $\mathbf{R}_{e,i}$ receives attention at at most 2^{2i} stages s such that $t_0 \leq s \leq t_1$.

To prove the claim, suppose that $\mathbf{R}_{e,i}$ receives attention at stage $s_1 + 1$ such that $t_0 \leq s_1 + 1 \leq t_1$. Let $s_2 + 1 \leq t_1$ be the least stage, if any, beyond $s_1 + 1$ such that A F-permits i at $s_2 + 1$. Then by induction on s, for all s such that $s_1 + 1 \leq s < s_2 + 1$, we have $n(e,s) \geq i$, $d_{j,s_1} = d_{j,s_2}$ for all $j \leq i$, and $z(e,i,s) = z(e,i,s_1+1)$. Furthermore $z(e,i,s_1+1)$ is enumerated in T at stage s_2+1. By the same argument applied to j such that $i < j \leq n(e,s_1+1)$, we have that each value $z(e,j,s_1+1)$ defined at stage s_1+1 is enumerated in T at or before stage s_2+1. Thus, for the y enumerated in Q_e at stage s_1+1, we have enumerated into T_{s_2+1} all the elements the basic strategy for $\mathbf{R}_e$ would have immediately enumerated into T. Thus, applying the arguments of Lemmas 4.10 and 4.11, we see that requirement $\mathbf{R}_{e,i}$ receives attention at most 2^{2i} times between stages t_0 and t_1.

To see that the claim is enough to prove the lemma, first note that the claim and the choice of f imply that $\mathbf{R}_{e,i}$ receives attention at most $f(i)$ times. Thus, since $|F_i| > f(i)$, an element of $F_i - V_e$ is always available for $\mathbf{R}_{e,i}$ if it requires attention. Therefore, if $V_e \cap F_i = A \cap F_i$ and s is a stage such that no higher priority requirement that $\mathbf{R}_{e,i}$ receives attention after s and such that $V_{e,s} \cap F_i = V_e \cap F_i$ and $A_s \cap F_i = A \cap F_i$, then

$\mathbf{R}_{e,i}$ never receives attention after stage s. This implies both that $\mathbf{R}_e$ is satisfied and that $\mathbf{R}_e$ never requires attention after stage s because of case (b). Thus $\mathbf{R}_e$ receives attention only finitely often. $\square$

(if) Suppose that T is a Martin–Pour-El theory. We construct A array nonrecursive such that $A \leq_{\mathrm{T}} T$. Let a very strong array $\{F_n\}_{n\in N}$ be given. As usual we have the requirements

$$\mathbf{R}_e: \quad (\exists n)[W_e \cap F_n = A \cap F_n].$$

We reserve $F_{\langle e,0\rangle}, F_{\langle e,1\rangle}, \ldots$ for meeting $\mathbf{R}_e$. The proof is quite similar to that of Theorem 4.4, (b) implies (a). As in that theorem, we will first give the construction of $A \leq_{\mathrm{T}} T$ and then show that if the construction fails to meet $\mathbf{R}_e$, then there is an r. e. theory $V \supset T$ which is not principal over T.

As in the proof of the other direction of this theorem, let $d_{0,s} < d_{1,s} < \cdots$ list the p_i in order of increasing i such that neither p_i nor $\neg p_i$ is in T_s. Define a recursive function $g : N^2 \to N$ as follows.

$$\begin{aligned} g(e,0) &= 1 + |F_{\langle e,1\rangle}| \quad \text{for every } e \in N \\ g(e,i) &= g(e,i-1) + 1 + |F_{\langle e,i+1\rangle}| \quad \text{for every } e \in N \text{ and } i > 0. \end{aligned}$$

To insure that $A \leq_{\mathrm{T}} T$, we require that $x \in A_{s+1} - A_s$ and $x \in F_{\langle e,i\rangle}$ implies that $d_{g(e,i),s} \neq d_{g(e,i),s+1}$.

CONSTRUCTION.

Stage $s+1$

For every e and i, if $W_{e,s} \cap F_{\langle e,i\rangle} \neq A_s \cap F_{\langle e,i\rangle}$ and $d_{g(e,i),s} \neq d_{g(e,i),s+1}$, then enumerate all of $W_{e,s+1} \cap F_{\langle e,i\rangle}$ into A at stage $s+1$.

To see that each requirement $\mathbf{R}_e$ is satisfied, suppose that e is a counterexample; i. e., that $W_e \cap F_n \neq A \cap F_n$ for all $n \in N$. We construct an r. e. theory V such that $V \supset T$, V is consistent, but V is not a principal extension of T. To construct V, we shall produce an infinite sequence (not necessarily recursive) z_i of elements such that

(28) $$V = (T \cup \{z_i \mid i \in N\})^*.$$

We will ensure that $z_i \notin (T \cup \{z_j \mid j \neq i\})^*$.

Let $d_n = \lim_s d_{n,s}$ for all n, let $D_0 = \{d_0, \ldots, d_{g(e,0)}\}$, and let $D_i = \{d_{g(e,i-1)+1}, \ldots, d_{g(e,i)}\}$ for all $i \in N$. We will define for each $i \in N$ a nonempty finite set $E_i \subseteq D_i$. Given any finite set $X \subset \{p_i \mid i \in N\}$, let $\hat{X}$ denote the subset of X resulting from removing the literal p_i of greatest index. Then for every $i \in n$, z_i is defined by

(29) $$z_i = \bigvee(\bigcup_{j<i} \hat{E}_j) \vee \bigvee E_i.$$

It is easy to see that if the z_i are defined in this way then V as defined in (28) has the desired properties except possibly for recursive enumerability.

At each stage we will have defined approximations to finitely many of the sets E_i. $E_{i,s}$, if defined, will denote the approximation to E_i at stage s. We will have that $E_{i,s} \subseteq D_{i,s} = \{d_{g(e,i-1)+1,s}, \ldots, d_{g(e,i),s}\}$. The two-state diagram is as follows. Of course

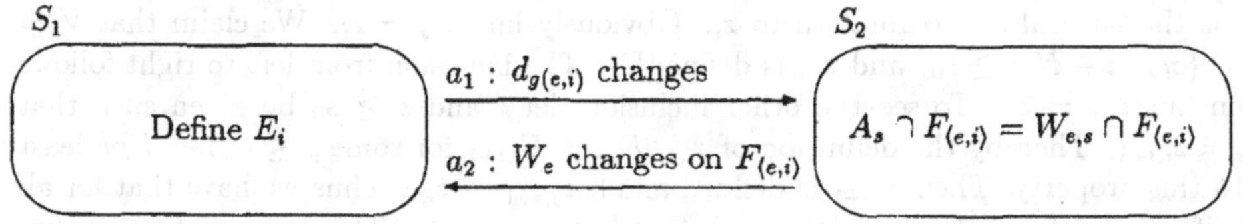

Figure 7: State diagram of the construction.

the i-module pictured in Figure 7 works at a stage s under the assumption that the j-modules for $j < i$ are in state S_1 for every stage after stage s.

The construction of the sets E_i is in stages as follows. First, let s_0 be the least stage such that $D_{0,s_0} = D_0$ and $W_{e,s_0} \cap F_{\langle e,0\rangle} \neq A_{s_0} \cap F_{\langle e,0\rangle}$.

CONSTRUCTION.

Stage s_0
Let $E_{0,s_0} = D_{0,s_0} = D_0$ and E_{i,s_0} be undefined if $i > 0$.
Stage $s+1 > s_0$.
There are three cases.

Case (a): (Arrow a_1.) There exists i such that $E_{i,s}$ is defined and $d_{g(e,i),s} \neq d_{g(e,i),s+1}$. In this case, let i be least with this property. Let $E_{j,s+1}$ be undefined for $j \geq i$. Let $E_{j,s+1} = E_{j,s}$ for $j < i-1$. Let $E_{i-1,s+1} = \hat{E}_{i-1,s}$.

Case (b): (Arrow a_2.) If i is least such that $E_{i,s}$ is not defined, then $W_{e,s+1} \cap F_{\langle e,i\rangle} \neq A_{s+1} \cap F_{\langle e,i\rangle}$. (Note that this case cannot happen if case (a) happens.) In this case define $E_{i,s+1} = D_{i,s+1}$ and $E_{j,s+1} = E_{j,s}$ for all $j < i$.

Case (c): Otherwise, let $E_{i,s+1} = E_{i,s}$ for all i such that $E_{i,s}$ is defined.

The following facts can easily be shown by induction on $s \geq s_0$. For each s there is i such that $E_{j,s}$ is defined for $j \leq i$ and undefined for $j > i$. Further, if $E_{j,s}$ is defined, then $W_{e,s} \cap F_{\langle e,j\rangle} \neq A_s \cap W_{\langle e,j\rangle}$ and $E_{j,s} \subseteq D_{j,s}$. Finally, if $E_{j,s+1}$ and $E_{j,s}$ are both defined then $E_{j,s+1} = E_{j,s}$ unless case (a) applies at stage $s+1$ with $i = j+1$. In this case $E_{j,s+1} \subset E_{j,s}$ and $|E_{j,s} - E_{j,s+1}| = 1$.

Fix i and let s_i be the least stage such that for all $s \geq s_i$, E_{i,s_i} is defined. (This is consistent with our definition of s_0.) At s_i, E_{i,s_i} is defined by case (b) (unless $i = 0$) and $E_{i,s_i} = D_{i,s_i}$. Thus $|E_{i,s_i}| = |F_{\langle e,i+1\rangle}| + 1$. Furthermore, $d_{g(e,i),s_i} = d_{g(e,i)}$ so that $D_{i,s_i} = D_i$. Now for $s \geq s_i$, $E_{i,s+1} \neq E_{i,s}$ only if $E_{i+1,s}$ is defined but becomes undefined at stage $s+1$ because case (a) applies to $i+1$. This can happen at most $|F_{\langle e,i+1\rangle}|$ times since it corresponds to the traversal of arrow a_1 in the $(i+1)$-module. But if this happens at stage $s+1$, we have that $E_{i,s+1} \subset E_{i,s}$ and $|E_{i,s} - E_{i,s+1}| = 1$. Thus we have that $\lim_s E_{i,s} = E_i$ exists and is nonempty. Therefore, if z_i is defined as in (29), we have that V as defined in (28) is a nonprincipal extension of T. We need only show that V is r. e.

We now give an enumeration procedure for V. For any s such that $E_{i,s}$ is defined, let

$$z_{i,s} = \bigvee(\bigcup_{j<i} \hat{E}_{j,s}) \vee \bigvee E_{i,s}.$$

$z_{i,s}$ is the natural approximation to z_i. Obviously $\lim_s z_{i,s} = z_i$. We claim that $V = (T \cup \{z_{i,s} \mid i \in N, s \geq s_0, \text{ and } z_{i,s} \text{ is defined}\})^*$. The inclusion from left to right follows from $\lim_s z_{i,s} = z_i$. To see the other inclusion, let i and $s \geq s_0$ be given such that $z_{i,s} \neq z_{i,s+1}$. Then by the definition of $z_{i,s}$, $E_{j,s} \neq E_{j,s+1}$ for some $j \leq i$. Let j be least with this property. Then $z_{j,s+1}$ is defined and $\vdash z_{j,s+1} \rightarrow z_{i,s}$. Thus we have that for all i, s, if $z_{i,s}$ is defined there is $j \leq i$ such that $\vdash z_{j,s+1} \rightarrow z_{i,s}$. This immediately implies that for every i, s such that $z_{i,s}$ is defined, there is $j \leq i$ such that $\vdash z_j \rightarrow z_{i,s}$. This establishes the desired inclusion. ■

References

[AJSS] K. Ambos-Spies, C. G. Jockusch Jr., R. A. Shore, and R. I. Soare, An algebraic decomposition of the recursively enumerable degrees and the coincidence of several degree classes with the promptly simple degrees, *Trans. Amer. Math. Soc.* **281** (1984) 109-128.

[C] S. B. Cooper, Minimal pairs and high recursively enumerable degrees, J. Symbolic Logic **39** (1974) 655–660.

[D1] R. G. Downey, Maximal theories, *Journal of Pure and Applied Logic* **33** (1987) 245–282.

[D2] R. G. Downey, Array nonrecursive degrees and lattice embeddings of the diamond, to appear.

[DJ] R. Downey and C. G. Jockusch Jr., T-degrees, jump classes and strong reducibilities, *Trans. Amer. Math. Soc.* **301** (1987) 103–136.

[J1] C. G. Jockusch Jr., Semirecursive sets and positive reducibility, *Trans. Amer. Math. Soc.* **131** (1968) 420–436.

[J2] C. G. Jockusch Jr., Degrees in which the recursive sets are uniformly recursive, *Canadian Journal of Mathematics* **24** (1971) 1092–1099.

[JS] C. G. Jockusch Jr. and R. I. Soare, Π^0_1 classes and degrees of theories, *Trans. Amer. Math. Soc.* **173** (1972) 33-56.

[K] A. Kučera, An alternative, priority-free solution to Post's problem, *Proceedings MFCS '86*, Lecture Notes in Computer Science, Springer-Verlag.

[L] A. H. Lachlan, Embedding nondistributive lattices in the recursively enumerable degrees, *Conference in Mathematical Logic — London 1970*, Lecture Notes in Mathematics, vol 255, (1972) Springer-Verlag 149–157.

[MSS] W. Maass, R. A. Shore, and M. Stob, Splitting properties and jump classes, *Israel J. Math.* **39** (1981) 210–224.

[M] D. A. Martin, Classes of recursively enumerable sets and degrees of unsolvability, *Z. Math. Grundlag. Math.* **12** (1966) 295–310.

[MP] D. A. Martin and M. B. Pour-El, Axiomatizable theories with few axiomatizable extensions, *J. Symbolic Logic* **35** (1970) 205–209.

[R] R. W. Robinson, Simplicity of recursively enumerable sets, *J. Symbolic Logic* **32** (1967) 162–172.

[S] R. I. Soare, *Recursively enumerable sets and degrees* (1987) Springer-Verlag.

SUPERBRANCHING DEGREES[1]

Rod Downey
Mathematics Department
Victoria University
PO Box 600
Wellington
NEW ZEALAND

Joe Mourad
Mathematics Department
National University of Singapore
Kent Ridge, 0511
SINGAPORE

§1 INTRODUCTION

All sets, degrees, etc will be r.e. unless explicitly stated otherwise. A degree $\mathbf{a} \neq \mathbf{0}'$ is called *branching* if there exist $\mathbf{c}|\mathbf{b}$ with $\mathbf{c} \cap \mathbf{b} = \mathbf{a}$. By constructing a minimal pair (i.e. $\mathbf{c}|\mathbf{d}$ with $\mathbf{c} \cap \mathbf{d} = \mathbf{0}$), Lachlan [7] and Yates [13] showed that branching degrees exist, thereby giving a negative solution to Shoenfield's conjecture [10]. The technique introduced is now called the minimal pair method and relied on the conscious use of nested strategies. It can be viewed to provide the conceptual framework that lead to the tree method and ultimately to the much of our understanding of **R**, the usl of r.e. degrees.

Various other results concerning branching degrees have been proven. Notably Lachlan [7] showed that not all r.e. degrees are branching and later Fejer showed that the nonbranching degrees are dense in **R**([6]) and above every low r.e degree there exists a branching degree ([5]). This second result was improved by Slaman [11] who used a rather difficult $\mathbf{0}'''$ argument to show that the branching degrees are dense. We should point out that these were the first elementary classes of degrees shown to be dense in **R**. Obvious consequences (via Sacks splitting theorem) are that the branching and nonbranching degrees each generate the r.e. degrees under join. Another consequence of the Fejer density theorem (in [4] see Soare [12, chIX, 4.5]) is that the first Centor-Bendixon derivative of **R** has no isolated points.

Interest has begun to focus on fragments of **R**, and, in particular, upon intervals in **R**. A nice open question here is whether there exist $\mathbf{a} < \mathbf{b}$ with the first order theory of $[\mathbf{a}, \mathbf{b}]$

1. This research partially supported by a Victoria University IGC grant enabling Downey to visit NUS.

decidable. a natural approach to such questions is to try to transport the machinery used to analyse **R** globally into various intervals. As **0** is branching and there are nonbranching degrees already we know that various intervals must be different.

We approached the above with the question of what sorts of constructions can be carried out relative to a given r.e. degree. A good and natural candidate was Lachlan's nonbounding construction [8] where Lachlan showed the existence of an r.e. degree $\mathbf{a} \neq \mathbf{0}$ such that for all $\mathbf{c}, \mathbf{d}$ with $\mathbf{0} < \mathbf{c}, \mathbf{d} < \mathbf{a}$, $\mathbf{c} \cap \mathbf{d} \neq \mathbf{0}'$. In effect, permitting and the minimal pair method are incompatible. The more general question was whether if $\mathbf{a} < \mathbf{0}'$ there exists $\mathbf{b} > \mathbf{a}$ such that for all $\mathbf{c}, \mathbf{d}$ if $\mathbf{a} < \mathbf{c}, \mathbf{d} < \mathbf{b}$ then $\mathbf{c} \cap \mathbf{d} \neq \mathbf{a}$.

In this paper we shall answer this question negatively by introducing a new class of r.e. degrees which we call the *superbranching degrees*. We define $\mathbf{a} \neq \mathbf{0}'$ to be superbranching if for all $\mathbf{b} > \mathbf{a}$ there exist $\mathbf{c}$ and $\mathbf{d}$ with $\mathbf{b} > \mathbf{c}, \mathbf{d} > \mathbf{a}$ and $\mathbf{c} \cap \mathbf{d} = \mathbf{a}$. The construction uses a concept we call *streaming* where nodes on a strategy tree process numbers into a well-behaved stream. We have found this viewpoint rather useful in other constructions (eg [1, 3] and believe it is a good way to view many tree arguments. Our argument blends with permitting. In fact, the first author has shown [2] that the superbranching degrees are dense in **R** and hence generate the r.e. degrees under join.

By Lachlan's results we know **0** is branching but not superbranching. We remark that it is possible (but remarkably tedious) to construct a nonzero branching-but-not-superbranching degree, using a variation on Lachlan's construction. It seems conceivable that the branching but not superbranching degrees are dense in **R**.

Notation is standard and follows Soare [12]. All uses, etc are bounded by s and stage s.

§2. We shall construct a low superbranching degree. In fact we will supress the lowness requirements as the argument will be seen to be easily combined with (eg) permitting. We build $A = \bigcup_s A_s$, $C_e = \bigcup_s C_{e,s}$ and $B_e = \bigcup_s B_{e,s}$ to satisfy the requirements that $C_e, B_e \leq_T A \oplus W_e$ and

$$P_{e,i} : (W_e \leq_T A) \vee (\Phi_i (A) \neq C_e)$$

$$\hat{P}_{e,i} : (W_e \leq_T A) \vee (\Phi_i (A) \neq B_e)$$

$$N_{e,i} : \Phi_i (\hat{C}_e) = \Phi_i (\hat{B}_e) = f \text{ total} \Rightarrow f \leq_T A, \text{ where } \hat{C}_e = A \oplus C_e \text{ and } \hat{B}_e = A \oplus B_e.$$

To meet the $N_{e,i}$ we first attempt a minimal pair type strategy. That is, let

$$l(e,i,s) = \max\left\{x : (\forall y < x)\left[\Phi_{i,s}(\hat{C}_{e,s}; y) = \Phi_{i,s}(\hat{B}_{e,s}; y)\right]\right\}$$

$$ml(e,i,s) = \max\{l(e,i,t) : t < s\}.$$

We say a stage s is (e, i) - *expansionary* if l(e,i,s) > ml(e,i,s). In a minimal pair type strategy we only allow one side of a computation to change between (e,i) - expansionary stages. Hence we know that one side always preserves computations we have seen. (We assume the reader familiar with the minimal pair technique.)

Should the minimal pair strategy fail, we will code the atomic fact that "both sides have changed" by enumeration into A along the lines of Lachlan [9]. As we will see this involves 'processing' the various follows of $P_{f,j}$, $\hat{P}_{f,j}$ by $N_{e,i}$ so that such a coding procedure coheres with the construction.

The basic idea used to meet the $P_{e,i}$ (and dually $\hat{P}_{e,i}$) is to define a *stream* of followers x(e,i,0,s), x(e,i,l,s), ... as follows. Pick x(e,i,0,s). Wait till L(e,i,s) > x(e,i,0,s) where $L(e,i,s) = \max\left\{x : (\forall y < x)\left[\Phi_{i,s}(A_s : y) = C_{e,s}(y)\right]\right\}$. We then appoint a follower x(e,i,l,s) so that $x(e,i,l,s) > u(\Phi_{i,s}(A_s; x(e,i,0,s)))$. Continue this in the obvious way so that (for a single requirement $P_{e,i}$ alone) we have a stream x(e,i,0,s), x(e,i,l,s), ..., x(e,i,n,s) with $x(e,i,j+1,s) > u(\Phi_{i,s}(A_s; x(e,i,k,s)))$ for $k \leq j$.

We then wait till we see W_e permit some j and then act by enumerating x(e,i,j,s) into C_e and restrain A on $u(\Phi_{i,s}(A_s; x(e,i,j,s)))$. Note that for a *single* $P_{e,i}$ by the above Friedberg action, W_e must be recursive: To compute $W_e[j]$ simply wait until x(e,i,j,s) is defined.

In the context of more than one $P_{e,i}$ ($\hat{P}_{e,i}$), the fact that we cannot know if (eg) W_e is recursive causes some complications. The most naive approach would be to allow $P_{e,i}$ to initialise all lower priority $P_{f,j}$ whenever it acts. In that way the (e,i) - stream would be a permanent set-up and would be sure of winning $P_{e,i}$. However, if $W_e = \emptyset$ (say) such a procedure would cause us to initialise $P_{f,j}$ cofinally in the construction.

Our solution to this dilemma is to allow $P_{f,j}$ of lower priority to upset some of the x(e,i,k,s) - set ups by enumerating these numbers into A, whilst we await W_e-permissions. We do this by asking that $P_{f,j}$ ($\hat{P}_{f,j}$) of lower priority than $P_{e,i}$ choose as their followers only numbers from $P_{e,i}$'s stream, so that they will only "refine" $P_{e,i}$'s stream. That is, if $P_{f,j}$ is a version guessing that $P_{e,i}$ has eventually infinitely many followers, $P_{f,j}$ will only select followers of $P_{e,i}$ as its followers.

The reader should note that there is no real conflict between $P_{e,i}$ and $P_{f,j}$ here since one targets such x for C_e and the other for C_f. The problems will occur due to the interaction of $P_{e,i}$, $P_{f,j}$ and $N_{f,k}$.

Consider $P_{f,j}$ and $\hat{P}_{f,g}$ cohereing with $N_{f,k}$ which is infinitely active. What $N_{f,k}$ does is to process numbers in the same way as $P_{e,i}$ above. It will be given a stream y_0, y_1 ..., and will process them into $z_{0,s}$, $z_{1,s}$, ..., so that at any stage s,

$$z_{i+1,s} > \max\left\{u(\Phi_{k,s}(\hat{C}_{f,s};\ z_{i,s})),\ u(\Phi_{k,s}(\hat{B}_{f,s}\ ;\ z_{i,s}))\right\}.$$

It will then hand these numbers to lower priority nodes to use. In particular let us suppose that $\hat{P}_{f,g}$ has lower priority then $P_{f,j}$ but is guessing $P_{f,j}$ is finitely active. Then this version of $\hat{P}_{f,g}$ will be initialised each time $P_{f,j}$ acts, but at any stage we may have refined the z_i - stream to look like

$$x(f,j,0,s),\ x(f,j,1,s),\ \ldots\ x(f,j,n,s),\ \hat{x}(f,g,0,s),\ \ldots,\ \hat{x}(f,g,m,s),$$

where x ($\hat{x}$) follows $P_{f,j}$ ($\hat{P}_{f,j}$). If W_f permits i < n then $P_{f,j}$ will act and initialise all the $\hat{x}$. However if $\hat{P}_{f,g}$ acts first, we will enumerate some $\hat{x}(f,g,t,s)$ into B_f. Before the next (f,k) - expansionary stage it may be that W_f permits i < n and so $P_{f,j}$ wishes to act to enumerate x(f,j,i,s) into C_f. This act will cause a conflict with $N_{f,k}$ which is resolved by enumerating $\hat{x} = \hat{x}(f,g,t,s)$ into A. The conflict is that both sides have changed at $\hat{x}$. By the order of events, since $N_{f,k}$ has processed the z_i, we know that this makes $N_{f,k}$ happy and also $\hat{x} > u(\Phi_{j,s}(C_{f,s}, x(f,j,i,s)))$ and so $P_{f,j}$ is also happy.

If additionally $N_{f,k}$ was guessing infinitary $P_{e,i}$ activity, we know that the z_i - stream is a refinement of the (e,i) - stream. The process above will injure the (e,i) - stream since for some q, $x(e,i,q,s) = \hat{x}(f,j,t,s)$. Thus we will need to pick a new x(e,i,q,s+1). The whole point is that if $x(e,i,q,s) \neq x(e,i,q,s+1)$, then $x(e,i,q,s) \in A$. Hence A can decide if x(e,i,q,s) is permanent. Care is taken to ensure that $\lim_s x(e,i,q,s) = x(e,i,q)$ for all q (should $P_{e,i}$ be infinitely active) but, once this is done we see that $W_e \leq_T A$.

<u>**Formal Details Now Follow.**</u>

- First we will always *dump*. Thus if $x \in A_{s+1} - A_s$ then

$(\forall z)(x \leq z \leq s \rightarrow z \in A_{s+1} - A_s)$.

2.1 The Priority Tree Let T be the tree of all $\sigma \in \{f, 0, 1\}^{<\omega}$ with $f < 0 < 1$ such that if i is even then $\sigma(i) \in \{0, 1\}$ and if i is odd then $\sigma(i) \in \{f, 0, 1\}$. We refer to $\sigma \in T$ as *guesses*, with $lh(\sigma)$ denoting the length of σ. Let $\sigma \leq_L \tau$ denote lexicographical ordering. Assign priorities as follows.

<u>$lh(\sigma) = 1$</u> Let $e(\sigma) = i(\sigma) = 0$. Let

$$L_1(\sigma\hat{\ }j) = \begin{cases} \omega - \{\langle 0, 0\rangle\} & \text{if } j = 1 \text{ or } j = f, \\ \omega - \{\langle 0, k\rangle : k \in \omega\} & \text{if } j = 0. \end{cases}$$

$$L_3(\sigma\hat{\ }j) = L_2(\sigma\hat{\ }j) = \begin{cases} \omega & \text{if } j = 1 \text{ or } j = f, \\ \omega - \{\langle 0, k\rangle ; k \in \omega\} & \text{if } j = 0. \end{cases}$$

<u>$lh(\sigma) > 1$ and $lh(\sigma)$ odd</u>

Case 1 $lh(\sigma) = 4n + 1$ for some n.

Let $\langle e(\sigma), i(\sigma)\rangle = \mu z(z \in L_1(\sigma))$. Define

$$L_1(\sigma\hat{\ }m) = \begin{cases} L_1(\sigma) - \{\langle e(\sigma), i(\sigma)\rangle\} & \text{if } m \neq 0, \\ L_1(\sigma) - \{\langle e(\sigma), k\rangle : k \in \omega\} & \text{if } m = 0. \end{cases}$$

For p = 2,3, define

$$L_p(\sigma\hat{\ }m) = \begin{cases} L_p(\sigma) & \text{if } m \neq 0, \\ L_p(\sigma) - \{\langle e(\sigma), k\rangle : k \in \omega\} & \text{if } m = 0. \end{cases}$$

Case 2 $lh(\sigma) = 4n+3$ for some n.

Let $\langle e(\sigma), i(\sigma)\rangle = \mu z(z \in L_2(\sigma))$. Define

$$L_2(\sigma\hat{\ }m) = \begin{cases} L_2(\sigma) - \{\langle e(\sigma), i(\sigma)\rangle\} & \text{if } m \neq 0 \\ L_2(\sigma) - \{\langle e(\sigma), k\rangle : k \in \omega\} & \text{if } m = 0. \end{cases}$$

For p = 1, 3, define

$$L_p(\sigma\hat{}m) = \begin{cases} L_p(\sigma) \text{ if } m \neq 0 \\ L_p(\sigma) - \{\langle e(\sigma), k\rangle : k \in \omega\} \text{ if } m = 0. \end{cases}$$

<u>lh(σ) even</u> If $lh(\sigma) = 0$ let $e(\sigma) = i(\sigma) = 0$. Otherwise let $\langle e(\sigma), i(\sigma)\rangle = \mu z\,[z \in L_3(\sigma)]$. Set $L_3(\sigma\hat{}j) = L_3(\sigma) - \{\langle e(\sigma), i(\sigma)\rangle\}$ and $L_p(\sigma\hat{}j) = L_p(\sigma)$ for $p \in \{1, 2\}$. This concludes the priority assignment.

2.2 Definition Let $\alpha \in T$.

i) We say a stage s is an *α-stage* if s=0 or α=σ(t,s) at some substage t of stage s. (σ(t,s) is defined later)

ii) If $lh(\alpha) = 2e$, we say a stage q is *α-expansionary* if q=0 or q is an α-stage and (a) and (b) below hold

a) $l(e(\alpha), i(\alpha), q) > \max\{l(e(\alpha), i(\alpha), \hat{q}) : \hat{q} \text{ is an } \alpha\text{-stage} < q\}$.

b) If there exist followers with guesses $\rho \supseteq \alpha\hat{}0$ not yet α-confirmed and $x(\rho, k, q)$ is the least such, then $l(e(\alpha), i(\alpha), q) > x(\rho, k, q)$.

iii) If $lh(\alpha) \equiv 1 \pmod 4$ and $e = e(\alpha)$, $i = i(\alpha)$, we say that q is α–expansionary if q=0 or q is an α-stage where

a) $L(e,i,q) > \max\{L(e,i,t) : t \text{ is an } \alpha\text{-stage} < q\}$

b) $L(e,i,q) > \max\left\{\begin{array}{l} x(\alpha,k,\hat{q}) : x(\alpha,k,\hat{q}) \text{ is a follower of } \alpha \text{ at any} \\ \text{stage } \hat{q} < q \end{array}\right\}$

c) If there exists $\gamma\hat{}0 \subseteq \alpha$ with $lh(\gamma)$ odd, let ρ denote the longest such γ. Then $x(\rho,g,q)$ is defined for $g=r(\alpha,q)$ and there exists a follower $x(\rho,k,q)$ with $k > g$ such that.

ci) $x(\rho,k,q)$ has never been assigned to α

cii) $x(\rho,k,q) > u(\Phi_{i,s}(A_s ; \hat{x}))$ where

$$\hat{x} = \max\left\{\begin{array}{l} x(\alpha,j,q),\ r(\alpha,q) : x(\alpha,j,\hat{q}) \text{ is the largest} \\ \text{follower of } \alpha \text{ at any } \hat{q} < q \end{array}\right\}$$

ciii) For all $\gamma\hat{}0$ with $lh(\gamma)$ even and $\rho \subseteq \gamma$ if ρ exists, and for all j with $x(\alpha,j,q)$ existing, if $m = m(\gamma, x(\alpha,j,q), q)$ is defined then $x(\rho,k,q) > m$. (r and m are defined in the construction)

Remark

The point of (b) and (c) of (iii) above is to guarantee that at α^0-stages we will get a new follower to add to the $P_{e,i}$-stream at guess α. The point of (ciii) is to ensure that $x(\rho,k,q)$ respects γ^0 in the sense that $x(\rho,k,q)$ can be used to trace $x(\alpha,j,q)$ for the sake of γ and yet not interfere with $x(\alpha,j,q)$'s γ^0-computations. As we will see this is achieved by our confirmation machinery. Note that (ii)(b) ensures that we don't believe a stage is α-expansionary until we can confirm some follower, should there exist some not yet confirmed follower.

iv) If $lh(\alpha) \equiv 3 \pmod 4$ we proceed as in (iii) above but with $\hat{L}$ in place of L and $\hat{P}$ in place of P (where $\hat{L}$ is used for the length of agreement between A and B.)

v) If $lh(\alpha)$ is odd, we say that α *requires attention at stage q via* $x(\alpha, k+1,q)$ if

a) q is an α-stage

b) If $\hat{q}$ is the stage where $x(\alpha, k+1, q) = x(\alpha, k+1, \hat{q})$ was appointed to α then

$$W_{e,q}[x(\alpha,k,q)] \neq W_{e,q}[x(\alpha,k,\hat{q})]$$

(The reader should note the use of k in place of k+1 here)

c) $L(e(\alpha), i(\alpha), q) > \max\{x, L(e(\alpha), i(\alpha), \hat{q}) : \hat{q} \text{ is an } \alpha\text{-stage} < q\}$.

Remark

There is a lot of room for variation here. For instance, we could simply wait for *any* stage where $x(\alpha,k,q)$ was still alive and we see $x(\alpha,k,q)$ permitted by W_e. Our version is more useful in the first author's proof of the density of superbranching degrees. The trade-off is that we need to work a little harder to get $C_e, B_e \leq_T W_e \oplus A$.

In the construction to follow we use the phrase "initialise α". As usual, this means reset $F(\alpha,s)$ (the current state of the α-module) to be 1 for $lh(\alpha)$ odd, cancel all followers of α, set $r(\alpha,s) = lh(\alpha) + s \geq \max\{r(\tau,s) : \tau \leq_L \alpha\}$ and cancel all $m(x,\alpha,s)$.

Construction

Stage 0 Set $r(\alpha,0)=0$ all $\alpha \in T$ and $\sigma_0 = \lambda$. (Here λ denotes the empty string).

Stage s+1 Refer to substage t of stage s+1 as stage (t, s+1). A parameter $Q \neq \sigma$ at stage (t, s+1) is denoted by Q_t.

Stage (0, s+1) Define $\sigma(0, s+1) = \lambda$.

Stage (t+1, s+1) We are given $\sigma(t, s+1)$ and for all α with $lh(\alpha)$ odd we are given $F_t(\alpha) = F_t(\alpha, s+1)$. Let $\alpha = \sigma(t, s+1)$. Adopt the first case below to pertain.

Case 1 $lh(\alpha)$ even.

Subcase 1 s+1 is not α-expansionary.

Action Let $\sigma(t+1, s+1) = \alpha$^1. If t=s set $\sigma_{s+1} = \alpha$ ^1 and initialise all τ with $\tau \nleq_L \sigma_{s+1}$. If $t \neq s$ go to stage (t+1, s+1).

Subcase 2 s+1 is α-expansioanry.

Step 1 If there is a least follower $x = x(\rho,k,s)$ with $\rho \supset \alpha$^0 not yet α^0-confirmed set $m(\alpha$^0,x,s+1) = s+1 and cancel all followers $x(\gamma,n,s)$ with $\rho \leq_L \gamma$ and $x(\rho,k,s) < x(\gamma,n,s)$. The reader should carefully note that some of these $x(\gamma,n,s)$ may also be $x(\eta,g,s)$ for some $\eta \leq_L \rho$. **The cancellation procedure will only cancel $x(\gamma,n,s)$ as a follower of γ and NOT of η.** On the other hand once $x = x(\rho,k,s)$ is confirmed by α^0, if x is later assigned to some $\beta \supsetneq \rho$ we regard it as also confirmed as a β-follower. (The reason being that x has done its duty to α^0.) We will do this automatically.

If no such x exists, let q denote the last α^0-stage < s. If no such q exists or α has been initialised at any stage u with $q \leq u \leq s$ go to step 2. Otherwise for each $x(\alpha,k,q) \in A_q - A_{q-1}$ define $m(\alpha, x(\alpha,k,q), s+1) = s+1$.

Step 2 Set $\sigma(t+1, s+1) = \alpha$^0 and go to stage (t+2, s+1) unless t=s. If t=s set $\sigma_{s+1} = \alpha$^0 and initialise all $\tau \nleq_L \sigma_{S+1}$.

Case 2 $lh(\alpha) \equiv 1 \pmod 4$. First reset $r_t(\alpha, s+1) = \max\{r(\alpha,s), lh(\alpha)\}$.

Subcase 1 $F_t(\alpha, s+1) = f$.

Action Define $\sigma(t+1, s+1) = \alpha$^f and go to stage (t+2, s+1).

Subcase 2 $F_t(\alpha, s+1) = 1$ and α requires attention via $x = x(\alpha,k,s)$.

Action Define $F_{t+1}(\alpha, s+1) = F(\alpha, s+1) = \alpha^f= \sigma_{s+1}$, set $A_{s+1} = A_s \cup \{z : x \leq z \leq s\}$ (dumping) set $C_{e(\alpha), s+1} = C_{e(\alpha), s} \cup \{x(\alpha, k-1, s)\}$. Initialise all $\gamma \leq_L \alpha$^f. Cancel all (other) followers of α. (Here of course we mean cancel them *as followers* of α. Perhaps they follow $\delta \subsetneq \alpha$. We would not cancel them as followers of δ.) Set $r(\alpha, s+1) = r(\alpha, s) + s+1$.

Subcase 3 $F_t(\alpha, s+1) = 1$ and α does not require attention yet s+1 is α-expansionary. We know by (2.2)(c) that there is a follower x ready to be assigned to α if there exists $\rho\hat{}0 \subset \alpha$ with $lh(\rho)$ odd. If such a ρ exists and $x(\alpha,0,s)$ is not yet defined set $x(\alpha,0,s) = x$. Otherwise, if ρ exists, let $x(\alpha,j,s)$ denote the largest currently defined member of α's stream. Assign $x(\alpha, j+1, s+1) = x$.

If no such $\rho\hat{}0$ exists assign $x(\alpha, j+1, s+1) = r(\alpha, s) + s+1$ (or $x(\alpha, 0, s+1)$ as the case may be).

Set $\sigma(t+1, s+1) = \alpha\hat{}0$ and go to stage $(t+2, s+1)$.

Subcase 4 Otherwise.

Action Define $\sigma(t+1, s+1) = \alpha\hat{}1$. If t=s set $\sigma_{s+1} = \alpha\hat{}1$ and initialise all $\tau \not\leq_L \sigma_{s+1}$. If $t \neq s$ go to stage $(t+2, s+1)$.

Case 3 $lh(\alpha) \equiv 3 \pmod 4$.

Action Proceed as in Case 2 with B_e in place of C_e.

End of construction

__**Verification**__

Let β denote the left most path.

__**Lemma**__ *Let* $\alpha \subseteq \beta$. *Then*

(i) *α receives attention only finitely often,*

(ii) *for all* $\gamma \leq_L \alpha$, $lim_s\, r(\gamma,s) = r(\gamma)$ *exists,*

(iii) $\left|\{s : \sigma_s \leq_L \alpha\}\right| < \infty$, *and*

(iv) *If* $lh(\gamma)$ *odd and* $\gamma\hat{}0 \subseteq \alpha$ *then* $lim_s\, x(\gamma,k,s) = x(\gamma,k)$ *exists.*

__**Proof**__

Assume (i), (ii) and (iii) for all $\hat{\alpha} \subsetneq \alpha$. Let $\alpha = \alpha^{+\hat{}j}$. Let s_0 be a stage where, for all $s \geq s_0$ and $\gamma \leq_L \alpha^+$

a) $r(\gamma,s) = r(\gamma)$

b) $\hat{\alpha} \leq_L \alpha \rightarrow \hat{\alpha}$ does not receive attention at s, and

c) $\sigma_s \nleq_L \alpha^+$.

Note that (ii) $\Rightarrow$ (iii) since $\alpha \subset \beta$ and after stage s_0, $\sigma \subseteq \alpha$ only if α receives attention (via some $x=x(\alpha, k+1, s)$). If such a stage occurs we would enumerate $x(\alpha, k-1, s)$ into C_e or B_e (as the case may be) and initialise all $\gamma \nleq_L \alpha$. Choice of s_0 ensure that all numbers to enter A after s must exceed s and so, in particular, exceed x. A number z can enter

A only if some $x(\eta, m, s) \leq z$ simultaneously enters A and η receives attention. Such η can only have $\alpha \leq_L \eta$ after stage s. Thus to see that α never again receives attention it suffices to show that

(2.3) $x(\alpha, k+1, s) > u(\Phi_{i(\alpha), s}(A_s ; x(\alpha, k, s)))$.

If (2.3) holds then $P_{e(\alpha), i(\alpha)}$ (or $\hat{P}_{e(\alpha), i(\alpha)}$ as the case may be) is met since by clause (c) of (2.2) we only attack α at $(e(\alpha), i(\alpha))$ - expansionary stages.

We claim that (2.3) holds as x is still alive at stage s. To see this let s_1 be the stage where $x = x(\alpha, k+1, s) = x(\alpha, k+1, s_1)$ was appointed to α. By initialisation, $s_1 \geq s_0$. We know s_1 is α-expansionary and $L(e(\alpha), i(\alpha), s_1) > x(\alpha, k, s)$. By the way we appoint, it follows that (2.3) holds at $s=s_1$. Because we dump since $x(\alpha, k+1, s) \notin A_s - A_{s1}$ no number $< x$ can have entered $A_s - A_{s1}$. Hence at s the relevant computations in (2.3) are the same since $A_s[x] = A_{s1}[x]$. Thus (2.3) holds. Thus (i), (ii) and (iii) hold for α.

Finally, to see that (iv) holds note that if $\gamma \nleq_L \beta$ then $r(\gamma,s) \rightarrow \infty$ monotonically. We claim that for all $k \leq r(\alpha)$ and all $\gamma \subset \alpha$ with $\alpha\hat{}0 \subset \beta$ and $lh(\alpha)$ odd $\lim_s x(\gamma,k,s) = x(\gamma,k)$ exists. (This will suffice to give (iv).)

For an induction, additionally assume that this is true at stage s_0. That is $x(\gamma,k,s) = x(\gamma,k,s_0)$ for all $k \leq r(\hat{\gamma})$ and $s \geq s_0$ where $\hat{\gamma}$ is the longest γ with $\gamma\hat{}0 \subseteq \gamma$. Let $s_1 > s_0$ be the stage where α ceases receiving attention. Then $r(\alpha, s_1) = r(\alpha)$. If $\gamma\hat{}0 \subset \beta$ with $\gamma \subset \alpha$ then there exist infinitely many γ-expansionary stages. Choice of s_1 ensures that no follower $x < r(\gamma)$ can be cancelled except by confirmation. As any x is only confirmed finitely often we see that $x(\gamma,k,s)$ will eventually all exist in a confirmed state for all $k \leq r(\gamma)$. These are now never enumerated or cancelled, giving the result.

Lemma 2 $C_e \leq_T W_e \oplus A$, $B_e \leq_T W_e \oplus A$.

Proof

Let x be given. To decide if $x \in C_e$ or not compute a stage s where

$W_{e,s}[x] = W_e[x]$. If x does not yet follow some $P_{e,i}$ then $x \in C_e$ iff $x \in C_{e,s}$. If x follows some $P_{e,i}$ then $x = x(\alpha,k,s)$ for some k. If $x(\alpha,k+1,s)$ is currently undefined or $x(\alpha,k+1,s) \notin A$ then $x \notin C_e$. If $x(\alpha,k+1,s) \in A$ compute u where $x(\alpha,k+1,s) \in A_u$. Then $x \in C_e$ iff $x \in C_{e,u}$. Hence $C_e \leq_T W_e \oplus A$ and $B_e \leq W_e \oplus A$ is the same.

Lemma 3 $(\forall e)(P_e)$ *and* $(\forall e)(\hat{P}_e)$.

Proof

(For P_e) Suppose that for some i, $P_{e,i}$ fails to be met by a disagreement. Then α^0 $\subset \beta$ with $e = e(\alpha)$ and $i = i(\alpha)$ for some α. By lemma 1, there are eventually infinitely many stable followers $x(\alpha,k,s) = x(\alpha,k)$ following α with $x(\alpha,k) \notin A$. Once $x(\alpha,k,s)$ is confirmed, $x(\alpha,k,s+1) \neq x(\alpha,k,s)$ iff $x(\alpha,k,s) \in A$. Thus, as we outlined in the introduction, to compute $W_e[k]$ simply find a stage s($> s_0$ of lemma 1) where $x = x(\alpha,k+1,s)$ is defined, confirmed and $x \notin A$. Then $W_{e,s}[k] = W_e[k]$ and so $W_e \leq_T A$.

Lemma 4 $(\forall e,i)(N_{e,i})$

Proof

Let α^0 $\subset \beta$ with lh(α) even and e=e(α), i=i(α). Let $\Phi_i(\hat{B}_e) = \Phi_i(\hat{C}_e) = f$. Let z be given. Let s_0 be a stage good for α in the sense of Lemma 1.

To compute f(z) find an α^0-stage $s > s_0$ where $l(e,i,s) > z$. Now compute an a α^0-stage $\hat{s} > s$ with $l(e,i,\hat{s}) > s$ and $A_{\hat{s}}[s] = A[s]$. We claim that $f_{\hat{s}}(z) = f(z)$. Indeed, we claim that for all stages $u > \hat{s}$.

$$\text{(2.4)} \quad \begin{cases} \text{one of } \Phi_{i,u}(\hat{B}_{e,u};\ z) = \Phi_{i,\hat{s}}(\hat{B}_{e,\hat{s}};\ z) \text{ or} \\ \Phi_{i,u}(\hat{C}_{e,u};\ z) = \Phi_{i,\hat{s}}\ \hat{C}_{e,\hat{s}};\ z) \text{ holds.} \end{cases}$$

If (2.4) is to fail, there must exist α^0-stages $s_1 > s_2 \geq \hat{s}$ with s_2 the preceding α – expansionary stage before s_1 such that two numbers y_1 and y_2 entered respectively the $\hat{C}_e$ and $\hat{B}_e$ sides below the z-uses at s_2. We argue that this cannot happen.

We know that the construction ensures that we can take y_1 and y_2 as followers. By the way we appoint (and cancel) at α^0-stages y_1 and y_2 must have both been present at stage $\hat{s}$. This means that if y_1 enters first (wlog) then y_2 must have higher priority than y_1 (let it be cancelled). By (c)(iii) of (2.2) we know that y_1 must respect y_2's confirmations. That is, when y_1 was assigned

$$s > y_1 > m(\alpha\hat{\ }0, y_2, s).$$

In particular, $y_1 > u(\Phi_{i,p}(\hat{B}_{e,p}; y_2)), u(\Phi_{i,p}(\hat{B}_{e,p}; y_2))$ where p is the stage that y_1 was assigned to its node. By the dump in the construction it follows that the $(\Phi_{i,p}; y_2)$-computations are unchanged at stage s and indeed at stage s_2. But now we are done. When y_2 enters y_1 is automatically put into A causing $A_{\hat{s}}[s] \neq A[s]$ as $s_2 > s$. ⊗

REFERENCES

1. Downey, R.G., A contiguous nonbranching degree, Z. Math. Logik. Grund. Math. (to appear).

2. Downey, R.G., Notes on the **0'''** priority method with special attention to density arguments, these proceedings.

3. Downey, R.G and T. Slaman, Completely mitotic r.e degrees, Annals Pure and Applied Logic *41* (1989) 119-152.

4. Fejer, P., *On the Structure of Definable Subclasses of the Recursively Enumerable Degrees,* Ph.D. Diss., University of Chicago, 1980.

5. Fejer, P., Branching degrees above low degrees, Trans. Amer. Math. Soc. *273* (1982) 157-180.

6. Fejer, P., The density of the nonbranching degrees, Ann. Pure and Applied Logic *24* (1983) 113-130.

7. Lachlan, A.H., Lower bounds for pairs of recursively enumerable degrees, Proc. London Math. Soc. *16* (1966) 537-569.

8. Lachlan, A.H., Bounding minimal pairs, J. Symb. Logic *44* (1979) 626-642.

9. Lachlan, A.H., Decomposition of recursively enumerable degrees, Proc. Amer. Math. Soc. *79* (1980) 629-634.

10. Shoenfield, J., Applications of model theory to degrees of unsolvability, in *Symposium on the Theory of Models* (ed. Addison, Henkin, and Tarski), North-Holland Amsterdam (1966) 359-363.

11. Slaman, T., *The recursively Enumerable Branching Degrees are Dense in the Recursively Enumerable Degrees,* handwritten notes, Univ. of Chicago (1980).

12. Soare, R.I., *Recursively Enumerable Sets and Degrees,* Springer-Verlag, New York 1987.

13. Yates, C.E.M., A minimal pair of recursively enumerable degrees, J.Symbolic Logic, *31* (1966) 159-168.

A Direct Construction of A Minimal Recursively Enumerable Truth-Table Degree

Peter A. Fejer
Department of Mathematics and Computer Science
University of Massachusetts at Boston
Boston, MA 02125
U.S.A.

Richard A. Shore*
Department of Mathematics
Cornell University
Ithaca, NY 14853
U.S.A.
and
MSRI
1000 Centennial Drive
Berkeley, CA 94720
U.S.A.

1 Introduction

Given a reducibility r between sets of natural numbers, one can form the structures $\mathcal{D}_r$ of all the r-degrees, $\mathcal{D}_r(\leq 0')$ of the r-degrees less than or equal to the r-degree of $\emptyset'$, and $\mathcal{R}_r$ of the recursively enumerable (r.e.) r-degrees. An obvious question about any of these structures is whether they contain minimal degrees. Besides being of intrinsic interest, a positive answer to this question is usually a first step towards showing the undecidability of the structure.

Lachlan has shown for both one-one (1) reducibility ([6]) and many-one (m) reducibility ([7]) that there is an r.e. degree which is minimal among the r.e. degrees, hence among all the degrees, since for these reducibilities the r.e. degrees are closed downwards. (See Rogers [14] and Odifreddi [11] and [12] for the definitions of and more information on reducibilities stronger than Turing reducibility.) Thus for these two reducibilities, there are minimal degrees in all three of the structures mentioned above.

For Turing (T) reducibility, one of the earliest results is Spector's construction in [19] of a minimal degree. Since this construction uses total trees, it is not hard to see that it

*Research partially supported by NSF grants DMS-8601048 and DMS-8912797.

produces a set whose degrees with respect to both weak truth table (*wtt*) and truth table (*tt*) reducibility are also minimal. However, the minimal degree produced is not below $\mathbf{0}'$, so Spector's result does not produce minimal degrees in any of the structures $\mathcal{D}_r(\leq \mathbf{0}')$ or $\mathcal{R}_r$. (Note that by the density of the r.e. Turing degrees ([16]) and r.e. *wtt*-degrees ([8]), there are no minimal degrees in $\mathcal{R}_T$ and $\mathcal{R}_{wtt}$.) In [15], Sacks constructed a minimal degree in $\mathcal{D}_T(\leq \mathbf{0}')$ and later Shoenfield ([17]) gave a simpler oracle construction of this type of degree. Both of these constructions use partial trees and hence do not produce minimal degrees below $\mathbf{0}'$ in any of the reducibilities stronger than Turing reducibility.

In 1973, Degtev [1] showed that there is a minimal degree in $\mathcal{R}_{tt}$ and in 1975 Marchenkov [9] showed that there is an r.e. degree in $\mathcal{D}_{tt}(\leq \mathbf{0}')$ which is minimal. Both of these results were indirect in that they have the flavor of the 1978 result of Kobzev [5] who showed that the degree of a nonrecursive η-maximal semirecursive set (a type of set shown to exist by Degtev [1]) is minimal among all *tt*-degrees. In this paper we give a direct construction of a minimal r.e. *tt*-degree. The obvious next step after giving such a direct construction would be to extend it to embed other lattices as initial segments. If a large enough class of lattices (e.g., the finite distributive lattices) can be so embedded, then the theory of the r.e. *tt*-degrees is undecidable. We tried to embed the finite distributive lattices as initial segments using our methods, but found that we could not even embed the four element Boolean algebra. However, an extension of our construction has been used by Haught and Shore [3] to show the undecidability of the theory of the r.e. *tt*-degrees. They showed that for every finite partition lattice $\mathcal{L}_0$, if $\mathcal{L}$ is the lattice obtained from $\mathcal{L}_0$ by taking the dual and adding an extra element below all the others, then $\mathcal{L}$ is embeddable in $\mathcal{R}_{tt}$. The reason why we could not embed the four element Boolean algebra as an initial segment, and the reason why Haught and Shore had to use such unusual lattices in their proof, has been made clear by Harrington and Haught. They show in [2] that in every nontrivial finite initial segment of $\mathcal{R}_{tt}$, the elements other than $\mathbf{0}$ have a least element. Exactly which finite lattices can be obtained as initial segments of $\mathcal{R}_{tt}$ is an interesting open question.

Another interesting open question (Problem 16 in [11]) is whether $\mathcal{R}_{tt}$ and $\mathcal{D}_{tt}(\leq \mathbf{0}')$ are elementarily equivalent. In [10], Marchenkov shows that there is an incomplete r.e. *tt*-degree which bounds all minimal r.e. *tt*-degrees. One might hope that our construction could be used to show that the minimal *tt*-degrees below $\mathbf{0}'$ are not bounded away from $\mathbf{0}'$ and thus obtain an elementary difference in the two structures. However, Shore has remarked that Marchenkov's proof can be modified to show that there is an incomplete *tt*-degree below $\mathbf{0}'$ which bounds all minimal *tt*-degrees below $\mathbf{0}'$, so some other approach will be needed to show that the two structures are not elementarily equivalent.

In our construction of a recursively enumerable set A whose *tt*-degree is minimal, if we put a number x into A at stage s, then we also put into A at stage s all numbers y with $x \leq y \leq s$. This ensures that the set we construct is semirecursive. We do not know if our set is η-maximal or if it can be made so.

Our construction can be modified to show that there is a minimal degree in $\mathcal{D}_{wtt}(\leq \mathbf{0}')$. In [4], Haught and Shore extend the construction of a minimal *wtt*-degree below $\mathbf{0}'$ (in ways analogous to those in [3]) to show that the theory of $\mathcal{D}_{wtt}(\leq \mathbf{0}')$ is undecidable.

To summarize then, for $r = 1, m$, and *tt*, there are r.e. r-degrees which are minimal among all the r-degrees. For $r = wtt$ and T, there are no minimal r.e. degrees, but there are minimal r-degrees below $\mathbf{0}'$. In each case, the minimal degree construction is a first step towards showing undecidability.

We complete the introduction by giving the definitions and notations we will use.

Our notation is for the most part standard. (See, e.g., [18].) By "set" we mean set of natural numbers and by "number" we mean natural number. We identify a set with its characteristic function. A *string* is an element of $2^{<\omega}$. If $\sigma \in 2^n$, then the *length* of σ, denoted $|\sigma|$, is n. For two strings σ and τ, $\sigma \subseteq \tau$ means that σ is an initial segment of τ while $\sigma \preceq \tau$ means that σ comes before τ in lexicographic ordering where 0 is considered to be less than 1. We say that two strings σ and τ *split* a string ρ if σ and τ are incompatible extensions of ρ. We write $\{e\}$ for the eth Turing reduction.

A *truth table* is a pair $\alpha = (\langle x_0, \ldots, x_{n-1}\rangle, t)$ consisting of a strictly increasing sequence $\langle x_0, \ldots, x_{n-1}\rangle$ of numbers and a function $t : 2^n \rightarrow 2$ for some number n. For a truth table α as in the previous sentence, the *size* of α is n and the *length* of α, denoted $|\alpha|$, is $x_{n-1}+1$ if $n > 0$ and is 0 if $n = 0$. We take $\{\alpha_n\}_{n\in\omega}$ to be an effective enumeration of all truth tables such that $|\alpha_n| \leq n$.

If A is a set and $\alpha = (\langle x_0, \ldots, x_{n-1}\rangle, t)$ is a truth table, then we say that A *satisfies* α if $t(A(x_0), \ldots, A(x_{n-1})) = 1$, in which case we write $\alpha^A = 1$. Otherwise, we write $\alpha^A = 0$ and similarly for finite strings σ with $|\sigma| \geq |\alpha|$.

If A and B are sets, then we say that A is truth table (tt) reducible to B ($A \leq_{tt} B$) if for some recursive function f, for every x, $x \in A \leftrightarrow \alpha^B_{f(x)} = 1$. Then $\equiv_{tt}$ and the notion of truth table (tt)-degree are defined as usual. By a result of Nerode (see [14, p. 143]) $A \leq_{tt} B$ iff $A = \{e\}^B$ for some e such that $\{e\}^X$ is total for every X.

If $\{e\}(x)\downarrow$, we write $[e](x)$ for $\alpha_{\{e\}(x)}$. For a number e and set A, we define a partial function $[e]^A$ by

$$[e]^A(x) = \begin{cases} ([e](x))^A & \text{if } \{e\}(x)\downarrow \\ \uparrow & \text{otherwise.} \end{cases}$$

Then $A \leq_{tt} B$ iff $(\exists e)(A = [e]^B)$. If σ is a string, we define a partial function $[e]^\sigma$ by

$$[e]^\sigma(x) = \begin{cases} ([e](x))^\sigma & \text{if } \{e\}(x)\downarrow \text{ and } |\sigma| \geq |[e](x)| \\ \uparrow & \text{otherwise.} \end{cases}$$

We also define $[e]^A_s$ by

$$[e]^A_s(x) = \begin{cases} ([e](x))^A & \text{if } \{e\}_s(x)\downarrow \\ \uparrow & \text{otherwise.} \end{cases}$$

$[e]^\sigma_s$ is defined similarly.

If σ and τ are strings, then say that σ and τ *e-split* if for $t = \min\{|\sigma|, |\tau|\}$, there is some x such that $[e]^\sigma_t(x)$ and $[e]^\tau_t(x)$ both converge and give different answers. Since we adopt the usual convention that if $\{e\}_s(x)\downarrow$ then $x < s$, given two strings σ and τ, we can effectively tell whether or not they e-split.

2 Construction of a Minimal R.E. *tt*-Degree

We will construct an r.e. set A. In order to make A have minimal tt-degree, we have for all $e \geq 0$ the requirements

$$P_e : A \neq \overline{W_e}$$

and

$$Q_e : \text{ If } [e]^A \text{ is total, then either } [e]^A \text{ is recursive or } A \leq_{tt} [e]^A.$$

If we meet all of these requirements, then A will in fact have minimal tt-degree.

We base our construction on the so-called "full-approximation" (no oracle) construction of a minimal Turing degree below $\mathbf{0}'$ and we assume familiarity with this method. (Posner [13] is an excellent introduction to this area.) Of course there must be some modifications to suit our current situation. First of all, we seek to build an r.e. set, something that does not happen with the usual full approximation construction. Second, we want our set to have minimal *tt*-degree and not just minimal Turing degree. As mentioned in the introduction, constructions using partial trees do not produce *tt*-reductions so we need a device to deal with this problem as well.

When thinking about trees, we envision 0 as being drawn to the left of 1. Since we are building A to be r.e., we only add elements to A, never remove them. Hence A can only move to the right as it is constructed. This means that there is no reason to keep track of any branchings to the left of A's current value. In fact, in our construction, rather than keep track of a full binary tree of branchings, we just keep track of a single infinite sequence of branchings. For each branching, one half of the branching (the "left" half) is contained in the current value of A, and the other half can be obtained from the left half by changing 0's to 1's. (Thus if we wish to move A to the right half of one of the branchings in order to meet one of the P_e requirements, we can do so by only adding elements to A.) Each branching extends the left half of the previous branching.

For a given e, if on the eth tree we are able eventually to find on each level of the tree an e-splitting to put up as the next branching and $[e]^A$ is total, then we need to show that $A \leq_{tt} [e]^A$. When we put up an e-splitting on some level of the eth tree, we have no way of knowing effectively whether diagonalization will ever move A to the right half of the splitting. If we are just trying to get a T-reduction of A to $[e]^A$ this is not a problem since we can ask the $[e]^A$ oracle if A is ever moved to the right side of the splitting, and if it does get moved, we can wait for this to happen. For a *tt*-reduction however, we must give an answer no matter what oracle we are given and hence we cannot have as part of our reduction procedure to wait for something that might never happen. Instead, we think of the right halves of the e-splittings we put up as being extended by 1's. If at stage s we move A to the right side of a splitting, then we also put into A every number from the length of the splitting up to the stage number s. Thus as we go through more stages, we are committing ourselves more and more as to what A will look like if we do eventually move A to the right side of the e-splitting. This allows us to get a *tt*-reduction of A to $[e]^A$. At the same time, when diagonalization occurs, we have committed ourselves to only a finite amount of information about A, so we still have room to work above this commitment to look for more e-splittings.

With this as motivation, we now formally define the kind of trees we will be using.

Let $\Omega = \{\emptyset, 0, 1, 00, 01, 000, 001, \ldots\} = \{\sigma \in 2^{<\omega} : \forall n(n < |\sigma| - 1 \rightarrow \sigma(n) = 0)\}$. A *tree* is a function $T : \Omega \rightarrow 2^{<\omega}$ such that for all k, $T(0^k0)$ and $T(0^k1)$ split $T(0^k)$ and $|T(0^k0)| = |T(0^k1)|$. We write $T(\sigma)(\tau)$ for the pair of strings $T(\sigma), T(\tau)$. $T(0^k0), T(0^k1)$ are called *T-partners*. If T is a tree and $q = T(0^k)$, then the *left successor* of q is $q_0 = T(0^k0)$ and the *right successor* of q is $q_1 = T(0^k1)$. $T(0^k)$ is the *predecessor* of both $T(0^k0)$ and $T(0^k1)$.

If T is a tree, then we adopt the convention that R denotes the range of T. Similarly, R_1 is the range of T_1 and so on. If T is a tree, then we define

$$R^- = \{T(0^k) : k \geq 0\}$$

and

$$R^* = R^- \cup \{\sigma \in 2^{<\omega} : \exists k, k', s[|\sigma| = |T(0^k)| \wedge \sigma = T(0^{k'}1)*1^s]\}.$$

Thus for any tree T, $R^- \subseteq R \subseteq R^*$.

If T is a tree and k a natural number, we define *level* k of T to consist of those strings in R^* which have the same length as $T(0^k)$. Thus there are $k+1$ strings in level k of T. These are $T(0^k)$ together with all strings obtained from strings $T(0^{k'}1)$ with $k' < k$ by adding enough 1's at the end to get the correct length. Note that any string on level k of a tree T must have length $\geq k$.

We can now state the computation lemmas which are used to construct a minimal *tt*-degree using trees.

Lemma 1 *Let* $\langle T_t : t \geq 0\rangle$ *be a recursive sequence of trees and let* A *be a set. Suppose that for each* $t \geq 0$,

1. *A contains one of the strings on level* t *of* T_t.
2. *There are no e-splittings on level* t *of* T_t.

Then if $[e]^A$ *is total, it is recursive.*

Proof: Suppose that $[e]^A$ is total. Given an x, find a $t \geq x$ such that $\{e\}_t(x)\downarrow$ and such that $t \geq |[e](x)|$. If σ is a string on level t of T_t, then $|\sigma| \geq t$, so $[e]^\sigma(x)\downarrow$. Since there are no e-splittings on level t of T_t, this answer must be the same for all strings σ on level t. Since A contains one of the strings on level t this answer is $[e]^A(x)$. ■

Lemma 2 *Let* $\langle T_t : t \geq 0\rangle$ *be a recursive sequence of trees and let* A *be a set. Suppose that for each* $t \geq 0$,

1. *A contains one of the strings on level* t *of* T_t.
2. *For each* $k < t$, $T_t(0^k0)(0^k1)$ *e-split.*

Then if $[e]^A$ *is total,* $A \leq_{tt} [e]^A$.

Proof: For a given x, consider T_t with $t = x+1$. All of the strings on level t of T_t have length greater than x and since T_t e-splits up through level t, it is possible to distinguish between the strings on level t using a truth table. More precisely, for $0 \leq k < t$, let q_k be the extension of $T_t(0^k1)$ on level t of T_t, and let $q_t = T_t(0^t)$. Then $q_0, \ldots, q_t$ are all the strings on level t of T_t. By assumption, for each $k < t$, $T_t(0^k0)(0^k1)$ e-split, so for each k with $1 \leq k \leq t$, there are y_k, a_k, b_k with $[e]^{T_t(0^k)}(y_k) = a_k$, $[e]^{T_t(0^{k-1}1)}(y_k) = b_k$, and $a_k \neq b_k$. If $r = \max\{r' : 1 \leq r' \leq t \wedge [e]^A(y_{r'}) = a_{r'}\}$ (taking $\max\emptyset = 0$), then $A(x) = q_r(x)$. It follows that it is possible to write a truth table (of length $\max\{y_k : 1 \leq k \leq t\}$) such that $[e]^A$ satisfies the truth table if and only if $x \in A$. Since the sequence of trees is recursive, given x, the numbers y_k and a_k can be found effectively. Thus the truth table can be found effectively given x, and $A \leq_{tt} [e]^A$. ■

We next define the appropriate notion of subtree for the kind of trees we are using. If T and T_1 are trees, then we call T_1 a *subtree* of T and write $T_1 \subseteq^* T$ if $R_1^- \subseteq R^-$ and $R_1 \subseteq R^*$. The following Lemma is an easy consequence of this definition.

Lemma 3 *For all trees* T_1, T_2, T_3,

1. *If* $T_1 \subseteq^* T_2$, *then* $R_1^* \subseteq R_2^*$.
2. *If* $T_1 \subseteq^* T_2 \subseteq^* T_3$, *then* $T_1 \subseteq^* T_3$.

■

We now consider the format of the construction and define the e-state of a string on the trees we will construct. At stage s, we will define a finite set A_s and trees $T_{-1,s}, T_{0,s}, T_{1,s}, \ldots$. We will have $A_s \subseteq A_{s+1}$ for all $s \geq 0$ and we will set $A = \bigcup_s A_s$, so, since the construction will be effective, A will be r.e.

We will also have $T_{e,s} \subseteq^* T_{e-1,s}$ for all $e \geq 0$. Thus for every e', e with $-1 \leq e' \leq e$, by the second part of Lemma 3, $T_{e,s} \subseteq^* T_{e',s}$ and hence by the first part of the Lemma, $R^*_{e,s} \subseteq R^*_{e',s}$. For $q \in R^*_{e,s}$,we define $U_{e,s}(q)$, *the e-state of q at stage s* as follows. First, for each $e \geq 0$ we define $S_{e,s} : R_{e,s} \to \{0,1\}$ by

$$S_{e,s}(T_{e,s}(\emptyset)) = 1$$

$$S_{e,s}(T_{e,s}(0^k0)) = S_{e,s}(T_{e,s}(0^k1)) = \begin{cases} 1 & \text{if } T_{e,s}(0^k0)(0^k1) \text{ } e\text{-split} \\ & \text{and have length} \leq s \\ 0 & \text{otherwise.} \end{cases}$$

Next we extend $S_{e,s}$ to $R^*_{e,s}$ by defining for $\sigma \in R^*_{e,s}$

$$S_{e,s}(\sigma) = S_{e,s}(\tau) \text{ where } \tau \in R^-_{e,s} \text{ and } |\tau| = |\sigma|.$$

Note that the value of $S_{e,s}(\sigma)$ is the same for all σ on a given level of $R^*_{e,s}$. Finally, for $e \geq -1$, we define $U_{e,s} : R^*_{e,s} \to 2^{e+1}$ by

$$U_{e,s}(\sigma) = (S_{0,s}(\sigma), S_{1,s}(\sigma), \ldots, S_{e,s}(\sigma))$$

(This definition makes sense since for all e' with $0 \leq e' \leq e$, $R^*_{e,s} \subseteq R^*_{e',s}$.) Note that $U_{-1,s}(\sigma) = \emptyset$ for all $\sigma \in R^*_{-1,s}$ and that if $e' < e$ and $\sigma \in R^*_{e,s} \subseteq R^*_{e',s}$, then $U_{e',s}(\sigma) \subseteq U_{e,s}(\sigma)$. Also, for a given $e \geq -1$, the definition of $U_{e,s}$ does not depend on any of the trees $T_{e',s}$ with $e' > e$, so we may compute $U_{e,s}$ once $T_{-1,s}, \ldots, T_{e,s}$ are defined. If σ, τ are two strings in $R^*_{e,s}$ which have the same value under $U_{e,s}$, then we write $U_{e,s}(\sigma, \tau)$ for this value. In particular, suppose that σ and τ are two strings in $R^*_{e,s}$ of the same length. Then (since strings in different levels must have different lengths), σ and τ are on the same level of each $R_{e',s}$ with $e' \leq e$ and hence (since the value of $S_{e',s}$ depends only on the level of the argument), $U_{e,s}(\sigma) = U_{e,s}(\tau)$.

Before we can give the conditions that our construction will meet, there is one technicality we must discuss since it does not come up in the usual full approximation constructions. This is what we call "stretching". Suppose that, at stage s, string q is on $T_{e,s}$, the left and right successors of q are q_0 and q_1, and the successors of q_0 are q_{00} and q_{01}. If at the beginning of stage $s+1$ we move A over to q_{01} to diagonalize, it might seem that, while it is obvious that the splitting q_{00}, q_{01} cannot be left on the eth tree, there is no reason to take q, q_0, and q_1 off of this tree unless it is possible to improve e-state. However, although q_{00} and q_{01} are incompatible extensions of q_0, the first disagreement between these two strings need not be until well past the length of q_0. Thus if q_0, q_1 are left as the successors of q, then in looking for a splitting to put up as the successor of q_0, we might find a splitting one half of which is contained in q_{01} (and hence is contained in the new A) and the other half of which is to the right of q_{01}. A further diagonalization could then push A to the right half of this new splitting and hence off of $T^*_{e,s}$. The way to avoid this undesirable situation is to not leave q_0, q_1 as the successors of q but instead to make the new successors of q be extensions of q_0, q_1 which have been stretched out long enough to prevent A from going off $T^*_{e,s}$. This stretching adds complication to the construction, but does have the advantage of making it correct.

Before we give the construction, we will list conditions which the construction will satisfy, and then we will derive consequences from these conditions. The intuitive meanings of the conditions will be given after we list them. For all $e \geq -1$ and $s \geq 0$ the following will be true.

C1) For all $q \in R^{-}_{e,s}$, $q \subset A_s$.

C2) If $e \geq 0$, $q \in R^{-}_{e,s}$, q_0, q_1 are the $T_{e,s}$ successors of q, and r_0, r_1 are the $T_{e-1,s}$ successors of q, then

$$U_{e-1,s}(q_0, q_1) = U_{e-1,s}(r_0, r_1).$$

C3) If $e \geq 0$, $q \in R^{-}_{e,s}$, q_0, q_1 are the $T_{e,s}$-successors of q and there exist r_0, r_1 with

a) $r_0, r_1 \in R^{*}_{e-1,s}$,

b) $r_0 \in R^{-}_{e-1,s}$,

c) $|r_0| = |r_1| \leq s$,

d) $r_0, r_1 \supseteq q$,

e) r_0, r_1 e-split,

f) $U_{e-1,s}(r_0, r_1) = U_{e-1,s}(q_0, q_1)$,

then q_0, q_1 e-split.

C4) If $s > 0$, $q \in R^{-}_{e,s} \cap R^{-}_{e,s-1}$, q_0, q_1 are the $T_{e,s-1}$- successors of q, q_{00}, q_{01} are the $T_{e,s-1}$-successors of q_0, r_0, r_1 are the $T_{e,s}$-successors of q, and $q_0 \subset A_s$, then $U_{e,s}(r_0, r_1) \succeq U_{e,s-1}(q_0, q_1)$. Further, if $U_{e,s}(r_0, r_1) = U_{e,s-1}(q_0, q_1)$, then $q_0 \subseteq r_0, q_1 \subseteq r_1$ and the following hold.

a) If $q_{00} \subset A_s$, then $q_0 = r_0, q_1 = r_1$.

b) If $q_{00} \not\subset A_s$ and $U_{e,s-1}(q_{00}) = U_{e,s-1}(q_0)$, then $|r_0| \geq s$.

C5) For all x, if x enters A at stage s and $x \leq y \leq s$, then y enters A at stage s.

Condition C1 simply states that the left half of every branching on all of the trees built at stage s is on A_s. Condition C2 says that if q is in $R^{-}_{e,s}$, then in searching for the successors of q on $T_{e,s}$, we must look only among those strings on $T_{e-1,s}$ which have the same $e-1$-state as the successors of q on $T_{e-1,s}$ do. This is the usual nesting of e-states strategy used in the full-approximation construction to make sure that the construction settles down. Condition C3 says that if there are strings on $T_{e-1,s}$ which are suitable for use as successors of some string q on $T_{e,s}$ and which e-split, then the successors of q on $T_{e,s}$ do e-split. Condition C4 is somewhat complicated due to stretching, but it says that if a string q is on both the trees $T_{e,s}$ and $T_{e,s-1}$, then, unless diagonalization interferes, the e-state of the successors of q on $T_{e,s}$ is at least as high as the e-state of the successors of q on $T_{e,s-1}$ and if the e-state has not improved then the successors have not changed (unless stretching interferes in which case the new successors of q are extensions of the old ones). The final condition C5 ensures that when diagonalization occurs, A remains on the $T^{*}_{e,s}$ for those e which are less than or equal to the priority of the diagonalizing requirement.

We now prove some lemmas which follow from the just-given conditions which our construction will meet. The first lemma says that as you go away from the root of $T_{e,s}$, the e-state gets lower.

Lemma 4 *Suppose that $s \geq 0, e \geq -1$ and condition C3 holds for all e' with $0 \leq e' \leq e$. Then for all $q, r \in R^-_{e,s}$, $q \subseteq r \rightarrow U_{e,s}(q) \succeq U_{e,s}(r)$.*

Proof: We use induction on e. For $e = -1$ the result is trivial ($U_{e,s}(q) = \emptyset = U_{e,s}(r)$).

Suppose that $e \geq 0$, that the result holds for $e-1$, that condition C3 holds for all e' with $0 \leq e' \leq e$, and that $q, r \in R^-_{e,s}$ with $q \subseteq r$. Since $T_{e,s} \subseteq^* T_{e-1,s}$, $R^-_{e,s} \subseteq R^-_{e-1,s}$ and thus by induction hypothesis, $U_{e-1,s}(q) \succeq U_{e-1,s}(r)$. Suppose for a contradiction that $U_{e,s}(q) \prec U_{e,s}(r)$. Then $U_{e-1,s}(q) = U_{e-1,s}(r)$, $S_{e,s}(q) = 0$, and $S_{e,s}(r) = 1$. Since $S_{e,s}(q) = 0$, $q \neq T_{e,s}(\emptyset)$. Let q^- be the $T_{e,s}$-predecessor of q, and let q', r' be the $T_{e,s}$-partners of q, r respectively. Then by definition of $S_{e,s}$, r and r' e-split and have length less than or equal to s. Since $q \subseteq r$, $|q| \leq |r| \leq s$, so since $S_{e,s}(q) = 0$, q and q' must not e-split. This contradicts condition C3 because $q^- \in R^-_{e,s}$, q, q' are the $T_{e,s}$-successors of q^-, r, r' are in $R_{e,s} \subseteq R^*_{e-1,s}$, $r \in R^-_{e,s} \subseteq R^-_{e-1,s}$, $|r| = |r'| \leq s$, $r, r' \supseteq q^-$, r, r' e-split, and $U_{e-1,s}(r, r') = U_{e-1,s}(q, q')$. Thus $U_{e,s}(q) \succeq U_{e,s}(r)$. ■

Our next lemma is a somewhat technical extension of condition C4. It states that if a string q is on both $T_{e,s}$ and $T_{e,s-1}$ and the e-state of the successors of q does not change from stage $s-1$ to s (call this e-state u), then the whole part of $T_{e,s-1}$ which is above q and has e-state u at stage $s-1$ is still on $T_{e,s}$ and has e-state u, unless diagonalization or stretching interfere.

Lemma 5 *Suppose that $s > 0$, $e \geq -1$, conditions C3 and C4 hold for all e', s' with $s' \in \{s, s-1\}$ and $-1 \leq e' \leq e$, $q \in R^-_{e,s} \cap R^-_{e,s-1}$, q_0, q_1 are the $T_{e,s}$ successors of q, $r \in R^-_{e,s-1}$, $q \subset r \subset A_s$, and $U_{e,s-1}(r) = U_{e,s}(q_0, q_1) = u$. Let r_0, r', r^- be the $T_{e,s-1}$ left successor, partner, and predecessor of r, respectively.*

1. *If $r_0 \subset A_s$, then $r, r' \in R_{e,s}$ and $U_{e,s}(r) = U_{e,s-1}(r)$.*
2. *If $r_0 \not\subset A_s$, then $r^- \in R^-_{e,s}$, and if p, p' are the $T_{e,s}$ successors of r^-, then $U_{e,s}(p, p') = U_{e,s-1}(r)$ and $r \subseteq p, r' \subseteq p'$. Further, if $U_{e,s-1}(r_0) = U_{e,s-1}(r)$, then $|p| \geq s$.*

Proof: 1) Let $q = p_0 \subset p_1 \subset \ldots \subset p_t = r$ be all strings between q and r on $T_{e,s-1}$ (so $t > 0$). Let p'_i be the $T_{e,s-1}$-partner of p_i, $1 \leq i \leq t$. We show by induction on i, $0 \leq i \leq t-1$, that

1. $p_i \in R^-_{e,s}$,
2. the $T_{e,s}$-successors of p_i are p_{i+1}, p'_{i+1},
3. $U_{e,s}(p_{i+1}, p'_{i+1}) = u = U_{e,s-1}(p_{i+1}, p'_{i+1})$.

For $i = 0$ we have $p_i = p_0 = q_0 \in R^-_{e,s}$ by hypothesis. Also

u	$=$	$U_{e,s}(q_0, q_1)$	by definition of u
	$\succeq$	$U_{e,s-1}(p_1, p'_1)$	by C4 for $T_{e,s}$ since $p_1 \subset r \subset A_s$
	$\succeq$	$U_{e,s-1}(r)$	by Lemma 4 for $T_{e,s-1}$ since $p_1 \subseteq r \in R^-_{e,s-1}$
	$=$	u	by hypothesis

Thus by C4 for $T_{e,s}$ (since $p_1 \subseteq r \subset r_0 \subset A_s$), $q_0 = p_1, q_1 = p'_1$ and hence $U_{e,s}(p_1, p'_1) = U_{e,s}(q_0, q_1) = u = U_{e,s-1}(p_1, p'_1)$.

Now suppose that the result holds for $i-1$ with $0 < i \leq t-1$. Then by induction hypothesis, $p_i \in R^-_{e,s}$. Let s_0, s_1 be the $T_{e,s}$-successors of p_i. We have

$$\begin{array}{rcll} u & = & U_{e,s}(p_i, p_i') & \text{by induction hypothesis} \\ & \succeq & U_{e,s}(s_0, s_1) & \text{by Lemma 4 for } T_{e,s} \\ & \succeq & U_{e,s-1}(p_{i+1}, p_{i+1}') & \text{by C4 for } T_{e,s} \text{ since } p_{i+1} \subseteq r \subset A_s \\ & \succeq & U_{e,s-1}(r) & \text{by Lemma 4 for } T_{e,s-1} \\ & = & u & \text{by hypothesis} \end{array}$$

Thus $u = U_{e,s}(s_0, s_1) = u_{e,s-1}(p_{i+1}, p_{i+1}')$, so by C4 for $T_{e,s}$ (since $p_{i+1} \subseteq r \subset r_0 \subset A_s$), $s_0 = p_{i+1}, s_1 = p_{i+1}'$ and $U_{e,s}(p_{i+1}, p_{i+1}') = U_{e,s}(s_0, s_1) = u = u_{e,s-1}(p_{i+1}, p_{i+1}')$.

Taking $i = t-1$ gives the result.

2) If $r^- = q$, then $r^- \in R^-_{e,s}$ and $u = U_{e,s}(p,p') = U_{e,s-1}(r,r')$, so the result follows from C4.

If $r^- \neq q$, then $u = U_{e,s-1}(r) \preceq U_{e,s-1}(r^-) \preceq U_{e,s-1}(p_1, p_1') \preceq U_{e,s}(q_0, q_1) = u$ (where p_1, p_1' are the $T_{e,s-1}$-successors of q), so $U_{e,s-1}(r^-) = u$. Thus by part (1), $r^- \in R^-_{e,s}$ and $U_{e,s}(r^-) = U_{e,s-1}(r^-) = u$. We now have

$$\begin{array}{rcll} u & = & U_{e,s}(r^-) & \\ & \succeq & U_{e,s}(p,p') & \text{by Lemma 4 for } T_{e,s} \\ & \succeq & U_{e,s-1}(r,r') & \text{by C4 for } T_{e,s} \text{ since } r \subset A_s \\ & = & u & \text{by hypothesis} \end{array}$$

Thus $U_{e,s}(p,p') = U_{e,s-1}(r,r')$, so the result follows from C4 for $T_{e,s}$. ■

The next lemma is just an extension of the previous lemma to all of $T^*_{e,s-1}$.

Lemma 6 *Suppose that $s > 0$, $e \geq -1$, conditions C3 and C4 hold for all e', s' with $s' \in \{s, s-1\}$ and $-1 \leq e' \leq e$, $q \in R^-_{e,s} \cap R^-_{e,s-1}$, q_0, q_1 are the $T_{e,s}$ successors of q, $r \in R^-_{e,s-1}$, $q \subset r \subset A_s$, $U_{e,s-1}(r) = U_{e,s}(q_0, q_1) = u$, $p^* \in R^*_{e,s-1}$, $|p^*| = |r|$, and $q \subseteq p^*$. Let r_0 be the $T_{e,s-1}$ left successor of r.*

1. *If $r_0 \subset A_s$, then $p^* \in R^*_{e,s}$ and $U_{e,s}(p^*) = U_{e,s-1}(p^*)$.*
2. *If $r_0 \not\subset A_s$, then some extension p_1 of p^* is in $R^*_{e,s}$ with $U_{e,s}(p_1) = U_{e,s-1}(p^*)$, and if $U_{e,s-1}(r_0) = U_{e,s-1}(r)$, then we may take $|p_1| \geq s$.*

Proof: 1) If $p^* = r$, then this is just a restatement of Lemma 5. Suppose that $p^* \neq r$. Then since $q \subset p^*$, for some $s_0 \in R^-_{e,s-1}$, $q \subset s_0 \subseteq r$ and $p^* = s_0' * 1^t$ for some t where s_0' is the $T_{e,s-1}$-partner of s_0. By the usual argument, if p_1, p_1' are the $T_{e,s-1}$-successors of q, then $U_{e,s-1}(r) \preceq U_{e,s-1}(s_0) \preceq U_{e,s-1}(p_1, p_1') \preceq U_{e,s}(q_0, q_1) = U_{e,s}(r)$ and $s_0 \subseteq r \subset r_0 \subset A_s$, so by part 1 of Lemma 5 applied to s_0, $s_0' \in R_{e,s}$. By the same Lemma applied to r, $r \in R^-_{e,s}$. Thus $p^* \in R^*_{e,s}$. Finally $U_{e,s}(p^*) = U_{e,s}(r) = U_{e,s-1}(r) = U_{e,s-1}(p^*)$.

We omit the proof of (2) which follows from the second part of Lemma 5 in much the same way that the argument just given used the first part of that Lemma. ■

We now give the construction and along the way show that all of the conditions C1 to C5 are met. At any stage during the construction, a requirement P_e may receive attention. At most one P_e receives attention at a given stage. P_e can receive attention at stage s only if it has never received attention at any earlier stage, so each P_e receives attention at most once.

The construction has infinitely many stages. At stage $s \geq 0$, we construct a sequence of trees $\langle T_{-1,s}, T_{0,s}, T_{1,s}, \ldots \rangle$ with $T_{e,s} \subseteq^* T_{e-1,s}$ for all $e \geq 0$.

Stage s.

(1) If $s > 0$, find the least $e \leq s-1$, if any, such that

a) P_e has not received attention at any stage $s' < s$.

b) $x \in W_e$ where x is minimal so that $T_{e,s-1}(0)(x) \neq T_{e,s-1}(1)(x)$.

If any such e exists, take the least such and put the corresponding x into A_s. In addition, for any y with $x \leq y \leq s$, put y into A_s. (Thus condition C5 is met for s.) We say that P_e receives attention at stage s.

(2) We now construct the trees $T_{e,s}$ by induction on $e \geq -1$. We assume that each of the conditions C1 to C5 is met by all trees $T_{e',s'}$ with $s' < s$ and we will verify that the trees constructed at stage s also meet these conditions.

First we define $T_{-1,s}$. Let $\overline{A_s} = \{a_0^s < a_1^s < \ldots\}$. We define

$$T_{-1,s}(\emptyset) = \emptyset$$

and for $k \geq 0$

$$\begin{aligned} T_{-1,s}(0^k 0) &= A_s \lceil a_{k+1}^s \\ T_{-1,s}(0^k 1) &= \begin{cases} T_{-1,s}(0^k 0)(y) & \text{if } y < a_{k+1}^s,\ y \neq a_k^s \\ 1 & \text{if } y = a_k^s. \end{cases} \end{aligned}$$

It is easy to see that $T_{-1,s}$ is a tree and that C1 is met, while C2 and C3 are vacuously true. Condition C4 is partly vacuous and the nontrivial part follows from the way $T_{-1,s}$ is defined. (We need to assume of course that when $s > 0$, $T_{-1,s-1}$ was defined the same way.) We leave this verification to the reader. (The fact that when stretching occurs the stretched string has length greater than or equal to s follows from C5.)

Note that since we will have each later $T_{e,s}$ a subtree of $T_{e-1,s}$, we have by Lemma 3 that $T_{e,s} \subseteq^* T_{-1,s}$ and hence C1 will automatically be met for $T_{e,s}$ since it is met by $T_{-1,s}$.

Now suppose that $e \geq 0$ and that for all e' with $-1 \leq e' < e$ we have defined $T_{e',s}$ satisfying conditions C1 to C4. We define $T_{e,s}$ inductively.

First we set

$$T_{e,s}(\emptyset) = T_{e-1,s}(0).$$

Then, assuming that we have defined $T_{e,s}(0^k) \in R^-_{e-1,s}$, say $T_{e,s}(0^k) = q$, we define $T_{e,s}(0^k 0)(0^k 1)$. Let u be the $e-1$ state at stage s of the $T_{e-1,s}$-successors of q.

Case 1. $s > 0$, $q \in R^-_{e,s-1}$, u is the $e-1$-state at stage $s-1$ of the $T_{e-1,s-1}$-successors of q, and, if q_0, q_1 are the $T_{e,s-1}$-successors of q and q_{00} is the $T_{e,s-1}$ left successor of q_0, $q_{00} \subset A_s$.

Then by C2 for $T_{e,s-1}$, $U_{e-1,s-1}(q_0, q_1) = u$, and hence by Lemmas 5 and 6 for $T_{e-1,s}$, $q_0 \in R^-_{e-1,s}$, $q_1 \in R^*_{e-1,s}$, and $U_{e-1,s}(q_0, q_1) = u$. Thus q_0, q_1 are still available at stage s to be used as the successors of q. We do this unless we can improve the e-state. That is, if $S_{e,s-1}(q_0, q_1) = 0$ and there are strings $r_0, r_1 \in R_{e-1,s}$ such that $r_0 \in R^-_{e-1,s}$, $|r_0| = |r_1| \leq s$, $r_0, r_1 \supseteq q$, r_0, r_1 e-split, and $U_{e-1,s}(r_0) = U_{e-1,s}(r_1) = u$, then we take the first such strings in $R^*_{e-1,s}$ (in some effective ordering of pairs of strings) and let $T_{e,s}(0^k 0)(0^k 1)$ be this pair. Otherwise, we let $T_{e,s}(0^k 0)(0^k 1) = (q_0, q_1)$.

It is now easy to see that $T_{e,s}(0^k)(0^k 1)$ split $T(0^k)$ and have the same length, that $T_{e,s}(0^k 0) \in R^-_{e-1,s}$, $T_{e,s}(0^k 1) \in R^*_{e-1,s}$ and that C2, C3, C4 all remain true up through level $k+1$ of $T_{e,s}$.

Case 2. Same as Case 1 except that $q_0 \subset A_s$, $q_{00} \not\subset A_s$.

Then by C2 for $T_{e,s-1}$, $U_{e-1,s-1}(q_0, q_1) = u$. The $T_{e-1,s-1}$ left successor of q_0 may or may not be $\subset A_s$, but in either case one part or the other of Lemma 6 says that some extension p_1 of q_1 is on $R^*_{e-1,s}$ and $U_{e-1,s}(p_1) = U_{e-1,s-1}(q_1) = u$. Then we use this p_1 to define $T_{e,s}(0^k0)(0^k1)$ unless we can improve the e-state. That is, if $S_{e,s-1}(q_0, q_1) = 0$ and there are strings $r_0, r_1 \in R^*_{e-1,s}$ meeting the conditions as given in Case 1, then we define $T_{e,s}(0^k0)(0^k1)$ to be the first such pair. Otherwise, we define $T_{e,s}(0^k0)(0^k1)$ as follows. We will take $T_{e,s}(0^k1)$ to be an extension p_1 of q_1 such that $p_1 \in R^*_{e-1,s}$ and $U_{e-1,s}(p_1) = u$ (such p_1 exists as noted above) and then we will take $T_{e,s}(0^k0)$ to be the element p in $R^-_{e-1,s}$ with $|p| = |p_1|$ (so $q_0 \subseteq T_{e,s}(0^k0)$ and $U_{e-1,s}(T_{e,s}(0^k0)) = U_{e-1,s}(p_1) = u$). To choose p_1, take the shortest possible p_1 with length $\geq s$, if such p_1 exists; otherwise, take the shortest possible p_1.

We now verify one part of condition C4. The other needed verifications are easy. Suppose that $U_{e,s-1}(q_{00}) = U_{e,s-1}(q_0)$. We must show that there is an extension p_1 of q_1 on $R^*_{e-1,s}$ with $|p_1| \geq s$ and $U_{e-1,s}(p_1) = u$. Take $r \in R^-_{e-1,s-1}$ such that $q_0 \subseteq r \subset q_{00}$, $r \subset A_s$, and $r_0 \not\subset A_s$ where r_0 is the $T_{e-1,s-1}$ left successor of r. Since $U_{e,s-1}(q_{00}) = U_{e,s-1}(q_0)$, we have $U_{e-1,s-1}(q_{00}) = U_{e-1,s-1}(q_0) = u$. Thus by Lemma 4, $u = U_{e-1,s-1}(q_{00}) \preceq U_{e-1,s-1}(r_0) \preceq U_{e-1,s-1}(r) \preceq U_{e-1,s-1}(q_0) = u$. Hence $U_{e-1,s-1}(r) = U_{e-1,s-1}(r_0) = u$. Let p^* be the extension of q_1 by 1's with length $|r|$, so $p^* \in R^*_{e-1,s-1}$. By Lemma 6, there is an extension p_1 of p^* in $R^*_{e-1,s}$ with $U_{e-1,s}(p_1) = U_{e-1,s-1}(p^*) = u$ and $|p_1| \geq s$, as desired.

Case 3. Neither Case 1 nor Case 2 holds.

In this case either there is no $T_{e,s-1}$ since $s = 0$ or $q \notin R_{e,s-1}$ or q is in $R_{e,s-1}$ but the successors of q on $T_{e,s-1}$ do not have the right $e-1$ state as dictated to make C2 true. In this case we will ignore what happened at stage $s-1$ and just do the obvious thing. That is, if there are strings r_0, r_1 satisfying the properties given in Case 1, then we define $T_{e,s}(0^k0)(0^k1)$ to be the first such pair of strings. If no such pair exists, then we just define $T_{e,s}(0^k0)(0^k1)$ to be the $T_{e-1,s}$-successors of q.

We must verify that condition C4 is met; the other verifications are routine. Suppose that $s > 0$, $q \in R^-_{e,s-1}$, and with q_0, q_1, q_{00} as in the statement of C4 suppose that $q_0 \subset A_s$. Let p_0, p_1 be the $T_{e-1,s-1}$-successors of q, and let $u' = U_{e-1,s-1}(p_0, p_1)$. Since neither Case 1 nor Case 2 applies, we must have $u' \neq u$. Now $q \in R^-_{e-1,s} \cap R^-_{e-1,s-1}$ and since $T_{e,s-1} \subseteq^* T_{e-1,s-1}$, $p_0 \subseteq q_0 \subset A_s$. Thus by C4 for $T_{e-1,s}$, $u \succeq u'$. But $u \neq u'$, so $u \succ u'$. By C2 for $T_{e,s-1}$, $U_{e-1,s-1}(q_0, q_1) = u'$, and by construction, $U_{e-1,s}(r_0, r_1) = u$. Thus $U_{e,s}(r_0, r_1) \succ U_{e,s-1}(q_0, q_1)$, so C4 holds.

This completes the construction.

The next Lemma is the key to showing that the construction settles down. It states that if $T_{e,s}(0^k)$ has settled down by stage s_0 and u is the greatest e-state that the successors of $T_{e,s}(0^k)$ have at any stage later than s_0, then any splitting with e-state u put onto a tree $T_{e,s}$ with $s \geq s_0$ above $T_{e,s}(0^k)$ is permanent unless diagonalization or stretching interfere.

Lemma 7 *Suppose that for all $s \geq s_0$, $T_{e,s}(0^k) = T_{e,s_0}(0^k)$. Let*

$$u = \max\{U_{e,s}(T_{e,s}(0^k0)(0^k1)) : s \geq s_0\}.$$

Suppose that $k' \geq k$, $s_1 \geq s_0$, $U_{e,s_1}(T_{e,s_1}(0^{k'}0)) = u$, and $T_{e,s_1}(0^{k'}0) \subset A_{s_1+1}$. Then

$$U_{e,s_1+1}(T_{e,s_1+1}(0^{k'}0)(0^{k'}1)) = u.$$

(Hence by C4, $T_{e,s_1+1}(0^{k'}0) \supseteq T_{e,s_1}(0^{k'}0)$, and $T_{e,s_1+1}(0^{k'}1) \supseteq T_{e,s_1}(0^{k'}1)$. Further, if $T_{e,s_1}(0^{k'}00) \subset A_{s_1+1}$, then $T_{e,s_1+1}(0^{k'}0)(0^{k'}1) = T_{e,s_1}(0^{k'}0)(0^{k'}1)$, while if $T_{e,s_1}(0^{k'}00) \not\subset A_{s_1+1}$ and $U_{e,s_1}(T_{e,s_1}(0^{k'}00)) = u$, then $|T_{e,s_1+1}(0^{k'}0)| \geq s_1 + 1$.)

Proof: We use induction on k'. First suppose that $k' = k$. Let $q = T_{e,s_1}(0^k) = T_{e,s_1+1}(0^k)$. Then

$$\begin{array}{rcll} u & = & U_{e,s_1}(T_{e,s_1}(0^k0)) & \text{by hypothesis, since } k = k' \\ & \preceq & U_{e,s_1+1}(T_{e,s_1+1}(0^k0)) & \text{by C4 since } T_{e,s_1}(0^k0) \subset A_{s_1+1} \\ & \preceq & u & \text{by definition of } u. \end{array}$$

Thus $u = U_{e,s_1+1}(T_{e,s_1+1}(0^k0)(0^k1))$, as desired.

Now suppose that the result holds for some $k' \geq k$ and that for some $s_1 \geq s_0$, $U_{e,s_1}(T_{e,s_1}(0^{k'+1}0)) = u$ and $T_{e,s_1}(0^{k'+1}0) \subset A_{s_1+1}$. Then

$$\begin{array}{rcll} u & = & U_{e,s_1}(T_{e,s_1}(0^{k'+1}0)) & \text{by hypothesis} \\ & \preceq & U_{e,s_1}(T_{e,s_1}(0^{k'}0)) & \text{by Lemma 4} \\ & \preceq & U_{e,s_1}(T_{e,s_1}(0^{k}0)) & \text{by Lemma 4} \\ & \preceq & u & \text{by definition of } u \end{array}$$

Thus $u = U_{e,s_1}(T_{e,s_1}(0^{k'}0))$ and $T_{e,s_1}(0^{k'}00) = T_{e,s_1}(0^{k'+1}0) \subset A_{s_1+1}$. So by induction hypothesis, $U_{e,s_1+1}(T_{e,s_1+1}(0^{k'}0)(0^{k'}1)) = u$ and $T_{e,s_1+1}(0^{k'}0)(0^{k'}1) = T_{e,s_1}(0^{k'}0)(0^{k'}1)$.

We now have

$$\begin{array}{rcll} u & = & U_{e,s_1+1}(T_{e,s_1+1}(0^{k'}0)) & \\ & \succeq & U_{e,s_1+1}(T_{e,s_1+1}(0^{k'+1}0)) & \text{by Lemma 4} \\ & \succeq & U_{e,s_1}(T_{e,s_1}(0^{k'+1}0)) & \text{by C4 since } T_{e,s_1}(0^{k'+1}0) \subset A_{s_1+1} \text{ and} \\ & & & T_{e,s_1+1}(0^{k'}0) = T_{e,s_1}(0^{k'}0) \\ & = & u. & \end{array}$$

Thus $u = U_{e,s_1+1}(T_{e,s_1+1}(0^{k'+1}0)(0^{k'+1}1))$ as desired. ■

Lemma 8 *For all $e \geq -1$ and $s, k \geq 0$, $T_{e,s}(0^k) \subseteq T_{e+k,s}(\emptyset)$.*

Proof: By induction on k. For $k = 0$ the result is trivial. Suppose that the result is true for k. Then $T_{e+k+1,s}(\emptyset) = T_{e+k,s}(0) \supset T_{e+k,s}(\emptyset) \supseteq T_{e,s}(0^k)$, so $T_{e+k+1,s}(\emptyset) \supset T_{e,s}(0^k)$. But $T_{e+k+1,s} \subseteq^* T_{e,s}$, so $T_{e+k+1,s}(\emptyset) \in R^-_{e,s}$. Thus $T_{e+k+1,s}(\emptyset) \supseteq T_{e,s}(0^{k+1})$. ■

Lemma 9 *For all $e \geq 1$ and all $\sigma \in \Omega$, $\lim_s T_{e,s}(\sigma)$, $\lim_s U_{e,s}(T_{e,s}(\sigma))$ both exist.*

Proof: By induction on e. Suppose that the result is true for all $e' < e$. We proceed by induction on $|\sigma|$ to show that the desired limits exist. First suppose $\sigma = \emptyset$. If $e = -1$, then $T_{e,s}(\sigma) = \emptyset$ and $U_{e,s}(T_{e,s}(\sigma)) = \emptyset$ for all s, so the desired limits certainly exist. If $e > -1$, then for all s we have $T_{e,s}(\emptyset) = T_{e-1,s}(0)$ and $U_{e,s}(T_{e,s}(\emptyset)) = U_{e-1,s}(T_{e-1,s}(0))*1$ so by induction hypothesis the limits exist.

Now suppose that $\lim_s T_{e,s}(0^k)$ exists for some $k \geq 0$. Take s_0 such that

1. For all $s \geq s_0$, $T_{e,s}(0^k) = T_{e,s_0}(0^k)$.

2. For all $s \geq s_0$, for all $n \leq e + k + 1$, P_e does not receive attention at stage s.

Note that by (2) and Lemma 8, for all $s \geq s_0$, $T_{e,s}(0^k00) \subset A_{s+1}$. Let $u = \max_{s \geq s_0} U_{e,s}(T_{e,s}(0^k0)(0^k1))$ and take $s_1 \geq s_0$ with $U_{e,s_1}(T_{e,s_1}(0^k0)(0^k1)) = u$. Then by induction on s, using Lemma 7 and the fact noted above, for all $s \geq s_1$,

$$T_{e,s}(0^k0)(0^k1) = T_{e,s_1}(0^k0)(0^k1)$$

and

$$U_{e,s}(T_{e,s}(0^k0)(0^k1)) = u.$$

■

For all e, σ, let $T_e(\sigma) = \lim_s T_{e,s}(\sigma)$, so T_e is a tree. Define U_e by $U_e(T_e(\sigma)) = \lim_s U_{e,s}(T_{e,s}(\sigma))$. By Lemma 4, for all $q, r \in R_e^-$, if $q \subseteq r$, then $U_e(q) \succeq U_e(r)$. Thus, for fixed e, as k increases, $U_e(T_e(0^k))$ decreases, so has a limit, say u_e. Then $u_e \subset u_{e+1}$ for all $e \geq -1$.

For all $e, s \geq 0$, $T_{e,s} \subseteq^* T_{-1,s}$ and hence for any k, if

$$x = \mu y(T_{e,s}(0^k0)(y) \neq T_{e,s}(0^k1)(y))$$

then for all z

$$x \leq z \leq |T_{e,s}(0^k1)| \rightarrow T_{e,s}(0^k1)(z) = 1.$$

Lemma 10 *For all $e \geq 0$, P_e is met.*

Proof: If P_e ever receives attention, then for some x, $W_e(x) = 1 = A(x)$. If P_e never receives attention, let

$$x = \mu y(T_e(0)(y) \neq T_e(1)(y)).$$

Then $x \notin W_e$ (else P_e would eventually receive attention), so $W_e(x) = 0 = T_e(0)(x) = A(x)$. ■

Lemma 11 *For all $e \geq 0$, Q_e is met.*

Proof: Fix $e \geq 0$. We wish to use Lemmas 1 and 2 to show that Q_e is met and hence we need to define a sequence $\langle T'_{e,t} : t \geq 0 \rangle$ satisfying the hypotheses of one of these two Lemmas. Take $\overline{k} \geq 1$ so that $U_e(T_e(0^{\overline{k}-1})) = u_e$. Then for all $k \geq \overline{k}$ we also have $U_e(T_e(0^k)) = u_e$. Fix t_0 such that for all $s \geq t_0$,

$$\begin{aligned} &T_{e,s}(0^{\overline{k}-1}) = T_e(0^{\overline{k}-1}),\\ &T_{e,s}(0^{\overline{k}}) = T_e(0^{\overline{k}}),\\ &U_{e,s}(T_{e,s}(0^{\overline{k}-1})) = u_e = U_{e,s}(T_{e,s}(0^{\overline{k}})). \end{aligned}$$

Given $t \geq 0$, let $k = \overline{k} + t$ and find $s_0 \geq t_0$ such that

$$\begin{aligned} &s_0 \geq |T_{e,s_0}(0^k)|,\\ &U_{e,s_0}(T_{e,s_0}(0^k)) = u_e. \end{aligned}$$

Define $T'_{e,t}$ to be the subtree of T_{e,s_0} above $T_{e,s_0}(0^{\overline{k}})$. Once the values of u_e, $\overline{k}$, and t_0 are fixed, for a given t one can effectively find k and s_0, so the sequence $\langle T'_{e,t} : t \geq 0 \rangle$ is recursive.

Note that level t of $T'_{e,t}$ consists of the leftmost $t+1$ strings (in the usual left to right ordering) in level $\overline{k} + t = k$ of T_{e,s_0}, i.e., those strings in level k of T_{e,s_0} which extend $T_{e,s_0}(0^{\overline{k}})$. Let these strings be $q_{\overline{k}}, \ldots, q_k$ where $q_k = T_{e,s_0}(0^k)$ and for $\overline{k} \leq k' < k$, $q_{k'}$ is $T_{e,s_0}(0^{k'}1)$ extended by enough 1's to have the right length.

Suppose that $u_e(e) = 1$. Then we claim that there are no e-splittings on level t of $T'_{e,t}$, i.e., no e-splittings among $q_{\overline{k}}, \ldots, q_k$. For if there were an e-splitting among these strings then since the strings are all of the same length and we are considering

tt-reductions, there will be an e-splitting involving $q_k = T_{e,s_0}(0^k)$. Further, the strings on level k of T_{e,s_0} have length less than or equal to s_0 and have the same $e-1$-state at stage s_0 as $T_{e,s_0}(0^{\overline{k}})(0^{\overline{k}-1}1)$, the T_{e,s_0} successors of $T_{e,s_0}(0^{\overline{k}-1})$ (namely, u_{e-1}). Thus by C3, $T_{e,s_0}(0^{\overline{k}})(0^{\overline{k}-1}1)$ must e-split and they have length $\leq s_0$, so they cannot have e-state u_e at stage s_0, contradicting our choice of $s_0 \geq t_0$.

Now suppose that $u_e(e) = 0$. Then by Lemma 4, for each k' with $\overline{k} \leq k' < k$,

$$U_{e,s_0}(T_{e,s_0}(0^{k'}0)(0^{k'}1)) = u_e$$

and hence $T_{e,s_0}(0^{k'}0)(0^{k'}1)$ e-split. That is, for $0 \leq k' < t$, $T'_{e,t}(0^{k'}0)(0^{k'}1)$ e-split, as demanded by Lemma 2.

Thus, we will be done if we can show that one of the strings on level t of $T'_{e,t}$ (i.e., one of $q_{\overline{k}}, \dots, q_k$) is on A. If $t = 0$ (so $k = \overline{k}$) then this fact is immediate since level 0 of $T'_{e,0}$ is $T_{e,s_0}(0^{\overline{k}}) = T_e(0^{\overline{k}}) \subset A$. Thus we may assume that $t > 0$. By induction on $s \geq s_0$, we show that there is some $q_{k'}$ such that $A_s \supset q_{k'}$. In fact we show that for every $s \geq s_0$ there is a k_s, $\overline{k} \leq k_s \leq k-1$ such that

1. $T_{e,s}(0^{k_s}0) \supseteq q_{k_s+1}$, $T_{e,s}(0^{k_s}1) \supseteq T_{e,s_0}(0^{k_s}1)$,
2. $(\forall k')(\overline{k} \leq k' \leq k_s - 1 \rightarrow T_{e,s}(0^{k'}0)(0^{k'}1) = T_{e,s_0}(0^{k'}0)(0^{k'}1))$,
3. $(\forall k')(\overline{k} \leq k' \leq k_s \rightarrow U_{e,s}(T_{e,s}(0^{k'}0)(0^{k'}1)) = u_e)$.

The intuition is that at stage s the eth tree is the same up through level k_s as it was at stage s_0 and at level k_s+1 the only change has been stretching through the corresponding q_{k_s+1}.

We may take $q_{s_0} = k-1$. (Then (3) follows from Lemma 4.) Suppose that $s \geq s_0$ and we have k_s. Let k_{s+1} be the largest number $\leq k_s$ such that $T_{e,s}(0^{k_{s+1}}0) \subset A_{s+1}$. Since $T_{e,s}(0^{\overline{k}}) = T_{e,s+1}(0^{\overline{k}}) \subset A_{s+1}$, $\overline{k}-1 \leq k_{s+1}$. (We will see later that $\overline{k} \leq k_{s+1}$.)

Suppose that $\overline{k} \leq k' \leq k_{s+1}$. Then by induction hypothesis (since $k_{s+1} \leq k_s$), $U_{e,s}(T_{e,s}(0^{k'}0)(0^{k'}1)) = u_e$ and

$$T_{e,s}(0^{k'}0) \subseteq T_{e,s}(0^{k_{s+1}}0) \subset A_{s+1},$$

so by Lemma 7

$$U_{e,s+1}(T_{e,s+1}(0^{k'}0)(0^{k'}1)) = u_e.$$

and

$$T_{e,s+1}(0^{k'}0) \supseteq T_{e,s}(0^{k'}0),$$

$$T_{e,s+1}(0^{k'}1) \supseteq T_{e,s}(0^{k'}1).$$

If, in addition, $k' \leq k_{s+1} - 1$, then

$$T_{e,s}(0^{k'}00) \subseteq T_{e,s}(0^{k_{s+1}}0) \subset A_{s+1}$$

so by Lemma 7,

$$T_{e,s+1}(0^{k'}0)(0^{k'}1) = T_{e,s}(0^{k'}0)(0^{k'}1).$$

Thus (2) and (3) hold for k_{s+1}.

To see that (1) holds, first suppose that $k_{s+1} = k_s$. Then

$$\begin{aligned}
T_{e,s+1}(0^{k_{s+1}}0) &\supseteq T_{e,s}(0^{k_{s+1}}0) && \text{as established above}\\
&= T_{e,s}(0^{k_s}0)\\
&\supseteq q_{k_s+1} && \text{by induction hypothesis}\\
&= q_{k_{s+1}+1}
\end{aligned}$$

and

$$\begin{aligned}
T_{e,s+1}(0^{k_{s+1}}1) &\supseteq T_{e,s}(0^{k_{s+1}}1)\\
&= T_{e,s}(0^{k_s}1)\\
&\supseteq T_{e,s_0}(0^{k_s}1)\\
&= T_{e,s_0}(0^{k_{s+1}}1)
\end{aligned}$$

so (1) holds and $\overline{k} \leq k_s = k_{s+1}$, so we are done.

Hence, for the rest of the proof, we suppose that $k_{s+1} < k_s$. Then $T_{e,s}(0^{k_{s+1}+1}0) \not\subset A_{s+1}$, and, by induction hypothesis, since $\overline{k} \leq k_{s+1}+1 \leq k_s$, we have $U_{e,s}(T_{e,s}(0^{k_{s+1}+1}0)) = u_e$. Thus, by Lemma 7,

$$|T_{e,s+1}(0^{k_{s+1}}0)| \geq s+1 > s_0 \geq |q_{k_{s+1}+1}|.$$

If we now show that $q_{k_{s+1}+1} \subset A_{s+1}$, then we will have $T_{e,s+1}(0^{k_{s+1}}0) \supseteq q_{k_{s+1}+1}$.

To see that $q_{k_{s+1}+1} \subset A_{s+1}$, first note that since $T_{e,s}(0^{k_{s+1}+1}0) \not\subset A_{s+1}$, some $P_{e'}$ receives attention at stage $s+1$. If $e' < e$, then $T_{e',s}(0) \subseteq T_{e,s}(\emptyset)$ and a number $< |T_{e,s}(\emptyset)|$ enters A at stage $s+1$, contradicting $T_{e,s}(0^{\overline{k}}) \subset A_{s+1}$. Thus $e \leq e'$ and hence $T_{e',s}(0) \in R^-_{e,s}$, $T_{e',s}(1) \in R^*_{e,s}$. Further, since $T_{e,s}(0^{k_{s+1}+1}) \subset A_{s+1}$, we have $T_{e,s}(0^{k_{s+1}+1}) \subseteq T_{e',s}(0), T_{e',s}(1)$ and since $T_{e,s}(0^{k_{s+1}+1}0) \not\subset A_{s+1}$, $T_{e,s}(0^{k_{s+1}+1}0) \not\subseteq T_{e',s}(1)$. Thus,

$$\begin{aligned}
T_{e,s}(0^{k_{s+1}+1}0) &\subseteq T_{e',s}(0)\\
T_{e,s}(0^{k_{s+1}+1}1) &\subseteq T_{e',s}(1).
\end{aligned}$$

Hence,

$$\mu y(T_{e',s}(0)(y) \neq T_{e',s}(1)(y)) = \mu y(T_{e,s}(0^{k_{s+1}+1}0) \neq T_{e,s}(0^{k_{s+1}+1}1))$$

and if we let x_0 be this number, then x_0 as well as any number z with $x_0 \leq z \leq s+1$ is put into A_{s+1}.

First suppose that $k_{s+1}+1 < k_s$. Then by induction hypothesis,

$$T_{e,s}(0^{k_{s+1}+1}0)(0^{k_{s+1}+1}1) = T_{e,s_0}(0^{k_{s+1}+1}0)(0^{k_{s+1}+1}1).$$

Since $s+1 > s_0 \geq |q_{k_{s+1}+1}|$, adding to A_{s+1} all numbers z, $x_0 \leq z \leq s+1$ and no others causes $q_{k_{s+1}+1} \subset A_{s+1}$, as desired.

Now suppose that $k_{s+1}+1 = k_s$. Then by induction hypothesis,

$$T_{e,s}(0^{k_{s+1}+1}0) \supseteq q_{k_s+1} \supseteq T_{e,s_0}(0^{k_s}0)$$

and

$$T_{e,s}(0^{k_{s+1}+1}1) \supseteq T_{e,s_0}(0^{k_s}1)$$

so $x_0 = \mu y(T_{e,s_0}(0^{k_s}0)(y) \neq T_{e,s_0}(0^{k_s}1)(y))$. Thus $q_{k_{s+1}+1} = q_{k_s} \subset A_{s+1}$.

Thus we now have $T_{e,s+1}(0^{k_{s+1}}0) \supseteq q_{k_{s+1}+1}$ and hence $k_{s+1} \neq \overline{k}-1$ (else $T_{e,s+1}(0^{\overline{k}}) \neq T_{e,s_0}(0^{\overline{k}})$), so $\overline{k} \leq k_{s+1}$.

Finally,

$$\begin{aligned} T_{e,s+1}(0^{k_{s+1}}1) &\supseteq T_{e,s}(0^{k_{s+1}}1) && \text{by Lemma 7} \\ &= T_{e,s_0}(0^{k_{s+1}}1) && \text{by induction hypothesis,} \\ & && \text{since } \overline{k} \leq k_{s+1} \leq k_s - 1. \end{aligned}$$

■

References

[1] A. N. Degtev. *tt*- and *m*-degrees. *Algebra and Logic*, 12:78–89, 1973.

[2] L. Harrington and C. A. Haught. Limitations on initial segment embeddings in the r.e. *tt*-degrees. To appear.

[3] C. A. Haught and R. A. Shore. Undecidability and initial segments of the (r.e.) *tt*-degrees. To appear, Journal of Symbolic Logic.

[4] C. A. Haught and R. A. Shore. Undecidability and initial segments of the *wtt*-degrees $\leq \mathbf{0}'$. This Volume.

[5] G. N. Kobzev. On *tt*-degrees of r.e. *T*-degrees. *Mathematics of the USSR, Sbornik*, 35:173–180, 1978.

[6] A. H. Lachlan. Initial segments of one-one degrees. *Pacific Journal of Mathematics*, 29:351–366, 1969.

[7] A. H. Lachlan. Recursively enumerable many-one degrees. *Algebra and Logic*, 11:186–202, 1972.

[8] R. E. Ladner and L. P. Sasso, Jr. The weak truth table degrees of recursively enumerable sets. *Annals of Mathematical Logic*, 8:429–448, 1975.

[9] S. S. Marchenkov. The existence of recursively enumerable minimal truth-table degrees. *Algebra and Logic*, 14:257–261, 1975.

[10] S. S. Marchenkov. On the comparison of the upper semilattices of recursively enumerable *m*-degrees and truth-table degrees. *Mathematical Notes of the Academy of Sciences of the USSR*, 20:567–570, 1976.

[11] P. Odifreddi. Strong reducibilities. *Bulletin (New Series) of the American Mathematical Society*, 4:37–86, 1981.

[12] P. Odifreddi. *Classical Recursion Theory*. Studies in Logic and the Foundations of Mathematics. North-Holland, Amsterdam, 1989.

[13] D. B. Posner. A survey of non-r.e. degrees $\leq \mathbf{0}'$. In F. R. Drake and S. S. Wainer, editors, *Recursion Theory: Its Generalizations and Applications, Proceedings of Logic Colloquium '79, Leeds, August 1979. London Mathematical Society Lecture Note Series, Volume 45*, pages 52–109, Cambridge, 1980. Cambridge University Press.

[14] H. Rogers, Jr. *Theory of Recursive Functions and Effective Computability*. MIT Press, Cambridge, 1987.

[15] G. E. Sacks. A minimal degree less than $\mathbf{0}'$. *Bulletin of the American Mathematical Society*, 67:416–419, 1961.

[16] G. E. Sacks. The recursively enumerable degrees are dense. *Annals of Mathematics (2)*, 80:300–312, 1964.

[17] J. R. Shoenfield. A theorem on minimal degrees. *Journal of Symbolic Logic*, 31:539–544, 1966.

[18] R. I. Soare. *Recursively Enumerable Sets and Degrees: The Study of Computable Functions and Computably Generated Sets.* Perspectives in Mathematical Logic, Ω-Series. Springer-Verlag, Berlin, 1987.

[19] C. Spector. On degrees of recursive unsolvability. *Annals of Mathematics (2)*, 64:581–592, 1956.

Σ_2-INDUCTION AND THE CONSTRUCTION OF A HIGH DEGREE

Marcia Groszek
Dept. of Math. and C.S., Dartmouth College, Bradley Hall
Hanover, NH, 03755, U.S.A.

Michael Mytilinaios
P.O. Box 470, Dept. of Math., University of Crete
71409 Iraklion, Greece

ABSTRACT:

We show that the subsystem of Peano Arithmetic containing only induction for Σ_2 formulas suffices to prove the existence of a recursively enumerable set of high, incomplete degree. By a result of Mytilinaios and Slaman, bounding for Σ_2 formulas does not suffice to prove that such a set exists. In contrast, by a result of Groszek and Slaman, induction for Σ_1 formulas suffices to carry out any Friedberg-style finite injury priority argument.

INTRODUCTION:

Since partial recursive functions can be coded as integers, a relatively weak subsystem of Peano Arithmetic suffices to discuss questions about Turing computability. (The statement of Post's problem, "there exists a recursively enumerable set of intermediate degree", can be phrased in the first-order language of Peano Arithmetic.) This allows us to consider the proof-theoretic strength of certain theorems and techniques of classical recursion theory.

One extremely important technique, used in particular to study the partial order of Turing degrees of recursively enumerable sets (the r.e. degrees), is the priority construction. Priority constructions are well understood to form a hierarchy of increasing complexity (finite injury, infinite injury, and beyond; or, 0' constructions, 0'' constructions, 0''' constructions, etc.) The complexity of a proposition about the r.e. degrees tends to be related in a natural way to the complexity of the priority construction used to prove it, and in many cases, one feels that this complexity is necessary; a simpler construction could not possibly prove the theorem.

We are indebted, directly and indirectly, to Steve Simpson for suggesting a proof-theoretic analysis of this hierarchy. The natural proof that a finite injury argument succeeds appeals to induction for Σ_2 formulas. (One must show, inductively, that each requirement is injured only finitely often; to say the injury set is finite takes a Σ_2 formula.) Simpson showed that the proof of the Friedberg-Muchnik theorem, which is a simple finite injury argument, can actually be carried out with only induction for Σ_1 formulas.

Mytilinaios [My] showed that the Sacks splitting theorem, which was originally proven using a more complicated finite injury argument [Sk], also holds in the presence of induction for Σ_1 formulas. However, Mytilinaios did not use Sacks's original proof, but a modification depending on ideas from alpha recursion theory (namely, Shore blocking, [Sr].) (Joe Mourad has since shown that the splitting theorem is in fact equivalent to induction for Σ_1 formulas. [Mo]) Mytilinaios and Ted Slaman also proved a negative result, regarding a theorem whose proof uses a simple infinite injury argument: They showed that bounding for Σ_2 formulas does not suffice to prove the existence of an r.e. set of high, incomplete degree ([MS].)

Groszek and Slaman ([GS]) showed that simple finite injury constructions akin to that of the Friedberg-Muchnik theorem (Π_1 constructions) can always be carried out using induction for Σ_1 formulas. In contrast, they showed that the more complicated finite injury constructions, such as that used in the original proof of the Sacks splitting theorem (Σ_2 constructions) generally require induction for Σ_2 formulas. (This is in fact an equivalence, in the spirit of the "reverse mathematics" program of Harvey Friedman, Simpson, and others: Over an appropriate base theory, induction holds for all Σ_1 formulas if and only if all Π_1 constructions can be carried out. Induction holds for all Σ_2 formulas if and only if all Σ_2 constructions can be carried out.) In particular, combined with the result of Mytilinaios and Slaman mentioned above, this shows that a Friedberg-style finite injury (Π_1)

argument cannot possibly prove the existence of a high incomplete r.e. degree.

In the result presented in this paper, Groszek and Mytilinaios showed that the construction of an r.e. set of high incomplete degree can be carried out using induction for Σ_2 formulas. (The argument that works in this setting is essentially the original argument; see [Sd].) This is an example of a basic infinite injury argument.

Groszek and Slaman, subsequently, isolated the notion of simple infinite injury in the definition of Π_2 construction, and showed, as above, that all Π_2 constructions can be carried out iff induction holds for all Σ_2 formulas. Thus the level of the theorem "there exists an r.e. set of high incomplete degree" in our hierarchy of theories is exactly the level of the technique originally used to prove the theorem. (This is in contrast to the Sacks splitting theorem.)

There has been much other work on these questions in the last several years.

The basic idea behind an infinite injury argument is as follows: There is an infinite list of requirements, arranged according to priority. Each requirement has the possibility of acting infinitely often, and hence, a priori, of imposing infinitely much restraint on the rest of the construction. Some technique must be used to keep a single requirement from taking over the whole construction.

A single requirement can be kept under control as follows: There are two possible cases for the action of the requirement. If it needs to act only finitely often, it will impose only finitely much restraint on the rest of the construction, which we can deal with as in a finite injury argument. If, on the other hand, it needs to act infinitely often, we can arrange so that each time it acts it first temporarily abandons (most of) its restraint on the rest of the construction. The result is that, in either case, the lim inf of the restraint imposed by a single requirement is finite.

Another idea is called for in combining two or more requirements. If two requirements each act (and therefore temporarily stop restraining the rest of the construction) infinitely often, but at different times, their actions may combine to impose infinitely much restraint on the rest of the construction. They must abandon their restraint simultaneously if anything interesting is to happen. Therefore requrements must be interleaved in such a way that the lim inf of the combined restraint of any finite collection of requirements remains finite. Proving that this works involves an argument which may appear as a "window lemma" or an analysis of what happens along the "true path" of the construction.

Notice that the action of each requirement is guided by the truth or falsity of a Σ_2 proposition, "there are only finitely many stages at which the requirement needs to act." In the case of the construction of a high degree, the Σ_2 proposition guiding the action of each requirement is of a particularly simple form, in that it does not refer to the construction. Specifically, the guiding proposition for the eth requirement is "$e \in 0''$ ". Thus our "window lemma", or "true path analysis", is accomplished by guiding the construction according to a "true path analysis" of $0''$. In fact, in combination with the recursion (fixed point) theorem, this organization can be made to work for any similarly simple construction.

In essence, the construction of a high incomplete r.e. set we present here is no different from the usual construction. Our analysis shows that induction for Σ_2 constructions suffices to carry out this argument.

BASIC DEFINITIONS:

We work in subsystems of Peano Arithmetic. Our base theory consists of P^- (finitely many axioms about the behavior of <, +, and $\cdot$) together with induction for Σ_0 formulas. (See [My].) In this base theory one can prove basic facts of number theory.

Over the base theory, there is a hierarchy of subsystems of Peano Arithmetic, identified by limited bounding or induction schema. In increasing strength, these theories are $B\Sigma_1$, $I\Sigma_1$, $B\Sigma_2$, $I\Sigma_2$, etc. (See [KP].)

In any model M of the base theory, we may use the exponential function to interpret single elements as codes of sets or sequences. (Any standard coding will do; see [E].) A subset of M is called M-finite (we may say simply finite) if it has a code in M.

We define the columns of a set A by interpreting the elements of A in any standard way as codes for ordered pairs; then the eth column of A, $A^{(e)}$, is

$$\{ x \mid (e, x) \in A \}.$$

The recursively enumerable subsets of M are those with a Σ_1 definition in M. The recursive subsets are those with a Δ_1 definition. (In each case, parameters of M are allowed in the definition.) The power to code and decode finite sequences allows us to define a complete Σ_1 set, that is, a recursively enumerable set, 0', such that any Σ_1 set A can be represented as a column of 0': for some fixed e,

$$x \in A \iff (e, x) \in 0'.$$

Furthermore, the number e can be found recursively from the Σ_1 definition for A and the parameters used in that definition.

Similarly, there is a complete Σ_2 set, 0''. As Σ_2 sets are Σ_1 relative to the complete Σ_1 set 0', we can view 0'' as the complete $\Sigma_1(0')$ set. In the same way, for any set A, we can define A' to be the complete $\Sigma_1(A)$ set. (Then 0'' is, in fact, (0')'.)

We will often identify a set with its characteristic function, thus, instead of $i \in A$, we may say $A(i) = 1$.

ENUMERATIONS AND TURING REDUCTIONS, IN $I\Sigma_1$:

Induction for sigma-one formulas is enough to show that a set A is recursively enumerable iff there is a recursive sequence, $(A[s] \mid s \geq 0)$, of finite sets whose union is A. We can also demand that:

(i) $A[0] = \emptyset$

(ii) $t > s \Rightarrow A[t] \supseteq A[s]$.

A sequence such as this is called a recursive enumeration of A.

Induction for sigma-one formulas also suffices to show that A is amenable, specifically, that for every n there is a stage s such that A agrees with A[s] up to n, i.e.,

$$(\forall x < n) [x \in A \iff x \in A[s]].$$

A reduction procedure is a recursively enumerable set Φ of 4-tuples, (P, N, i, j), where P and N are disjoint finite sets and i and j are numbers. We may also assume our reduction procedures are monotone and well-defined, i.e.,

$$[(P, N, i, j) \in \Phi \ \& \ P' \supseteq P \ \& \ N' \supseteq N \ \& \ P' \cap N' = \emptyset] \Rightarrow$$
$$[(P', N', i, j') \in \Phi \Leftrightarrow j' = j].$$

We say that $\Phi^A(i) = j$ provided that there are finite sets, P included in A, and N disjoint from A, such that $(P, N, i, j) \in \Phi$. The 4-tuple $(P, N, i, j) \in \Phi$ is said to be a computation in Φ that applies to A. The sets P and N are the positive and negative neighborhood conditions of the computation.

We say B is recursive in A ($B \leq_T A$) iff for some reduction procedure Φ, the characteristic function of B is equal to Φ^A. B can be shown to be Σ_1 relative to A iff B is the domain of some relation R which is recursive in A ($R \leq_T A$.)

We let Φ_e designate the eth reduction procedure (or, more precisely, the reduction procedure given by the eth recursive enumeration $(\Phi_e[s] \mid s \geq 0)$.) We say $\Phi_{e,s}{}^A(i) = j$ iff some computation $(P, N, i, j) \in \Phi[s]$ applies to A.

Induction for sigma-one formulas also implies a kind of amenability for reductions from recursively enumerable sets. Namely, if A is a recursively enumerable set given by $(A[s] \mid s \geq 0)$, then $\Phi^A(i) = j$ iff, for some s, $\Phi_{e,s}{}^A(i) = j$. This holds iff, for some s, some computation $(P, N, i, j) \in \Phi[s]$ applies to A[s], (so $\Phi_{e,s}{}^{A[s]}(i) = j$), and furthermore $N \cap A = \emptyset$.

We define the use of a computation (P, N, i, j) to be the least number larger than every element of $P \cup N$. We define use($\Phi^A(i)$) to be the use of the least computation (P, N, i, j) in Φ which applies to A. ($I\Sigma_1$ guarantees that there is in fact a least such computation, if A is recursively enumerable.) The amenability of reductions from recursively enumerable sets can then be rephrased as follows:

$$\Phi^A(i) = j \Leftrightarrow \text{ for some } s,\ \Phi_{e,s}{}^{A[s]}(i) = j, \text{ and } A \text{ agrees with } A[s] \text{ up to use}(\Phi_{e,s}{}^{A[s]}(i)).$$

REPRESENTING 0'' AS A LEFTMOST PATH IN $I\Sigma_2$:

Definition: Let (T, <) be the tree of (M)-finite sequences from {0, 1}, ordered by inclusion.

For $\alpha \in T$, $|\alpha|$ denotes the length of α.

For α and β in T, say α is to the left of β iff, for all i in their common domain, $\alpha(i) = 1 \Rightarrow \beta(i) = 1$, and there is some i such that $\beta(i) = 1$ & $\alpha(i) = 0$. We say α is to the right of β iff α is not to the left of β and α and β do not agree on their common domain. (Note the somewhat counter-intuitive asymmetry in this definition: As α is to the right of β iff there is some i such that $\beta(i) = 0$ & $\alpha(i) = 1$, it is quite possible that α is to the right of β and also β is to the right of α.)

Define an ordering $<^*$ on T by $\alpha <^* \beta$ iff either α is to the left of β or $\beta \supset \alpha$.

Claim: There is a recursive sequence $(\alpha_s \mid s \geq 0)$ from T such that $|\alpha_s| = s$, and for all n, $0''\lceil n$ is the unique $\alpha \in T$ of length n such that

(i) $(\forall t)\,(\exists s > t)\,[\,\alpha_s\lceil n = \alpha\,]$.

(ii) $(\exists t)\,(\forall s > t)\,[\,\alpha <^* \alpha_s\,]$.

Remark: This is the usual description of 0'' as the "leftmost path visited infinitely often" on a recursive walk through the tree T.

We will show this claim is provable from $I\Sigma_2$. As it implies the amenability of 0'', hence of all Σ_2 sets, it is therefore equivalent to $I\Sigma_2$.

Definition: Let 0'' have the Σ_2 definition:

$$(x \in 0'') \Leftrightarrow (\exists y)\,(\forall z)\,[\,\phi(x, y, z)\,].$$

Define from this functions:

$$f_s(x) = \begin{cases} \text{the least } y < s \text{ such that } (\forall z < s)\,[\,\phi(x, y, z)],\\ \quad \text{if such a } y \text{ exists} \\ s \text{ otherwise} \end{cases}$$

$$g_s(x) = \max(x, f_s(x))$$

$$g(x) = \lim_{s \to \infty} g_s(x)$$

Remark: Since $s < t \Rightarrow g_s(x) \leq g_t(x)$, we have

$$g(x) = y \Leftrightarrow [(\exists s)\,[\,g_s(x) = y]\ \&\ (\forall s)\,[\,g_s(x) \leq y]].$$

This is a fairly simply defined function. It is easy to see in $I\Sigma_2$ that

$$x \in 0'' \Rightarrow \quad g(x) \text{ exists}$$
$$x \notin 0'' \Rightarrow \quad \lim_{s \to \infty} g_s(x) = \infty.$$

For any n, we can find a stage t such that, for all $x < n$,

$$x \in 0'' \Rightarrow g(x) = g_t(x) < t.$$

(This is an essential use of $I\Sigma_2$, to show that the partial Π_1 function which takes x to the least stage t at which $g(x) = g_t(x)$, when restricted to a finite domain, has bounded range.) Using the fact that $g_s(x) \geq x$, we can find a further stage t' such that, for all x,

$$g(x) > t \text{ or } x \notin 0'' \Rightarrow g_{t'}(x) > t$$
$$g(x) \leq t \Rightarrow g_{t'}(x) = g_t(x) = g(x).$$

<u>Definition:</u> Define $(\alpha_s \mid s \leq 0)$ recursively:

$$\alpha_0 = \emptyset$$
$$\alpha_{s+1}(s) = 0$$

for $x < s$,

$$\alpha_{s+1}(x) = \begin{cases} 0 \text{ if } (\exists x' < s)\,[\,g_s(x') \leq g_s(x)\ \&\ g_{s+1}(x') \neq g_s(x')\,] \\ 1 \text{ otherwise.} \end{cases}$$

<u>Proposition:</u> This sequence approximates 0'' as claimed.

<u>Proof:</u> Let n be given. As remarked above, there are t and t' such that:

$$x < n\ \&\ x \in 0'' \Rightarrow g(x) < t$$
$$g(x) \leq t \Rightarrow g(x) = g_{t'}(x)$$
$$[g(x) \geq t \text{ or } g(x) \text{ is undefined}] \Leftrightarrow g_{t'}(x) \geq t.$$

Clearly, for $s > t'$ and $x < n$, if $x \in 0''$, then $\alpha_s(x) = 1$. So $0''\lceil n <^* \alpha_s$.

For any $s > t'$, we can choose $x < s$, $x \notin 0''$, such that $g_s(x)$ is a minimum. (This uses the amenability of 0''.) Let s' be least such that $g_{s'}(x) > g_s(x)$. Then $\alpha_{s'}\lceil n = 0''\lceil n$.

This proves the claim.

<u>FRIEDBERG REQUIREMENTS, IN $I\Sigma_1$:</u>

Definition: Let $(A[s] \mid s \geq 0)$ and $(B[s] \mid s \geq 0)$ be recursive enumerations of sets A and B. We say these enumerations respect the requirement

$$\Phi_e^A(i) \neq B(i)$$

at stage t iff one of

(i) $i \notin B[t+1] - B[t]$ &
$i \in B[t] \Rightarrow$ for the unique s such that $i \in B[s+1] - B[s]$, and for the least computation $(P, N, i, 0) \in \Phi_e[s]$ applying to $A[s]$, $(A[t+1] - A[t]) \cap N = \varnothing$.

(ii) $i \in B[t+1] - B[t]$ & $\Phi_{e,t}^{A[t]}(i) = 0$ &
for the least computation $(P, N, i, 0) \in \Phi_e[t]$ applying to $A[t]$, $(A[t+1] - A[t]) \cap N = \varnothing$.

We say the requirement is acted on at stage t iff it is respected at stage t and

$[i \notin B[t]$ & $\Phi_{e,t}^{A[t]}(i) = 0] \Rightarrow$ case (ii) above holds.

Proposition: Suppose the requirement $\Phi_e^A(i) \neq B(i)$ is respected at all stages and acted on at unboundedly many stages. Then $\Phi_e^A(i) \neq B(i)$.

Proof: We consider two cases.

If $i \notin B$, then consider any stage t at which the requirement is acted on. Since $i \notin B[t+1]$, the requirement is not respected according to case (ii); since $i \notin B[t]$, it must be the case that $\Phi_{e,t}^{A[t]}(i) \neq 0$. This is true for unboundedly many stages t, and therefore $\Phi_e^A(i) \neq 0$. By assumption, however, $B(i) = 0$, giving us the desired inequality.

If $i \in B$, consider the unique stage s such that $i \in B[s+1] - B[s]$. Since the requirement is respected at every stage t, then for every $t \geq s$, for the least computation $(P, N, i, 0) \in \Phi_e[s]$ applying to $A[s]$, $(A[t+1] - A[t]) \cap N = \varnothing$. But then, since $A[s] \cap N = \varnothing$, in fact $A \cap N = \varnothing$. Therefore the computation $(P, N, i, 0)$ still applies to A, and so $\Phi_e^A(i) = 0$. By assumption, however, $B(i) = 1$, giving us the desired inequality.

THE CONSTRUCTION:

We work inside a model M of $I\Sigma_2$.

We describe a recursive enumeration of a set H, $(H[s] \mid s \geq 0)$, and a recursive enumeration of a set B, $(B[s] \mid s \geq 0)$.

We will guarantee that H is incomplete by arranging that H does not compute B, using Friedberg strategies. We will guarantee that H is high ($0'' \leq H'$) by arranging that

$i \in 0'' \Rightarrow$ column i of H is finite

$i \notin 0'' \Rightarrow$ column i of H is cofinite.

The basic strategy for guaranteeing this is by noticing that $i \in 0''$ iff g(i), the limit of $g_s(i)$, exists; this is true iff, for all but finitely many s, $\alpha_s(i) = 1$. Therefore, we can accomplish our purpose by enumerating (i, j) into H at stage s iff $(i, j) < s$ and $\alpha_s(i) = 0$. For each i, there will be finitely much additional restraint imposed by the Friedberg strategies, but this is harmless.

We use the leftmost-path approximation to 0'' to guide our construction, not just in building H to be high, but in deciding which Friedberg requirements to act on. Basically, at each stage s we act as though α_s is the correct approximation to 0''; we freely destroy action taken on the assumption that 0'' is to the right of α_s, and preserve action taken on the assumption that 0'' is to the left of α_s (but take no new action on that assumption.) Therefore if α is an initial segment of 0'', eventually it will always be the case that $\alpha <^* \alpha_s$ (so there is only finitely much action taken to the left of α to be preserved by action at α) and infinitely often α_s will agree with α (so action taken to the right of α can be ignored, and action will be taken at α infinitely often.)

There is an additional reason for explicitly guiding our construction by $(\alpha_s \mid s \geq 0)$, besides ease of exposition, in this setting. The only use of $I\Sigma_2$ in this argument, when organized in this way, is in the leftmost-path approximation of 0''. This will have some consequences regarding models in which $I\Sigma_2$ fails.

<u>Definition:</u> The construction is defined recursively as follows.

<u>Stage 0:</u> $H[0] = \varnothing$

$B[0] = \varnothing$.

<u>Stage s+1:</u> For $\alpha \in T$ such that $|\alpha| \leq s$:

If α is left of α_s:

$r_{s+1}(\alpha) = r_s(\alpha)$

$H(s+1, \alpha) = H(s, \alpha)$

$\Phi_{f(s+1, \alpha)} = \Phi_{f(s, \alpha)}$
$i_{s+1}(\alpha) = i_s(\alpha)$.

If α is right of α_s:

$r_{s+1}(\alpha)$ is undefined
$H(s+1, \alpha)$ is undefined
$\Phi_{f(s+1, \alpha)}$ is undefined
$i_{s+1}(\alpha)$ is undefined.

If $\alpha_s \supseteq \alpha$:

Let β_s be the smallest initial segment β of α_s such that:

$i_s(\beta) \notin B[s]$
$(\Phi_{f(s, \beta), s}{}^{H[s]})(i_s(\beta)) = 0$.

If there is such a β_s we set

$B[s+1] = B[s] \cup \{i_s(\beta_s)\}$.

We set

$$r_{s+1}(\alpha) = \begin{cases} r_s(\alpha) & \text{if } \beta_s \supset \alpha \text{ or there is no such } \beta_s \\ & \text{and } r_s(\alpha) \text{ is defined} \\ \text{use } ((\Phi_{f(s, \alpha), s}{}^{H[s]})(i_s(\alpha))) & \text{if } \alpha = \beta_s \\ 0 & \text{if } \alpha \supset \beta_s \text{ or } r_s(\alpha) \text{ is undefined.} \end{cases}$$

We set

$$H[s+1] = H[s] \cup \{(i, j) \mid (i, j) \le s \ \& \ \alpha_s(i) = 0 \ \& \ (\forall \gamma <^* \alpha_s \lceil i) \ [r_{s+1}(\gamma) < (i, j)]\}.$$

We set

$$H(s+1, \alpha) = \{(i, j) \mid i < |\alpha| \ \& \ [\text{either } (i, j) \in H[s+1] \text{ or } [\alpha(i) = 0 \ \& \ (\forall \gamma <^* \alpha) \ [r_{s+1}(\gamma) < (i, j)]\ ]]\}.$$

We set $\Phi_{f(s+1, \alpha)}$ to be the reduction prodecure defined by:

$$\Phi_{f(s+1, \alpha)}{}^X = \Phi_{|\alpha|}{}^{X \cup H(s+1, \alpha)}.$$

We set

$$i_{s+1}(\alpha) = \begin{cases} i_s(\alpha) & \text{if } \beta_s \supset \alpha \text{ or there is no such } \beta_s \\ & \text{and } i_s(\alpha) \text{ is defined} \\ \max \{i_{s'}(\gamma) \mid s' \le s\} + |\alpha| + 1 & \\ & \text{otherwise.} \end{cases}$$

This defines the construction.

Claim: Suppose $|\alpha| < t$ and $t < t'$ are stages such that, for $t \le s < t'$, $\alpha <^* \alpha_s$, and β_s is not a proper initial segment of α. Then, for $t \le s < t'$,

$i_{s+1}(\alpha) = i_t(\alpha)$
$H(s+1, \alpha) = H(t, \alpha)$

$\Phi_{f(s+1,\,\alpha)} = \Phi_{f(t,\,\alpha)}$.

Furthermore, there is at most one s such that $t \leq s < t'$ and $\beta_s = \alpha$, and if there is no such s, then also for $t \leq s < t'$,

$r_{s+1}(\alpha) = r_t(\alpha)$.

<u>Proof:</u> This, being a recursive proposition about s, can be proven by induction on s.

Since $i_s(\alpha)$ is not redefined at stages when either α is to the left of α_s or α is an proper initial segment of β_s, clearly for all s between t and t', $i_{s+1}(\alpha) = i_s(\alpha)$.

For the remainder of the claim, first consider the case where it is also the case that β_s is never equal to α for $t \leq s < t'$. Then we also know that, for all such s, $r_{s+1}(\alpha) = r_s(\alpha)$, by the same reasoning that we used for $i_{s+1}(\alpha)$.

For any α in question, then, we know that for all $\gamma <^* \alpha$, and all s such that $t \leq s < t'$, $r_{s+1}(\gamma) = r_s(\gamma)$. What is more, for all such stages s, we know that for any i in domain(α), $\alpha_s(i) = 0 \Rightarrow \alpha(i) = 0$ (by definition of $<^*$.) This implies, in particular, that if i is in domain(α) and (i, j) is in H[t'] but not in H[t], then $\alpha(i) = 0$. (This is because (i, j) is enumerated into H at stage s only in the case that $\alpha_s(i) = 0$, by the recursive definition of H.)

Now, we show that $H(s+1, \alpha) = H(s, \alpha)$ for all relevant s. Suppose (i, j) is in $H(s+1, \alpha)$ but not in $H(s, \alpha)$. In particular, then, (by the definition of $H(s, \alpha)$), (i, j) is not in H[s]; also, either $\alpha(i) = 1$ or $(\exists\gamma <^* \alpha)[r_s(\gamma) < (i, j)]$. If (i, j) is in $H(s+1, \alpha)$, this must be because (i, j) is in H[s+1] -- by the above paragraph, if $(\exists\gamma <^* \alpha)[r_s(\gamma) < (i, j)]$ then also $(\exists\gamma <^* \alpha)[r_{s+1}(\gamma) < (i, j)]$. But that means that (by definition of H[s+1]) $\alpha_s(i) = 0$, so that $\alpha(i) = 0$. And that means that (again by definition of $H(s, \alpha)$), since it is not the case that (i, j) is in $H(s, \alpha)$, $(\exists\gamma <^* \alpha)[r_s(\gamma) \geq (i, j)]$. But now, since $\alpha <^* \alpha_s$, $(\exists\gamma <^* \alpha_s)[r_s(\gamma) \geq (i, j)]$, which contradicts our conclusion above that (i, j) must be in H[s+1] but not in H[s] (by the definition of H[s+1].)

Finally, since $f(s, \alpha)$ is defined from $H(s, \alpha)$, it remains only to show that there can be at most one s such that $\beta_s = \alpha$. So suppose that $\beta_s = \alpha$, where s is chosen to be the least such s between t and t'. By construction, at this stage $i_s(\alpha)$ is enumerated into B. For s' > s, between t and t', we have shown that $i_{s'}(\alpha) = i_s(\alpha)$. In particular, this means that $i_{s'}(\alpha) \in B[s']$. But by definition of $\beta_{s'}$, this means $\beta_{s'}$ cannot equal α.

This completes the proof of this claim.

Claim: Suppose $|\alpha| < t$ and $t < t'$ are stages such that, for $t \le s < t'$, $\alpha <^* \alpha_s$. Then there are at most $2^{|\alpha|}$ stages s such that $t \le s < t'$ and $\alpha \supseteq \beta_s$.

Proof: This follows easily from the above claim, by induction on $|\alpha|$.

Corollary: Suppose α is an initial segment of 0''. (By the amenability of 0'', α is in M.) Then there is a stage t such that, for $s \ge t$, $\alpha <^* \alpha_s$, β_s is not a proper initial segment of α, $i_{s+1}(\alpha) = i_t(\alpha)$, $H(s+1, \alpha) = H(t, \alpha)$, and $\Phi_{f(s+1, \alpha)} = \Phi_{f(t, \alpha)}$.

Proof: This uses the fact that, in $I\Sigma_1$, a partial recursive function on a finite set has bounded range. The relevant partial recursive function is the following: Let t' be a stage such that, for $t' \le s$, $\alpha <^* \alpha_s$. (There is such a t' by the properties of the leftmost-path approximation of 0''.) Then define f(i), for $i < 2^{|\alpha|}$, to be the ith stage $s > t'$ such that $\alpha \supseteq \beta_s$. Choose t to be any number greater than everything in the range of f.

Claim: For this stage t, we have:

$$H(t, \alpha) = \{(i, j) \in H \mid i < |\alpha|\}$$

and $\Phi_{f(t, \alpha)}{}^H = \Phi_{|\alpha|}{}^H$.

Proof: The second part of this claim follows from the first. To prove the first, let (i, j) be any pair such that $i < |\alpha|$.

First, suppose that $(i, j) \in H(t+1, \alpha)$ $(= H(t, \alpha).)$ If $(i, j) \in H[t+1]$, then clearly $(i, j) \in H$. If not, then by definition of $H(t+1, \alpha)$, $\alpha(i) = 0$, and $(\forall\gamma <^* \alpha)\ [r_{t+1}(\gamma) < (i, j)]$. By choice of t, and the above claims, this last fact still holds for any stage $t' > t$: $(\forall\gamma <^* \alpha)\ [r_{t'+1}(\gamma) < (i, j)]$. By properties of the leftmost-path approximation of 0'', we can choose $t' > t$, $t' < (i, j)$, such that $\alpha_{t'} \supseteq \alpha$; then, by definition of H[t'+1], $(i, j) \in H[t'+1]$, so again $(i, j) \in H$.

Conversely, suppose that $(i, j) \in H$. Then, for some stage $t' > t$, $(i, j) \in H(t', \alpha)$; but this, by choice of t, is the same as $H(t, \alpha)$.

Claim: Suppose α is an initial segment of 0''. Consider the least stage t such that, for all $s \ge t$, $\alpha <^* \alpha_s$, and β_s is not a proper initial segment of α. There is a least such t by $I\Sigma_1$. (For this stage t, by the above, $i_{s+1}(\alpha) = i_t(\alpha)$, $H(s+1, \alpha) = H(t, \alpha)$, and $\Phi_{f(s+1, \alpha)} = \Phi_{f(t, \alpha)}$.) Then, for all stages s, the Friedberg requirement $\Phi_{f(t, \alpha)}{}^H(i_t(\alpha)) \neq B(i_t(\alpha))$ is respected at stage s; furthermore, if $s \geq t$ is a stage such that $\alpha_s \supseteq \alpha$, then this requirement is acted on at stage s.

<u>Proof:</u> First, note that by definition of t (and by the construction), t must be the least stage s such that $i_s(\alpha) = i_t(\alpha)$. (I.e., $i_s(\alpha)$ is redefined at stage t.) So, by construction, $i_t(\alpha)$ is not in B[t], and is different from any other $i_{s'}(\gamma)$ (for any $\gamma \neq \alpha$ and any stage s'.)

If $i_t(\alpha)$ is not in B, then the requirement is respected at all stages. To show it is acted on at stages $s \geq t$ such that $\alpha_s \supseteq \alpha$, it suffices to show that, for such stages s, $\Phi_{f(t, \alpha),s}{}^{H[s]}(i_t(\alpha)) \neq 0$, i.e., that $\Phi_{f(s, \alpha),s}{}^{H[s]}(i_s(\alpha)) \neq 0$. But this must be the case, otherwise (since we know that β_s is not a proper initial segment of α) we must have $\beta_s = \alpha$, which would mean that $i_s(\alpha) \in B[s+1]$.

If $i_t(\alpha)$ is in B, that means that for some stage $t' \geq t$, we have $i_t(\alpha) \in B[t'+1] - B[t']$, and $i_{t'}(\alpha) = i_t(\alpha)$. The same reasoning as above applies to $s < t'$. At stage t', we must have $\beta_{t'} = \alpha$. This means that we have $\Phi_{f(t', \alpha),t'}{}^{H[t']}(i_{t'}(\alpha)) = 0$. Also, for this stage t', we have $r_{t'+1}(\alpha) =$ use $((\Phi_{f(t', \alpha), t'}{}^{H[t']})(i_{t'}(\alpha))$, and since the reduction procedure $\Phi_{f(t', \alpha)}{}^X$ does not depend on the columns $X^{(i)}$ for $i < |\alpha|$, we can assume that for the computation $(P, N, i_{t'}(\alpha), 0)$ in $\Phi_{f(t', \alpha)}$ determining that use, the negative neighborhood condition N does not contain any elements (i, j) for $i < |\alpha|$. But, since for all further stages s' we have $\alpha <^* \alpha_s$, and β_s is not an initial segment (proper or otherwise) of α, we have for all $s' \geq t'$, $r_{t'+1}(\alpha) = r_{s'+1}(\alpha) =$ use $((\Phi_{f(t', \alpha), t'}{}^{H[t']})(i_{t'}(\alpha))$, and no $(i, j) < r_{s'+1}(\alpha)$ for $i \geq |\alpha|$, (in particular, no (i, j) in N), is enumerated into H at stage s'. This shows that the requirement is acted on at stage t' (the stage at which $i_t(\alpha)$ is enumerated into B), and at all stages $s' \geq t'$.

<u>Claim:</u> B is not recursive in H.

<u>Proof:</u> For any e, choose $\alpha = 0''\ulcorner e$, and choose a stage t as above. Then $\Phi_{f(t, \alpha)}{}^H = \Phi_{|\alpha|}{}^H$. Furthermore, for all stages s, the Friedberg requirement $\Phi_{f(t, \alpha)}{}^H(i_t(\alpha)) \neq B(i_t(\alpha))$ is respected at stage s; furthermore, if $s \geq t$ is a stage such that $\alpha_s \supseteq \alpha$, then this requirement is acted on at stage s. Since there are unboundedly many stages s such

that $\alpha_s \supseteq \alpha$, this suffices to show that $\Phi_{f(t,\alpha)}{}^H(i_t(\alpha)) \neq B(i_t(\alpha))$, thus that $\Phi_e{}^H(i_t(\alpha)) \neq B(i_t(\alpha))$.

<u>Claim:</u> 0" is recursive in H'.

<u>Proof:</u> We show that, for any i,

$i \in 0'' \Rightarrow$ column i of H is finite

$i \notin 0'' \Rightarrow$ column i of H is cofinite.

The first statement is clear, because if i is in 0", then by the leftmost path approximation to 0", there is a stage t such that, for all $s > t$, $\alpha(s) = 1$. But then no (i, j) for $j > t$ is ever enumerated into H (by construction, i.e., by the definition of H[s+1].)

If i is not in 0", choose $\alpha = 0''\lceil i+1$, and choose a stage t such that, for $s \geq t$, $\alpha <^* \alpha_s$, and β_s is not an initial segment of α. By the above claims, for $s \geq t$, and $\gamma <^* \alpha$, $r_s(\gamma) = r_t(\gamma)$. Consider any j large enough so that $(i, j) > \max\{r_t(\gamma) \mid \gamma <^* \alpha\}$. Then, we know there is a stage $s > t$, $s > (i, j)$, such that $\alpha_s \supseteq \alpha$; by construction, $(i, j) \in H[s+1]$.

<u>Corollary:</u> H is a recursively enumerable set of incomplete high degree.

SUMMARY:

We have proved the following:

<u>Theorem:</u> $I\Sigma_2$ suffices to prove there is an r.e. set of incomplete high degree.

As pointed out, the only use of $I\Sigma_2$ was in the leftmost path approximation to 0". The existence of such an approximation to a Σ_2 set A is equivalent (given $B\Sigma_2$) to the amenability of A. Therefore we see:

<u>Proposition:</u> In any model of $B\Sigma_2$, if A is an amenable Σ_2 set, then there is an incomplete r.e. set B such that $A \leq_T B'$.

This does not necessarily mean that there are many options for the jump of an r.e. set: The model of Mytilinaios and Slaman, in which there is no high incomplete r.e. degree, actually has the property that any incomplete r.e. set A is low ($A' \leq_T 0'$.) A way to interpret this is that $B\Sigma_2$ is not very close to $I\Sigma_2$, in that $I\Sigma_2$ is equivalent to the amenability of every Σ_2 set. (In another sense, $B\Sigma_2$ captures most of $I\Sigma_2$; $I\Sigma_2$ can be phrased as $I\Sigma_1(0')$, and by a result of Groszek, Mytilinaios, and Slaman, $B\Sigma_2$ implies $I\Sigma_1(A')$ for every incomplete r.e. degree A.)

As we pointed out above, the existence of an r.e. set of incomplete high degree is an example of a theorem whose proof-theoretic strength (as measured by the standard hierarchy of $B\Sigma_1$ $I\Sigma_1$, $B\Sigma_2$ $I\Sigma_2$, etc.), is approximately the same as that of the priority argument used in the standard proof; the usual proof works, as above in $I\Sigma_2$, and the theorem cannot be proven from $B\Sigma_2$. The Sacks splitting theorem, in contrast, has weaker proof-theoretic strength than that of the priority argument used in its standard proof.

Question: Is there a theorem whose usual proof is via a (Sacks-style) finite injury argument, which can be shown to require the full strength of this technique ($I\Sigma_2$), or at least to be unprovable in $B\Sigma_2$?

REFERENCES:

[E] Enderton, H. B., A Mathematical Introduction to Logic, Academic Press, New York, 1972.

[GS] Groszek, M. J., and Slaman, T. A., "Foundations of the Priority Method I: Finite and Infinite Injury," to appear.

[KP] Kirby, L. A. S., and Paris, J. B., "Σ_n-collection schemas in arithmetic", in Logic Colloquium '77, North Holland, Amsterdam, 1978, 199-209.

[Mo] Mourad, J., Ph.D. thesis, University of Chicago, 1989.

[My] Mytilinaios, M. E., "Finite injury and Σ_1-induction", Jour. Sym. Log. 54 (1989), 38-49.

[MS] Mytilinaios, M. E., and Slaman, T. A., "Σ_2-collection and the infinite injury priority method", Jour. Sym. Log. 53 (1988), 212-221.

[Sk] Sacks, G. E., "On degrees less than 0''", Ann. of Math. (2) 77 (1963), 211-231.

[Sd] Shoenfield, J. R., "Undecidable and creative theories", Fundamenta Mathematicae 49 (1961), 171-179.

[Sr] Shore, R. A., "The recursively enumerable α-r.e. degrees are dense", Annals Math. Logic 9 (1976), 123-155.

This paper has not appeared or been submitted for publication elsewhere.

Undecidability and Initial Segments of the wtt-Degrees $\leq 0'$

CHRISTINE ANN HAUGHT
RICHARD A. SHORE

§1. INTRODUCTION

A notion of reducibility $\leq_r$ between sets is specified by giving a set of procedures for computing one set from another. We say that a set A is r-reducible to one B, $A \leq_r B$, if one of the procedures applied to B gives A. Associated with any such reducibility notion is the structure $\mathcal{D}_r$ of r-degrees, the equivalence classes of sets with respect to this reducibility, with the induced ordering. The most general notion of a computable reducibility is that of Turing, $\leq_T$. Here we say that $A \leq_T B$ if there is a Turing machine, φ_e which, when equipped with an oracle for B, computes A: $\varphi_e^B = A$. Such Turing degree computations are characterized by the phenomena that only during the computation itself do we discover how much information about B is needed to determine $A(x)$.

In contrast, for nearly all other computable reducibilities the amount of information, and indeed the set of questions to be asked of the oracle, is given in advance by a recursive procedure. Perhaps the most common example of such a procedure is many-one reducibility, $\leq_m$: $A \leq_m B$ if there is a recursive function f such that $x \in A \iff f(x) \in B$. Reducibilities with the property that the output, $A(x)$, is determined directly (by a truth table) by the answers that B gives to a set of questions calculated recursively from x are said to be of tabular type. The most general tabular reducibility is called *truth-table* reducibility, $\leq_{tt}$.

The reducibility we are concerned with here, weak truth table reducibility, $\leq_{wtt}$, lies between Turing reducibility and the tabular reducibilities in strength. In a wtt-computation, the output, $A(x)$, is determined by a recursive calculation based upon the answers that the oracle, B, gives to a set of questions calculated recursively from x. The recursive calculation based on answers to oracle queries need not be a truth table calculation; in fact it need not be total. Thus the wtt reduction procedures are sometimes partial (depending upon the oracle), whereas the tt reduction procedures are always total (regardless of which oracle set is used.)

The procedures $\{e\}$ associated with weak truth table reducibility are specified by a recursive function $f(= \{e\})$ which, for each x, gives a set of n questions about the oracle and, for each of the possible 2^n sets of answers, gives the corresponding output, if the computation converges, otherwise the computation is undefined. As usual this defines

---

The authors' research was supported by NSF grants DMS-8705818 and DMS-8601048, respectively and MSRI. In addition, Shore was supported by a grant from the U.S.-Israel Bi-national Science Foundation.

$A \leq_{wtt} B$ as "there is an e such that $\{e\}^B = A$". Basic information on several such strong reducibilities can be found in Rogers [**7**]. For more information we recommend the survey articles by Odifreddi [**5**] and Degtev [**2**] as well as Odifreddi's book [**6**].

Some of the methods used to prove structural results about the r.e. truth table degrees can be adapted to prove the analogous theorem about the wtt degrees below $0'$, $\mathcal{D}_{wtt}(\leq 0')$. In particular, an adaptation of Fejer and Shore's construction of a minimal r.e. tt-degree [**3**] provides the only known proof of the existence of a minimal degree in the wtt-degrees below $0'$. In this paper we prove that the first order theory of the wtt-degrees below $0'$ (in the language with $<$) is undecidable. The proof is an adaptation of the corresponding proof for the r.e. tt-degrees [**4**]. We prove that any finite partition lattice can be embedded as an interval in $\mathcal{D}_{wtt}(\leq_{wtt} 0')$ (the wtt-degrees below $0'$.)

This kind of translation is possible because of the similarity between tt- and wtt-reductions. One difference between tt- and wtt-reducibility is that, while still utilizing a recursively bounded use, wtt-reductions may be or appear to be partial. However, in general, if a reduction is really partial then it can be ignored in our proofs. So, although diagonalizations and the searches for e-splits are not as easy as they were in the tt-case where everything converged immediately, they can be handled by moving to a string which is at least as long as the necessary use but at the present time appears not to give a convergent computation. If the computation remains partial with this string as an initial segment of the oracle then this reduction procedure may be ignored. Otherwise it converges and we have a way to force the reduction procedures to become total, i.e. to act like tt-reductions. Notice that in order to achieve this control over a single wtt-reduction procedure, only a recursively bounded amount of the set we are constructing is used.

To see that embedding the class of finite partition lattices as initial segments of a structure suffices to prove the undecidability of the structure, we refer the reader to the corresponding discussion in [**4**]. The proof relies on results proved by Burris and Sankappanavar [**1**].

We will make use of the following notational conventions. The recursive use function for the wtt-reduction procedure with index e will be denoted by $\{e\}$. A wtt-reduction procedure with index e and oracle X will be written as $\{e\}^X$.

In section 2 we give a brief sketch of the construction in [**4**] for embedding the finite partition lattices in the r.e. tt-degrees. In section 3 we discuss the modifications of this construction which need to be made for the wtt-degrees below $0'$. Section 4 contains the proof of the main theorem; conclusions are in section 5.

§2. R.E. TT-DEGREES

The embedding of the finite partition lattices in the r.e. tt-degrees above a minimal degree builds upon Fejer and Shore's construction of a minimal r.e. tt-degree. Throughout the rest of the paper we will assume that the reader is familiar with the Fejer-Shore construction; it appears in this volume, [3].

We modify the basic minimal degree construction to embed the finite partition lattices. A single right branch in the minimal degree construction is replaced by an n-tuple of right branches in this construction.

The trees used for the r.e. tt-degree embeddings were maps,

$$T : \{0^m \star i : i \in \{0, 1, 2, ..., n\}, m \in \omega\} \longrightarrow 2^{<\omega}$$

with the property that

for each $m \in \omega$ there is a unique x such that for all $i \in \{1, 2, ..., n\}$

$$T(0^m \star i)(nx + i) = 1, \text{ and}$$
$$\forall j \in \{1, 2, ..., n\}, \text{ if } j \neq i \text{ then } T(0^m \star i)(nx + j) = 0, \text{ and}$$
$$\forall x' \text{ such that } x < x' \leq s, \forall j \in \{1, 2, ..., n\}, T(0^m \star i)(nx' + j) = 1$$

The lattices embedded are the duals of the lattices of all equivalence relations on the finite set $\{1, 2, ..., n\}$. The following notation is used.

$$\Pi_n = \{\rho : \rho \text{ is an equivalence relation on } \{1, 2, ..., n\}\}$$

The elements of Π_n are ordered by the dual of refinement:

$$\rho \leq \tau \iff (\forall i, j \in \{1, 2, ..., n\} \, (i\tau j \Rightarrow i\rho j))$$

The reversed order is used since it is the order produced by our trees. Note that the embedding of these dual lattices also allows us to deduce the undecidability of $\mathcal{D}_{tt}(\leq 0')$ and $\mathcal{R}_{tt}$ as a corollary.

Consider the possibilities for e-splittings among the n-tuple of right branches at a particular level in our newly configured trees. In the minimal degree construction we were only interested in whether or not two strings e-split. In this construction we want to analyze which sets of branches at a particular level do or do not e-split. This corresponds almost exactly to the set of all equivalence relations on the n right branches. In this way, the tree shape yields an embedding of Π_n above a minimal degree.

The r.e. tt-degree construction enumerates an r.e. set $A = \bigcup_{s \in \omega} A_s$; A_s will lie on each tree of a nested sequence of trees, viz.

$$T_{-1,s} \subseteq T_{0,s} \subseteq ... \subseteq T_{e,s} \subseteq ... \subseteq T_{s,s}$$

For each $\rho \in \Pi_n$, we define a set A_ρ from A in a positive way so that A_ρ is also r.e.:

$$A_\rho = \{nx + i : \exists j \leq n \, (nx + j \in A \text{ and } j\rho i)\}.$$

Congruence modulo ρ ($\equiv_\rho$) on strings in $2^{<\omega}$ is defined by

$$\sigma_1 \equiv_\rho \sigma_2 \iff \{nx+i : \exists j \leq n(\sigma_1(nx+j) = 1 \text{ and } j\rho i\} = \{nx+i : \exists j \leq n(\sigma_2(nx+j) = 1 \text{ and } j\rho i\}$$

The shape of the trees now ensures that for all $e, s, k \in \omega$, all $\rho \in \Pi_n$ and all $i, j \in \{1, 2, ..., n\}$,

$$T_{e,s}(0^k \star 0) \not\equiv_\rho T_{e,s}(0^k \star i), \text{ and}$$

$$i\rho j \iff T_{e,s}(0^k \star i) \equiv_\rho T_{e,s}(0^k \star j)$$

Note that for any e and s the visible pattern of e-splittings among the n-tuple of right branches at a particular level of a tree $T_{e,s}(0^m)$ at stage s defines an equivalence relation on $\{1, 2, ..., n\}$ as follows: For $i, j \in \{1, 2, ..., n\}$, define the relation $i\sigma j \iff [e]^{T_{e,s}(0^m \star i)}$ and $[e]^{T_{e,s}(0^m \star j)}$ are compatible as strings. Our trees will have the property that for any $i, j \in \{1, 2, ..., n\}$, $|T_{e,s}(0^k \star i)| = |T_{e,s}(0^k \star j)| = s$, and so our notational conventions will guarantee that for $i \in \{1, 2, ..., n\}$ the strings $[e]^{T_{e,s}(0^m \star i)}$ all have the same length, and so it is easy to see that σ is an equivalence relation. Thus, the e-states correspond to a sequence of elements of Π_n of length e.

Next consider the interaction between the leftmost branch and the n-tuple of right branches. There are two possibilities.

Case 1. For all $k \geq m$, $T_e(0^k \star 0)$ e-splits with each of $T_e(0^k \star i)$ for each $i \in \{1, 2, ..., n\}$. Let ρ be such that in the limiting tree, T_e, for all but finitely many k, for each $i, j \in \{1, 2, ..., n\}$

$$\Big(\big(T_e(0^k \star i) \text{ and } T_e(0^k \star j) e\text{-split}\big) \iff \neg(i\rho j)\Big)$$

Fix m so that this holds for all $k \geq m$.In this case we argue that $[e]^A \equiv_{tt} A_\rho$, since all but finitely many paths in T_e are congruent modulo $[e]^A$ if and only if they are congruent modulo A_ρ.

Case 2. For some $k \geq m$ and some $j \in \{1, 2, ..., n\}$, $T_e(0^k \star 0)$ does not e-split with $T_e(0^k \star j)$. Take i to be the least j such that $T_e(0^k \star 0)$ does not e-split with $T_e(0^k \star j)$. In this case we argue that $[e]^A$ is recursive. Even though the tree T_e is not recursive, with knowledge of the correct final e-state we can recover enough of T_e to give a recursive computation of $[e]^A$. We assign to this configuration the e-state i. We claim that there can never be an e-split between A and $T_e(0^k \star i) \star 1^s$ for any s. (Suppose that there were such a split. Let s and k'' be large enough so that $A \supseteq T_{e,s}(0^{k''})$ and $T_{e,s}(0^{k''})$ e-splits with $T_{e,s}(0^k \star i)$ and $k'' \geq k+1$ and $T_{e,s}(0^k \star i) = T_e(0^k \star i)$. This would be visible during the construction at stage $s+1$ and the definition $T_{e,s+1}(0^k \star 0) \supseteq T_{e,s}(0^{k''})$

and $T_{e,s+1}(0^k \star j) = T_{e,s}(0^k \star j)$ for all $j \in \{1,2,...,n\}$ would increase the value of the e-state at e of $T_{e,s+1}(0^k)$ to something larger than i. Since $T_{e,s+1}$ is built by maximizing e-states, this change will occur during the construction. Thus, $T_e(0^k \star 0)$ really would e-split with $T_e(0^k \star i)$, a contradiction. This produces a recursive computation of $[e]^A$: $[e]^A(x) = [e]^{T_e(0^k \star i) \star 1^h}(x)$, where $h = |\,[e](x)| - |T_e(0^k \star i)|$.

This coding and tree structure preserve the order relations holding in Π_n since if $\rho \leq \tau$ then

$$nx + i \in A_\rho \iff \exists j \in \{1,2,...,n\}\,(nx + j \in A \text{ and } j\rho i)$$
$$\iff \exists k \in \{1,2,...,n\}\,(nx + k \in A_\tau \text{ and } k\rho i)$$

and so $A_\rho \leq_{tt} A_\tau$.

The coding and tree structure also allow room for diagonalization. If $\rho \not\leq \tau$, then there exist i and j such that $i\tau j$ and $\neg i\rho j$, and so for each e, s, $T_{e,s}(i) \equiv_\tau T_{e,s}(j)$ and $T_{e,s}(i) \not\equiv_\rho T_{e,s}(j)$. Thus the opportunity to make A extend $T_{e,s}(i)$ or $T_{e,s}(j)$ allows us to satisfy $A_\rho \neq [e]^{A_\tau}$.

§3. Changes necessary for the wtt-degrees

We isolate and discuss four problems (and their solutions) which arise when translating the proof from the r.e. tt-degree setting to the Δ_2^0 wtt-degree setting. These problems are all related to the partialness of wtt-reductions.

(1) Because of partialness, the diagonalization step might not be carried out instantaneously. Suppose we are working on the requirement

$$P_k : \{e\}^{G_\rho} \neq G_\tau.$$

As in the tt-degree proof, we reserve the first level of the tree T_k for work on requirement P_k. There will be $i, j \in \{1,...,n\}$ such that $T_k(i) \equiv_\rho T_k(j)$ and there is an x such that $x \in (T_k(i))_\tau$ and $x \notin (T_k(i))_\tau$. In the tt-case we could assume that $[e]^{(T_k(i))_\rho}(x)$ converged since it was a tt-reduction. Now, it might be the case that at stage s when we are acting for P_k, $\{e\}_s^{(T_k(i))_\rho}(x)$ diverges. If the computation is truly divergent, then by making G extend $T_k(i)$ we will have won the diagonalization requirement. But, it may be the case that $\{e\}^{(T_k(i))_\rho}(x)$ later does converge and gives the same answer as $(T_k(i))_\tau(x)$. In this case we can win the diagonalization requirement by making G extend $T_k(j)$. Of course we cannot recognize which of these two cases holds during a recursive construction. The solution is as follows: If we act for P_k at stage s, then we make G extend the right branch which gives a disagreement (visible at stage s), if such a branch exists. Otherwise, we make G extend $T_k(h)$, where h is the least of i, j such that $\{e\}^{(T_k(h))_\rho}(x)$ diverges. If at a later stage we see that $\{e\}^{(T_k(h))_\rho}(x)$ converges and agrees with $(T_k(h))_\tau(x)$, then we move G to extend $T_k(j)$ (w.l.o.g. $h \neq j$) and win the diagonalization requirement.

(2) Apparent partialness may temporarily conceal a better e-state. This could cause a problem if the better e-state is not visible until a stage in the construction when some action for the sake of a diagonalization requirement has made it impossible to make use of this improved e-state. However, during a recovery procedure, the splitting may become apparent and destroy our assumptions about the existence and non-existence of e-splitting in the tree. This potential problem is solved by the fact that G is Δ^0_2. We can undo action taken for the sake of a diagonalization requirement and move back to a setting where the better e-state is available. A priority ordering among the positive requirements and the indices of the tree guarantees that the strategies for the positive requirements will still be able to succeed.

(3) The argument used in the tt-degree proof that either we get e-splits with the left path or $\{e\}^G$ is recursive no longer works because the computation procedure we had hoped to use is not necessarily total. The solution to this problem is an active one: we move to the path which is threatening to make $\{e\}^G$ partial. If it stays partial then the degree of $\{e\}^G$ is irrelevant. If it converges on a predetermined number then we will be able to carve out an improved e-state. We must now have a method for imposing a restriction for T_e that G extend a certain branch until an improved e-state is achieved. The nesting of the trees is used to help ensure this.

(4) The e-state (i.e. its correspondence with an equivalence relation) might not be well-defined. Since the wtt-reduction may be partial, some branches may not give convergence on numbers where other branches are exhibiting an e-split. Thus, one branch may appear (because of partialness) to be equivalent to two distinct branches which are inequivalent. This is factored into the definition of e-state; equivalence relations on a subset of $\{1, 2, ..., n\}$ are also possible values for the e-state. As in item 3 above, we can move to the branch which appears to make $\{e\}^G$ partial and conclude that $\{e\}^G$ is partial or the universe of the equivalence relation will increase in size and so we gain an improved e-state.

§4. Finite partition lattices in the wtt-degrees $\leq 0'$

We start with notation and definitions.

We will sometimes need to refer to intervals in a characteristic function, C, or in a string, γ. The symbol $C[a, b]$ denotes the string, σ, of length $b - a + 1$ such that for all $x \leq b - a$, $\sigma(x) = C(a + x)$. Similarly, the symbol $\gamma[a, b]$ denotes the string σ of length $b - a + 1$ such that for all $x \leq b - a$, $\sigma(x) = \gamma(a + x)$.

We say that a string σ is "left of" a string τ the value of σ at their first point of difference is less than the value of τ at that point of difference. More precisely, $\sigma <^L \tau$ if $\exists x < |\sigma|, |\tau|(\sigma(x) < \tau(x)$ and $\forall y \leq x(\sigma(y) = \tau(y)))$.

We embed the same lattices as in the tt-case - Π_n for $n \in \omega$, the duals of the finite partition lattices. The same notation and definitions for the lattice and the decoding of sets from the top set will be used. The Δ_2^0 set constructed will be named G.

Our process of building trees is more general now. We include the congruences in the definition of a tree in order to emphasize the need for maintaining them.

There is a further and more basic change in our notion of a tree. In the tt-degree construction the r.e. set constructed, A, always lay along the leftmost path of our trees. In light of the new diagonalization strategies, we no longer attempt to maintain this property. A diagonalization step for requirement P_k no longer destroys $T_{k,s}(i)$ for $i \in \{0,1,2,...,n\}$. We need to preserve this level of the tree so that we can later complete the diagonalization step if necessary. At a stage when a requirement P_k is in this intermediate state, i.e. it has acted but has not yet completed its diagonalization, we say that "P_k is active". The diagonalization step now moves G_s but preserves the node on the tree $T_{k,s}$. Thus, G_s might extend $T_{k,s+1}(i)$ for $i \neq 0$. Our trees will still have a single distinguished path (sometimes referred to as the trunk.) This will be the path that G_s takes through the tree.

Because we may need to return to old branchings when their splittings on lower priority trees become visible, we keep old branchings (ones which we have moved away from for the sake of diagonalization) on the tree in case they later have a better e-state for a lower priority tree. We do not erect new splits along paths where G_s does not lie. This means that the tree will be bushier than before, but will still have a single infinite path.

DEFINITION. A tree, T, is a partial map $T : \{0,1,2,...,n\}^{<\omega} \longrightarrow 2^{<\omega}$ with a single distinguished path, called the trunk, satisfying

(1) $\forall \sigma, \tau \in dom(T)(\sigma \subseteq \tau \leftrightarrow T(\sigma) \subseteq T(\tau))$.

(2) $\tau \in dom(T) \implies \forall \sigma \subseteq \tau(\sigma \in dom(T))$.

(3) $\forall \sigma \in dom(T)(\exists i \in \{0,1,2,...,n\}(\sigma \star i \in dom(T)) \implies \forall i \in \{0,1,2,...,n\}(\sigma \star i \in dom(T)))$

(4) For $\sigma \in dom(T)$, define $l(\sigma) = min\{|T(\sigma \star k)| : k \in \{0,1,...,n\}\}$. Then $\forall i,j \in \{1,2,...,n\}, \forall p \in \Pi_n, ((T(\sigma \star i) \restriction l(\sigma) \equiv_p T(\sigma \star j) \restriction l(\sigma) \longleftrightarrow ipj)$ and $T(\sigma \star 0) \not\equiv_p T(\sigma \star i))$

We now have a more liberal definition of a subtree. Whenever possible, the trunk of a tree is unchanged by the operation of taking a subtree. But, in some cases, when it is necessary to improve e-state, we allow the subtree operation to move the trunk in a restricted way. We use a relaxed definition of a subtree (rather than the one used in the r.e. tt-degree construction); necessary restrictions will be enforced by the construction.

DEFINITION. T_2 is a subtree of T_1 if

(1) T_1 and T_2 are trees, and

(2) There is an $i \in \{0,1,2,...,n\}$ such that $T_2(\emptyset) \supseteq T_1(i)$, and

(3) For all $\sigma \in dom(T_2)$ there are η and $\mu \in dom(T_1)$ such that ($\mu \supset \eta$ and $\forall i \in \{1,2,...,n\}(T_2(\sigma \star i) = T_1(\eta \star i)$ and $T_2(\sigma \star 0) = T_1(\mu)))$.

Because of the potentially partial nature of wtt-computations, we can no longer assert that the pattern of e-splitting among an n-tuple of right branches forms an equivalence relation. Instead we get an equivalence relation on a subset of the n-tuple of right branches. This complication is reflected in the definition of the e-state. We will consider equivalence relations on subsets of the universe, $\{1,2,...,n\}$.

DEFINITION. Let ρ be an equivalence relation on a subset of $\{1,2,...,n\}$. This subset is the universe of ρ, and is denoted as X_ρ.

DEFINITION. Let $\mathcal{E}_n$ be the structure $\langle E_n, \leq_{E_n}\rangle$, where

$$E_n = \{\rho : \rho \text{ is an equivalence relation on a subset of } \{1,2,...,n\}\}$$

and $\leq_{E_n}$ is a partial order on E_n defined as follows for ρ and $\tau \in E_n$:

$$\rho \leq_{E_n} \tau \longleftrightarrow (\forall x,y \in X_\rho \cap X_\tau(x\tau y \implies x\rho y))$$
$$\text{and}$$
$$(\text{if } \forall x,y \in X_\rho \cap X_\tau(x\tau y \leftrightarrow x\rho y) \text{ then } X_\tau \supseteq X_\rho).$$

In building splitting trees, we give first priority to establishing splittings between the left path and each member of an n-tuple of right branches. There are two situations where there is no e-split between the left path and some member of the n-tuple of right branches. In the first case there is no split between the left path and any member of the n-tuple of right branches. In this case we prove that if $\{e\}^G$ is total then it is recursive. To handle this case we allow as a possible value of the e-state at e, the values -1 and 0, signifying that there is no splitting between the left branch and any of the n-tuple of right branches. This might arise in one of two ways. There could be splittings among the right branches, but none with the left path. In this case we use the value 0, and can prove that if this configuration is permanent then $\{e\}^G$ is partial. Otherwise, there are no splittings among the right branches and no splittings with the left branch. In this case we use the value -1 and can prove that if this configuration is permanent and if $\{e\}^G$ is total then it is recursive.

The second case is where there is an e-split between the left path and some member of the n-tuple of right branches. We will take action during the construction to ensure that if this situation is permanent, then $\{e\}^G$ is partial. To handle this case we allow as possible values of the e-state at e, the symbols i, for each $i \in \{1,2,...,n\}$. The value i will be assigned to the configuration where i is the smallest number such that $T_e(\sigma \star 0)$ does not e-split with $T_e(\sigma \star i)$, and there is a splitting between the left branch and some member of the n-tuple of right branches.

The correspondence between e-state and sequences of members of $E_n \cup \{-1,0,1,2,...,n\}$ warrants some discussion. We would like to match up the possible patterns of e-splitting at a node on the tree with equivalence relations on $\{1,2,...,n\}$ (possibly partial). In order to make this correspondence meaningful, we must restrict the set of points at which we are looking for e-splittings. For a node $T_{k,s}(\sigma)$, an upper bound on this set of points will be denoted by $X(e,k,s,\sigma)$. $X(e,k,s,\sigma)$ is defined inductively (on σ) as follows:

$X(e,k,s,\sigma) = max\,\{X_{i,j}(e,k,s,\sigma) : i,j \in \{0,1,2,...,n\}\}$, where $X_{i,j}(e,k,s,\sigma)$ is defined as follows:

For the sake of notational convenience, we take $X(e,k,s,\emptyset^-) = -1$.

$X_{i,j}(e,k,s,\sigma)$ = the least x such that $\forall y \leq X(e,k,s,\sigma^-)$ $(\{e\}^{T_{k,s}(\sigma\star i)}(y)\downarrow$ and $\{e\}^{T_{k,s}(\sigma\star j)}(y)\downarrow$ and $\{e\}^{T_{k,s}(\sigma\star i)}(x)\downarrow \neq \{e\}^{T_{k,s}(\sigma\star j)}(x)\downarrow$, if there is such an x, otherwise $X_{i,j}(e,k,s,\sigma) = -1$.

Next we define a member, α, of $E_n \cup \{-1,0,1,2,...,n\}$ to correspond to the e-splitting among $T_{k,s}(\sigma \star i)$ for $i \in \{0,1,2,...,n\}$.

If for all $i,j \in \{0,1,2,...,n\}$, $X_{i,j}(e,k,s,\sigma) = -1$, then $\alpha = -1$.

If for all $i \in \{0,1,2,...,n\}$, $X_{0,i}(e,k,s,\sigma) = -1$, but for some $i,j \in \{1,2,...,n\}$, $X_{i,j}(e,k,s,\sigma) \neq -1$, then $\alpha = 0$.

If there is an $i \in \{1,2,...,n\}$ such that $X_{0,i}(e,k,s,\sigma) = -1$, but for some $j \in \{1,2,...,n\}$, $X_{0,j}(e,k,s,\sigma) \neq -1$, then α = the least such i.

Otherwise we assume that $X_{0,i}(e,k,s,\sigma) > -1$ for all $i \in \{1,2,...,n\}$, and thus $X(e,k,s,\sigma) > -1$. Define $Y = \left\{i : \forall x \leq X(e,k,s,\sigma), \{e\}^{T_{k,s}(\sigma\star i)}(x)\downarrow\right\}$.

α will be an equivalence relation on Y. For $i,j \in Y$, define $i\alpha j$ if and only if there is no $x \leq X(e,k,s,\sigma)$ such that $\{e\}^{T_{k,s}(\sigma\star i)}(x)\downarrow \neq \{e\}^{T_{k,s}(\sigma\star j)}(x)\downarrow$.

The j-state of a node $T_{k,s}(\sigma)$, for $k \geq j$, is a sequence of length $j+1$ from $E_n \cup \{1,2,...,n\}$. The value of this sequence at e is the α defined above.

We define a total order on the potential values for the e-states so that we can build the trees by maximizing e-states.

DEFINITION. Let $\leq$ be a total order on $\{-1,0,1,2,...,n\} \cup E_n$ satisfying

(1) $\forall \rho,\tau \in E_n\,(\rho \leq_{E_n} \tau \implies \rho \leq \tau)$.
(2) $-1 < 0 < 1 < 2 < ... < n$
(3) $\forall \rho \in E_n (n \leq \rho)$

Now each possible pattern of e-splitting among an n-tuple of right branches corresponds to an element of E_n. So we use the elements of E_n as the values for the e-state. We will arrange our construction so that if $\{e\}^G$ is total we can prove that no element of $E_n - \Pi_n$

can be the value of the final e-state at e. So, as in the tt-degree proof, if $\{e\}^G$ is total and the value of the final e-state at e is ρ, we will show that $\{e\}^G \equiv_{wtt} G_\rho$.

We must modify the construction in order to make this work. We want to be able to argue that if one of the "partial" equivalence relations appears in the final e-state, then in fact $\{e\}^G$ is partial. This is accomplished by moving G to extend a branch which appears to be divergent. If in fact the branch does later give a computation at the number(s) in question, then we have increased the size of the universe of the equivalence relation corresponding to e-splittings at this level. We move G back to the left path it was on when we started considering this branching and gain an increased e-state. Otherwise G extends a branch which never gives a computation at at least one of the numbers in question and so $\{e\}^G$ is partial.

Next we consider the case where the left path does e-split with some right branch but not with all of them. In the tt-degree proof we could argue in this case that the right path gave a recursive computation of $\{e\}^G$, since the right path extended trivially by $1's$ was guaranteed to give a convergent computation which must agree with $[e]^G$. This is no longer the case with wtt-reducibility; the reduction procedure $\{e\}$ using the right path extended by $1's$ may not be total. First assume that there is some e-split between the left path and one of the n-tuple of right branches. Our solution to the problem in this case has an ingredient similar to the solution used above. We need only worry about this case if this is the final e-state, so we may assume that we have at least two levels with this e-state. For the sake of definiteness, say these two levels are $T_e(0^m)$ and $T_e(0^{m+1})$. Further suppose that it is the ith right branch which fails to e-split with the left branch and that $T_e(0^{m+1}\star 0)$ does e-split with $T_e(0^{m+1}\star j)$. Let x_{m+1} be such that $\{e\}^{T_e(0^{m+1}\star 0)}(x_{m+1}) \neq \{e\}^{T_e(0^{m+1}\star j)}(x_{m+1})$. The use of this computation is no larger than $|T_e(0^{m+1}\star 0)|$. Now consider the computation $\{e\}^{T_e(0^m\star i)\star G_s[|T_e(0^m\star i)|,|T_e(0^{m+1}\star 0)|]}(x_{m+1})$. If this diverges, then we would like to make $T_e(0^m\star i)\star G_s[|T_e(0^m\star i)|,|T_e(0^{m+1}\star 0)|]$ an initial segment of G and thus satisfy the e requirement by making $\{e\}^G$ partial. If it converges, then it must differ from one of $\{e\}^{T_e(0^{m+1}\star 0)}(x_{m+1})$ and $\{e\}^{T_e(0^{m+1}\star j)}(x_{m+1})$. Thus we could improve the e-state of $T_e(0^m)$ by making the left path either $\{e\}^{T_e(0^{m+1}\star 0)}$ or $\{e\}^{T_e(0^{m+1}\star j)}$, whichever gives the e-split. Until now such moves of the left path to improve e-state have not been needed in our initial segment constructions. We do allow it now, although it will complicate the proof that the limit tree exists. Another problem is that during a recursive construction we will not be able to recognize if the computation $\{e\}^{T_e(0^m\star i)\star G_s[|T_e(0^m\star i)|,|T_e(0^{m+1}\star 0)|]}(x_{m+1})$ diverges. The solution to that problem is to assume it will diverge until it actually converges. So, if at stage s we think that $\{e\}^{T_{e,s}(0^m\star i)\star G_s[|T_e(0^m\star i)|,|T_e(0^{m+1}\star 0)|]}(x_{m+1})$ diverges, then we move G to extend $T_e(0^m\star i)\star G_s[|T_e(0^m\star i)|,|T_e(0^{m+1}\star 0)|]$ until such a stage when it converges, if it ever does. If it does not converge, we are done. If it does converge, then we move G back at the stage when the convergence is apparent and gain an increased e-state. Such moving around of G can happen at most finitely often for a given node on the tree because with each such cycle the e-state of the node increases.

The other case to consider is where there is no e-split between the left path and any of the right branches. In this case we take no special action; we will be able to prove that if $\{e\}^G$ is total then it is recursive.

Let $\{P_i : i \in \omega\}$ be an effective listing of all requirements of the forms

$$\{e\}^{G_\rho} \neq G_\tau, \text{ for all } \rho, \tau \in \Pi_n \text{ such that } \tau \not\leq \rho$$

and

$$\{e\} \neq G_\psi, \text{ where } \psi \text{ is the least element of } \Pi_n$$

$G_{s,s}$ will be an intermediate value for G_s; it is the value of G_s after the trees have been built at stage s but before any diagonalization step has taken place at stage s.

We say that $P_k(= \{e\}^{G_\rho} \neq G_\tau)$ requires attention to begin diagonalization at stage s if $T_{e,s}(0) \subset G_{s,s}$ and x, i, and j are such that $T_{e,s}(i) \equiv_\rho T_{e,s}(j)$, $(T_{e,s}(i))_\tau(x) \neq (T_{e,s}(j))_\tau(x)$, $|T_{e,s}(i)| = |T_{e,s}(j)| \leq \{e\}(x)$, and $\{e\}^{(T_{e,s}(0))_\rho}(x) = (G_{s,s})_\tau(x)$.

We say that $P_k(= \{e\}^{G_\rho} \neq G_\tau)$ requires attention to complete a diagonalization at stage s if $x \in \omega$ and $i, j \in \{1, 2, ..., n\}$ are such that $T_{e,s}(i) \subseteq G_{s,s}$ and $T_{e,s}(i) \equiv_\rho T_{e,s}(j)$, $(T_{e,s}(i))_\tau(x) \neq (T_{e,s}(j))_\tau(x)$, $|T_{e,s}(i)| = |T_{e,s}(j)| \leq \{e\}(x)$, and $\{e\}^{(T_{e,s}(i))_\rho}(x) = (G_{s,s})_\tau(x)$.

At stage s of the construction only strings of length less than or equal to s will be in the domain of the trees.

CONSTRUCTION.

Stage 0: $G_0 = \emptyset$; $T_{0,0}(\emptyset) = \emptyset$; all requirements are inactive at stage 0.

Stage s: The stage will be carried out in two phases. During Phase 1 we build trees by maximizing e-states. During Phase 2 we act for diagonalization requirements.

PHASE 1. We will have substages e for $e \leq s+1$. For each such e we will define $G_{e,s}$ and $T_{e,i,s}$ for i such that $e \leq i \leq s$. $G_{e,s}$ will be the version of G believed to be correct at the end of substage e. $T_{e,e,s}$ will be our initial definition of the e-splitting tree at stage s, built by maximizing e-states. This tree may need to be (non-essentially) modified by stretching instigated by lower priority trees; stretching caused by the i-splitting tree appears in $T_{e,i,s}$. Stretching may also be instigated by action taken for the sake of one of the diagonalization requirements. This will be handled during Phase 2 and will appear in the trees $T_{e,s}$ for $e \leq s$.

Substage -1: We define $T_{-1,-1,s}$ and $G_{-1,s}$. For $\sigma \in domain(T_{-1,s-1})$, define $T_{-1,-1,s}(\sigma) = T_{-1,s}(\sigma)$. Fix σ, the $\subseteq$-longest string in $domain(T_{-1,s})$ such that $T_{-1,s}(\sigma) \subset G_{-1,s}$. We will define $T_{-1,-1,s}(\tau \star i)$ for $i \in \{0, 1, 2, ..., n\}$ and $\tau = \sigma \star 0^m$, for $m \in \omega$ by induction on m. Inductively assume that $T_{-1,-1,s}(\tau)$ has been defined.

Fix x the least such that $nx > |T_{-1,-1,s}(\tau)|$. Define $T_{-1,-1;s}(\tau \star 0) = T_{-1,-1,s}(\tau) \star 0^{(n+1)x - |T_{-1,-1,s}(\tau)|}$, and for $i \in \{1, 2, ..., n\}$ and $y \leq (n+1)x$ define

$$T_{-1,-1,s}(\tau \star i)(y) = \begin{cases} T_{-1,-1,s}(\tau)(y), & \text{if } y < |T_{-1,-1,s}(\tau)| \\ 1, & \text{if } y = nx + i \\ 0, & \text{if } y = nx + j \text{ and } j \neq i \\ 0, & \text{if } |T_{-1,-1,s}(\tau)| \leq y \leq nx. \end{cases}$$

Define $G_{-1,s}$ to be the set extending $T_{-1,-1,s}(\sigma \star 0^m)$ for all $m \in \omega$.

Substage e: We will define $T_{e,e,s}$, $G_{e,s}$, and $T_{j,e,s}$ for $j < e$. $T_{e,e,s}$ is built by maximizing e-states. $G_{e,s}$ reflects the possible changes made to G by the building of $T_{e,e,s}$. As in the r.e. tt-degree proof, the definition of $T_{e,e,s}(\sigma)$ is complicated by the process of making splitting trees. $T_{e,e,s}(\sigma)$ plays a double role: its e-state is determined by the splittings among $T_{e,e,s}(\sigma \star i)$ for $i \in \{0, 1, 2, ..., n\}$, but the splitting among $T_{e,e,s}(\sigma)$ and $T_{e,e,s}(\sigma^- \star i)$ for $i \in \{0, 1, 2, ..., n\}$ determines the e-state of $T_{e,e,s}(\sigma^-)$. For this reason $T_{e,e,s}(\sigma)$ is defined in two steps. First it is defined for the sake of maximizing the e-state of $T_{e,e,s}(\sigma^-)$. Later it may be "lifted" for the sake of maximizing the e-state of $T_{e,e,s}(\sigma)$. The first (temporary) definition is called $T_{e,e,s-1/2}(\sigma)$, the second (permanent) definition is called $T_{e,e,s}(\sigma)$.

Assume that $T_{i,e-1,s}$ and $G_{e-1,s}$ have been defined for all $i < e$. (In general, if $G_{i,s}$ is not defined during substage i then we take $G_{i,s} = G_{i-1,s}$.) We define $T_{e,e,s}$ as a (modified) subtree of $T_{e-1,e-1,s}$. Its domain contains only strings of length less than or equal to s.

$T_{e,e,s}(\sigma)$ is defined by induction on σ. When considering extensions of σ, we will define $T_{e,e,s-1/2}(\sigma \star i)$ and $T_{e,e,s}(\sigma)$.

First we define $T_{e,e,s}(\emptyset)$ and $T_{e,e,s-1/2}(i)$ for $i \in \{1, 2, ..., n\}$. If P_e is not currently active, (requirements become active or inactive during Phase 2 of the construction) then fix $\sigma \in 2^{<n}$ the $\subseteq$-shortest such that $|\sigma| \geq 1$ and $T_{e-1,e-1,s}(\sigma\star 0) \subset G_{e-1,s}$ and for all $i \leq e-1$ the value of the $e-1$-state of $T_{e-1,e-1,s}(\sigma)$ at i is not in $\{0, 1, 2, ..., n\}$. Here we are avoiding all nodes where a tree with index $i < e$ might prevent the requirement P_e from changing G. Note that such a σ exists because there can be at most finitely many ineligible nodes on $T_{e-1,e-1,s}$. This is the case because for each $i \leq e-1$, at most finitely many node on $T_{i,e-1,s}$ can be in the i-state with value at i in $\{0, 1, 2, ..., n\}$. Define $T_{e,e,s}(\emptyset) = T_{e-1,e-1,s}(\sigma)$. Define $T_{e,e,s-1/2}(i) = T_{e-1,e-1,s}(\sigma \star i)$ for $i \in \{0, 1, 2, ..., n\}$. (We do not attempt to maximize the e-state of $T_{e,e,s}(\emptyset)$; it is simply a set-up for diagonalization.)

Next, assume that $T_{e,e,s-1/2}(\sigma)$ has been defined. We consider $T_{e,e,s-1/2}(\sigma \star i)$ for $i \in \{0, 1, 2, ..., n\}$. If $T_{e,e,s-1/2}(\sigma \star i)$ are about to be defined then it will be by one of the following three cases:

Case I. First assume that $G_{e,s}$ has not yet been defined, and that $T_{e,e,s-1/2}(\sigma) \subset G_{e-1,s}$. Fix η such that $T_{e-1,e-1,s}(\eta) = T_{e,e,s-1/2}(\sigma)$. Choose $\gamma \supseteq \mu \supseteq \eta$ so that defining $T_{e,e,s-1/2}(\sigma \star 0) = T_{e-1,e-1,s}(\gamma)$ and $T_{e,e,s-1/2}(\sigma \star i) = T_{e-1,e-1,s}(\mu \star i)$ for $i \in \{1, 2, ..., n\}$ will make the e-state of $T_{e,e,s-1/2}(\sigma)$ as large as possible.

Define $T_{e,e,s}(\sigma) = T_{e-1,e-1,s}(\mu)$, $T_{e,e,s-1/2}(\sigma \star 0) = T_{e-1,e-1,s}(\gamma)$, and $T_{e,e,s-1/2}(\sigma \star i) = T_{e-1,e-1,s}(\mu \star i)$ for $i \in \{1, 2, ..., n\}$.

If the definition of $T_{e,e,s-1/2}(\sigma \star k)$ has caused one of the following four events to occur, then we will at this point define $G_{e,s}$, $T_{e,e,s}(\sigma \star i)$ for $i \in \{1,2,...,n\}$ and $T_{j,e,s}$ for $j < e$ and complete the definition of $T_{e,e,s}$ by making dummy extensions.

(1) There is no $k \in \{0,1,2,...,n\}$ such that $G_{e-1,s} \supseteq T_{e,e,s-1/2}(\sigma \star k)$.

Again fix γ so that $T_{e,e,s-1/2}(\sigma \star 0) = T_{e-1,e-1,s}(\gamma)$. If there is a $\gamma' \supset \gamma$ in $dom(T_{e-1,e-1,s})$, let γ'' be the $\subseteq$-longest and $<^L$-leftmost such string. Fix τ so that $(G_{e-1,s} \upharpoonright |T_{e-1,e-1,s}(\gamma'')|) \star \tau = G_{e-1,s} \upharpoonright s$. Define $T_{e,e,s}(\sigma \star 0) = T_{e-1,e-1,s}(\gamma'') \star \tau$. Define $G_{e,s}=$ the set extending $T_{e,e,s}(\sigma \star 0^m)$ for all $m \in \omega$. Define $T_{e,e,s}(\sigma \star i) = T_{e,e,s-1/2}(\sigma \star i)$ for $i \in \{1,2,...,n\}$.

There is no need to stretch any nodes on $T_{j,e-1,s}$, so define $T_{j,e,s}(\tau) = T_{j,e-1,s}(\tau)$ for all $j < e$ and all $\tau \in dom(T_{j,e-1,s})$.

(2) There is a $k \in \{1,2,...,n\}$ such that $G_{e-1,s} \supseteq T_{e,e,s-1/2}(\sigma \star k)$, and the e-state of $T_{e,e,s}(\sigma)$ is higher than the e-state of $T_{e,e,s-1}(\sigma)$.

Fix τ so that $(G_{e-1,s} \upharpoonright |T_{e,e,s-1/2}(\sigma \star k)|) \star \tau = G_{e-1,s} \upharpoonright s$. Define $T_{e,e,s}(\sigma \star i) = T_{e,e,s-1/2}(\sigma \star i) \star \tau$ for $i \in \{1,2,...,n\}$. Let τ_0 be the final segment of τ of length $s - |T_{e,e,s-1/2}(\sigma \star 0)|$. Define $T_{e,e,s}(\sigma \star 0) = T_{e,e,s-1/2}(\sigma \star 0) \star \tau_0$. Define $G_{e,s} =$ the set extending $T_{e,e,s}(\sigma \star 0^m)$ for all $m \in \omega$.

Finally we stretch the nodes corresponding to $T_{e,e,s}(\sigma \star i)$ on trees $T_{j,e,s}$ for $j < e$. For each $j < e$, fix η_j so that $T_{e,e,s}(\sigma) = T_{j,e-1,s}(\eta_j)$. For each $j < e$, for each $i \in \{1,2,...,n\}$, define $T_{j,e,s}(\eta_j \star i) = T_{j,e-1,s}(\eta_j \star i) \star \tau$. For $\alpha \neq \eta_j \star i$, define $T_{j,e,s}(\alpha) = T_{j,e-1,s}(\alpha)$.

(3) $G_{e-1,s} \supseteq T_{e,e,s-1/2}(\sigma \star 0)$ and the value of the e-state of $T_{e,e,s}(\sigma)$ at e is $\rho \in E_n - \Pi_n$, and for all $i < e$ the tree T_i is not controlling G at this node.) (We say that the tree T_i is controlling G at a node $T_{e,e,s}(\sigma)$ if the value of the e-state at i is in $\{0,1,2,...,n\} \cup E_n - \Pi_n$ and G has been moved to extend one of the right branches for the sake of gaining a convergence and improving the i-state, or if σ is the $\subseteq$-least string such that the value of the e-state at i is $\{0,1,2,...,n\}$.

Let i be the smallest number such that $i \notin X_\rho$. Fix τ so that $(G_{e-1,s} \upharpoonright |T_{e,e,s-1/2}(\sigma \star k)|) \star \tau = G_{e-1,s} \upharpoonright s$. Define $T_{e,e,s}(\sigma \star k) = T_{e,e,s-1/2}(\sigma \star k) \star \tau$, for $k \in \{1,2,...,n\}$. Define $G_{e,s} =$ the set extending $T_{e,e,s}(\sigma \star 0^m)$ for all $m \in \omega$.

Finally we stretch the nodes corresponding to $T_{e,e,s}(\sigma \star i)$ on trees $T_{j,e,s}$ for $j < e$. For each $j < e$, fix η_j so that $T_{e,e,s}(\sigma) = T_{j,e-1,s}(\eta_j)$. For each $j < e$, for each $i \in \{1,2,...,n\}$, define $T_{j,e,s}(\eta_j \star i) = T_{j,e-1,s}(\eta_j \star i) \star \tau$. For $\alpha \neq \eta_j \star i$, define $T_{j,e,s}(\alpha) = T_{j,e-1,s}(\alpha)$.

(4) $G_{e-1,s} \supseteq T_{e,e,s-1/2}(\sigma \star 0)$ and $\sigma = \sigma^- \star 0$ and the values of the e-states of $T_{e,e,s}(\sigma^-)$ and $T_{e,e,s}(\sigma)$ are both $i \in \{1,2,...,n\}$, and for all $i < e$ the tree T_i is not controlling G at this node.) Let $j \in \{1,2,...,n\}$ be such that $T_{e,e,s}(\sigma)$ and $T_{e,e,s}(\sigma^- \star j)$ e-split.

Define $T_{e,e,s}(\sigma^- \star k) = T_{e,e,s-1/2}(\sigma^- \star k) \star G_{e-1,s}\left[|T_{e,e,s-1/2}(\sigma \star k)|, s\right]$ for $k \in \{1,2,...,n\}$. Define $G_{e,s}$ to be the set extending $T_{e,e,s}(\sigma \star 0^m)$ for all $m \in \omega$.

Fix τ so that $(G_{e-1,s} \upharpoonright |T_{e,e,s-1/2}(\sigma \star k)|) \star \tau = G_{e-1,s} \upharpoonright s$. Define $T_{e,e,s}(\sigma^- \star k) =$

$T_{e,e,s-1/2}(\sigma^- \star k)\star\tau$ for $k \in \{1, 2, ..., n\}$. Define $G_{e,s}$ = the set extending $T_{e,e,s}(\sigma\star 0^m)$ for all $m \in \omega$.

Finally we stretch the nodes corresponding to $T_{e,e,s}(\sigma^- \star i)$ on trees $T_{j,e,s}$ for $j < e$. For each $j < e$, fix η_j so that $T_{e,e,s}(\sigma^-) = T_{j,e-1,s}(\eta_j)$. For each $j < e$, for each $i \in \{1, 2, ..., n\}$, define $T_{j,e,s}(\eta_j \star i) = T_{j,e-1,s}(\eta_j \star i) \star \tau$. For $\alpha \neq \eta_j \star i$, define $T_{j,e,s}(\alpha) = T_{j,e-1,s}(\alpha)$.

Now the definition of $T_{e,e,s}$ and substage e are completed by making dummy extensions. Fix $i \in \{0, 1, 2, ..., n\}$ such that $G_{e,s} \supseteq T_{e,e,s}(\sigma \star i)$ and τ such that $T_{e,e,s}(\sigma \star i) = T_{e-1,e-1,s}(\tau)$. If $T_{e-1,e-1,s}$ has an infinite path above $T_{e-1,e-1,s}(\tau)$, then copy this path of nodes into $T_{e,e,s}$. Otherwise we erect an infinite path of dummy extensions on $T_{e,e,s}$ above $T_{e,e,s}(\sigma \star i)$ as follows: Inductively assume that $T_{e,e,s}(\tau \star k)$ have been defined for $k \in \{0, 1, 2, ..., n\}$ and $\tau = \sigma\star i0^m$. We define $T_{e,e,s}(\tau\star 0k)$ for $k \in \{0, 1, 2, ..., n\}$. Fix x the least such that $nx > |T_{e,e,s}(\tau\star 0)|$. Define $T_{e,e,s}(\tau\star 00) = T_{e,e,s}(\tau\star 0)\star 0^{(n+1)x-|T_{e,e,s}(\tau\star 0)|}$, and for $i \in \{1, 2, ..., n\}$ and $y \le (n+1)x$ define

$$T_{e,e,s}(\tau \star 0i)(y) = \begin{cases} T_{e,e,s}(\tau \star 0)(y), & \text{if } y < |T_{e,e,s}(\tau \star 0)| \\ 1, & \text{if } y = nx + i \\ 0, & \text{if } y = nx + j \text{ and } j \neq i \\ 0, & \text{if } |T_{e,e,s}(\tau \star 0)| \le y \le nx. \end{cases}$$

Case II. Next assume that $G_{e,s}$ has not yet been defined and $T_{e,e,s-1/2}(\sigma) \not\subseteq G_{e-1,s}$.

If $G_{e-1,s} <^L T_{e,e,s-1/2}(\sigma)$, then we define no extensions of $T_{e,e,s-1/2}(\sigma)$, and define $T_{e,e,s}(\sigma) = T_{e,e,s-1/2}(\sigma)$.

If $T_{e,e,s-1/2}(\sigma) <^L G_{e-1,s}$, then we copy over old splittings into the tree. If there is an α such that $T_{e,e,s-1/2}(\sigma) = T_{e,e,s-1}(\alpha)$, then for all β such that $\alpha \star \beta \in dom(T_{e,s-1})$, define $T_{e,e,s-1/2}(\sigma \star \beta) = T_{e,e,s-1}(\alpha \star \beta)$.

There is no need to stretch any nodes on $T_{j,e-1,s}$, so define $T_{j,e,s}(\tau) = T_{j,e-1,s}(\tau)$ for all $j < e$ and all τ in $dom(T_{j,e-1,s}(\tau)$.

Case III. $G_{e,s}$ has already been defined.

If $G_{e,s} <^L T_{e,e,s-1/2}(\sigma)$, then we define no extensions of $T_{e,e,s-1/2}(\sigma)$, and define $T_{e,e,s}(\sigma) = T_{e,e,s-1/2}(\sigma)$.

Otherwise we finish off the definition of $T_{e,e,s}$ by copying over old nodes which are to the left of $G_{e,s}$ as follows.

If $T_{e,e,s-1/2}(\sigma) <^L G_{e,s}$, then we copy over old splittings into the tree. If there is an α such that $T_{e,e,s-1/2}(\sigma) = T_{e,s-1}(\alpha)$, then for all β such that $\alpha \star \beta \in dom(T_{e,s-1})$, define $T_{e,e,s-1/2}(\sigma \star \beta) = T_{e,,s-1}(\alpha \star \beta)$.

There is no need to stretch any nodes on $T_{j,e-1,s}$, so define $T_{j,e,s}(\tau) = T_{j,e-1,s}(\tau)$ for all $j < e$ and all $\tau \in dom(T_{j,e-1,s}(\tau)$.

PHASE 2.

In this phase of the construction we act for diagonalization requirements. Let P_k be the highest priority requirement requiring attention at stage s. If there is no such k, then define $T_{e,s} = T_{e,s,s}$ for all $e \leq s$, and $G_s = G_{s,s}$.

If there is such a k, fix σ such that the action called for by P_k is to make G_s extend $T_{k,s,s}(\sigma)$. Fix τ such that $(G_{s,s} \upharpoonright |T_{k,s,s}(\sigma)|) \star \tau = G_{s,s} \upharpoonright s$. Define $T_{k,s}(\sigma^- \star j) = T_{k,s,s}(\sigma^- \star j) \star \tau$ for $j \in \{1, 2, ..., n\}$. For $\alpha \in dom(T_{k,s,s})$ not of the form $\sigma^- \star j$, define $T_{k,s}(\alpha) = T_{k,s,s}(\alpha)$. Define $G_s =$ the set extending $T_{k,s}(\sigma \star 0^m)$ for all $m \in \omega$.

Next we do one final stretching of nodes corresponding to $T_{k,s}(\emptyset)$ on trees $T_{e,s,s}$ for $e < k$. The trees obtained after this stretching will be called $T_{e,s}$. For each $e < k$, fix η_e so that $T_{k,s}(\sigma^-) = T_{i,s,s}(\eta_e)$. For each $e < k$, for each $i \in \{1, 2, ..., n\}$, define $T_{e,s}(\eta_e \star i) = T_{e,s,s}(\eta_e \star i) \star \tau$. For $\alpha \neq \eta_e \star i$, define $T_{e,s}(\alpha) = T_{e,s,s}(\alpha)$.

For $e > k$, define $T_{e,s} = T_{e,s,s}$.

end of construction.

Verifications.

LEMMA 1. *For all* $e, s \in \omega$ *and* $\sigma \in n^{<\omega}$,
(1) $lim_s T_{e,s}(\sigma)$ *exists, and*
(2) P_e *acts at most finitely often.*

Note that "undefined" is a possible value for $lim_s T_{e,s}(\sigma)$, since our trees are partial. We will show that if $T_{e,s}(\sigma)$ is undefined infinitely often then $T_{e,s}(\sigma)$ is undefined co-finitely often.

PROOF: We prove a variation of the statement of the lemma in order to make the induction proceed more smoothly. The proof that the pointwise limit of the trees exists is complicated by the process of "lifting" that is sometimes used in defining $T_{e,s}(\sigma)$. This occurs when we define extensions of $T_{e,s}(\sigma)$, $T_{e,s}(\sigma \star j)$ for $j \in \{0, 1, 2, ..., n\}$. When this happens, $T_{e,s}(\sigma)$, is lifted up to the base of the node used for $T_{e,s}(\sigma \star j)$. We will prove that $\forall \sigma \in n^{<\omega}$, the limit exists for $T_{e,s}(\sigma)$ modulo lifting. From this we can easily deduce the existence of the limit: To see that $lim_s T_{e,s}(\sigma)$ exists, wait for a stage s_0 large enough so that for all $i \in \{0, 1, 2, ..., n\}$, $T_{e,s}(\sigma \star i)$ changes after stage s_0 only by lifting. Then $\forall s \geq s_0 T_{e,s}(\sigma) = T_{e,s_0}(\sigma)$, since lifting of $T_{e,s}(\sigma \star i)$ does not change $T_{e,s}(\sigma)$.

The statement we actually prove (by induction on k) is $\forall k Q(k)$, where $Q(k)$ is

$$
\begin{aligned}
Q(k) \iff & \forall e \forall \sigma \text{ such that } k = e + |\sigma| \exists s_0 \forall s' \geq s_0 \\
& (T_{e,s'}(\sigma) = T_{e,s'+1}(\sigma) \text{ or } T_{e,s'+1}(\sigma) \text{ differs from } T_{e,s'}(\sigma) \text{ only by lifting}) \\
& \text{and } P_k \text{ acts at most finitely often, and} \\
& lim_s G_s \upharpoonright l(k, s) \text{ exists}
\end{aligned}
$$

Where

$$l(k,s) = max\left\{|T_{k,t+1}(0)| : \begin{array}{l} T_{k,t+1}(0) \neq T_{k,t}(0) \text{ and } t < s \\ \text{and the change was not occassioned by lifting} \end{array}\right\}.$$

The idea is that the limit of G exists up to the essential part of $T_{k,s}(0)$. Since we are disregarding lifting in the existence of limits in the trees, we disregard it here to make the induction proceed more smoothly.

$Q(0)$. If $e+|\sigma| = 0$, then $e = 0$ and $\sigma = \emptyset$, so we only need consider $lim_s T_{0,s}(\emptyset)$ modulo lifting, and P_0. $T_{0,s}(\emptyset) = \emptyset$ for all s, and in particular is never even changed by lifting. If the requirement P_0 ever acts, it can be injured only if the 0-state of $T_{0,s}(\emptyset)$ increases. The 0-state of $T_{0,s}(\emptyset)$ is bounded and non-decreasing with s, and so can increase at most finitely often. After all injury to P_0 stops, then P_0 can act at most twice. Thus, P_0 acts at most finitely often. After P_0 stops acting, $G_s \restriction l(0,s)$ can change only for the sake of improving the 0-state of $T_{0,s}(\emptyset)$. As discussed above, this can occur at most finitely often. Thus $Q(0)$ holds.

$Q(k+1)$. If we prove that $lim_s T_{k+1,s}(\emptyset)$ exists then it will be relatively easy to show that P_{k+1} acts at most finitely often. $T_{k+1,s}(\emptyset)$ is never changed by lifting, and after a stage large enough so that $T_{k+1,s}(\emptyset)$ does not change, P_{k+1} can only be injured at stages when the $k+1$-state of $T_{k+1,s}(\emptyset)$ increases. As in the proof for $Q(0)$, this can occur at most finitely often (and in fact will correspond to an increase in the k-state of $T_{k+1,s}(\emptyset)$)(assuming that $T_{k+1,s}(\emptyset)$ has settled down.) Thus P_{k+1} will act at most finitely often. So, we first prove that $lim_s T_{e,s}(\sigma)$ exists, modulo lifting for all e and σ such that $e+|\sigma| \leq k+1$.

We show that $lim_s T_{e,s}(\sigma)$ exists, modulo lifting by a subinduction on $|\sigma|$. Consider first $|\sigma| = 0$, i.e., $\sigma = \emptyset$ and $e = k+1$. By the induction hypothesis, let s_0 be a stage large enough so that $\forall s \geq s_0$, $\forall e, \sigma$ such that $e+|\sigma| \leq k$, $T_{e,s}(\sigma) = T_{e,s+1}(\sigma)$ or $T_{e,s+1}(\sigma)$ differs from $T_{e,s}(\sigma)$ only by lifting and for all $j \leq k$, P_j does not act at stage s, and $G_s \restriction l(k,s) = G_{s+1} \restriction l(k,s+1)$.

If $s \geq s_0$ and $T_{k+1,s}(\emptyset) \neq T_{k+1,s+1}(\emptyset)$ then an event of one of the following two types must have occurred:

(1) $G_s \restriction |T_{k+1,s+1}(\emptyset)| \neq G_s \restriction |T_{k+1,s}(\emptyset)|$, or

(2) For σ such that $T_{k+1,s}(\emptyset) = T_{k,s}(\sigma)$, $T_{k,s}(\sigma) \neq T_{k,s+1}(\sigma)$.

Consider first events of type (1). Since $s \geq s_0$, the change in G must appear on the tree $T_{k,s}$ and be above $|T_{k,s}(0)|$. Since the change is below $|T_{k+1,s}(0)|$, and no requirement P_j could have acted at stage s, the change in G must have occasioned the improvement of the k-state of a node on $T_{k,s}$. Furthermore, there are a finite number of nodes whose k-state can improve in conjunction with such a change in G_s. Let τ be such that $T_{k+1,s_0}(\emptyset) = T_{k,s_0}(\tau)$. Let $s_1 > s_0$ be the least such that $G_{s_1} \restriction |T_{k+1,s_1}(0)| \neq G_{s_0} \restriction |T_{k+1,s_0}(0)|$. If $G_{s_1} \restriction |T_{k+1,s_1}|$ ever changes after stage s_1, then at the least such stage, call it s_2,

$G_{s_2} \supset T_{k,s_0}(\tau \star 0)$, and so $T_{k+1,s_2}(\emptyset) = T_{k,s_2}(\tau)$. The point is that at stages $s \geq s_0$ such that $G_s \supset T_{k,s}(\tau \star 0)$, $T_{k+1,s}(\emptyset) = T_{k,s}(\tau)$. At stages $t \geq s_0$ such that $G_t \supset T_{k,s}(\tau \star i)$ for some $i \neq 0$, then either P_{k+1} is active at stage t, or $G_t \supset T_{k,t}(\tau \star i \star 0)$. Thus $G_s \upharpoonright |T_{k+1,s}(0)|$ can change for the sake of improving the k-state of at most two consecutive nodes on the tree $T_{k,s}$. This can happen at most finitely often, and so events of type (1) can occur at most finitely often.

Next assume that $s_1 > s_0$ is a stage large enough so that no events of type (1) occur after stage s_1. We will show that after stage s_1 events of type two can occur at most finitely often. Fix τ such that $T_{k+1,s_1}(\emptyset) = T_{k,s_1}(\tau)$. Then for all $s \geq s_1$, $T_{k+1,s}(\emptyset) = T_{k,s}(\tau)$. And so events of type (2) can occur only if the k-state of $T_{k,s}(\tau)$ improved. As discussed above, this can occur at most finitely often.

Thus $lim_s T_{k+1,s}(\emptyset)$ exists.

Now assume that $lim_s T_{i,s}(\eta)$ exists modulo stretching for all i, η such that $i + |\eta| \leq k$ or $i + |\eta| = k + 1$ and $|\eta| \leq m$. We show that $lim_s T_{k-m,s}(\sigma)$ exists modulo stretching, where $|\sigma| = m + 1$.

By the induction hypothesis, let s_0 be a stage large enough so that $\forall s \geq s_0$, $\forall i, \eta$ such that $i + |\eta| \leq k$ or $i + |\eta| = k + 1$ and $|\eta| \leq m$, $T_{i,s}(\eta) = T_{i,s+1}(\eta)$ or $T_{i,s+1}(\eta)$ differs from $T_{i,s}(\eta)$ only by lifting and for all $j \leq k$, P_j does not act at stage s, and $G_s \upharpoonright l(k,s) = G_{s+1} \upharpoonright l(k, s+1)$.

Suppose that $s > s_0$ and $T_{k-m,s}(\sigma) \neq T_{k-m,s+1}(\sigma)$. Then one of the following events must have occurred:

(1) The changed was caused by lifting, or
(2) a requirement P_j with $j < e + |\sigma|$ acted at stage s, or
(3) for some τ such that $|\tau| < |\sigma|$, $T_{k-m,s}(\tau) \neq T_{k-m,s+1}(\tau)$, or
(4) the $(k-m)$-state of $T_{k-m,s+1}(\sigma^-)$ is higher than the $(k-m)$-state of $T_{k-m,s}(\sigma^-)$, or
(5) the change was caused by stretching.

Events of type (1) are disregarded by $Q(k+1)$. By the induction hypothesis, since $s > s_0$, we may assume that events of type (2) and (3) do not occur at stage s. After stage s_0 the $(k-m)$-state of $T_{k-m,s}(\sigma^-)$ is monotone increasing, and so events of type (4) can occur at most finitely often. Suppose an event of type (5) occurs at stage s. Then the stretching must have been instigated by an essential change (i.e. a change other than lifting) in a node on a tree $T_{i,s}$ with $i > k - m$. Furthermore, the node must be of the form $T_{i,s}(\tau)$, where $i + |\tau| \leq (k-m) + |\sigma| = k+1$. Since $i > k - m$, we get that $|\tau| < |\sigma|$, and the sub-induction hypothesis tells us that such changes are impossible after stage s_0. So $T_{k-m,s}(\sigma)$ can change at most finitely often (other than by lifting) after stage s_0. The subinduction is complete, and we conclude that $lim_s T_{e,s}(\sigma)$ exists modulo lifting for all e, σ such that $e + |\sigma| \leq k + 1$. As discussed above, from this we can conclude that P_{k+1} acts at most finitely often.

Next we consider $lim_s G_s \upharpoonright |T_{k+1,s}(0)|$. After the requirement P_{k+1} stops acting, $G_s \upharpoonright$

$|T_{k+1,s}(0)|$ can change only for the sake of improving the $(k+1)$-state of $T_{k+1,s}(\emptyset)$. Again, this can occur at most finitely often, and so the limit exists.

LEMMA 2. *Each requirement P_k is satisfied.*

PROOF: We use induction on k. Assume that for all $i < k$, requirement P_i is satisfied. Let s_0 be a stage large enough so that for all $s \geq s_0$, for all $i < k$, P_i is satisfied at stage s and for all $i \leq k$, P_i is not active at stage s. Consider requirement $P_k : \{e\}^{G_\rho} \neq G_\tau$. Let $s_1 > s_0$ be a stage large enough so that for all $s \geq s_1$, for all $i \in \{0,1,2,...,n\}$, $T_{e,s}(i) = T_{e,s_1}(i)$, and $G_s \upharpoonright |T_{e,s}(0)| = G_{s_1} \upharpoonright |T_{e,s_1}(0)|$. Suppose that P_k fails and that $\{e\}^{G_\rho} = G_\tau$. Let i and j be such that $i \equiv_\rho j$ and $i \not\equiv_\tau j$, and x be such that $(T_{e,s_1}(i))_\tau(x) \neq (T_{e,s_1}(j))_\tau(x)$. Since P_k fails, we know that $\{e\}^{G_\rho}$ is total, and in particular that $\{e\}^{G_\rho}(x)$ converges. Let $s_2 \geq s_1$ be large enough so that $\{e\}^{G_\rho}(x)$ converges by stage s_2. But then P_k would require attention at stage s_2, and become active, a contradiction to our assumptions about $s_2 > s_1$.

LEMMA 3. *For each tree T_e, there is a final e-state.*

PROOF: We begin by making a definition of the "final e-state".

DEFINITION. The *final e-state of a (partial) tree T_e is a string $\varphi \in (E_n \cup \{-1,0,1,2,...,n\} \cup \Pi_n)^{<\omega}$ of length $e+1$ such that for all but finitely many σ in the domain of T_e, there is a stage s_σ such that either for all $s \geq s_\sigma$ the e-state of $T_{e,s}(\sigma)$ is φ or for all $s \geq s_\sigma$, $T_{e,s}(\sigma) <^L G_s$.*

Note that for all strings σ and τ in $\{0,1,2,...,\}^{<\omega}$ and all $s \in \omega$, if $T_{e,s}(\sigma) \subset G_s$ and $\sigma \subseteq \tau$, then either the e-state of $T_{e,s}(\tau)$ is no larger than the e-state of $T_{e,s}(\sigma)$, or $T_{e,s}(\tau) \not\subset G_s$. Since the e-state cannot decrease infinitely often, there will be some node on the tree beyond which the e-state never decreases; this will be our final e-state.

LEMMA 4. *For all e, if the value of the final e-state of T_e at e is in $E_n - \Pi_n$ then $\{e\}^G$ is partial or recursive.*

PROOF: Fix e and σ $\subseteq$-minimal for e such that for all sufficiently large s, the value of the e-state at e of $T_{e,s}(\sigma)$ is in $E_n - \Pi_n$ and no tree T_i for $i < e$ is controlling G at the node $T_{e,s}(\sigma)$. Such a σ exists because each tree T_i for $i < e$ controls at most finitely many nodes, and since we are dealing with the final e-state, there are infinitely many nodes with the appropriate e-state. Fix s_0 and $\alpha \in E_n$ such that for all $s \geq s_0$, for all $i \in \{0,1,2,...,n\}$, $T_{e,s}(\sigma) = T_{e,s_0}(\sigma)$, $T_{e,s}(\sigma \star i) = T_{e,s_0}(\sigma \star i)$, $G_s \upharpoonright T_{e+|\sigma|+1,s}(0) = G_{s_0} \upharpoonright T_{e+|\sigma|+1,s_0}(0)$ and the value of the e-state of $T_{e,s}(\sigma)$ is α. We show that $\{e\}^G$ is partial. Let $i \in \{1,2,...,n\}$ be the least such that $i \notin X_\alpha$. Our construction guarantees that for all $s \geq s_0$, $G_s \supset T_{e,s}(i)$, and so $G \supset T_{e,s}(i)$. There is an x in the set of difference points for this node such that $\{e\}^{T_{e,s_0}(i)}(x)$ diverges (otherwise i would be in X_α.) $|T_{e,s_0}(i)| > |\{e\}(x)|$, and so we see that $\{e\}^G(x)$ diverges.

LEMMA 5. *For all e, if the value of the final e-state of T_e at e is $i \in \{1, 2, ..., n\}$, then $\{e\}^G$ is partial.*

PROOF: Fix e and σ ($\subseteq$-minimal for e) such that for all sufficiently large s, the values of the e-states at e of $T_{e,s}(\sigma)$ and $T_{e,s}(\sigma \star 0)$ are both $i \in \{1, 2, ..., n\}$.

Such a σ must exist for the following reasons: The value of the final e-state at e is i, so there is an infinite path of nodes on T_e with e-state i. The problem now is to find two such nodes of the form σ and $\sigma \star 0$. But the only way such nodes could fail to appear is if G is moved to extend $\sigma \star i$ for some $i > 0$. This can only happen for the sake of positive requirements of priority higher than e (of which there are only finitely many). Thus there will be an ample number of nodes of the form σ and $\sigma \star 0$ with e-state at e equal to i.

Fix s_0 and $i, j \in \{1, 2, ..., n\}$ such that for all $s \geq s_0$, for all $l \in \{0, 1, 2, ..., n\}$, $T_{e,s}(\sigma) = T_{e,s_0}(\sigma)$, $T_{e,s}(\sigma \star l) = T_{e,s_0}(\sigma \star l)$, $G_s \restriction T_{e+|\sigma|+2,s}(0) = G_{s_0} \restriction T_{e+|\sigma|+2,s_0}(0)$ and the value of the e-state of $T_{e,s}(\sigma)$ is i.

We show that $\{e\}^G$ is partial. Our construction guarantees that for all $s \geq s_0$, $G_s \supset T_{e,s}(\sigma^- \star i)$, and so $G \supset T_{e,s}(\sigma^- \star i)$. There is an x in the set of difference points for the node $T_{e,s}(\sigma)$ such that $\{e\}^{T_{e,s_0}(\sigma^- \star i)}(x)$ diverges. If this were not the case, then $\{e\}^{T_{e,s_0}(\sigma^- \star i)}(x)$ would necessarily differ from one of $\{e\}^{T_{e,s_0}(\sigma \star 0)}(x)$ and $\{e\}^{T_{e,s_0}(\sigma^- \star j)}(x)$. But in that case we would have acted to improve the e-state of the node $T_{e,s}(\sigma^-)$ by using the e-split between $T_{e,s}(\sigma^- \star i)$ and $T_{e,s}(\sigma^- \star h)$, where $h = 0$ or j. $|T_{e,s_0}(\sigma^- \star i)| > |\{e\}(x)$, and so we see that $\{e\}^G(x)$ diverges.

LEMMA 6. *For all e, if the value of the final e-state of T_e at e is -1 and $\{e\}^G$ is total, then $\{e\}^G$ is recursive.*

PROOF: Assume that $\{e\}^G$ is total; we prove that $\{e\}^G$ is recursive. Let α be such that for all $\sigma \supseteq \alpha$, for all sufficiently large s the value of the e-state at e of $T_{e,s}(\sigma)$ is -1, or $T_{e,s}(\sigma) \not\supset G_s$. Then to compute $\{e\}^G(x)$, simply search for σ and an $s \geq s_0$ such that $\{e\}^{T_{e,s}(\sigma)}(x)$ converges and $T_{e,s}(\sigma) \subset G_s$. Then $\{e\}^{T_{e,s}(\sigma)}(x) = \{e\}^G(x)$. If they were unequal, then a split would appear in the tree and it could be used to improve the e-state.

LEMMA 7. *For all e, if the value of the final e-state of T_e at e is 0 then $\{e\}^G$ is partial.*

PROOF: Fix a node σ and a stage s_0 such that for all $s \geq s_0$, for all $i \in \{0, 1, 2, ..., n\}$, $T_{e,s}(\sigma \star i) = T_{e,s_0}(\sigma \star i)$ and the value of the e-state at e of $T_{e,s}(\sigma)$ is 0. There are $i, j \in \{1, 2, ..., n\}$ such that $T_{e,s}(\sigma \star i)$ and $T_{e,s}(\sigma \star j)$ e-split, say $\{e\}^{T_{e,s}(\sigma \star i)}(x) \neq \{e\}^{T_{e,s}(\sigma \star i)}$. Then $\{e\}^G(x)$ diverges. This is true since if $\{e\}^G(x)$ converged, then eventually $\{e\}^{T_{e,s}(\sigma \star 0)}(x)$ would converge (the use is less than $|T_{e,s}(\sigma \star 0)|$), and the value of the e-state at e would not be 0, since $T_{e,s}(\sigma \star 0)$ e-splits with at least one of $T_{e,s}(\sigma \star i)$ or $T_{e,s}(\sigma \star j)$.

LEMMA 8. *If the value of the final e-state at e is $\rho \in \Pi_n$ and $\{e\}^G$ is total, then $\{e\}^G \equiv_{wtt} G_\rho$.*

PROOF: We must prove that $G_\rho \leq_{wtt} \{e\}^G$ and $\{e\}^G \leq_{wtt} G_\rho$.

$G_\rho \leq_{wtt} \{e\}^G$. We first describe a wtt-computation procedure and then verify that it succeeds.

Fix σ and s_0 such that $G \supset T_e(\sigma)$ and for all $\sigma' \supseteq \sigma$ there is a $s_{\sigma'} \geq s_0$ such that either for all $s \geq s_{\sigma'}$ the value of the final e-state at e of $T_{e,s}(\sigma')$ is ρ or for all $s \geq s_{\sigma'}$, $T_{e,s}(\sigma') <^L G_s$ and $\forall s \geq s_0$, $T_{e,s}(\sigma) = T_{e,s'}(\sigma)$ and $G \supset T_{e,s_0}(\sigma)$.

To compute $G_\rho(x)$:

If $T_{e,s_0}(\sigma)$ determines $G_\rho(x)$, then give this answer. Otherwise, search for an $s > s_0$ and a $\tau \supseteq \sigma$ such that $|T_{e,s}(\tau)| > x$ and $s > x$ and $G_s \supseteq T_{e,s}(\tau)$ and for all $\sigma' \in dom(T_{e,s})$ such that $|\sigma'| < |\tau|$, the e-state of $T_{e,s}(\sigma')$ at e is ρ and $\{e\}^{T_{e,s}(\tau)} \subseteq \{e\}^G$.

Claim. $G_\rho(x) = (T_{e,s}(\tau))_\rho(x)$.

Suppose not. Fix η in $dom(T_{e,s})$ such that $T_{e,s}(\eta) \subset G$ and $\{e\}^{T_{e,s}(\eta)}(x)$ converges and $\eta \supseteq \sigma$. Such an η exists because G extends a branch of $T_{e,s}$ and $G \supseteq T_{e,s}(\sigma)$. Then $\{e\}^{T_{e,s}(\tau)}$ and $\{e\}^{T_{e,s}(\eta)}$ are compatible, but $T_{e,s}(\tau)$ and $T_{e,s}(\eta)$ are ρ-split.

Fix γ the longest common initial segment of η and τ. $\gamma \supseteq \sigma$ since $\eta \supseteq \sigma$ and $\tau \supseteq \sigma$. Then there must be $i, j \in \{1, 2, ..., n\}$ such that $\gamma \star i \subseteq \tau$ and $\gamma \star j \subseteq \eta$. Neither i nor j can be 0 because $T_{e,s}(\tau)$ and $T_{e,s}(\eta)$ do not e-split, but the e-state of $T_{e,s}(\gamma)$ is ρ, which indicates that $T_{e,s}(\gamma\star 0)$ e-splits with each of $T_{e,s}(\gamma\star i)$ for $i \in \{1, 2, ..., n\}$. We know that $T_{e,s}(\gamma)$ has e-state ρ because $\gamma \supseteq \sigma$ and $T_{e,s}(\gamma) \subset G$.

At stage s, $G_s \supset T_{e,s}(\gamma \star i)$, but $G \supset T_{e,s}(\eta) \supset T_{e,s}(\gamma \star j)$. So, there is a least stage $t > s$ such that $G_t \supset T_{e,t}(\gamma \star i) = T_{e,s}(\gamma \star i)$, but $G_{t+1} \not\supseteq T_{e,t}(\gamma \star i)$. At this stage $t+1$, the then current G_t was copied across all of the $T_{e,t+1}(\gamma \star k)$ for $k \in \{0, 1, 2, ..., n\}$, in particular on $T_{e,s}(\gamma \star j) = T_{e,t}(\gamma \star j)$, and so $G_t \restriction t = T_{e,t+1}(\gamma \star j) \restriction t = T_{e,t+1}(\eta) \restriction t$ (if defined). But, since $T_{e,t+1}(\eta) \restriction t \not\equiv_\rho T_{e,t+1}(\tau) \restriction t$, we get that $T_{e,t+1}(\gamma\star i) \restriction t = G_t \restriction t \not\equiv_\rho T_{e,t+1}(\tau) \restriction t = T_{e,t+1}(\gamma \star j) \restriction t$. Since the final e-state of $T_{e,s}$ indicates that branches on $T_{e,s}$ e-split if and only if they ρ-split, we get that G_t and $T_{e,t+1}(\tau)$ are e-split. But then $T_{e,t}(\gamma \star i)$ and $T_{e,t}(\gamma \star j)$ e-split, but do not ρ split; contrary to our assumption that the final e-state is ρ.

$\{e\}^G \leq_{wtt} G_\rho$. Again we describe a wtt-computation procedure and then verify that it succeeds.

Fix σ and s_0 such that $G \supset T_e(\sigma)$ and for all $\sigma' \supseteq \sigma$ there is a $s_{\sigma'} \geq s_0$ such that either for all $s \geq s_{\sigma'}$ the value of the final e-state at e of $T_{e,s}(\sigma')$ is ρ or for all $s \geq s_{\sigma'}$, $T_{e,s}(\sigma') <^L G_s$, and for all $s \geq s_0$, $T_{e,s}(\sigma) = T_{e,s_0}(\sigma)$ and $G_s \supseteq T_{e,s}(\sigma)$.

To compute $\{e\}^G(x)$:

If $|T_e(\sigma)| < \{e\}(x)$, then $\{e\}^G(x) = \{e\}^{T_e(\sigma)}(x)$. Otherwise, search for an $s \geq s_0$ and a $\tau \supset \sigma$ such that $|T_{e,s}(\tau)| > \{e\}(x)$ and $s > \{e\}(x)$ and $G_s \supset T_{e,s}(\tau)$ and for all σ' such that $\sigma \subseteq \sigma' \subseteq \tau$, the value of the e-state of $T_{e,s}(\sigma')$ at e is ρ and $(T_{e,s}(\tau))_\rho \subseteq G_\rho$.

Claim: $\{e\}^G(x) = \{e\}^{T_{e,s}(\tau)}(x)$.

Suppose not. Then, as above, there will be two branches on T_e, both extending $T_e(\sigma)$ that are e-split but are not ρ split. This contradicts our assumption that the final e-state at e is ρ.

§5. Conclusions

We have now shown that every non-zero degree below $deg_{wtt}(G)$ coincides with $deg(G_\rho)$ for some $\rho \in \Pi_n$. Since all of the positive requirements have been satisfied, we see that the lattice Π_n has been embedded as an initial segment of the Δ_2^0 wtt-degrees above a minimal degree. In fact the embedding is in the wtt-degrees $\leq_{wtt} 0'$. To see that this is true, we need only check that there is a recursive function $b(x)$ such that $|\{s : G_{s+1}(x) \neq G_s(x)\}| \leq b(x)$. But this is the case because of the nesting of the trees. Only action taken for the sake of satisfying P_0 can change $G_s(0)$. This can happen at most twice. The only possible actions which can change $G(x)$ are taken for the sake of satisfying some P_e where e $\leq x$, or improving the e-state of some $T_{e,s}(\sigma)$, where $e+|\sigma| \leq x$. A simple induction shows that this is bounded by $(2+n+|E_n|+2)^{2^{x+1}-1}$.

REFERENCES

1. Burris, S. and Sankappanavar, A.P., *Lattice theoretic decision problems in universal algebra*, **Algebra Univ. 5** (1975), 163-177.
2. Degtev, A., *On truth-table-like reducibilities in the theory of algorithms*, **Usp. Mat. Nauk 34** (1979), 137-168, 248 (Russian); **Russ. Math. Surv. 34**, 155-192 (English translation).
3. Fejer, P. and Shore, R., *Minimal tt- and wtt- degrees*, in this volume.
4. Haught, C. A. and Shore, R., *Undecidability and initial segments of the r.e. tt-degrees*, to appear in **Journal of Symbolic Logic**.
5. Odifreddi, P., *Strong reducibilities*, **Bull. Amer. Math. Soc. (N.S.) 4** (1981), 37-86.
6. Odifreddi, P., "**Classical Recursion Theory**," North-Holland, Amsterdam, New York, Oxford, 1989.
7. Rogers, H., "**Theory of Recursive Functions and Effective Computability**," McGraw-Hill, New York, 1967.

(Haught) Department of Mathematics, The University of Chicago, Chicago, IL 60637 and Department f Mathematical Sciences, Loyola University of Chicago, Chicago, IL 60626

(Shore) Department of Mathematics, Cornell University, Ithaca, New York, 14853

RANDOMNESS AND GENERALIZATIONS OF FIXED POINT FREE FUNCTIONS

Antonín Kučera
Department of Computer Science, Charles University
Malostranské náměstí 25, 118 00 Praha 1, Czechoslovakia

1. INTRODUCTION

The study of functions which diagonalize all partial recursive functions or, similarly, all r.e. sets has brought several interesting results. There arose a question whether more complicated ways of a diagonalization or a diagonalization of more complicated objects can give us straightforwardly yet stronger results. We use measure arguments to show a substantial limitation in this direction.

A total function g is called a fixed point free function (an FPF function) if $W_e \neq W_{g(e)}$ for all e. A total function f is called a diagonally nonrecursive function (a DNR function) if $f(e) \neq \{e\}(e)$ for all e. A DNR (FPF) degree is a degree containing a DNR (FPF) function. DNR degrees coincide with FPF degrees ([6], Lemma 4.1).

It is clear that $\underset{\sim}{0}$ is not an FPF degree while $\underset{\sim}{0}'$ is an FPF degree. M.M.Arslanov ([1], [2]) proved the following result known as a completeness criterion. If A is an r.e. set then A is complete (i.e. $deg(A) = \underset{\sim}{0}'$) iff there is a DNR function recursive in A. Let us note that this completeness criterion was further extended in [6] from r.e. sets to all finite levels of n-REA hierarchy introduced in [7].

From another point of view, DNR (or FPF) degrees turned out to play an important role for constructions of r.e. degrees. Namely, every FPF degree $\leq \underset{\sim}{0}'$ bounds a nonzero r.e. degree [11] and the technique to prove that does not use the priority method and, thus, yields an alternative, priority-free, solution to Post's problem as well as alternative constructions to a considerable part of the finite injury priority results. In [12] it is shown how to extend the use of DNR functions to infinite injury results. The resulting method of constructions of r.e. sets is different from the standard priority method but, roughly speaking, it is at the same level of complexity. In this connection it was an open question whether in the case of a stronger diagonalization we could still have reasonable existential theorems for r.e. degrees with a significant simplification of the resulting constructions of r.e. sets (in question).

The first natural candidate was the class of *-FPF functions. A total function h is called a *-FPF function if $W_e \neq^* W_{h(e)}$ for all e, where $=^*$ means equality modulo finite sets.

Concerning a generalization of the completeness criterion, the following restricted version of it was proved by M.M.Arslanov [1] (see also [17]). If A is a Σ^0_2 set (i.e. A is r.e. in $\emptyset'$) and $\emptyset' \leq_T A$ then there is a *-FPF function recursive in A iff A is Σ^0_2 complete (i.e. deg(A) = $\underset{\sim}{0}''$). A further generalization to n-REA hierarchy was proved in [6] (see also [17]). Nevertheless, it was an open question whether this generalization can be extended to all Σ^0_2 sets (i.e. whether the assumption $\emptyset' \leq_T A$ is really necessary).

Concerning a connection to constructions of r.e sets it was not known whether every degree $\underset{\sim}{a} \leq \underset{\sim}{0}''$ (or, at least, every Σ^0_2 degree $\underset{\sim}{a}$) which contains a *-FPF function bounds a nonzero r.e. degree.

We use measure arguments, more precisely, we use 2-random sets (or, equivalently, $\emptyset'$-NAP sets - see below) to answer the above questions negatively.

Our notation and terminology are standard. We deal with sets and functions over N = {0,1,2, ... }. We frequently identify subsets of N with their characteristic functions. $\{e\}^A$ is the eth Turing reduction functional (in some standard enumeration of all such functionals), {e} denotes $\{e\}^{\emptyset}$. For A ⊆ N let deg(A) denote the Turing degree of A. By a degree we mean a Turing degree. If a degree contains a set or a function with a certain property we say that the degree itself has that property.

A string is a finite sequence of 0's and 1's. Strings may also be viewed as functions from finite initial segments of N into {0,1}. By $\sigma \subseteq A$ we denote that the characteristic function of A extends σ and we use also other usual notation concerning strings. We apply notions of recursion theory to strings via Gödel numbering.

Let Ext(σ) denote the class of all sets A for which $\sigma \subseteq A$, for A ⊆ N let Ext(A) be the union of Ext(σ) for σ ∈ A. A class $\mathfrak{C}$ of subsets of N is a $\Sigma^{0,A}_1$ class and x is its index if $\mathfrak{C}$ = Ext(W^A_x). We write Σ^0_1 for $\Sigma^{0,\emptyset}_1$. Further, $\Pi^{0,A}_1$ classes are complements (in $\{0,1\}^N$) of $\Sigma^{0,A}_1$ classes. We write Π^0_1 for $\Pi^{0,\emptyset}_1$. A Π^0_1 class can be thought of as just the class of all infinite branches of a binary recursive tree. A string σ is called $\mathfrak{C}$-extendible for a class $\mathfrak{C}$ of sets if Ext(σ) ∩ $\mathfrak{C}$ ≠ ∅.

Let |A| denote the cardinality of A and for every n let $(A)_n$ denote {x : <n,x> ∈ A}. By μ we denote Lebesgue measure. [17] is a general reference for unexplained terminology.

2. RANDOMNESS

The concept of randomness was extensively studied from several points of view and for different purposes by Kolmogorov, Chaitin, Solovay, Martin-Löf, Kurtz, Demuth and others ([15], [4], [5], [3], [13], [14], [10], [12] - is a partial list of basic references). Two basic approaches use either the concept of Kolmogorov complexity or measure theory. Nevertheless, various definitions of 1-randomness (sometimes under different names) turned out to be equivalent (an unpublished result of Solovay, [13], [14], [3] etc.). From this reason and since 1-randomness is now a generally accepted notation we will call from now on by a 1-random set what was called in [10], [12] etc. by a NAP set and we will follow this style at higher levels of the arithmetical hierarchy too.

We reformulate the definition introduced in [10] together with a straightforward relativization.

Definition. 1) Let A be a set.
a) A class $\mathfrak{C} \subseteq \{0,1\}^N$ is of $\Sigma_1^{0,A}$ measure zero if there is a recursive sequence of $\Sigma_1^{0,A}$ indices of $\Sigma_1^{0,A}$ classes $\subseteq \{0,1\}^N$ $\mathcal{B}_0, \mathcal{B}_1, \mathcal{B}_2, \ldots$ such that
$\forall n(\mu(\mathcal{B}_n) < 2^{-n})$ and $\mathfrak{C} \subseteq \bigcap_n \mathcal{B}_n$.
b) A set B is A-random if the class {B} is not of $\Sigma_1^{0,A}$ measure zero.
2) For $n \geq 0$ $\emptyset^{(n)}$-random sets are called n+1-random sets.

Remark. 1) n-random sets ($n \geq 1$) arise, roughly speaking, by a diagonalization of all effective approximations in measure by $\Sigma_1^{0,\emptyset^{(n-1)}}$ classes (or, equivalently, by Σ_n^0 classes).
2) n-random sets ($n \geq 1$) are just $\emptyset^{(n-1)}$-NAP sets in an alternative terminology (however, we will not use it).

P.Martin-Löf proved [15] that there is a universal class which has Σ_1^0 measure zero, i.e. there is a recursive sequence of Σ_1^0 indices of Σ_1^0 classes $\subseteq \{0,1\}^N$ $\mathcal{U}_0, \mathcal{U}_1, \ldots$ such that $\mathcal{U}_n \supseteq \mathcal{U}_{n+1}$ for all n, $\forall n(\mu(\mathcal{U}_n) < 2^{-n})$ and for every class $\mathfrak{C}$ of Σ_1^0 measure zero $\mathfrak{C} \subseteq \bigcap_n \mathcal{U}_n$.
Analogously, for every set A we have a universal class of $\Sigma_1^{0,A}$ measure zero. For later purposes we need to fix some standard way of a construction of such universal approximating sequences.

Let A be a set. Let U_n^A denote an A-r.e. set constructed as follows: for each e, e > n, take y = {e}(e) (if it exists) and for all s put all elements of $W_{y,s}^A$ into U_n^A under the condition $\mu(\mathrm{Ext}(W_{y,s}^A)) < 2^{-e}$. Let $\mathcal{U}_n^A$ denote $\mathrm{Ext}(U_n^A)$. It is easy to see that we can fix a recursive sequence $\{u_n\}_{n \in N}$ such that for each n u_n is a $\Sigma_1^{0,A}$ index of a $\Sigma_1^{0,A}$ class $\mathcal{U}_n^A$ (for every set A).

The class $\bigcap_n \mathfrak{U}_n^A$ is obviously the required universal class of $\Sigma_1^{0,A}$ measure zero.

Let $\mathfrak{P}_n^A$ denote the $\Pi_1^{0,A}$ class which is the complement (in $\{0,1\}^N$) of $\mathfrak{U}_n^A$. Obviously $\mathfrak{P}_0^A \subseteq \mathfrak{P}_1^A \subseteq \mathfrak{P}_2^A \subseteq \ldots$ and the class of all A-random sets is just the class $\bigcup_n \mathfrak{P}_n^A$. Thus, the class of all A-random sets is a $\Sigma_2^{0,A}$ class. On the other hand, A-random degrees are exactly determined by degrees of members of some $\Pi_1^{0,A}$ class of sets. In fact, the same argument as that used to prove Lemma 3 in [10] shows that the classes of degrees $\deg(\mathfrak{P}_n^A)$ coincide for all n and are equal to the class of all A-random degrees, i.e. we have the following.

Lemma 1. For every set A, the class of all A-random degrees is equal to $\deg(\mathfrak{P}_n^A)$ for all n.

Analogously to Corollary 3 of Lemma 3 from [10], for every nonempty $\Pi_1^{0,A}$ class $\mathfrak{C}$ (of sets) of positive measure degrees of members of $\mathfrak{C}$ contain all A-random degrees. Obviously, the class of all A-random sets has measure one.

Basic methods for working with A-random sets (from the point of view of recursion theory) include methods for dealing with $\Pi_1^{0,A}$ classes (like applications of Low Basis Theorem from [8]), measure arguments and, due to a certain effective way of diagonalization of $\Sigma_1^{0,A}$ objects, also a certain kind of self-reference. For example, the following generalization of Lemma 8 from [10] presents a weak version of "incompleteness phenomena", roughly speaking, every nonempty $\Pi_1^{0,A}$ class which is a subclass of $\mathfrak{P}_n^A$ is still of a positive measure and we can even effectively find a lower bound of the measure.

Lemma 2. There are recursive functions F,G such that for all A,x,σ,n we have $G(\sigma,n,x) > 0$ and if $\mathcal{B} = \mathrm{Ext}(\sigma) \cap (\mathfrak{P}_n^A - \mathrm{Ext}(W_x^A)) \neq \emptyset$ then

a) $\mu(\mathcal{B}) \geq G(\sigma,n,x)$,

b) $F(\sigma,n,x) \geq$ the least number at which the left-most and right-most member (in the standard lexicographic ordering) of $\mathcal{B}$ differ each other.

Proof. Let σ,n,x be given. Recall that $\mathfrak{P}_n^A$ is a $\Pi_1^{0,A}$ class which is the complement of $\mathrm{Ext}(W_{u_n}^A)$, for every set A.

We can find e, $e > n$, such that for all j $\{e\}(j)!$ and, for every set A, $\{e\}(j)$ is a $\Sigma_1^{0,A}$ index of a $\Sigma_1^{0,A}$ class $\mathfrak{C}_j$, where

a) $\mathfrak{C}_j = \mathrm{Ext}(\sigma) - (\ \mathrm{Ext}(W_{u_n,t}^A) \cup \mathrm{Ext}(W_{x,t}^A)\ )$ if there is an s such that $\mu(\mathrm{Ext}(\sigma) - (\ \mathrm{Ext}(W_{u_n,s}^A) \cup \mathrm{Ext}(W_{x,s}^A)\ )) < 2^{-j}$, in which case t denotes the least such s,

b) $\mathfrak{C}_j = \emptyset$ otherwise.

Clearly $\mathfrak{C}_e \subseteq \mathfrak{U}_n^A$, i.e. if $\mu(\mathcal{B}) < 2^{-e}$, where $\mathcal{B} = \mathrm{Ext}(\sigma) \cap (\mathfrak{P}_n^A - \mathrm{Ext}(W_x^A))$,

then $\mathcal{B} = \emptyset$.
Thus, it is sufficient to let $F(\sigma,n,x) = e$ and $G(\sigma,n,x) = 2^{-e}$. $\square$

The above principle was used in [10] to show that 1-random degrees contain the upper cone $\{\underset{\sim}{a} : \underset{\sim}{a} \geq \underset{\sim}{0}'\}$. We use Lemma 2 to prove a more general result.

Theorem 3. Let A be a set, $\underset{\sim}{a} = \deg(A)$. For every degree $\underset{\sim}{b} \geq \underset{\sim}{a}'$ there is an A-random degree $\underset{\sim}{d}$ such that $\underset{\sim}{d} \cup \underset{\sim}{a} = \underset{\sim}{b}$, i.e.
$\{\underset{\sim}{d} \cup \underset{\sim}{a} : \underset{\sim}{d} \text{ is A-random}\} \supseteq \{\underset{\sim}{b} : \underset{\sim}{b} \geq \underset{\sim}{a}'\}$.

Proof. We will construct a total tree T recursive in A' such that
i) all infinite branches of T are A-random sets,
ii) for every set $B \geq_T A'$ $\deg(T(B)) \cup \deg(A) = \deg(B)$.
We start with the class $\mathfrak{P}_0^A$. Applying the previous Lemma we can construct a recursive function H such that for all strings σ
$H(\sigma) \geq$ the least number at which the left-most and right-most member of $\mathfrak{P}_0^A$ extending σ differ each other (if σ is $\mathfrak{P}_0^A$-extendible).
Let $T(\emptyset) = \emptyset$. Assume $T(\sigma)$ has been defined. Using an oracle A' we can find two strings τ_0, τ_1 extending $T(\sigma)$ of length $H(T(\sigma))$ which are beginnings of the left-most, resp. right-most member of $\mathfrak{P}_0^A$ extending $T(\sigma)$. It follows that $\tau_0 \neq \tau_1$ and τ_0,τ_1 are incompatible.
Let $T(\sigma*0) = \tau_0$ and $T(\sigma*1) = \tau_1$.
Obviously, T is a total tree recursive in A'. Its infinite branches are A-random sets (from the class $\mathfrak{P}_0^A$).
Clearly, for every set B $T(B) \leq_T B \oplus A'$. Hence, if $B \geq_T A'$ we have $T(B) \leq_T B$.
It remains to show that for every set B $B \leq_T T(B) \oplus A$.
Assume we already know $\sigma \subseteq B$. We have to show how to find j such that $\sigma*j \subseteq B$. From T(B), using the function H, we can find $T(\sigma)$ and compute the length k of both $\tau_0 = T(\sigma*0)$ and $\tau_1 = T(\sigma*1)$. Let ρ be a string of length k, $\rho \subseteq T(B)$. Now we have to determine whether $\rho = \tau_0$ or $\rho = \tau_1$. It is easy to see that what we need is to decide (by means of an oracle A) which of the following two $\Sigma_1^{0,A}$ conditions is true:
case 0: all strings of length k extending $T(\sigma)$ which lexicographically precede ρ are not $\mathfrak{P}_0^A$-extendible,
case 1: analogously but with "lexicographically follow" instead of "lexicographically precede".
Since $\mathfrak{P}_0^A$ is a $\Pi_1^{0,A}$ class of sets and due to the construction of τ_0,τ_1 just one of the previous two conditions is true and an oracle A can be used to find which one, i.e. whether $\sigma*0 \subseteq B$ or $\sigma*1 \subseteq B$. Thus, we have $B \leq_T T(B) \oplus A$. $\square$

The class of 1-random degrees is not closed upwards [10]. We now show that neither the class of 2-random degrees (or, C-random degrees for $C \geq_T \emptyset'$) is closed upwards; it even does not contain any nonempty upper cone of degrees.

<u>Theorem 4</u>. Let A be a nonrecursive set, $A \leq_T \emptyset'$. Then the class $\{B : B \geq_T A\}$ is of $\Sigma_1^{0,\emptyset'}$ measure zero.

<u>Proof</u>. It is sufficient to show that for all z the class $\{B : \{z\}^B = A\}$ is of $\Sigma_1^{0,\emptyset'}$ measure zero.
Let z be given. G.Sacks [16] proved that the class $\{B : B \geq_T A\}$ is of measure zero. Thus, for every n there exists k such that

$$\mu(\{B : \{z\}^B \upharpoonright k = A \upharpoonright k\}) < 2^{-n} . \qquad (1)$$

Observe that for every string σ $\{B : \{z\}^B \upharpoonright lh(\sigma) = \sigma\}$ is a Σ_1^0 class and, thus, we can use an oracle $\emptyset'$ to compute the measure of this class (which is a real recursive in $\emptyset'$).
Since $A \leq_T \emptyset'$ it is easy to see that we can use an oracle $\emptyset'$ to find for each n some k satisfying (1).
It now easily implies that $\{B : \{z\}^B = A\}$ is of $\Sigma_1^{0,\emptyset'}$ measure zero.□

We immediately have the following.

<u>Corollary 1</u>. The class $\{B : \exists A \ (B \geq_T A \ \& \ A \leq_T \emptyset' \ \& \ \emptyset <_T A)\}$ is of $\Sigma_1^{0,\emptyset'}$ measure zero.

<u>Corollary 2</u>. For every 2-random degree $\underset{\sim}{a}$
$\underset{\sim}{a} \notin \{\underset{\sim}{b} : \exists \underset{\sim}{d} \ (\underset{\sim}{d} \neq \underset{\sim}{0} \ \& \ \underset{\sim}{d} \leq \underset{\sim}{0}' \ \& \ \underset{\sim}{b} \geq \underset{\sim}{d})\}$, (i.e. 2-random degrees avoid the upper cone of degrees generated by nonzero degrees below $\underset{\sim}{0}'$).

By the same method we can prove also the following.

<u>Corollary 3</u>. For every C-random degree $\underset{\sim}{a}$, where $C \geq_T \emptyset'$,
$\underset{\sim}{a} \notin \{\underset{\sim}{b} : \exists \underset{\sim}{d} \ (\underset{\sim}{b} \geq \underset{\sim}{d} \ \& \ \underset{\sim}{d} \neq \underset{\sim}{0} \ \& \ \underset{\sim}{d} \leq deg(C))\}$.

3. A GENERALIZATION OF FPF FUNCTIONS

Every 1-random degree is also an FPF degree ([10]). There is an analogous relation between various kinds of randomness and generalizations of FPF functions. We start with 2-random degrees and we consider the situation at higher levels of the arithmetical hierarchy later.

<u>Theorem 5</u>. Every 2-random degree is also a *-FPF degree.

Idea of the proof. We think of W_x as a set of pairs. Let us consider a set M of indices of finite columns in W_x, i.e. a set of indices satisfying a Σ_2^0 property. Every 2-random set A has to differ from such set and we can even find (A-effectively) a finite string $\sigma \subseteq A$ such that $M \neq \{i : \sigma(i) = 1\}$. This is enough to construct an r.e. set $\neq^* W_x$.

<u>Proof</u>. Let $\underset{\sim}{a}$ be a 2-random degree. Due to Lemma 1, we can fix a

2-random set $A \in \underset{\sim}{a}$ such that $A \in \mathfrak{P}_0^{\emptyset'}$. Let h be a recursive function such that

$W_{h(x)}^{\emptyset'} = \{i : (W_x)_i \text{ is finite}\}$.

Let e be a recursive function such that for every x,j $\{e(x)\}(j)!$, $e(x) > 0$ and $\{e(x)\}(j)$ is a $\Sigma_1^{0,\emptyset'}$ index of a $\Sigma_1^{0,\emptyset'}$ class $\mathfrak{C}_j^x$, where

i) $\mathfrak{C}_j^x = \emptyset$ if $|W_{h(x)}^{\emptyset'}| < j+1$,

ii) $\mathfrak{C}_j^x = \{B : B \supseteq \{a_0, \ldots, a_j\}\}$ if $|W_{h(x)}^{\emptyset'}| \geq j+1$, where a_i $(i = 0, \ldots, j)$ is the ith element of $W_{h(x)}^{\emptyset'}$ found in a standard $\emptyset'$-recursive search of all elements of this set.

It is easy to see that for all x,j $\mathfrak{C}_j^x$ is a $\Sigma_1^{0,\emptyset'}$ class the measure of which is $< 2^{-j}$.

Due to the construction of $\mathfrak{U}_0^{\emptyset'}$, obviously $\mathfrak{C}_{e(x)}^x \subseteq \mathfrak{U}_0^{\emptyset'}$, i.e. $\mathfrak{C}_{e(x)}^x \cap \mathfrak{P}_0^{\emptyset'} = \emptyset$ for every x.

On the other hand, $A \in \mathfrak{P}_0^{\emptyset'}$ and, thus, $A \notin \mathfrak{C}_{e(x)}^x$, for every x.

For every x let σ_x be a string satisfying $|\{i : \sigma_x(i) = 1\}| = e(x)+1$ and $\sigma_x \subseteq A$.

We claim that $W_{h(x)}^{\emptyset'} \neq \{i : \sigma_x(i) = 1\}$ for every x.

If not, then obviously $A \in \mathfrak{C}_{e(x)}^x$ for some x, a contradiction.

Let f be an A-recursive function such that $W_{f(x)}$ is equal to the set $\{\langle i,y\rangle : (y \in N \,\&\, i \geq lh(\sigma_x)) \vee (y \in N \,\&\, i < lh(\sigma_x) \,\&\, \sigma_x(i) = 0)\}$.

We have $W_x \neq^* W_{f(x)}$ for all x. Really, the only finite columns in $W_{f(x)}$ are those whose indices belong to $\{i : \sigma_x(i) = 1\}$ and this set is different from $W_{h(x)}^{\emptyset'}$ (the set of indices of finite columns in W_x). Thus, for some i $(W_x)_i \neq^* (W_{f(x)})_i$.

Since the class of *-FPF degrees is closed upwards the proof is complete.□

Remark. The class of *-FPF degrees is closed upwards while the class of 2-random degrees does not contain any nonempty upper cone of degrees. Thus, the class of 2-random degrees is a proper subclass of the class of *-FPF degrees.

An oracle $\emptyset''$ can be used to construct some particular 2-random degrees.

Theorem 6. There is a 2-random degree which is also a Σ_2^0 degree.

Proof. The class $\mathfrak{P}_0^{\emptyset'}$ is a nonempty $\Pi_1^{0,\emptyset'}$ class of sets. Let A be the left-most member of it (in the standard lexicographic ordering). Obviously A is 2-random and a relativization of Theorem 3 from [9] shows that $A \equiv_T B$ for some Σ_2^0 set B, i.e. A is of a Σ_2^0 degree.□

Corollary. There is a Σ_2^0 degree $\underset{\sim}{a}$ which is a *-FPF degree such that there is no nonzero r.e. degree (even no nonzero degree $\leq \underset{\sim}{0}'$) below $\underset{\sim}{a}$; thus, $\underset{\sim}{a} < \underset{\sim}{0}''$ and $\neg(\underset{\sim}{a} \geq \underset{\sim}{0}')$.

Proof. It is sufficient to use Theorem 5, Theorem 6 and Corollary 2 of Theorem 4.□

The above Corollary shows that Arslanov completeness criterion does not generalize to Σ_2^0 and *-FPF degrees without an additional assumption. It also shows that the use of all *-FPF degrees below $\underline{0}''$ cannot help straightforwardly for constructions of r.e. degrees.

The relation between 2-random degrees and *-FPF degrees can be generalized to higher levels of the arithmetical hierarchy. Let us first remember a sequence of equivalence relations on r.e. sets (see [6],[17]).

Definition. For every r.e. sets A,B and n let $A \sim_n B$ denote
$A = B$ for $n = 0$,
$A =^* B$ for $n = 1$, and
$A^{(n-2)} \equiv_T B^{(n-2)}$ for $n \geq 2$.

Definition. A total function f is called an n-FPF function if for all x $\neg(W_x \sim_n W_{f(x)})$.

Let us note that FPF functions are just 0-FPF functions and *-FPF functions are just 1-FPF functions.

Theorem 7. Every n+1-random degree is also an n-FPF degree (for all n).

Proof. The case n = 0 is proved in [10] (Corollary 1 of Theorem 6), the case n = 1 is Theorem 5.
Let $n \geq 2$.
Let us take a set B such that B is r.e. in $\emptyset^{(n-2)}$, $B \geq_T \emptyset^{(n-2)}$ and $(B)_j \geq_T \emptyset^{(n-2)}$ for all j, B is low over $\emptyset^{(n-2)}$ (i.e. $B' \equiv_T \emptyset^{(n-1)}$) and that the sequence $(B)_0, (B)_1, (B)_2, \ldots$ is strongly independent, i.e. for all j $\neg((B)_j \leq_T (B)_{\neq j})$, where $(B)_{\neq j} = \{\langle i,k\rangle : \langle i,k\rangle \in B \ \&\ i \neq j\}$ (such set can be constructed by a standard finite injury method).
Since B is low over $\emptyset^{(n-2)}$, $W_x^{(n-2)} \leq_T (B)_{\neq j}$ is a $\Sigma_3^{0,\emptyset^{(n-2)}}$ condition (in x,j).
Thus, there exists a recursive function h such that
$W_{h(x)}^{\emptyset^{(n)}} = \{i : W_x^{(n-2)} \leq_T (B)_{\neq i}\}$.
Further, let e be a recursive function such that for all x,j $e(x) > 0$, $\{e(x)\}(j)!$ and $\{e(x)\}(j)$ is a $\Sigma_1^{0,\emptyset^{(n)}}$ index of a $\Sigma_1^{0,\emptyset^{(n)}}$ class $\mathbb{C}_j^x$, where

i) $\mathbb{C}_j^x = \emptyset$ if $|W_{h(x)}^{\emptyset^{(n)}}| < j+1$,

ii) $\mathbb{C}_j^x = \{S : S \supseteq \{a_0, \ldots, a_j\}\}$ if $|W_{h(x)}^{\emptyset^{(n)}}| \geq j+1$,

where a_i $(i = 0,\ldots,j)$ is the ith element of $W_{h(x)}^{\emptyset^{(n)}}$ found in a standard $\emptyset^{(n)}$-recursive search of all elements of this set.

Due to the construction of $\mathfrak{U}_0^{\emptyset^{(n)}}$, it is easy to see that $\mathfrak{C}_{e(x)}^{x} \subseteq \mathfrak{U}_0^{\emptyset^{(n)}}$, i.e. $\mathfrak{C}_{e(x)}^{x} \cap \mathfrak{P}_0^{\emptyset^{(n)}} = \emptyset$ for all x.

Let $\underset{\sim}{a}$ be an n+1-random degree. Due to Lemma 1 we can fix an n+1-random set $A \in \underset{\sim}{a}$, $A \in \mathfrak{P}_0^{\emptyset^{(n)}}$. We get from the above $A \notin \mathfrak{C}_{e(x)}^{x}$ for all x.

For every x let σ_x be a string satisfying $|\{i : \sigma_x(i) = 1\}| = e(x)+1$ and $\sigma_x \subseteq A$.

We claim that $W_{h(x)}^{\emptyset^{(n)}} \neq \{i : \sigma_x(i) = 1\}$ for all x.

If not, then for some x $A \in \mathfrak{C}_{e(x)}^{x}$, a contradiction.

Let g be an A-recursive function such that

$W_{g(x)}^{\emptyset^{(n-2)}} = \{<j,k> : <j,k> \in B \ \& \ (j \geq lh(\sigma_x) \vee (j < lh(\sigma_x) \ \& \ \sigma_x(j) = 0))\}$ for all x.

Now for all x we have $W_x^{(n-2)} \not\equiv_T W_{g(x)}^{\emptyset^{(n-2)}}$.

To see that observe that the opposite case implies $W_{h(x)}^{\emptyset^{(n)}} = \{i : \sigma_x(i) = 1\}$, a contradiction.

Obviously, the set $W_{g(x)}^{\emptyset^{(n-2)}}$ is r.e. in $\emptyset^{(n-2)}$ and $\geq_T \emptyset^{(n-2)}$ so by the uniformity of the iterated Sacks jump theorem (and since the function g is A-recursive) there is an A-recursive function f such that for all x $W_{f(x)}^{(n-2)} \equiv_T W_{g(x)}^{\emptyset^{(n-2)}}$. Thus, $\neg(W_x \sim_n W_{f(x)})$ for all x.

Since the class of n-FPF degrees is closed upwards the proof is complete.$\square$

Remark. Since the class of n+1-random degrees for $n \geq 1$ does not contain any nonempty upper cone of degrees it is a proper subclass of the class of n-FPF degrees.

We can also generalize Theorem 6 and its Corollary.

Theorem 8. For every $n \geq 1$ there is a degree $\underset{\sim}{a}$ which is both Σ_{n+1}^0 and n+1-random and, hence, $\underset{\sim}{a}$ is an n-FPF degree $\leq \underset{\sim}{0}^{(n+1)}$ and there is no nonzero degree $\leq \underset{\sim}{0}^{(n)}$ below $\underset{\sim}{a}$ (thus, $\neg(\underset{\sim}{a} \geq \underset{\sim}{0}^{(n)})$ and $\underset{\sim}{a} < \underset{\sim}{0}^{(n+1)}$).

Proof. We use the same idea as in the proof of Theorem 6 and its Corollary. Namely, let A be the left-most member of $\mathfrak{P}_0^{\emptyset^{(n)}}$ (a $\Pi_1^{0,\emptyset^{(n)}}$ class of sets). Then by a relativization of Theorem 3 from [9], A is of a Σ_{n+1}^0 degree. Obviously A is n+1-random. Then we apply Theorem 7 and Corollary 3 of Theorem 4 to complete the proof.$\square$

Acknowledgement. The author would like to thank to M.Lerman for proposing the problem and to R.Shore and C.Jockusch for helpful comments.

The paper is in final form and no similar paper has been or is being submitted elsewhere.

REFERENCES

[1] M.M.Arslanov (1981) On some generalizations of the theorem on fixed points, Izv. Vyssh. Uchebn. Zaved. Mat. 228, No 5, 9-16 (Russian).

[2] M.M.Arslanov, R.F.Nadirov, V.D.Solov'ev (1977) Completeness criteria for recursively enumerable sets and some general theorems on fixed points, Izv. Vyssh. Uchebn. Zaved. Mat. 179, No 4, 3-7, (Russian).

[3] G.J.Chaitin (1977) Algorithmic information theory, IBM Journal of Research and Development, July 1977, 350-359.

[4] O.Demuth (1975) On constructive pseudonumbers, Comment. Math. Univ. Carolinae 16, 315-331 (Russian).

[5] O.Demuth (1982) On some classes of arithmetical real numbers, Comment. Math. Univ. Carolinae 23, 453-465 (Russian).

[6] C.G.Jockusch,Jr., M.Lerman, R.I.Soare, R.M.Solovay (1989) Recursively enumerable sets modulo iterated jumps and extensions of Arslanov's completeness criterion, to appear in J. Symbolic Logic.

[7] C.G.Jockusch,Jr., R.A.Shore (1984) Pseudo jump operators. II: Transfinite iterations, hierarchies, and minimal covers, J. Symbolic Logic 49, 1205-1236.

[8] C.G.Jockusch,Jr., R.I.Soare (1972) Π_1^0 classes and degrees of theories, Trans. Amer. Math. Soc. 173, 33-56.

[9] C.G.Jockusch,Jr., R.I.Soare (1972) Degrees of members of Π_1^0 classes, Pacific J. Math., 40, 605-616.

[10] A.Kučera (1985) Measure, Π_1^0-classes and complete extensions of PA, In: Recursion Theory Week, Proceedings (Ebbinghaus, Müller, Sacks editors), Lect. Notes in Math. No. 1141, Springer-Verlag, 245-259.

[11] A.Kučera (1986) An alternative, priority-free, solution to Post's problem, In: Proceedings, MFCS'86 (Gruska, Rovan, Wiedermann editors.), Lect. Notes in Comp. Science, No. 233, Springer-Verlag, 493-500.

[12] A.Kučera (1989) On the use of diagonally nonrecursive functions, In: Proceedings, Logic Colloquium'87 (Ebbinghaus et al. editors), North-Holland, 219-239.

[13] S.A.Kurtz (1981) Randomness and genericity in the degrees of unsolvability, PhD thesis, Univ. of Illinois at Urbana-Champaign.

[14] S.A.Kurtz (ta) A note on random sets, to appear.

[15] P.Martin-Löf (1966) The definition of random sequences, Information and Control, 9, 602-619.

[16] G.E.Sacks (1963) Degrees of unsolvability, Annals of Math. Studies 55, Princeton University Press, Princeton, N.J.

[17] R.I.Soare (1987) Recursively enumerable sets and degrees: A study of computable functions and computably generated sets, Springer-Verlag.

Recursive Enumeration Without Repetition Revisited

Martin Kummer
Institut für Logik, Komplexität und Deduktionssysteme
Universität Karlsruhe
D-7500 Karlsruhe, Postfach 6980
F.R.G.

§0 Introduction

The study of uniformly r.e. classes of r.e. sets is a classical topic in recursion theory. In this context the problem to characterize the r.e. classes admitting a *recursive enumeration without repetition* or a *one-one numbering* had attracted a great deal of research effort. In 1958, Friedberg [F58], utilizing his insight into the "priority method", proved that the class of all r.e. sets and the class of all partial recursive functions can be enumerated without repetition. Pour-El [P60] rose the question if this could be done for every r.e. class, and answered it negatively [PP65] : Let K be the halting problem $\{i \mid \varphi_i(i)\downarrow\}$. The r.e. class

$$\{\{2x,2x+1\}\}_{x\in K} \cup \{\{2x\},\{2x+1\}\}_{x\notin K}$$

cannot be enumerated without repetition, for otherwise K would be co-r.e.

In the following years many sufficient conditions for the existence of one-one numberings were discovered by Howard and Pour-El [HP64], Pour-El and Putnam [PP65], Lachlan [La65], [La67], Wolf [Wo64], Mal'tsev [Mal65], Marchenkov [Mar71]. On the other hand, few necessary conditions have been found.

In the seventies and eighties research on one-one numberings has been confined nearly exclusively to russian authors. They studied them - under the direction of Yu.L. Ershov - in the course of their research program concerning the structure of the upper semilattices induced by computable numberings of r.e. classes. Goncharov [Go80] employed one-one numberings as a tool in constructive model theory.

In the present paper we want to reconsider the classical area of *sufficient* conditions for the existence of one-one numberings. We obtain some strong generalizations of Mal'tsev's criterion. In particular, we are dealing with classes S which may contain finite sets not having a proper superset in S, a case which is scarcely covered by previous criteria.

The results appearing here are part of the author's dissertation [Ku89c]. The paper is in final form and has not been submitted elsewhere.

§1 Notation and Definitions

Our notation is fairly standard and follows the textbooks of Rogers [Ro67] and Soare [So87]. We denote by $\mathcal{W}$ the class of all r.e. sets of natural numbers. $\subset$ denotes proper inclusion, \ denotes the settheoretical difference, and $|H|$ is the cardinality of the set H. $\langle .,. \rangle$ is a computable pairing function and $(.)_1$ is the decoding function for the first component. R_1 is the set of all total recursive unary functions, P_1 is the set of all partial recursive unary functions. Two partial functions f, g are called *incompatible* if there does not exist a common extension of f and g.

A set $M \subseteq \omega$ is called *increasing union of the sequence* $\{E_i\}_{i\in\omega}$ (of sets of natural numbers) if $M = \cup\{ E_i \mid i\in\omega \}$ and $E_i \subseteq E_{i+1}$ for all i. The following fact is obvious:

(F0) If M is increasing union of $\{E_i\}_{i\in\omega}$, L is increasing union of $\{F_i\}_{i\in\omega}$, and $\forall k\, \exists i,j \geq k\ [E_i = F_j]$, then M = L.

Let $S \subseteq \mathcal{W}$ be an r.e. class. A uniformly r.e. sequence $\{A_n\}_{n\in\omega}$ of sets from S which contains each $M \in S$ at least once is called a *numbering* of S. $\{A_n\}_{n\in\omega}$ is called a *one-one numbering* (henceforth abbreviated OON) of S if it is a numbering of S which contains each $M \in S$ exactly once. Note that this definition implies that only infinite classes can possess an OON.

For ease of conversation we introduce the following definitions:

Given two classes $L_1, L_2 \subseteq \mathcal{W}$. L_1 is called *strongly dense in* L_2 if every finite subset of a member of L_2 has infinitely many supersets in L_1. L_1 is called *weakly dense in* L_2 if every finite subset of a member of L_2 has a proper superset in L_1.

A class $S \subseteq \mathcal{W}$ is called *regular* if S is r.e. and there exists a canonically enumerable subclass $S_0 \subseteq S$ of finite sets such that each $M \in S$ is increasing union of a sequence from S_0. The following easy facts will be used: (Let S, S_0 be as in the above definition.)

(F1) S_0 coincides with the subclass of all finite sets from S. In particular, S_0 is uniquely determined by S.

(F2) If S is infinite then S_0 must be infinite, too.

(F3) ([Mal65], p. 157) S can be enumerated "on S_0" i.e. for every numbering $\{A_i\}_{i\in\omega}$ of S there exists a strong array $\{A_{i,j}\}_{i,j\in\omega}$ of sets from S_0 such that

$$\forall i,j\ [\, A_{i,j} \subseteq A_{i,j+1} \subseteq A_i \ \wedge\ A_i = \cup\{ A_{i,s} \mid s\in\omega\} \,].$$

The notions defined above also apply to classes of partial functions by identifying functions with their graphs. - In our algorithms we make use of the following comfortable notational convention :

Usually we have to enumerate some sets or functions in stages and use several parameters. We omit mentioning the stage when it is clear from the context. We write p[s] to explicitly refer to the value of the parameter (set, function) p at the end of stage s. The meaning of the assignment statement (x := y) is such as in conventional programming languages. If at one stage the value of a variable is not changed in an explicit way, it is assumed to be unchanged from the previous stage.

§2 On the criterion of Mal'tsev and Wolf

Our starting point is the following Theorem of Mal'tsev [Mal65], p.156 :

<u>Theorem M</u> Given an infinite r.e. class $S \subseteq \mathcal{W}$ containing a canonically enumerable subclass S_0 such that:

(1) S_0 is weakly dense in S.

(2) S is regular.

Then S has an OON.

Note that from $S_0 \subseteq S$ and (1) it follows that S_0 is strongly dense in S. Wolf [Wo64] independently announced a similar version of Theorem M (for a representative special case).

Mal'tsev's proof uses (F3) and is exactly analogous to Friedberg's.
In [Ku89a] we have introduced a powerful but easy method for organizing the construction of OONs : the "Extension Lemma", stating that an infinite r.e.class $S \subseteq \mathcal{W}$ has an OON if S can be partitioned into disjoint r.e. subclasses L_1, L_2 such that L_1 has an OON and L_1 is strongly dense in L_2. As a direct consequence we have shown that hypothesis (2) in Theorem M is not necessary, furthermore we got a priorityfree proof of the old and new results which is considerably simpler than Friedberg's original one [Ku89a],[Ku89b]. A number of further applications can be found in [Ku89c].
In Theorem 1 below we will show that hypothesis (1) in Theorem M is not necessary, too. Note that we may have to cope with $\subseteq$-maximal finite sets in S which constitutes the major difficulty in the proof of Theorem 1. In the following lines we want to point out why the previous criteria (including our own Extension Lemma) do not suffice for proving the result. Consider the following example: Let W_a be a simple set and define

$$S := \{ \{<i,0>, <i,s+1> \mid i \notin W_{a,s} \} \}_{i\in\omega} \cup \{ \{ <i,0>, <i,t+1> \mid t \leq s \wedge i \notin W_{a,t} \} \}_{i,s\in\omega} .$$

S is an infinite r.e. regular class. Thus, by Theorem 1 below, S has an OON. However, S contains finite sets, namely $\{<i,0>, <i,s+1> \mid i\notin W_{a,s} \}$ for $i\in W_a$, which do not have a proper superset in S. Therefore, we can not apply the criteria of Howard and Pour-El, Pour-El and Putnam, Mal'tsev (even in the stronger form without hypothesis (2)), Marchenkov. Lachlan's criterion requires the existence of a partial recursive function $\eta\in P_2$ such that for all i:

$$W_i \in S \rightarrow \{ W_{\eta(i,j)} \}_{j\in\omega} \text{ is a numbering of } S\backslash\{W_i\}.$$

Using the recursion theorem one can easily show that such an η does not exist.
Consider now a partitioning of S into disjoint r.e. classes L_1, L_2 such that L_1 has an OON and L_1 is strongly dense in L_2. Then, each $M\in L_2$ is contained in an infinite set from S, and, by W_a simple, we conclude that $\{i \mid \exists M\in L_2 [<i,0>\in M] \}$ is finite. In particular, S does not satisfy in a nontrivial way (i.e. such that we do not already presuppose that S has an OON) the hypothesis of our Extension Lemma. - So, we cannot use our easy method to prove Theorem 1. Interestingly, it turns out that the devices of Friedberg's original proof are helpful.

Theorem 1 Each infinite regular class of r.e. sets has an OON.

Proof

Let $S \subseteq \mathcal{W}$ be an infinite regular class, let S_0 be the infinite canonically enumerable subclass of all finite sets in S (cf. (F1),(F2)). Let $\{E_i\}_{i\in\omega}$ be a strong array which enumerates S_0 without repetition. According to (F3) choose a numbering $\{A_i\}_{i\in\omega}$ of S and a strong array $\{A_{i,s}\}_{i,s\in\omega}$ of sets from S_0 such that $\forall i,j\ [A_{i,j} \subseteq A_{i,j+1} \subseteq A_i \ \wedge\ A_i = \cup\{ A_{i,s} \mid s\in\omega\}]$.
The set $I := \{ i \mid \forall j<i\ \ A_i \neq A_j \}$ of the minimal A-indices is infinite and contains for each $M\in S$ exactly one index m such that $A_m = M$. Furthermore, $I\in\Sigma_2$, i.e. I is r.e. in K and we can choose an injective function $\bar{f}:\omega\to\omega$ such that $\bar{f} \leq_T K$ and $rg(\bar{f}) = I$. By the Limit Lemma (see e.g. [So87], p.57), there exists $f\in R_2$ such that $\lim_{s\to\infty} f(x,s) = \bar{f}(x)$ for all x.
We will give an algorithm for the construction of a numbering $\{M_x\}_{x\in\omega} \subseteq S \cup \{\emptyset\}$ which

contains each nonempty $M \in S$ exactly once. From this it follows that $S \setminus \{\emptyset\}$ has an OON and therefore S has an OON, too. Before giving the formal details we want to outline the main ideas of the construction.

Let i be fixed. We are trying to establish an index x such that $A_{\bar{f}(i)} = M_x$. During the construction there will be certain candidates z for such an x , and we try to enumerate $A_{f(i,s)}$ into M_z if z is candidate in stage s+1. At each step during the construction there will be at most one candidate. If it exists we call it (in accordance with Friedberg [F58] whose terminology we use) the follower of i. If i does not have a follower then a previously unused number (i.e. a number which has neither been a follower, nor has been chosen in step 3, substep (1)) will be apppointed to be a follower of i. Finitely often it may happen that $f(i,s+1) \neq f(i,s)$. If this happens we will release the follower z of i. z will not become a follower again for the rest of the construction and M_z will remain a finite set from S_0. A follower is called loyal if it is never released. Suppose that $f(i,s+1) = f(i,s)$ and x is follower of i in stage s+1. We want to enumerate $A_{f(i,s),t}$ into M_x for some $t \leq s$ such that $M_x[s] \subset A_{f(i,s),t}$ (call this a move of i). However, care must be taken that no duplications arise: i is not allowed to move if there exists a y such that $M_y = A_{f(i,s),t}$ and y is follower of some j<i, or if y has been displaced by x (see below) in any previous stage. Furthermore, it is required that E_s is a proper superset of $A_{f(i,s),t}$. Suppose that these conditions are satisfied and that for some y we have $M_y = A_{f(i,s),t} \subset E_s$ .If i is allowed to move then we will put $M_x := M_y$ and $M_y := E_s$, and in the future y will be regarded as "having been displaced by x". If y is a follower of some j > i then y is released. Of course, in each stage we have to decide among several candidates which i is allowed to move. This will be done by a "fair" procedure: with each follower x a "counter" p(x) is associated which is incremented if x moves. Now we let that i move for which the value of the counter of its follower is minimal. This rule guarantees that each i which infinitely often wants to move is infinitely often allowed to move.

Construction

Stage 0 For all x : $M_x := \emptyset$; $p(x) := \uparrow$.

Stage s+1

1.) $i := (s)_1$. If i does not have a follower then choose the least unused x, appoint x to be a follower of i and put $p(x) := 2s$.

2.) For all k = 0, ..., s :
If $f(k,s) \neq f(k,s+1)$ then release the follower of x, if there exists one.

3.) If there exists a pair (j,x) such that:
a.) x is a follower of j, and
b.) there exists $y > x$, $t \leq s$ such that:
- $M_x \subset M_y \subset E_s$,
- $M_y = A_{f(j,s),t}$,
- $| A_{f(j,s),t} | > x$,
- y is not a follower of some k < j ,
- y has not been displaced by x,

then choose the pair (j,x) such that p(x) is minimal; choose y such that b.) is satisfied and do the following:

(1) Choose the least unused z and put $M_z := M_x$;
(2) $M_x := M_y$; $p(x) := 2s+1$;
(3) $M_y := E_s$; release y if y is a follower
(from now on y is regarded as "having been displaced by x").

Otherwise choose the least unused z and put $M_z := E_s$. □

Verification

Lemma 0

(a) $\forall x,s\ [\ M_x[s] \in S_0 \cup \{\emptyset\} \wedge M_x[s] \subseteq M_x[s+1]\ ]$

(b) $\forall x,y,s\ [\ x \neq y \wedge M_x[s] \neq \emptyset \rightarrow M_x[s] \neq M_y[s]\ ]$

(c) $\forall x,s$ [x is follower of i at the end of stage s $\rightarrow M_x[s] \subseteq A_{f(i,s),s}$]

(d) $\forall x \quad M_x \in S \cup \{\emptyset\}$

Proof

(a)-(c) follow directly from the construction by induction on s.

(d) From (a) we get that each finite M_x is from $S_0 \cup \{\emptyset\}$. There are two cases left: (1) If x has never been a follower, or if x has been released as a follower, then there exists s_0 such that in each stage $s \geq s_0$ x is not a follower. For $s \geq s_0$ $M_x[s]$ changes only if x is displaced in stage s by some $z < x$. This can happen at most x times. It follows that M_x must be finite. (2) If x is a loyal follower of j and x has been appointed at stage s_0 then $f(j,s_0) = f(j,s) = \bar{f}(j)$ for all $s > s_0$. M_x is infinite only if (j,x) is chosen in infinitely many stages in step 3. It then follows that $M_x = A_{\bar{f}(j)} \in S$. □

Let $\bar{p}(y)[s+1]$ denote the value of p(y) at the end of step 2 in stage s+1. The following Lemma gives the important properties of p(x).

Lemma 1

(a) At each step of the algorithm all values p(y) which are defined are pairwisely different.

(b) $\forall n\, \exists s_0\, \forall s \geq s_0\, \forall j,y$ [If (j,y) is chosen in step 3 of stage s+1 then $\bar{p}(y)[s+1] > n$].

Proof

(a) follows directly form the construction.

(b) Consider the course of values of p(y) for some fixed y: If ever defined the value of p(y) will never decrease and it will increase whenever the pair (j,y) is chosen in step 3 of some stage s+1. - In each stage s+1 p(y) will be initialized for at most one y (and becomes equal to 2s); thus, for each n there exist only finitely many y such that p(y) is less than n. □

Lemma 2 $\forall x,y\ [\ x \neq y \wedge M_x \neq \emptyset \rightarrow M_x \neq M_y\ ]$

Proof Suppose that $x \neq y$ and $M_x \neq \emptyset$. If M_x is finite it follows from Lemma 0(b) that $M_x \neq$

M_y . If M_x is infinite then there exists an i such that x is loyal follower of i (cf. the proof of Lemma 0(d)). If in addition $M_x = M_y$ then there exists j such that y is loyal follower of j. In particular (i,x) and (j,y) are chosen infinitely often in step 3 of the algorithm. It follows that $A_{\bar{f}(i)} = M_x = M_y = A_{\bar{f}(j)}$.
From the fact that f enumerates minimal A-indices we get $\bar{f}(i) = \bar{f}(j)$, and, by f injective, i = j. This contradicts the fact that i has at most one follower in each stage. □

<u>Lemma 3</u> $\forall F \in S_0 \, \exists x \quad M_x = F$.

<u>Proof</u> Let $F \in S_0$ be given. There exists s such that $M = E_s$. It follows from the construction that for each stage $t \geq s+1$ there exists y_t such that $M_{y_t}[t] = E_s$. From condition b.) in step 3 we get that $y_t \neq y_{t+1}$ only if there exists a follower x in stage t+1 such that $x < |E_s|$ and $M_x[t] \subset M_x[t+1] \subset E_s$. E_s is finite and thus there can be only finitely many such stages. Therefore y_t is constant for almost all t. Let $y := \lim_{t \to \infty} y_t$. Then $M_y = E_s$. □

Recall that a follower x of i is called loyal if x is follower of i in almost every stage. If x is a loyal follower of i then x is uniquely determined and $M_x \subseteq A_{\bar{f}(i)}$.

<u>Lemma 4</u> Each i has exactly one loyal follower x_i . If $A_{\bar{f}(i)}$ is infinite then $A_{\bar{f}(i)} = M_{x_i}$.
In particular, each set from $S \setminus S_0$ is contained in $\{M_k\}_{k \in \omega}$.

<u>Proof</u> Suppose that the claim of the Lemma is true for all j < i. We will show that it is true for i. Choose s_0 such that for all $s \geq s_0$:

- $\bar{f}(k) = f(k,s)$ for all $k \leq i$,
- Each $j < i$ has a loyal follower x_j in stage s+1
- For all $t \geq 0, j < i$ we have:
  $A_{\bar{f}(i)}$ infinite $\rightarrow M_{x_j}[s] \neq A_{\bar{f}(i),t}$ and
  $A_{\bar{f}(j)}$ finite $\rightarrow M_{x_j}[s] = M_{x_j}$

(Use (F0) and the fact that all $A_{\bar{f}(k)}$ are pairwisely different.)

In each stage $s > s_0$ a follower of i cannot be released, neither in step 2 nor in step 3. Therefore, a loyal follower x_i of i must exist.
Suppose that $A_{\bar{f}(i)}$ is infinite. It remains to show that there are infinitely many stages s such that the pair (i,x_i) is chosen in step 3 of stage s. It then follows that $M_{x_i} = A_{\bar{f}(i)}$.
Assume for a contradiction that (i,x_i) is chosen only finitely often. Then there must exist $s_1 > s_0$ such that for all $s \geq s_1$ the following conditions are satisfied:

- x_i is follower of i in stage s ,
- (i,x_i) is not chosen in step 3 of stage s,

- $M_{x_i} = M_{x_i}[s_1] = M_{x_i}[s]$ and $p(x_i)[s_1] = p(x_i)[s]$,
- If some (j,z) is chosen in step 3 of stage s+1 then $p(x_i)[s_1] < \overline{p}(z)[s+1]$ (Use Lemma 1(b)).

Let t be minimal such that

- $M_{x_i} \subset A_{\overline{f}(i),t}$,
- $|A_{\overline{f}(i),t}| > x$,
- the unique y with $M_y = A_{\overline{f}(i),t}$ (whose existence is guaranteed by Lemma 3) satisfies: y is never displaced by x, y > x, and $y \notin \{x_j \mid j<i\}$.

(We use that there exist arbitrarily large sets $A_{\overline{f}(i),t} \supset M_{x_i}$, that $A_{\overline{f}(i),t} \in S_0$ for all t, and that only finitely many y can ever be displaced by x_i (for M_{x_i} is finite).)

Now choose y such that $M_y = A_{\overline{f}(i),t}$ (y is uniquely determined), and choose $s_2 > s_1$ such that $s_2 \geq t$ and $M_y = M_y[s_2]$. By choice of s_2 it follows that in stage s_2+1 the pair (i,x_i) satisfies the conditions a.), b.) in step 3. However (i,x_i) is not chosen because $s_2+1 > s_1$. Therefore, another pair (j,z) must have been chosen such that $\overline{p}(z)[s_2+1] < \overline{p}(x_i)[s_2+1] = p(x_i)[s_1]$. This contradicts the choice of s_1, which is our final contradiction. □

From Lemma 0(d), Lemma 3, Lemma 4 we obtain that $\{M_x\}_{x\in\omega}$ is a numbering of S or of $S\cup\{\emptyset\}$. Lemma 2 implies that every nonempty set from S is contained exactly once in $\{M_x\}_{x\in\omega}$. We conclude that S has an OON. □

Remarks

1.) The question may arise if the construction could be simplified by omitting the counter p(x) and instead just choosing in step 3 the pair (j,x) with j minimal. However, suppose that there exist i < j such that $A_{\overline{f}(i)} \subset A_{\overline{f}(j)}$ and $A_{\overline{f}(i)}$ is infinite. $A_{\overline{f}(i)}$ might be enumerated faster than $A_{\overline{f}(j)}$ such that each $E_s \subseteq A_{\overline{f}(j)}$ would be used for a move of i. Though the loyal follower y of j would infinitely often satisfy the condition in step 3, j would never move and $A_{\overline{f}(j)}$ would not be contained in $\{M_x\}_{x\in\omega}$.

2.) The conditon $|A_{f(j,s),t}| > x$ in step 3 b.) could be omitted. It is used only to simplify the verification.

Lachlan [La64], p.28 has introduced the notion of a "special standard class" (SSC) in order to describe r.e. classes of sets having an analog of a "Gödelnumbering". $S \subseteq \mathcal{W}$ is an SSC iff there exists $f\in R_1$ such that

$$S = \{W_{f(i)}\}_{i\in\omega} \ \wedge\ \forall i\,[\,W_i\in S \rightarrow W_i = W_{f(i)}\,] \ \wedge\ \forall i\ W_{f(i)} \subseteq W_i .$$

Lachlan [La64], Theorem 1.8 proved that an SSC is regular. Therefore we get:

Corollary 2 Any infinite SSC has an OON.

Another natural property for classes of r.e. sets is the closure under finite subsets. Call an r.e. class $S \subseteq \mathcal{W}$ normal if it is closed under finite subsets. Of course, the subclass S_0 of all finite sets of a normal class S is canonically enumerable, and each infinite set from S is increasing union of a sequence from S_0. Thus, a normal family is regular and we get:

Corollary 3 Any infinite normal family has an OON.

§ 3 One-one numberings of regular classes extended by total recursive functions

If $\{f_i\}_{i\in\omega}$ is a numbering of an r.e. class S of total recursive functions then the set I:= $\{i \mid \forall j<i\ f_i \neq f_j\}$ of the minimal f-indices is r.e. Thus, any infinite r.e. class of total recursive functions has an OON. This result is due to Liu [Li60] who obtained it for the representative special case of the primitive recursive functions. However, it is not clear whether each infinite r.e. class $L \cup F$ where $L \subseteq R_1$ and F is a canonically enumerable class of finite functions, does have an OON. Theorem 1 is not applicable because $L \cup F$ does not need to be regular. Liu´s result does not help because L does not need to be r.e. One might also consider the more general but related question whether each infinite r.e. class S, such that the subclass of all finite members of S is canonically enumerable, did have an OON. This question has a negative answer: Pour-El and Putnam [PP65], Theorem 6a have constructed an r.e. partitioning S of ω into infinitely many pairwisely disjoint r.e. nonrecursive sets. It follows that S cannot have an OON. - We will now answer the first question positively.

Theorem 4 Given a class $L \subseteq R_1$ and a canonically enumerable class F of finite functions such that $L \cup F$ is an infinite r.e. class. Then $L \cup F$ has an OON.

Proof

Let L, F satisfy the hypothesis of the Theorem and let $\{\nu_i\}_{i\in\omega}$ be a numbering of $L \cup F$. If F is finite then L must be r.e. and the conclusion of the Theorem follows from Liu´s result. For the following let F be infinite. Define $L_1 := \{f \in L \cup F \mid f$ is increasing union of a sequence from F $\}$. It is easy to show that L_1 is r.e. Thus, from Theorem 1 it follows that L_1 has an OON $\{\eta_i\}_{i\in\omega}$. By (F3) we choose a strong array $\{\eta_{i,s}\}_{i,s\in\omega}$ of functions from F such that:

$$\eta_i = \cup\{\eta_{i,s} \mid s\in\omega\} \wedge \eta_{i,s} \subseteq \eta_{i,s+1} \quad \text{for all i,s.}$$

Let $\{\nu_{i,s}\}_{i,s\in\omega}$ be a strong array of finite functions such that

$$\nu_i = \cup\{\nu_{i,s} \mid s\in\omega\} \wedge \nu_{i,s} \subseteq \nu_{i,s+1} \quad \text{for all i,s.}$$

Finally, let $\{\delta_i\}_{i\in\omega}$ be a strong array which enumerates F without repetition.

We will construct a numbering $\{\gamma_x\}_{x\in\omega} \subseteq L \cup F \cup \{\lambda x.\uparrow\}$ which contains each nonempty function from $L \cup F$ exactly once. From this it follows that $(L \cup F)\setminus\{\lambda x.\uparrow\}$ has an OON and therefore, $L \cup F$ has an OON, too. Before giving the formal details we want to outline the main ideas of the construction:

Let $L_2 := (L \cup F) \setminus L_1 = \{ f \in L \mid f$ is not increasing union of a sequence from $F \} \subseteq R_1$ (L_2 and even $L_2 \cup F$ need not be r.e.). We use two sorts of followers : follower$_1$ and follower$_2$.[1]
Followers$_1$ shall be used to enumerate L_1 , followers$_2$ to enumerate L_2 . Of course, we have to allow exceptions from this rule (for otherwise $L_2 \cup F$ would be r.e.) and it may be possible that a total function from L_1 will be enumerated by a follower$_2$. We require for all active followers$_2$ that their respective functions γ_x are pairwisely incompatible. Furthermore, if x is follower$_2$ of i in stage s+1 then $\nu_i \not\subseteq \delta_t$ for all $t \leq s$. If this condition cannot be verified for i (i.e. if i is "t-incorrect in stage s+1", see below) then i loses its follower$_2$. These properties are invariant over the whole construction, thus, in each stage s+1 at most one i loses its follower$_2$ (step 6). On the other hand, a problem arises when we want to appoint a follower$_2$ of i : First of all, one has to verify that ν_i is incompatible with all ν_j such that j already possesses a follower$_2$. This cannot be done effectively. Therefore, we put i into the set W where we record the "waiting" numbers (step 2), i gets a "time stamp" $q(i) := 2s$ (if we are in stage s+1). Now we wait until we can verify for all $j \in \omega$, such that $q(j) < q(i)$, that $\nu_i \neq \nu_j$. If this will ever occur then a follower$_2$ of i will be appointed. If it turns out that i is t-incorrect in stage s'+1 > s+1 then i will be deleted from W and q(i) gets undefined. In contrast to the previous construction we allow an x to be appointed as a follower$_2$ of i even when x is a released follower$_{1 \text{ or } 2}$. However, we take care that this happens for each x only finitely often (step 5 (4)). Thus x is a *loyal* follower$_2$ if x is follower$_2$ in almost every stage, though, x may have been released and re-appointed finitely often.
A follower$_1$ will move as in the previous construction (step 7), with the additional possibility to move if a follower$_2$ is appointed (step 5). On the other hand, we have to guarantee that a loyal follower$_1$ x moves infinitely often (and thus $\gamma_x \in R_1$) only if γ_x is different from all γ_y such that y is a loyal follower$_2$ (and thus $\gamma_y \in R_1$). Therefore, we have introduced condition (6) in step 5 and step 6, forcing each such x to eventually diagonalize every γ_y such that y is a loyal follower$_2$.
In this construction we do not need the "counter" p(x) from the previous algorithm because our infinite sets are graphs of total recursive functions and thus, the failure mentioned in Remark (1) above cannot occur.

Definition We call i <u>t-incorrect in stage s+1</u> iff there exists $r \geq s$ such that
$\nu_{i,r} \subseteq \delta_t$ and $\nu_{i,r} \in \{ \delta_u \mid u \leq r \}$.

Note that we can decide uniformly in i,s,t whether i is t-incorrect in stage s+1 : If $\nu_{i,r} \subseteq \delta_t$ for all $r \geq s$ then ν_i is a finite function from F, e.g. $\nu_i = \delta_v$, and for large enough $r \geq v$ we get $\nu_{i,r} = \nu_i \in \{ \delta_u \mid u \leq r \}$.

Construction
<u>Stage 0</u> $W := \emptyset$; for all x: $\gamma_x := \lambda x.\uparrow$; $q(x) := \uparrow$.

[1] Such a device was also used in [HP64].

Stage s+1

1.) If $i = (s)_1$ does not have a follower_1 then appoint the least unused x to be a follower_1 of i.

2.) If $i = (s)_1$ does not have a follower_2 and $i \notin W$ then put $W := W \cup \{i\}$; $q(i) := 2s$.

3.) For all i : If i has a follower_2 x then put $\gamma_x := v_{i,s}$.

4.) For all $i \in W$: If there exists $t \le s$ such that i is t-incorrect in stage s+1, then put $W := W \setminus \{i\}$; $q(i) := \uparrow$.

5.) [Appointment of a follower_2 / Move of a follower_1]

If there exists $i \in W$ such that

$\forall x < q(i)\ v_{i,s}(x)\downarrow$ and $\forall j\ [\ q(j)\downarrow < q(i) \rightarrow \exists u \le s\ \ v_{i,s}(u)\downarrow \ne v_{j,s}(u)\downarrow\ ]$

then choose that one whose q(i)-value is minimal, put $W := W \setminus \{i\}$ and perform the following actions:

If there exists a pair (j,y) and $t \le s$ such that:

(1) j has a follower_1 $x < y$,

(2) y is not follower_1 of some $k < j$,

(3) y is not follower_2 ,

(4) y has not been displaced by x,

(5) $[\ \gamma_x \subset \gamma_y = \eta_{j,t}\ ] \wedge |\text{dom}(\gamma_y)| > x$,

(6) $\forall k\ \ [\ q(k)\downarrow < t \wedge k$ has a $\text{follower}_2 \rightarrow \exists u < s\ \ \eta_{j,s}(u)\downarrow \ne v_{k,s}(u)\downarrow\ ]$,

(7) $\gamma_y \subseteq v_{i,s}$,

then choose the pair (j,y) with j minimal (x is the follower_1 of j) and perform the following actions in the given order:

- Choose the least unused z and put $\gamma_z := \gamma_x$;
- $\gamma_x := \gamma_y$;
- Release y if y is a follower_1 ;
- $\gamma_y := v_{i,s}$;
- Appoint y to be a follower_2 of i

  (from now on y is regarded as "having been displaced by x").

Otherwise choose the least unused z, put $\gamma_z := v_{i,s}$, and appoint z to be a follower_2 of i.

6.) [Release of a follower_2]

If there exists a pair (k,x) - which is uniquely determined - such that x is the follower_2 of k, and k is s-incorrect in stage s+1, then put $\gamma_x := \delta_s$; $q(k) := \uparrow$ and release x.

7.) [Move of a follower_1]

If the condition in step 6 was not satisfied, and there exists a pair (j,x) and a number y such that conditions (1)-(6) from step 5 hold for (j,y) and x, and $\gamma_y \subseteq \delta_s$, then choose a pair

(j,x) with minimal first coordinate and perform the following actions in the given order:
- Choose the least unused z and put $\gamma_z := \gamma_x$;
- $\gamma_x := \gamma_y$;
- $\gamma_y := \delta_s$;
- Release y if y is a follower_1
(from now on y is regarded as "having been displaced by x").

Otherwise choose the least unused z and put $\gamma_z := \delta_s$. □

Verification

In Lemma 0 we list some easy invariant properties of the algortihm.

Lemma 0

(a) $\forall x,s\ \ \gamma_x[s] \subseteq \gamma_x[s+1]$

(b) If x is not a follower_2 at the end of stage s then $\gamma_x[s] \in F \cup \{\lambda x.\uparrow\}$. Furthermore, the nonempty functions $\gamma_x[s]$ are pairwisely different for these x.

(c) Every i has exactly one loyal follower_1 .

(d) $\forall k,x,s$ [x is follower_2 of k at the end of stage s+1 $\rightarrow \gamma_x[s+1] = \nu_{k,s} \wedge \forall t{\le}s\ \ \nu_k \not\subseteq \delta_t$]

(e) $\forall x$ [[x is loyal follower_1 $\rightarrow \gamma_x \in L_1 \cup \{\lambda x.\uparrow\}$] $\wedge$ [x is loyal follower_2 $\rightarrow \gamma_x \in L$]]

(f) $\forall x$ [x is not a loyal $\text{follower}_{1 \text{ or } 2}$ $\rightarrow$ x is not $\text{follower}_{1 \text{ or } 2}$ in almost all stages $\wedge\ \gamma_x \in F \cup \{\lambda x.\uparrow\}$]

(g) If i has a loyal follower_2 , then q(i)[s] is defined for almost all s and is eventually constant. If i loses a follower_2 infinitely often then $\lambda s.q(i)[s]$ grows unbounded.

(h) For every n, there are only finitely many k such that $q(k)[t]\downarrow < n$ for some t. If $\lambda s.q(k)[s]$ is not eventually defined and constant, then there are only finitely many t such that $q(k)[t]\downarrow < n$.

(i) $\{\gamma_x\}_{x\in\omega} \subseteq L \cup F \cup \{\lambda x.\uparrow\}$

Proof

(a), (b) follow directly from the construction by induction on s.

(c) Each i has at most one follower_1 in each stage and therefore it has at most one loyal follower_1 . The proof that there exists at least one loyal follower_1 is the same as the proof of Lemma 4 in the previous verification (using the fact that $\{\eta_i\}_{i\in\omega}$ is an OON).

(d) Just note that x is appointed to be a follower_2 of i in stage s+1 only if $\nu_i \not\subseteq \delta_t$ for all t≤s. In this case $\gamma_x[s+1] = \nu_{i,s}$. In the following stages s'+1 > s+1 x will be released in step 6 if $\nu_k \subseteq \delta_{s'}$ (because then k is s'-incorrect in stage s'+1).

(e) If x is the loyal follower_1 of i then $\gamma_x[s] \subseteq \eta_i$ for all s. By (b) $\gamma_x[s]$ is from $F \cup \{\lambda x.\uparrow\} \subseteq L_1 \cup \{\lambda x.\uparrow\}$. This proves the claim if γ_x is finite. If γ_x is total then $\gamma_x = \eta_i \in L_1$.

If x is the loyal follower$_2$ of i then there exists s_0 such that x is follower$_2$ of i in every stage $s \geq s_0$. In step 3 we take action that $\gamma_x[s] = v_{i,s}$ for all $s>s_0$, thus, $\gamma_x = v_i$. From (d) it follows that $v_i \notin F$, thus, $\gamma_x = v_i \in L$.

(f) Each x can be appointed at most once to be a follower$_1$ and at most x+1 times to be a follower$_2$, namely, when x is unused or when x is displaced in step 5 by some $z < x$. Thus, if x is not a loyal follower$_{1 \text{ or } 2}$ then there exists s_0 such that x is neither follower$_1$ nor follower$_2$ in all stages $s \geq s_0$. Now, x can be displaced (by a follower$_1$ in step 7) at most x times in stages $s \geq s_0$. Thus, there must exist $s_1 > s_0$ such that $\gamma_x = \gamma_x[s] = \gamma_x[s_1]$ for all $s \geq s_1$. By (b) we get that $\gamma_x[s_1] \in F \cup \{\lambda x.\uparrow\}$.

(g), (h) follow from an examination of the course of values of q(i)[t] for fixed i : q(i)[t] is defined as long as $i \in W$ or a follower$_2$ of i exists. In such a phase q(i) does not change. q(i) can change only by temporarily becoming undefined. When reinitialized, it cannot attain an old value again. Also note that for each m there is at most one i such that q(i) can take on the value m.

(i) is a consequence of (e) and (f). □

<u>Lemma 1</u>

(a) If x, y are loyal followers$_2$ and $x \neq y$ then $\gamma_x \neq \gamma_y$.

(b) If x is a loyal follower$_2$ and y is a loyal follower$_1$ then $\gamma_x \neq \gamma_y$.

<u>Proof</u>

(a) Suppose that x, y ($x \neq y$) are loyal followers$_2$ of i and j, respectively. Each number can possess at most one loyal follower$_2$, thus, $i \neq j$. Before getting their final appointments, x and y must have been inserted (the last time) into W, say at stages s_0 and s_1. Then, $(s_0)_1 = j$ and $(s_1)_1 = i$, therefore $s_0 \neq s_1$, say $s_0 < s_1$. It follows that $q(j)[s] = q(j)[s_0] = 2s_0$ for all $s \geq s_0$ and $q(i)[s] = q(i)[s_1] = 2s_1$ for all $s \geq s_1$. According to the first condition in step 5 a follower$_2$ of i will be appointed in a stage $s \geq s_1$ only if there exists $u < s$ such that $v_i(u){\downarrow} \neq v_j(u){\downarrow}$ (by the way, this is the reason why all active followers$_2$ index pairwisely incompatible functions). In particular, x has got its final appointment in a stage $s \geq s_1$ and therefore $v_i \neq v_j$. From the loyality of x, y it follows that $\gamma_x = v_i$, $\gamma_y = v_j$, i.e. $\gamma_x \neq \gamma_y$.

(b) Suppose that x is loyal follower$_2$ of i and y is loyal follower$_1$ of j. Choose s_1 as above, i.e. $q(i)[s] = q(i)[s_1] = 2s_1$ for all $s \geq s_1$. From Lemma 0(e) it follows that $\gamma_x = v_i \in R_1$. Thus, $\gamma_x \neq \gamma_y$ if γ_y is a finite function. Now suppose that $\gamma_y \in R_1$. Then, (j,y) has been chosen infinitely often in step 5 or step 7, and $\gamma_y = \eta_j$. Let $s+1 > 2s_1$ be a stage, such that (j,y) is chosen in stage s+1, and, such that for all t which satisfy $\eta_{i,t} = \gamma_y[s+1]$ we have $t > 2s_1$. Then, in step 5 or step 7 of stage s+1 condition (6) must be satisfied for k=i. I.e. there must exist $u<s$ such that $\eta_j(u){\downarrow} \neq v_i(u){\downarrow}$, i.e. $\gamma_y = \eta_j \neq v_i = \gamma_x$. □

<u>Lemma 2</u> $\forall x,y\ [x \neq y \wedge \gamma_x \neq \lambda x.\uparrow \ \rightarrow \ \gamma_x \neq \gamma_y\ ]$

<u>Proof</u> Suppose for a contradiction that there exist $x \neq y$ such that $\gamma_x = \gamma_y \neq \lambda x.\uparrow$. The previous Lemma 1 implies that neither x nor y is a loyal follower_2. From the proof of Lemma 0 (f) it follows that there exists s_0 such that for all $s \geq s_0$ neither x nor y are followers_2 in stage s and $\gamma_x[s] \neq \gamma_y[s] \in F$ (Lemma 0 (b)). Thus, if γ_x is finite then $\gamma_x \neq \gamma_y$. Thus, by hypothesis, γ_x must be total. So we get that x, y are loyal followers_1, say of i and j, respectively, and $\gamma_x = \eta_i = \eta_j = \gamma_y$. Now, η is an OON and therefore i=j contradicting the fact that in each stage there exists at most one follower_1 of i. □

<u>Lemma 3</u> $\forall f \in F\ \exists x\ \gamma_x = f$.

<u>Proof</u> This is analogous to the proof of Lemma 3 in the verification of Theorem 1. Use condition (5) of step 5 and step 7 instead of condition b.) of step 3 of the previous algorithm. □

<u>Lemma 4</u> $\forall f \in L\ \exists x\ \gamma_x = f$.

<u>Proof</u> We distinguish two cases:
a.) Suppose that $f \in L_1 \cap L$. If there exists any loyal follower_2 y such that $\gamma_y = f$ then nothing remains to be proven. Now suppose that $\gamma_y \neq f$ for all loyal followers_2 y. Choose j such that $\eta_j = f$. By Lemma 0 (c) each k has a loyal follower_1 x_k, for ease of reference denote x_j by x. - It suffices to show that j is chosen infinitely often in step 5 or step 7. Suppose for a contradiction that there exists s_0 such that j is not chosen in step 5 or step 7 of stage s for all $s \geq s_0$, i.e. $\gamma_x = \gamma_x[s_0]$. Choose $t_0 \geq s_0$ such that:

- $\text{dom}(\eta_{j,t_0}) > x$, x is folloer_1 of j in stage t_0.
- $\forall t \geq t_0\ \forall k<j$ [x_k is follower_1 of k in stage t $\wedge$ [γ_{x_k} finite $\rightarrow$ $\gamma_{x_k} = \gamma_{x_k}[t_0]$] $\wedge$ [γ_{x_k} total $\rightarrow$ η_{j,t_0} and $\gamma_{x_k}[t]$ are incomparable]] (use that $\{\eta_k\}_{k \in \omega}$ is an OON)
- There exists $y > x$ such that
  - $\forall t \geq t_0$ $\gamma_x[t] \subset \eta_{j,t_0} = \gamma_y[t] = \gamma_y$ (use Lemma 3).
  - $\forall t \geq t_0$ y is not follower_1 of some $k<j$ in stage t (use Lemma 0 (c)),
  - y is never displaced by x (note that γ_x is finite and therefore only finitely many z are ever displaced by x),
  - $\forall t \geq t_0$ y is not follower_2 in stage t (use Lemma 0 (d)).

Condition (6) of steps 5, 7 is satisfied for $t=t_0$ in almost all stages s : Using Lemma 0 (g), (h), for s large enough, every k which has a follower_2 in stage s and satisfies $q(k)[s]\downarrow < t_0$, must have a loyal follower_2, say z, i.e. $\gamma_z = \nu_k \in R_1$. By hypothesis, $f \neq \gamma_z$, thus, a number u must exist such that $f(u) = \eta_j(u)\downarrow \neq \nu_k(u)\downarrow$. Hence, condition (6) can be satisfied for k. By Lemma 0 (h) there are only finitely many such k. Thus, condition (6) is eventually satisfied for $t=t_0$.- All in all, by choice of t_0, there exists $s_1 \geq t_0$ such that (j,y) and $t=t_0$ satisfy conditions (1)-(6) in step 5 and step 7 of stage s for all $s \geq s_1$. We distinguish two subcases each leading to a contradiction:

a1.) There exist only finitely many stages s at which a follower$_2$ z is appointed such that $\eta_{j,t_0} \subseteq \gamma_z[s]$: Then there are only finitely many z such that z is a follower$_2$ in stage t and $\eta_{j,t_0} \subseteq \gamma_z[t]$, for some t. This follows from the condition that $\nu_{i,s}(x)$ has to be defined for all $x < q(i)[s]$, if a follower$_2$ z of i is appointed in step 5 of stage s+1: for q(i) sufficiently large we find that η_{j,t_0} is already a subfunction of $\nu_{i,s}$ if it is a subfunction of ν_i. By the same argument, there can be only finitely many z such that there exists a stage s+1 at which $\eta_{j,t_0} \subseteq \delta_s$ and z is released as a follower$_2$ in step 6 of stage s+1 (i.e. $\gamma_z \subseteq \delta_s$ when z was appointed as follower$_2$.). Thus, using the fact that $\eta_{j,t_0} \subseteq \delta_s$, for infinitely many s, we find that there must exist a stage $s+1 > s_1$ such that the conditions in step 7 are satisfied for i, x, y as above. By choice of t_0 , j is the minimal candidate and is chosen in step 7 of stage s+1, contradicting the choice of s_0.

a2.) There exist infinitely many stages s at which followers$_2$ z are appointed such that $\eta_{j,t_0} \subseteq \gamma_z$ [s]: Consider the first such stage $s > s_1$. Then j is the minimal candidate such that conditions (1)-(7) of step 5 are satisfied, and thus, j is chosen, contradicting the choice of s_0.

b.) Suppose that $f \in L\setminus L_1$. We claim that there exists i such that $\nu_i = f$ and i possesses a loyal follower$_2$. Suppose for a contradiction that $f \neq \gamma_y$ for all loyal followers$_2$ y. First of all we consider an arbitrary i such that $\nu_i = f$. By $f \in L\setminus L_1$ it follows that there exists t_0 such that for all $t \geq t_0$ and all s: $\nu_{i,t} \neq \delta_s$. Thus , for all $s \geq t_0$ and all t : i is not t-incorrect in stage s+1. This implies that a follower$_2$ y of i cannot be released in any stage $s > t_0$ .Thus, by hypothesis, no follower$_2$ of i can be appointed in a stage $s > t_0$ (for otherwise this would be a loyal follower$_2$). We get: $i \in W[s]$ for almost all s and $q(i)[s]$ is eventually constant $=: \overline{q}(i)$.
Now consider that i such that $\nu_i = f$ and q(i) is minimal. There exists t_1 such that for all $s \geq t_1$: $i \in W[s]$, $\overline{q}(i) = q(i)[s]$, $\nu_{i,t_1}(x)\downarrow$ for all $x < \overline{q}(i)$, $\overline{q}(i)$ is less than all q-values of followers$_2$ which are appointed in stage s+1 (Lemma 0 (h)). In each such stage s+1 there must exist j satisfying :

(*) $j \in W[s+1] \wedge q(j)[s+1] < \overline{q}(i) \wedge \forall u \leq s\ [\ \nu_{i,s}(u)\downarrow \wedge \nu_{j,s}(u)\downarrow \rightarrow \nu_{i,s}(u) = \nu_{j,s}(u)\ ]$,
for otherwise a loyal follower$_2$ of i would be appointed in step 5.

There are only finitely many j such that $\exists s\ q(j)[s]\downarrow < \overline{q}(i)$. Thus, there exists a fixed j for which condition(*) is satisfied for infinitely many s. ν_j cannot be finite because of step 4. Thus, $\nu_j \in R_1$ and $\nu_j = \nu_i$, $\overline{q}(j) < \overline{q}(i)$, contradicting the choice of i. □

From Lemma 0 (i), Lemma 3, Lemma 4 we obtain that $\{\gamma_x\}_{x\in\omega}$ is a numbering of $L\cup F$ or $L\cup F\cup\{\lambda x.\uparrow\}$. Lemma 2 implies that each nonempty function occurs exactly once. Thus we conclude that $L\cup F$ has an OON. □

Theorem 4 can be easily generalized to obtain:

Theorem 5 Given an infinite regular class S of partial recursive functions and a class $L \subseteq R_1$, such that $S \cup L$ is r.e. Then $S \cup L$ has an OON.

Sketch of Proof
Let S, L be as in the hypothesis of the Theorem, and let F denote the canonically enumerable subclass of all finite functions from S. Let $S_1 := S \setminus R_1$, $S_2 := (S \cap R_1) \cup F$, $S_3 := S_2 \cup L$. It is easy to show that S_1, S_2, and S_3 are r.e. From Theorems 1, 4 it follows that each of these classes has an OON. Let $\{\bar{\eta}_i\}_{i \in \omega}$, $\{\bar{\rho}_i\}_{i \in \omega}$ be OONs of S_1 and S_2, respectively. Let $\bar{\nu}$ be a numbering of S_3. Apply the construction of Theorem 4 with the following modifications: Substitute η by $\bar{\eta} \oplus \bar{\rho}$ (where $(\bar{\eta} \oplus \bar{\rho})_{2k} = \bar{\eta}_k$, $(\bar{\eta} \oplus \bar{\rho})_{2k+1} = \bar{\rho}_k$) and substitute ν by $\bar{\nu}$. The counter p(x) is reintroduced for followers$_1$ x (thus, each follower$_1$ which infinitely often wants to move in step 5 or step 7 is infinitely often allowed to move). Furthermore, condition (6) in steps 5, 7 is applied only to $\bar{\rho}$-indices j but not to $\bar{\eta}$-indices j. The verification proceeds quite analogously as above. Lemma 1 remains true because $\gamma_x \in P_1 \setminus R_1$ for every loyal follower$_1$ x of an infinite function from S_1. For the proof of Lemma 0 (c) and of Lemma 2 note that an infinite function can have at most one $\bar{\eta} \oplus \bar{\rho}$-index. Part b) of the proof of Lemma 4 does not need to be modified. If $f \in S \setminus (L \cup F)$ then combine the argument of part a) with the argument of Lemma 4 in the verification of Theorem 1. □

Remark
One might suspect that for all $S \subseteq P_1$ which have an OON and for all $L \subseteq R_1$ such that $S \cup L$ is r.e. $S \cup L$ must possess an OON. This is false, however. In [Ku89a], Theorem 8 we have constructed an r.e. class F of finite functions, such that $R_1 \cup F$ is r.e. and does not have an OON. Of course, for every $f \in F$ there exists $g \in F$ such that $f \subset g$. Therefore, in the terminology of Howard and Pour-El [HP64], Definition 1, F has a partial recursive "height function", and has an OON by their Theorem 1.

§4 A further generalization of Theorem M

In this final section we want to present a generalization of Mal'tsev´s criterion, which will subsume both Theorem 1 above and Theorem 5 from [Ku89a].

Theorem 6 Let S be an infinite r.e. class and let $S_0 \subseteq S$ be any canonically enumerable subclass of finite sets from S such that S_0 is strongly dense in $S \setminus S_0$. Then S has an OON.

Note that $S \setminus S_0$ needs not be r.e. Clearly Theorem 1 follows from Theorem 6. As was stated above (following Theorem M) if $S_0 \subseteq S$ is weakly dense in S then S_0 is strongly dense in S and a fortiori S_0 is strongly dense in $S \setminus S_0$. Thus, from Theorem 6 it follows :

Corollary 7 [Ku89a] Hypothesis (2) in Theorem M is not necessary.

Note that an S satisfying the hypothesis of Theorem 6 may be "mixed" from classes as in

Theorem 1 and as in Corollary 7. Finally, we would like to mention that in the hypothesis of Theorem 6 "strongly dense" cannot be replaced by "weakly dense" , as is shown by the following counterexample :

$S = \{ \{2x\}, \{2x+1\} \}_{x \notin K} \cup \{ \{2x, 2x+1\} \}_{x \in K} \cup \{ \{0, 2x, 2x+1\} \}_{x \in \omega}$;

$S_0 = \{ \{0, 2x, 2x+1\} \}_{x \in \omega}$

This is just a trivial modification of the r.e. class of Pour-El and Putnam mentioned in section 0 above.

Proof of Theorem 6

Let S, S_0 satisfy the hypothesis of Theorem 6. It follows that S_0 and $L_1 := \{ A \in S \mid A$ is increasing union of a sequence from $S_0 \}$ are infinite classes. It is easy to show that L_1 is r.e. Thus, L_1 is a regular class and from Theorem 1 we get that L_1 has an OON, say $\{M_i\}_{i \in \omega}$. Let $\{E_s\}_{s \in \omega}$ be a strong array which enumerates S_0 without repetition. By (F3) we may choose a strong array $\{M_{i,s}\}_{i,s \in \omega}$ of finite sets from S_0 such that $M_i = \cup\{M_{i,s} \mid s \in \omega\}$ and $M_{i,s} \subseteq M_{i,s+1} \subseteq M_i$ for all i,s. Let $\{A_i\}_{i \in \omega}$ be an arbitrary numbering of S and let $\{A_{i,s}\}_{i,s \in \omega}$ be a strong array of finite sets such that $A_i = \cup\{A_{i,s} \mid s \in \omega\}$ and $A_{i,s} \subseteq A_{i,s+1} \subseteq A_i$. There exists an injective total function $\bar{f}: \omega \to \omega$, $\bar{f} \leq_T K$ such that $\bar{f}$ enumerates the set $\{i \mid \forall i<j \; A_i \neq A_j\}$ without repetition. By the Limit Lemma choose $f \in R_2$ such that $\bar{f}(x) = \lim_{s \to \infty} f(x,s)$, for all x.

We will now construct a numbering $\{V_k\}_{k \in \omega} \subseteq S \cup \{\emptyset\}$ which contains each nonempty $A \in S$ exactly once. Before giving the formal details of the construction we want to outline the main ideas. We are using two sorts of followers: follower$_1$ and follower$_2$. Followers$_1$ shall be used to enumerate L_1 , followers$_2$ to enumerate $L_2 := S \setminus L_1 = \{ A \in S \mid A$ is not increasing union of a sequence from $S_0 \}$. Note that every finite subset of a member from L_2 has infinitely many supersets in S_0. For the construction of L_1 basically the same strategy as in the proof of Theorem 1 is used, e.g. we will use again the counter p(x) for a follower$_1$ x. L_2 is enumerated in the following way: A follower$_2$ can be appointed in stage s+1 only if there exists $G \in S_0$ such that $A_{f(i,s),s} \subseteq G$. If $\bar{f}(i) = f(i,s)$ and $A_{\bar{f}(i)} \in L_2$ then, by hypothesis, there will be infinitely many such G's. We call G a "catcher of x" and require that any two catchers are pairwisely different and have not been used previously for the enumeration of L_1 .[2] In the parameter W (a finite set) we record the E-indices of the used members of S_0 ; in the parameter q(y) (a number) we record the E-index of the current catcher of y. Suppose, in stage t+1 > s+1 we find that $f(i,t) \neq f(i,t+1)$ or an u turns up such that $A_{f(i,t),x} \subseteq E_u \subseteq A_{f(i,t),t}$ then the follower$_2$ x of i will be released, and we put $V_x[s+1] := G$. To avoid duplication we have to require that the catcher of x is different from all $V_y[s+1]$ such that y is not a follower$_2$ (i.e. $V_y[s+1] \in S_0 \cup \{\lambda x.\uparrow\}$). Suppose that x is a loyal follower$_2$ of a finite set and $V_x \subset G$. Then G will not be contained in $\{V_x\}_{x \in \omega}$. In order to repair this defect we require that each loyal follower$_2$ has to change its catcher infinitely often. Furthermore, we have to avoid that for an infinite M_i almost all $E \in S_0$ such that $E \subseteq M_i$ are used as catchers. Therefore , we will respect the requirements of M_i (see below) when choosing new catchers.

---

[2] A similar device was used in [La65].

In intermediate steps of stage s+1 the abbreviation "(j,x) requires u" stands for the following condition B.):

B.) a.) x is $follower_1$ of j, and

b.) There exist y>x, t≤s such that:

- $V_x \subset V_y \subset E_u$,
- $V_y = M_{j,t} \wedge |M_{j,t}| > x$,
- y is not a $follower_2$,
- y is not a $follower_1$ of some k<j,
- y has not been displaced by x.

Construction

Stage 0 $a:= \lambda x.\uparrow$; $p:= \lambda x.\uparrow$; $W:= \emptyset$; $V_x := \emptyset$ for all x.

Stage s+1

1.) If $i = (s)_1$ does not have a $follower_1$ then appoint the least unused x to be a $follower_1$ of i, and put $p(x) := 2s$.

2.) If $i = (s)_1$ does not have a $follower_2$ then appoint the least unused x to be a $follower_2$ of i.

3.) [Release of $follower_2$]
For all i≤s :
If i has a $follower_2$ x and $f(i,s) \neq f(i,s+1)$ or $\exists t \leq s$ [$A_{f(i,s),x} \subseteq E_t \subseteq A_{f(i,s),s}$], then release x as $follower_2$ and, if in addition $a(x)\downarrow$ then put (in the given order):
$V_x := E_{a(x)}$; $W := W \cup \{a(x)\}$; $a(x) := \uparrow$.

4.) [Change of catcher]
If $i=(s)_1$ has a $follower_2$ x then search for the least number <r,u> such that I.) or II.) are satisfied:

I.) (1) $u \notin W \cup \{ k \mid \exists s' \leq s \ \ k \in rg(a[s']) \} \wedge A_{f(i,s),s} \subset E_u$,

(2) For all pairs (j,x): If "(j,x) requires u" then there exists v < u such that "(j,x) requires v" and $v \notin W \cup \{ k \mid \exists s' \leq s \ \ k \in rg(a[s']) \}$.

II.) $A_{f(i,s),s} \subseteq E_u \subseteq A_{f(i,s),r}$.

If case II.) occurs then release x as a $follower_2$ of i and if in addition $a(x)\downarrow$ then put (in the given order): $V_x := E_{a(x)}$; $W := W \cup \{a(x)\}$; $a(x) := \uparrow$.
If case II.) does not occur and case I.) occurs then put $V_x := A_{f(i,s),s}$; $a(x) := u$.

5.) [Move of $follower_1$]
Let $u := \mu r.\ r \notin (rg(a) \cup W)$.
If there exists a pair (j,x) such that "(j,x) requires u" then choose that pair for which p(x) is minimal, choose y such that condition B.)b.) is satisfied and execute the following actions in the given order:
- Choose the least unused z and put $V_z := V_x$;

- $V_x := V_y$; $p(x) := 2s+1$;
- $V_y := E_u$; release y as follower_1 if y is a follower_1
(from now on y is regarded as "having been displaced by x").

Otherwise choose the least unused z and put $V_z := E_u$.

In both cases put $W := W \cup \{u\}$. □

Verification

This is quite similar as the verification of Theorem 1.

Lemma 0

(a) $\forall x,s \ \ V_x[s] \subseteq V_x[s+1]$.

(b) The nonempty sets $V_x[s]$ such that x is not follower_2 in stage s are pairwisely different elements of S_0.

(c) V_x is finite if x is not a loyal $\text{follower}_{1 \text{ or } 2}$.

(d) The search in step 4 terminates.

Proof (a) and (b) follow by induction on s. Note that in stage s the active catchers are pairwisely different and are also different from all $V_x[s]$ where x is not a follower_2.
(c) follows from (b) and the fact that each x is displaced at most finitely often.
(d) Suppose that the search in step 4 of stage s+1 does not terminate by II.). Then there does not exist $G \in S_0$ such that $A_{f(i,s),s} \subseteq G \subseteq A_{f(i,s)}$ and it follows that $A_{f(i,s)} \notin L_1$. Thus, each finite subset of $A_{f(i,s)}$ has infinitely many supersets in S_0. In particular, this is true for $A_{f(i,s),s}$. In each stage there are at most finitely many j having a follower_1 , and $W[s] \cup \{ k \mid \exists s' \leq s \ k \in rg(a[s']) \}$ is finite. Therefore, among the infinitely many supersets E_u of $A_{f(i,s),s}$ there are at most finitely many such that u does not satisfy condition I.). All in all, the search must terminate by I.). □

Lemma 1 For all i : i has a loyal follower_2 x iff $A_{\bar{f}(i)} \in L_2$; if this is the case then $V_x = A_{\bar{f}(i)}$.

Proof Suppose that in all stages $s \geq s_0$, for some s_0 , x is follower_2 of i. As x is never released in step 3 it follows that: $f(i,s) = \bar{f}(i)$, for all $s \geq s_0$ and $\neg\exists G \in S_0 \ [A_{\bar{f}(i),x} \subseteq G \subseteq A_{\bar{f}(i)}]$. It follows that $A_{\bar{f}(i)} \notin L_1$ i.e. $A_{\bar{f}(i)} \in L_2$. On the other hand, suppose that $A_{\bar{f}(i)} \notin L_1$ and choose s_0 large enough such that $\bar{f}(i) = f(i,s)$ for all $s \geq s_0$ and $\neg\exists u \ [A_{\bar{f}(i),s_0} \subseteq E_u \subseteq A_{\bar{f}(i)}]$. Then no follower_2 $x \geq s_0$ of i can be released in any stage $s > s_0$. In particular, i has a loyal follower_2 x. By Lemma 0 (d) it follows that in all stages $s+1 > s_0$ such that $(s)_1 = i$, in step 4 case I.) must occur. Therefore, $V_x[s+1] = A_{\bar{f}(i),s}$, and so $V_x = A_{\bar{f}(i)}$. □

Lemma 2 For each $G \in S_0$ there exists x such that $V_x = G$.

Proof It suffices to show that for every u there exists s_0 such that $\forall s \geq s_0 \ u \notin rg(a[s])$. The rest follows as in the proof of Lemma 3 in the verification of Theorem 1. - Condition I.) (1) in step 4 ensures that for each u_0 there exists at most one x such that $u_0 = a(x)[s]$ for some $s \geq 0$. Suppose that this is the case, i.e. there exists a follower_2 x of some j such that $u_0 = a(x)[s]$ for some $s \geq 0$. Choose $t := \mu s' > s. (s')_1 = j$. Either x will be released in some stage $\leq t+1$, or in step 4 of stage t+1 case I.) occurs and a(x) is redefined. In both cases u_0 will disappear from rg(a) for ever. □

Let $\overline{p}(y)[s]$ denote the value of p(y) at the end of step 4 in stage s+1.

Lemma 3 Each i has a loyal follower$_1$ x, and, if M_i is infinite then $M_i = V_x$.

Proof A follower$_1$ y of i can be released only in step 5 of some stage s+1, namely, if there is a follower$_1$ x of some $j<i$ and $t \leq s$ such that $V_y[s] = M_{j,t} \wedge |M_{j,t}| > x \wedge V_x[s] \subset V_y[s]$.
Suppose the Lemma is true for all $j<i$ and denote by x_j the loyal follower$_1$ of j. Choose s_0 large enough such that for $s \geq s_0$:

- $\forall j<i$ x_j is follower$_1$ of j in stage s,
- $\forall j<i$ $[[M_j \text{ infinite} \rightarrow \forall t\ V_{x_j}[s] \neq M_{j,t}]$
  $\wedge [M_j \text{ finite} \rightarrow V_{x_j}[s] = V_{x_j}[s_0] = V_{x_j}]]$.

We are using the induction hypothesis, fact (F0) and the fact that $\{M_k\}_{k \in \omega}$ is an OON. Note that no follower$_1$ of i can be released in a stage $s > s_0$. Thus, a loyal follower$_1$ of i must exist.
Now suppose that M_i is infinite. If (i,x) is chosen infinitely often in step 5 of the algorithm then $V_x = M_i$. Suppose for a contradiction that (i,x) is chosen only finitely often. Then there exists $s_1 > s_0$ such that for all $s \geq s_1$:

- x is follower$_1$ of i in stage s_1,
- (i,x) is not chosen in step 5 of stage s,
- $V_x[s] = V_x[s_1] = V_x \wedge p(x)[s] = p(x)[s_1]$,
- If a pair (j,y) is chosen in step 5 of stage s+1 then $p(x)[s_1] < \overline{p}(y)[s+1]$.

Let t be minimal such that

- $M_{i,t} \supset V_x \wedge |M_{i,t}| > x$
- The unique y (whose existence is guaranteed by Lemma 2) with $V_y = M_{i,t}$ satisfies:

  y is never displaced by x , y > x, and $y \notin \{x_j \mid j<i\}$.

Choose this y and select $s_2 > s_1$ such that

- $s_2 \geq t \wedge V_y = V_y[s_2]$

By the choice of t,y,s_2, for all $s > s_2$ and all u such that "(i,x) requires u" in stage s, we get:

$M_{i,t} = V_y[s] \subset E_u$.

Let u be minimal such that:

- $u \notin W[s_2] \cup \{ k \mid \exists s' \leq s_2\ k \in rg(a[s']) \}$ and $M_{i,t} \subset E_u$.

(the existence of u follows from the fact that M_i is increasing union of a sequence from S_0).
"(i,x) requires u" is true in every stage $s > s_2$. From condition I.) (2) in step 4 and, by choice of u and s_2, it follows that $u \notin \cup\{rg(a[s]) \mid s \in \omega\}$. Thus, there must exist a stage $s > s_2$ in which u will be chosen at the beginning of step 5. By choice of $s > s_1$ (i,x) is the pair requiring u whose $\overline{p}(x)$-value is minimal. Thus, (i,x) will be chosen and $V_x[s_0] \subset V_x[s]$ contradicting the choice of s_1. This is our final contradiction. □

Lemma 4 (a) $\forall x,y\ [x \neq y \wedge V_x \neq \emptyset \rightarrow V_x \neq V_y]$.

(b) $S \subseteq \{V_x\}_{x\in\omega} \subseteq S \cup \{\emptyset\}$.

Proof (a) From Lemma 0 (b), (c) it follows that every two nonempty finite sets V_x, V_y ($x \neq y$) such that x is not a loyal follower_2, are pairwisely different. By Lemma 1, the sets V_x such that x is a loyal follower_2, are pairwisely different members of L_2. Finally, we have shown in Lemma 3 that the infinite sets V_x, such that x is a loyal follower_1, are pairwisely different members from $L_1 \setminus S_0$.

(b) $S \subseteq \{V_x\}_{x\in\omega}$ follows from Lemmas 1-3.

$\{V_x\}_{x\in\omega} \subseteq S \cup \{\emptyset\}$ follows from Lemma 0 (b), (c) and Lemmas 1,3. □

Directly from Lemma 4 we conclude that S has an OON. □

References

Friedberg, R.M.

[Fr58] *Three theorems on recursive enumeration. I. Decomposition II. Maximal set III. Enumeration without duplication.* J. Symbolic Logic 23, p. 309-316, 1958

Goncharov, S.S.

[Go80] *The problem of the number of non-self-equivalent constructivizations.* Sov. Math. Dokl. 21, 411-414, 1980

Howard, W.A. and Pour-El, M.B.

[HP64] *A structural criterion for recursive enumeration without repetition.* Z. Math. Logik Grundlag. Math. 10, 105-114, 1964

Kummer, M.

[Ku89a] *Numberings of* $R_1 \cup F$. in: CSL' 88, (Eds.: E. Börger, H. Kleine Büning, M. M. Richter), Lecture Notes in Computer Science 385, p. 166-186, Springer-Verlag, Berlin, 1989.

[Ku89b] *An easy priority-free proof of a theorem of Friedberg.* to appear in: Theoretical Computer Science

[Ku89c] *Beiträge zur Theorie der Numerierungen: Eindeutige Numerierungen.* Dissertation, Universität Karlsruhe, Fakultät für Informatik, 1989

Lachlan, A.H.

[La64] *Standard classes of recursively enumerable sets.* Z. Math. Logik Grundlag. Math. 10, 23-42, 1964

[La65] *On recursive enumeration without repetition.* Z. Math. Logik Grundlag. Math. 11, 209-220, 1965

[La67] *On recursive enumeration without repetition: A correction.* Z. Math. Logik Grundlag. Math. 13, 99-100, 1967

Liu, S.

[Li60] *An enumeration of the primitive recursive functions without repetition.* Tohoku Math. J. 12, 400-402, 1960

Mal'tsev, A.I.

[Mal65] *Algorithms and recursive functions.* English Translation: Wolters-Noordhoff Publishing, Groningen, 1970; Russian Edition: Izdatelstvo Nauka, Moskva, 1965; German Translation: Vieweg-Verlag, Braunschweig, 1974

Marchenkov, S.S.

[Mar71] *On minimal numerations of systems of recursively enumerable sets.* Sov. Math. Dokl. 12, 843-845, 1971

Pour-El, M.B.

[P60] Review of [Fr58], J. Symbolic Logic 35, 223-229, 1960

Pour-El, M.B. and Putnam, H.

[PP65] *Recursively enumerable classes and their application to sequences of formal theories.* Arch. math. Logik und Grundlagenforsch. 8, 104-121, 1965

Rogers, H. jr.

[Ro67] *Theory of recursive functions and effective computability.* McGraw-Hill Book Company, New York, 1967

Soare, R.I.

[So87] *Recursively enumerable sets and degrees.* Springer-Verlag, Berlin, 1987

Wolf, C.E.

[Wo64] *Standard recursively enumerable classes I.* Notices Amer. Math. Soc. 11, 386, 1964

PRIORITY ARGUMENTS USING ITERATED TREES OF STRATEGIES

Steffen Lempp and Manuel Lerman

Dept. of Mathematics, University of Wisconsin, Madison, WI 53706, USA
Dept. of Mathematics, University of Connecticut, Storrs, CT 06269, USA

Abstract. A general framework for priority arguments in classical recursion theory, using iterated trees of strategies, is introduced and used to present new proofs of four fundamental theorems of recursion theory.

1. Introduction. Only a few years after Cohen invented forcing in set theory in 1961, a quite general framework had been developed by Shoenfield, Solovay, Feferman, and others. This framework has since allowed set theorists to prove a wide variety of theorems by defining the appropriate partial order and appealing to the general lemmas on forcing. The lemmas could also be applied to extensions of forcing such as iterated forcing or class forcing.

Recursion theorists, unfortunately, have had a much harder time with the priority argument invented in 1956/57 by Muchnik and Friedberg. Even nowadays and even for finite-injury priority arguments, recursion theorists either reprove the combinatorics of finite injury or simply assume that the reader is familiar enough with the combinatorics to fill in the details. For the most complicated well-understood kind of priority argument, the $\mathbf{0}'''$-priority argument, the whole framework has to be reproved every time.

Of course, there have been numerous attempts at finding a framework: Lachlan [**7, 8**] tried a game-theoretical approach, and also a topological approach, using an effective version of the Baire Category Theorem. (The true stages method for some infinite-injury priority arguments has its origin there.) Lerman [**11**] devised the pinball machine model for infinite-injury priority arguments. Harrington conceived the tree of strategies method as a way to understand Lachlan's Nonsplitting Theorem, an approach that was then worked out in detail and popularized by Soare [**18, 19**]; this method is the most widely used today.

In the 1980's, Harrington introduced the "worker at level n" approach, later widely used in recursive model theory by Knight and others. Ash [**1, 2**] gave a more detailed version of this, working out a general framework in terms of iterated trees, which he and Knight then extended and used to prove results in recursive model theory. Groszek and Slaman [**4**] attempted a tree of trees approach in their work on reverse mathematics in recursion theory. Finally, Kučera [**5, 6**] introduced the construction of recursively

The authors would like to thank C. Ash, M. Groszek, J. Knight, A. Kučera, and T. Slaman for stimulating discussions and/or providing related preprints. The first author was partially supported by NSF grants DMS-8701891, DMS-8901529, a Binational NSF grant U.S.-West Germany, and post-doctoral fellowships of the Deutsche Forschungsgemeinschaft and the Mathematical Sciences Research Institute. The second author was partially supported by NSF grants DMS-8521843 and DMS-8900349 and by the Mathematical Sciences Research Institute.

enumerable degrees below degrees of diagonally nonrecursive functions as a way of eliminating one level of injury and of separating negative and positive strategies in the construction.

The framework that we would like to introduce here grew out of our work on the decidability of the existential theory of the recursively enumerable (r.e.) degrees with nth jump reducibility predicates [**9**], which, by the Shore Noninversion Theorem [**17**], requires a general $\mathbf{0}^{(n)}$-priority argument for arbitrarily large n. Of course, this framework was inspired by many of the approaches above, especially Ash's, and Groszek's and Slaman's. However, we see our approach as the most promising at this time to reach the goal set above, namely to prove combinatorial lemmas about the framework and to eliminate these from the individual priority arguments, which we hope to achieve in [**10**]. In this paper, however, we will prove the combinatorial lemmas in each instance separately to better give the intuition for the framework.

Our notation generally follows Chapter XIV of Soare [**19**]. We denote the *use* (the largest number *actually* used) of a partial recursive functional by the corresponding lower-case Greek letter, so e.g. φ is the use function of Φ. With the "opponent's" (i.e. the given) functionals, we will always assume that the use of a computation is bounded by the stage at which the computation first appears.

2. The Framework. We will define, for fixed $n > 0$, a sequence of trees of strategies $T_0, T_1, \ldots, T_n$ such that, roughly speaking, the construction taking place on T_i is recursive in $\mathbf{0}^{(i)}$, and for $i > 0$ is actually a finite-injury construction relative to $\mathbf{0}^{(i-1)}$, and such that each strategy $\alpha \in T_{i+1}$ is split up into ω many substrategies $\alpha_k \in T_i$, working on the kth instance of α's requirement. By the way the strategies will be arranged on the tree, the construction on T_i, which is recursive in $\mathbf{0}^{(i)}$, will work to satisfy not only (sub)requirements on T_i, but also higher-level (sub)requirements from $T_{i+1}, \ldots, T_n$. The assignment of (sub)requirements to nodes on the trees is thus at the heart of the framework. The splitting-up of requirements will automatically ensure the negative requirements and will follow the same pattern for all four theorems presented here. (For some other theorems, such as the Minimal Pair Theorem, the splitting-up would have to be modified slightly, however, to ensure that "at most one side of the minimal pair" is injured.)

We define the trees of strategies by induction as follows:

$$T_0 = \{\infty, 0\}^{<\omega},$$
$$T_{i+1} = \Big(\{\infty\} \cup \bigcup_{j \le i} T_j\Big)^{<\omega}. \tag{1}$$

Here ∞ and 0 are distinct symbols, and we tacitly assume that the nodes of each tree (including the empty node) are appropriately tagged so that the trees form pairwise disjoint sets, and we can thus tell which tree a particular node comes from. (Intuitively, for $\alpha \in T_i$, $\alpha\widehat{\ }\langle\infty\rangle$ denotes the Π_i-outcome of α, and $\alpha\widehat{\ }\langle\beta\rangle$ denotes a Σ_i-outcome witnessed by the Π_{i-1}-outcome of β on T_{i-1}.)

Suppose we are given two partial functions up: $T_i \to T_{i+1}$ and lev: $T_i \to \{j \mid j \le n\}$. (In the following, these will denote that $\alpha \in T_i$ is a substrategy of $\mathrm{up}(\alpha) \in T_{i+1}$ and works for a requirement of complexity level $\mathrm{lev}(\alpha)$. We do not split up a requirement in going from T_{i+1} to T_i if $i \geq \text{lev}(\alpha)$, but only if $i < \text{lev}(\alpha)$.) We define the *approximation function* $\lambda: T_i \to T_{i+1}$ by induction on m as follows (until $\lambda(\alpha)(m)\uparrow$):

$$\lambda(\alpha)(m) = \begin{cases} \alpha(k), & \text{if } \text{lev}(\lambda(\alpha) \upharpoonright m) \leq i \text{ and} \\ & \quad k = \mu k' < |\alpha| \, (\text{up}(\alpha \upharpoonright k') = \lambda(\alpha) \upharpoonright m), \\ \beta, & \text{if } \text{lev}(\lambda(\alpha) \upharpoonright m) > i \text{ and} \\ & \quad \beta = \mu\beta' \subset \alpha \, (\text{up}(\beta') = \lambda(\alpha) \upharpoonright m \;\&\; \beta'^\frown\langle\infty\rangle \subseteq \alpha), \\ \infty, & \text{if } \text{lev}(\lambda(\alpha) \upharpoonright m) > i \text{ and} \\ & \quad \forall\beta \subset \alpha \, (\text{up}(\beta) = \lambda(\alpha) \upharpoonright m \to \beta^\frown\langle\infty\rangle \not\subseteq \alpha) \text{ and} \\ & \quad \exists\beta \subset \alpha \, (\text{up}(\beta) = \lambda(\alpha) \upharpoonright m), \\ \uparrow, & \text{if } \forall\beta \subset \alpha \, (\text{up}(\beta) \neq \lambda(\alpha) \upharpoonright m). \end{cases} \tag{2}$$

(Intuitively, α believes that if α is an initial segment of the true path on T_i then $\lambda(\alpha)$ is an initial segment of the true path on T_{i+1}. We notice that, by the remark at the end of the preceding paragraph, we need not specify the i for T_i on up and λ; and we denote by up^j and λ^j the j-fold iteration of up and λ, respectively.)

The definition of the approximation function naturally extends to infinite paths $\Lambda \in [T_i]$; here $\lambda(\Lambda)$ can be either a path through T_{i+1} or a node on T_{i+1}.

Next, we define the concept of consistency. We say that *an instance of $\beta \in T_{i+1}$ is consistent at $\alpha \in T_i$* iff

$$\forall j < n - i\big(\text{up}^j(\beta) \subseteq \lambda^{j+1}(\alpha)\big), \tag{3.1}$$

$$\text{lev}(\beta) \leq i \to \forall\alpha' \subset \alpha\big(\text{up}(\alpha') \neq \beta\big), \text{ and} \tag{3.2}$$

$$\text{lev}(\beta) > i \to \forall\alpha' \subset \alpha\big(\alpha'^\frown\langle\infty\rangle \subseteq \alpha \to \text{up}(\alpha') \neq \beta\big). \tag{3.3}$$

(Intuitively, we state in (3.1) that $\text{up}^j(\beta)$ is consistent with α's guess at the true path through T_{i+j+1}; in (3.2) that β's requirement need not be split up from T_{i+1} to T_i and has not already been taken care of before α; and in (3.3) that along α, no substrategy working for β has had a Π_i-outcome and that therefore at α we must still search for a Σ_{i+1}-witness for β's requirement. There are theorems (not presented here) in the proofs of which this notion of consistency has to be defined in a less restrictive way.)

Next, we fix an effective 1–1 poset homomorphism par from $\bigcup_{i\in\omega} T_i$ *onto* ω with the following property:

$$\forall\alpha, \beta \in T_{i+1} \forall\gamma \in T_i\big(\beta^\frown\langle\gamma\rangle \subseteq \alpha \to \text{par}(\gamma) < \text{par}(\alpha)\big). \tag{4}$$

We call $\text{par}(\alpha)$ the *parameter* of α. (The intuition is that $\text{par}(\alpha)$ is a number reserved for α and "big" relative to α's predecessors.)

We now assign requirements to nodes. Fix n to be the highest (quantifier) complexity level of any requirement for the theorem at hand. Given an effective ω-ordering of all requirements $\{\mathcal{R}_k\}_{k\in\omega}$ for the theorem, we assign $\mathcal{R}_k$ to all nodes $\alpha \in T_n$ with $|\alpha| = k$ (or a proper subset thereof if certain outcomes contradict the hypotheses of the theorem, as in the Sacks Splitting and the Sacks Density Theorems, but always such that the nodes that are assigned requirements form a recursive subtree of T_n). We also define the complexity level $\text{lev}(\alpha)$ to be the (quantifier) complexity level of requirement $\mathcal{R}_k$. (This assignment of requirements on T_n will have to be modified slightly for the Sacks Splitting Theorem.)

We then proceed by reverse induction on $i < n$ and by induction on $|\alpha|$ for $\alpha \in T_i$. Given an effective ω-ordering $\preceq$ of $\{\, \beta \in T_{i+1} \mid \text{some requirement is assigned to } \beta \,\} \times \omega$ with the property

$$(\beta = \beta' \;\&\; k \leq k') \text{ or } (\beta \subset \beta' \;\&\; k \leq k' + \mathrm{par}(\beta')) \rightarrow \langle \beta, k \rangle \preceq \langle \beta', k' \rangle, \tag{5}$$

and given an assignment to all $\alpha' \subset \alpha$, we let $\langle \beta, k \rangle$ be $\preceq$-minimal such that

$$\exists^{\leq k} \alpha' \subset \alpha (\mathrm{up}(\alpha') = \beta), \text{ and} \tag{6.1}$$

$$\text{an instance of } \beta \text{ is consistent at } \alpha. \tag{6.2}$$

We then assign β's requirement to α (if $\mathrm{lev}(\beta) \leq i$) or the kth instance of β's requirement to α (if $\mathrm{lev}(\beta) > i$), respectively, and we set $\mathrm{up}(\alpha) = \beta$ and $\mathrm{lev}(\alpha) = \mathrm{lev}(\beta)$. If $\langle \beta, k \rangle$ does not exist then we assign no requirements to any $\alpha' \supseteq \alpha$.

We will now prove a general lemma about the splitting-up of requirements.

Lemma 1 (Splitting-Up Lemma). *Assume $\Lambda_0 \in [T_0]$ and $\Lambda_j = \lambda^j(\Lambda_0)$ for $0 < j \leq n$. Then:*

(i) *If $0 < j \leq n$ and $|\Lambda_j| < \infty$ then no requirement is assigned to the node Λ_j.*

(ii) *If $0 < j \leq n$ then for all t (with $t \leq |\Lambda_j|$ if Λ_j is a node) there is some t' such that $\lambda^j(\Lambda_0 \restriction t') = \Lambda_j \restriction t$ and for all $j' < j$, $\lambda^{j'}(\Lambda_0 \restriction t') \subseteq \Lambda_{j'}$.*

(iii) *If $\mathcal{R}$ is a requirement of complexity level i assigned to a (necessarily unique) node on Λ_n, say $\alpha_n^{\emptyset}$, then for all $j < n$:*

  (a) *If $j \geq i$ then $\mathcal{R}$ is assigned to a (necessarily unique) node on Λ_j, say $\alpha_j^{\emptyset}$.*

  (b) *If $j < i$ and $\alpha_{j+1}^{\sigma} {}^\frown \langle \beta \rangle \subseteq \Lambda_{j+1}$ for some $\beta \in T_j$ and some $\sigma \in \omega^{i-j-1}$ then the kth instance of α_{j+1}^{σ}'s requirement (for some k) is assigned to β with $\beta^\frown \langle \infty \rangle \subseteq \Lambda_j$, and no instance of α_{j+1}^{σ}'s requirement is assigned to any β' with $\beta \subset \beta' \subseteq \Lambda_j$. We denote by $\alpha_j^{\sigma^\frown \langle k' \rangle}$ the (necessarily unique) $\beta' \subset \Lambda_j$ to which the k'th instance of α_{j+1}^{σ}'s requirement is assigned (for $k' \leq k$).*

  (c) *If $j < i$ and $\alpha_{j+1}^{\sigma} {}^\frown \langle \infty \rangle \subseteq \Lambda_{j+1}$ for some $\sigma \in \omega^{i-j-1}$ then for all k, the kth instance of α_{j+1}^{σ}'s requirement is assigned to a (necessarily unique) node on Λ_j, say $\alpha_j^{\sigma^\frown \langle k \rangle}$.*

Proof. If no requirement is assigned to some $\alpha \subset \Lambda_0$ then we redefine Λ_0 to be the least such α, otherwise we leave α as is. This will not affect the definition of Λ_j for $j > 0$.

(i) We proceed by induction on j. Suppose (i) fails for some (least) $j \leq n$. Then Λ_j is finite, and some requirement is assigned to it. Since $\Lambda_j = \lambda(\Lambda_{j-1})$, we have, by the definition of λ, that $\Lambda_j = \lambda(\Lambda_{j-1} \restriction t)$ for all $t \geq t_0$ (for some t_0). Furthermore, by (3.1), $\mathrm{up}^l(\Lambda_j) \subseteq \Lambda_{j+l}$ for all $l \leq n-j$, and thus $\langle \Lambda_j, 0 \rangle$ must eventually be the least pair $\langle \beta, k \rangle$ used in splitting up T_j's requirements along Λ_{j-1}. Therefore (the 0th instance of) Λ_j's requirement must be assigned to some $\alpha \subset \Lambda_{j-1}$ whence $\Lambda_j \subset \lambda(\Lambda_{j-1})$, a contradiction.

(ii) We proceed by induction on j. Set $\alpha = \Lambda_j \restriction (t-1)$. If $\mathrm{lev}(\alpha) < j$ then there is a unique $\beta \subset \Lambda_{j-1}$ with $\mathrm{up}(\beta) = \alpha$, and we have $\Lambda_{j-1}(|\beta|) = \Lambda_j(|\alpha|)$. If $\mathrm{lev}(\alpha) \geq j$ and $\alpha^\frown \langle \beta \rangle = \Lambda_j \restriction t$ for some β then $\beta^\frown \langle \infty \rangle \subseteq \Lambda_{j-1}$. If $\mathrm{lev}(\alpha) \geq j$ and $\alpha^\frown \langle \infty \rangle = \Lambda_j \restriction t$ then there is a (least) $\beta \subset \Lambda_{j-1}$ with $\mathrm{up}(\beta) = \alpha$, and we have $\beta^\frown \langle \gamma \rangle \subseteq \Lambda_{j-1}$ for some γ. In all these cases, set $t'' = |\beta| + 1$. Since $\mathrm{up}(\beta) = \alpha$, we have

$\alpha \subseteq \lambda(\beta)$ and thus $\Lambda_j \restriction t \subseteq \lambda(\Lambda_{j-1} \restriction t'')$. By (3.2) and (3.3), it is also easy to see that $\Lambda_j \restriction t = \lambda(\Lambda_{j-1} \restriction t'')$. Now apply (ii) with t'' and $j - 1$ in place of t and j to get t' which will then make (ii) true for t and j.

(iii) We proceed by reverse induction on j. By the hypothesis, (a) holds for $\alpha_n^\emptyset$. Suppose one of (a), (b), or (c) holds for $j + 1$, and fix $\alpha_{j+1}^\sigma \subset \Lambda_{j+1}$ for some $\sigma \in \omega^{<i}$ in our notation. If $j < i$ and $\alpha_{j+1}^\sigma {}^\frown \langle \beta \rangle \subseteq \Lambda_{j+1}$ for some $\beta \in T_j$ then (b) holds since $\Lambda_{j+1} = \lambda(\Lambda_j)$. So suppose $j \geq i$ or $\alpha_{j+1}^\sigma {}^\frown \langle \infty \rangle \subseteq \Lambda_{j+1}$, and (a) or (c) fails, respectively. We distinguish two cases.

If Λ_j is a node (i.e. of finite length) then $\lambda^k(\Lambda_j) = \Lambda_{j+k}$ for all $k \leq n - j$. Since $\mathrm{up}^k(\alpha_{j+1}^\sigma) \subset \Lambda_{j+1+k}$ for all $k < n - j$ by hypothesis, (some instance of) α_{j+1}^σ's requirement would be assigned to Λ_j, contradicting (i) (or the first paragraph of this proof).

If Λ_j is an infinite path then by an application of (ii), there are infinitely many t such that for all $k < n - j$, $\lambda_{k+1}(\Lambda_j \restriction t) \supseteq \mathrm{up}^k(\alpha_{j+1}^\sigma)$. Thus an instance of α_{j+1}^σ is consistent at $\Lambda \restriction t$ for infinitely many t, contradicting the failure of (a) or (c).

This concludes the proof of Lemma 1. ∎

The general procedure to use this framework will then be as follows: For a theorem on r.e. objects, we define an ω-sequence of requirements of the form

$$(\rho \to \sigma) \,\&\, (\neg \rho \to \tau) \tag{7}$$

where ρ is a Σ_i-formula (or Π_i-formula), σ a Σ_{i+1}-formula (or Π_{i+1}-formula), and τ a Π_i-formula (or Σ_i-formula) if the *complexity level i* is odd (or even, respectively). We allow ρ, σ, and τ to contain free variables that will be substituted by parameters of nodes (or numbers computed from them), and we allow the innermost quantifier of ρ to be restricted to a set of "stages" $S = \{ \mathrm{par}(\xi) \mid \xi \subset \Lambda_0 \,\&\, \mathrm{up}^n(\xi) = \alpha \,\&\, \forall j \leq n(\mathrm{up}^j(\xi) \subset \Lambda_j) \}$ when a T_0-strategy works for this requirement with correct guesses at the true path on all trees. In splitting (sub)requirements into subrequirements, we bound the outermost unbounded quantifier in ρ, σ, and τ by k (for the kth instance of that requirement) or by $\mathrm{par}(\beta)$ (for β working on the kth instance of that requirement), proceeding from T_n down to T_i without splitting up, and then from T_i all the way down to T_0 while splitting up. The requirements assigned to T_0-nodes then determine the *true path* Λ_0 on T_0 (where we follow $\alpha^\frown \langle \infty \rangle$ if the $i$fold instance of ρ from the requirement for $\alpha \in T_0$ is true, and we follow $\alpha^\frown \langle 0 \rangle$ otherwise). We now have to define the action of an $\alpha \in T_0$ depending upon its "outcome" (∞ or 0). We thus obtain the (recursive) true path Λ_0 on T_0 by starting with $\emptyset \in T_0$ and, whenever α has been determined to be on Λ_0, letting α act and determine $\alpha^\frown \langle a \rangle$ to be on Λ_0 for α's outcome a. The construction consists of the actions of all $\alpha \subset \Lambda_0$. We finally define $\Lambda_i = \lambda^i(\Lambda_0)$ to be the *true path* on T_i (which is recursive in $0^{(i)}$) and verify that all $\alpha \in \Lambda_n$ satisfy their requirements.

3. The Sample Theorems. We now present the details of the construction in our framework for four well-known theorems of recursion theory. (We assume familiarity with the traditional proofs, as e.g. in Soare [19].)

Friedberg-Muchnik Theorem (Muchnik [**13**], Friedberg [**3**]). *There are two r.e. sets A and B of incomparable Turing degree.*

Proof. We need to satisfy $A \neq \Phi^B$ and $B \neq \Phi^A$ for all partial recursive (p.r.) functionals Φ. By symmetry, we assume throughout that we want to satisfy the former whenever we discuss an individual requirement. To ensure such an inequality, we need to ensure it at a number x. A requirement thus takes the form

$$\begin{aligned}\exists s \in S\,(\Phi_s^{B_s}(x)\downarrow = 0) \rightarrow \exists s \forall t \geq s (\Phi_s^{B_s}(x)\downarrow = 0 \,\&\, x \in A_{s+1} \,\&\\ B_s \upharpoonright (\varphi_s(x)+1) = B_t \upharpoonright (\varphi_s(x)+1)) \quad (8)\\ \neg\exists s \in S(\Phi_s^{B_s}(x)\downarrow = 0) \rightarrow \forall s\,(x \notin A_s).\end{aligned}$$

The complexity level is thus 1. We fix an effective ω-ordering $\{\mathcal{R}_e\}_{e\in\omega}$ of all requirements, assign requirement $\mathcal{R}_e$ to all nodes $\alpha \in T_1$ with $|\alpha| = e$, and substitute the x in α's requirement by $\mathrm{par}(\alpha)$ for all $\alpha \in T_1$.

The split-up requirements for $\beta \in T_0$ are obtained by restricting $s \leq \mathrm{par}(\beta)$ (where $B_{\mathrm{par}(\beta)}$ is that part of B enumerated before β acts). The action of $\beta \in T_0$ is to measure if

$$\exists s \leq \mathrm{par}(\beta)\,(s \in S \,\&\, \Phi_s^{B_s}(\mathrm{par}(\mathrm{up}(\beta)))\downarrow = 0)\,, \quad (9)$$

which, since all previous instances of α's requirement before β have had outcome 0, i.e. have been answered negatively, just means

$$\Phi_{\mathrm{par}(\beta)}^{B_{\mathrm{par}(\beta)}}(\mathrm{par}(\mathrm{up}(\beta)))\downarrow = 0. \quad (9')$$

If the answer is no then β has outcome 0, otherwise it has outcome ∞ and enumerates $\mathrm{par}(\mathrm{up}(\beta))$ into A.

We now verify that the construction satisfies the requirements.

We first observe a combinatorial fact about finite injury, namely that $\beta \subset \Lambda_0$ and $\mathrm{up}(\beta) \subset \Lambda_1$ implies that $\mathrm{up}(\beta)$ and $\mathrm{up}(\beta')$ are comparable for all β' with $\beta \subset \beta' \subset \Lambda_0$.

Suppose $A = \Phi^B$. Then some unique $\alpha \subset \Lambda_1$ was supposed to ensure $A(\mathrm{par}(\alpha)) \neq \Phi^B(\mathrm{par}(\alpha))$. First assume $\alpha\hat{\ }\langle\infty\rangle \subset \Lambda_1$. Then, by Lemma 1, there are infinitely many $\beta \subset \Lambda_0$ with $\mathrm{up}(\beta) = \alpha$, and for all such β we have $\beta\hat{\ }\langle 0\rangle \subset \Lambda_0$. Thus $\Phi_{\mathrm{par}(\beta)}^{B_{\mathrm{par}(\beta)}}(\mathrm{par}(\alpha))\downarrow = 0$ fails for all these β. This implies $\neg\Phi^B(\mathrm{par}(\alpha))\downarrow = 0$ and $\mathrm{par}(\alpha) \notin A$, a contradiction.

On the other hand, assume $\alpha\hat{\ }\langle\beta\rangle \subset \Lambda_1$ for some $\beta \in T_0$. Thus $\beta\hat{\ }\langle\infty\rangle \subset \Lambda_0$, $\Phi_{\mathrm{par}(\beta)}^{B_{\mathrm{par}(\beta)}}(\mathrm{par}(\alpha))\downarrow = 0$, and $\mathrm{par}(\alpha) \in A$. It suffices to show that no β' with $\beta \subset \beta' \subset \Lambda_0$ will enumerate a number $\leq \varphi_{\mathrm{par}(\beta)}(\mathrm{par}(\alpha))$ into B in order for us to establish that $B_{\mathrm{par}(\beta)} \upharpoonright (\varphi_{\mathrm{par}(\beta)}(\mathrm{par}(\alpha)) + 1) = B \upharpoonright (\varphi_{\mathrm{par}(\beta)}(\mathrm{par}(\alpha)) + 1)$. We analyze the different possibilities for the position of $\alpha' = \mathrm{up}(\beta')$ relative to α. By (3.1) and (3.3), $\alpha'\hat{\ }\langle\beta''\rangle \subseteq \alpha$ (for some β'') is impossible. By (3.1), $\alpha\hat{\ }\langle\infty\rangle \subseteq \alpha'$ is impossible. By $\alpha \subset \Lambda_1$, $\alpha'\hat{\ }\langle\infty\rangle \subseteq \alpha$ implies that β' does not enumerate any number. By (3.1), $\alpha\hat{\ }\langle\beta''\rangle \subseteq \alpha'$ (for some β'') implies $\beta = \beta''$, and thus $\mathrm{par}(\alpha') > \mathrm{par}(\beta) > \varphi_{\mathrm{par}(\beta)}(\mathrm{par}(\alpha))$ by (4). Suppose α and α' are incomparable, say they split at $\bar{\alpha}$. But then $\beta \subset \beta' \subset \Lambda_0$ is impossible by (3.1) and our observation. ∎

Sacks Splitting Theorem (Sacks [**14**]). *Any nonrecursive r.e. set A can be split into two r.e. subsets A_0 and A_1 of incomparable Turing degree.*

Proof. Given an r.e. set A, we need to satisfy (for all p.r. functionals Φ and $i = 0$ or 1) the requirements

$$A = A_0 \sqcup A_1, \text{ and} \tag{10.1}$$
$$A = \Phi^{A_i} \to A \leq_T \emptyset. \tag{10.2}$$

(Here $\sqcup$ denotes disjoint union.) We ensure $A_0 \cap A_1 = \emptyset$ simply by not enumerating the same number into both A_0 and A_1, and we ensure $A_0, A_1 \subseteq A$ simply by only enumerating elements of A into A_0 or A_1; so ensuring (10.1) is reduced to showing

$$A \subseteq A_0 \cup A_1. \tag{10.1*}$$

We try to show that A is recursive by building a partial recursive function Δ_α such that $A = \Delta_\alpha$ (for some $\alpha \in T_2$ working on (10.2)). Our requirements thus take the following form (for all x and all i and Φ):

$$\exists s \in S\,(x \in A_s) \to \exists s\big(x \in A_{0,s+1} \cup A_{1,s+1}\big), \text{ and} \tag{10.1'}$$
$$\forall x \exists s \in S\, \forall y \leq x\big(A_s(y) = \Phi_s^{A_{i,s}}(y)\downarrow\big) \to \forall x \exists s \forall t > s\big(A_t(x) = \Delta_{\alpha,t}(x)\downarrow\big). \tag{10.2'}$$

In the notation of (7), formulas τ are trivial, say $0 = 0$, so the requirements take the simpler form $\rho \to \sigma$. The complexity level of requirements (10.1′) and (10.2′) is 1 and 2, respectively. We index the requirements (10.1′) and (10.2′) by $\{\mathcal{P}_x\}_{x\in\omega}$ (for the x in (10.1′)) and $\{\mathcal{N}_e\}_{e\in\omega}$, respectively. We will assign requirements to nodes of T_2 in a different fashion than usual, namely as follows: We assign $\mathcal{N}_0$ to $\emptyset \in T_2$. Given assignments to all $\alpha' \subset \alpha$ for some $\alpha \in T_2$, we denote by α_0 the longest $\alpha' \subset \alpha$ to which a requirement (10.2′), say $\mathcal{N}_{e_0}$, has been assigned. If $\alpha_0{}^\frown\langle\infty\rangle \subseteq \alpha$ then no requirement is assigned to α (since α_0 has shown that A is recursive contrary to the hypothesis of the theorem). Otherwise let $\alpha_0{}^\frown\langle\beta\rangle \subseteq \alpha$ for some β; then $\mathcal{N}_{e_0+1}$ is assigned to α if (for all $x \leq \mathrm{par}(\beta)$) $\mathcal{P}_x$ has already been assigned to some $\alpha' \subset \alpha$; otherwise let x_0 be minimal such that $\mathcal{P}_{x_0}$ has not been assigned to some $\alpha' \subset \alpha$, and assign $\mathcal{P}_{x_0}$ to α. It is now easy to check that along any path $\Lambda \in [T_2]$, all requirements have been assigned unless $A \leq_T \emptyset$ is shown by $\alpha^\frown\langle\infty\rangle \subset \Lambda$ for some α working on a requirement (10.2′).

In going from T_2 to T_1, we do not split up requirements (10.1′), and we split up requirements (10.2′) by bounding $x \leq k$ (for $\beta \in T_1$ working on the kth instance of (10.2′)). The split-up requirements for T_0 are simply obtained by bounding $s \leq \mathrm{par}(\gamma)$ for $\gamma \in T_0$. (Here we fix an enumeration $\{A_s\}_{s\in\omega}$ of A, and $A_{i,\mathrm{par}(\gamma)+1}$ is the part of A_i enumerated by all $\gamma' \subseteq \gamma$, and likewise for Δ_α.)

The action of a $\gamma \in T_0$ working on a requirement (10.1′) is to measure if

$$\exists s \in S\,(s \leq \mathrm{par}(\gamma)\ \&\ x \in A_s), \tag{11}$$

which just amounts to measuring if

$$x \in A_{\mathrm{par}(\gamma)}. \tag{11'}$$

If the answer to (11′) is no then γ has outcome 0; otherwise it has outcome ∞, and, if now $x \notin A_0 \cup A_1$, then γ enumerates x into A_{1-i_0} where $\alpha_0 \subset \mathrm{up}^2(\gamma)$ is the longest strategy working on a requirement (10.2′), and α_0 uses $i = i_0$.

The action of a $\gamma \in T_0$ working on the kth instance of a requirement (10.2′) is to measure if

$$\forall x \leq k \exists s \leq \mathrm{par}(\gamma) \forall y \leq x \left(s \in S \,\&\, A_s(y) = \Phi_s^{A_{i,s}}(y)\right)\downarrow, \tag{12}$$

which, just as for (9), simply amounts to measuring if

$$A_{\mathrm{par}(\gamma)} \restriction (k+1) = \Phi_{\mathrm{par}(\gamma)}^{A_{i,\mathrm{par}(\gamma)}} \restriction (k+1)\downarrow. \tag{12′}$$

If the answer is no then γ has outcome 0; otherwise it has outcome ∞ and defines $\Delta_\alpha(k) = A_{\mathrm{par}(\gamma)}(k)$ (if $\Delta_\alpha(k)$ is undefined so far where $\alpha = \mathrm{up}^2(\gamma)$).

We now verify that the construction satisfies all the requirements (up to the first one, if any, that contradicts the hypothesis of the theorem by showing $A \leq_{\mathrm{T}} \emptyset$).

First, it is easy to see that any $\alpha \subset \Lambda_2$ working on a requirement $\mathcal{P}_x$ will ensure $\mathcal{P}_x$; so if Λ_2 is an infinite path then (10.1) is satisfied.

Next, we observe that $\gamma \subset \Lambda_0$ and $\mathrm{up}^2(\gamma) \subset \Lambda_2$ implies $\mathrm{up}(\gamma) \subset \Lambda_1$. Suppose not, say $\mathrm{up}(\gamma)$ and Λ_1 split at $\bar\beta$. By $\mathrm{up}(\gamma) \subseteq \lambda(\gamma)$, we have $\bar\beta^\frown\langle\infty\rangle \subseteq \mathrm{up}(\gamma)$ and $\bar\beta^\frown\langle\bar\gamma\rangle \subset \Lambda_1$ for some $\bar\gamma$ with $\gamma \subset \bar\gamma \subset \Lambda_0$. By the way requirements are assigned to nodes of T_2, we have $\mathrm{up}(\bar\beta) \subseteq \mathrm{up}^2(\gamma)$, and by (3.2) or (3.3), even $\mathrm{up}(\bar\beta) \subset \mathrm{up}^2(\gamma)$. But $\mathrm{up}^2(\gamma) \subset \Lambda_2$. So if $\mathrm{up}(\bar\beta)$ works on some requirement (10.1′) then $\mathrm{up}^2(\gamma)$ has a correct guess on $\bar\beta$'s outcome by $\bar\beta \subset \Lambda_1$. And if $\mathrm{up}(\bar\beta)$ works on some requirement (10.2′) then, by (3.1), we have $\mathrm{up}(\bar\beta)^\frown\langle\bar\beta\rangle \subseteq \mathrm{up}^2(\gamma)$, contradicting $\bar\beta^\frown\langle\bar\gamma\rangle \subset \Lambda_1$. This establishes the observation, and furthermore that $\mathrm{up}(\gamma) \subseteq \lambda(\gamma')$ and $\mathrm{up}^2(\gamma) \subseteq \lambda^2(\gamma')$ for all γ' with $\gamma \subseteq \gamma' \subset \Lambda_0$.

Now suppose that $\alpha^\frown\langle\infty\rangle \subset \Lambda_2$ for some strategy working on a requirement (10.2′). Then $\alpha^\frown\langle\infty\rangle = \Lambda_2$ by the assignment of requirements to nodes of T_2, and $\beta^\frown\langle\gamma\rangle \subset \Lambda_1$ (for some γ) for all $\beta \subset \Lambda_1$ with $\mathrm{up}(\beta) = \alpha$. Thus, by Lemma 1(iii)(c), Δ_α is total. So suppose $\Delta_\alpha(x)\downarrow \neq A(x)$ for some (least) x. Then $\Delta_\alpha(x)$ was defined by some $\gamma \subset \Lambda_0$, and later x entered A. Since $\varphi_{\mathrm{par}(\gamma)}(x) < \mathrm{par}(\gamma)$, it suffices to show that no $\tilde\gamma$ with $\gamma \subset \tilde\gamma \subset \Lambda_0$ will put any number $\leq \mathrm{par}(\gamma)$ into A_i.

So suppose there is such a $\tilde\gamma$. We set $\tilde\alpha = \mathrm{up}^2(\tilde\gamma)$. By our observation above and $\alpha \subset \Lambda_2$, α and $\tilde\alpha$ must be comparable. If $\tilde\alpha \subset \alpha$ then $\mathrm{up}(\tilde\gamma) \subset \mathrm{up}(\gamma)$ and, by $\gamma \subset \tilde\gamma$, $\mathrm{up}(\tilde\gamma)^\frown\langle\infty\rangle \subseteq \mathrm{up}(\gamma) \subset \Lambda_1$, and so $\tilde\gamma$ will not enumerate any number.

We conclude that $\alpha \subset \tilde\alpha$. Let $\alpha_0 \subset \tilde\alpha$ be the strategy determining if $\tilde\gamma$ enumerates its $\tilde x$ into A_0 or A_1 in the construction. Obviously, $\alpha \subseteq \alpha_0$; and since y enters A_i we even have $\alpha \subset \alpha_0$, say $\alpha^\frown\langle\beta_0\rangle \subseteq \alpha_0 \subset \tilde\alpha$. Then $\tilde x > \mathrm{par}(\beta_0)$ by the way requirements are assigned to nodes of T_2. But by our observation, $\mathrm{up}(\gamma) \subset \Lambda_1$ and so $\mathrm{up}(\gamma)^\frown\langle\gamma\rangle \subseteq \beta_0$; thus $\tilde x > \mathrm{par}(\beta_0) > \mathrm{par}(\gamma)$.

We have thus shown that $A >_{\mathrm{T}} \emptyset$ forces $\alpha^\frown\langle\beta\rangle \subset \Lambda_2$ (for some $\beta \in T_1$) for all $\alpha \subset \Lambda_2$ working on a requirement (10.2′). Thus Λ_2 is infinite, and all requirements are satisfied. ∎

Sacks Jump Inversion Theorem (Sacks [15]). *For any set $J \geq_T \emptyset'$ r.e. in $\emptyset'$, there is an r.e. set A with $A' \equiv_T J$.*

Proof. We need to build an r.e. set A and p.r. functionals Γ and Δ satisfying, for all x, the requirements

$$J(x) = \lim_y \Gamma^A(x,y) \tag{13}$$

(establishing $J \leq_T A'$ by the Limit Lemma), and

$$A'(x) = \Delta^{J \oplus \Lambda_1}(x) \tag{14}$$

(where $\Lambda_1 \leq_T \emptyset'$ is the true path on T_1, thus establishing $A' \leq_T J \oplus \emptyset'$). Since J is a Σ_2-set there is a recursive relation R such that $x \in J$ iff $\neg\forall y \exists s R(x,y,s)$. We ensure (13) by requiring

$$\forall y \exists s R(x,y,s) \rightarrow \forall y > y_0 \exists s \in S\big(\Gamma^{A_s+1}_{s+1}(x,y){\downarrow} = 0 \;\&\; \gamma_{s+1}(x,y) = -1\big), \text{ and} \tag{15.1}$$

$$\neg\forall y \exists s R(x,y,s) \rightarrow \exists y \forall z \geq y \forall s \geq s_z \big(s \in S \rightarrow \Gamma^{A_s+1}_{s+1}(x,z){\downarrow} = 1 \;\& \;\gamma_{s+1}(x,z) \leq \gamma_{s_z+1}(x,z)\big). \tag{15.2}$$

(Here y_0 and the s_z will be numbers determined by $\alpha \in T_2$ and the $\beta \in T_1$ working on (15). In the above, use -1 means that the oracle string is $\emptyset$.) The complexity level of this requirement is 2.

Recall that $A' = \{\, x \mid \Phi^A_x(x){\downarrow} \,\}$. We ensure (14) by requiring

$$\exists s \in S\,\big(\Phi^{A_s}_{x,s}(x){\downarrow}\big) \rightarrow \exists s \in S\, \forall t > s\big(\Delta^{\sigma \oplus (\beta^\frown\langle\gamma\rangle)}_{s+1}(x){\downarrow} = 1 \;\&\; \Phi^{A_s}_{x,s}(x){\downarrow} \;\& \; A_s \upharpoonright (\varphi_{x,s}(x)+1) = A_t \upharpoonright (\varphi_{x,s}(x)+1)\big), \text{ and} \tag{16.1}$$

$$\neg\exists s \in S\,\big(\Phi^{A_s}_{x,s}(x){\downarrow}\big) \rightarrow \forall s \in S\big(\Delta^{\sigma \oplus (\beta^\frown\langle\infty\rangle)}_{s+1}(x){\downarrow} = 0\big) \tag{16.2}$$

where $\sigma \subset J$ and $\beta^\frown\langle\gamma\rangle \subset \Lambda_1$ (or $\beta^\frown\langle\infty\rangle \subset \Lambda_1$, respectively) are determined by $\alpha \in T_2$ and the $\beta \in T_1$ working on (16), respectively. The complexity level of this requirement is 1.

We fix an effective ω-ordering $\{\, \mathcal{R}_e \,\}_{e\in\omega}$ of all the above requirements (in ascending order of x) and assign requirement $\mathcal{R}_e$ to all nodes $\alpha \in T_2$ with $|\alpha| = e$. For α working on (16), we set

$$\sigma(i) = \begin{cases} 0, & \text{if (for some } j)\ \alpha(j){\downarrow} = \infty \text{ and } \alpha \upharpoonright j \text{ works on (15) with } x = i, \\ 1, & \text{if (for some } j)\ \alpha(j){\downarrow} \neq \infty \text{ and } \alpha \upharpoonright j \text{ works on (15) with } x = i, \\ \uparrow, & \text{otherwise.} \end{cases}$$

In going from T_2 to T_1, we split up the left-hand sides of α's requirement (15) into the kth instance for $\beta \in T_1$ by bounding $y \leq \text{par}(\beta)$ in (15). (The splitting-up of the right-hand sides is a bit more complicated, there we bound y by a number computable from β.) We do not split up requirement (16) from T_2 to T_1, but we identify $\beta \in T_1$ with the β in (16) (and γ will be the unique γ' with $\beta^\frown\langle\gamma'\rangle \subset \Lambda_1$ if it exists). In going from T_1 to T_0, we just split up β's (sub)requirement (for $\beta \in T_1$) into the lth instance for $\gamma \in T_0$ by bounding $s, z \leq \text{par}(\gamma)$.

The action of a $\gamma \in T_0$ working on (the lth instance of the kth instance of) (15) is to measure if

$$\forall y \leq \text{par}(\text{up}(\gamma))\, \exists s \leq \text{par}(\gamma) R(x, y, s). \tag{17}$$

If the answer is no then γ has outcome 0 and sets $\Gamma^A(x, y) = 1$ for all $y \leq \text{par}(\gamma)$ (unless now defined to a different value). If the answer is yes then γ has outcome ∞; defines β_0 to be the longest $\beta' \subset \text{up}(\gamma)$ with $\text{up}(\beta') = \text{up}^2(\gamma)$ and γ_0 to be the longest $\gamma' \subset \gamma$ with $\text{up}(\gamma') = \beta_0$ (if β_0 exists); enumerates $\text{par}(\gamma^-)$ into A (if β_0 exists) where $\gamma^- \subseteq \gamma$ is minimal with $\text{up}^2(\gamma^-) = \text{up}^2(\gamma)$ and $\text{up}(\gamma^-) \supset \beta_0$; sets $\Gamma^A(x, y) = 0$ (unless now defined to a different value) for all y with $\text{par}(\gamma_0) < y \leq \text{par}(\gamma)$ (if γ_0 exists); and sets $\Gamma^A(x, y) = 1$ (unless now defined to a different value) for all other $y \leq \text{par}(\gamma)$. The use $\gamma(x, y)$ for setting $\Gamma^A(x, y) = 0$ here is -1 (i.e. oracle string $\emptyset$), and the use $\gamma(x, y)$ for setting $\Gamma^A(x, y) = 1$ here is $\text{par}(\gamma_*)$ where $\gamma_* \subseteq \gamma$ is minimal with $\text{par}(\gamma_*) \geq y$, $\text{up}(\gamma_*) \subseteq \text{up}(\gamma)$, and $\text{up}^2(\gamma_*) = \text{up}^2(\gamma)$. (When using oracle string $\emptyset$, we still get a p.r. functional if we adopt the convention that if two contradictory definitions would apply to a fixed oracle then the definition enumerated first is the one used. In our example, there will never be contradictory definitions applying to the r.e. oracle A.)

The action of a $\gamma \in T_0$ working on (the kth instance of) (16) is to measure if

$$\exists s \leq \text{par}(\gamma)\, (s \in S \;\&\; \Phi_{x,s}^{A_s}(x)\downarrow), \tag{18}$$

which, as for (9), just means

$$\Phi_{x,\text{par}(\gamma)}^{A_{\text{par}(\gamma)}}(x)\downarrow . \tag{18'}$$

(Here $A_{\text{par}(\gamma)}$ is the subset of A enumerated before γ's action.) If the answer is no then γ has outcome 0 and sets $\Delta^{\sigma\oplus(\beta^\frown\langle\infty\rangle)}(x) = 0$ (for its σ and β, unless previously set to 1 for a compatible initial segment of the oracle); otherwise γ has outcome ∞ and sets $\Delta^{\sigma\oplus(\beta^\frown\langle\gamma\rangle)}(x) = 1$ (for its σ and β, unless previously set to 0 for a compatible initial segment of the oracle).

We now verify that the construction satisfies all the requirements.

We first observe a combinatorial fact about infinite injury, namely that if $\gamma \subset \Lambda_0$, $\text{up}(\gamma) \subset \Lambda_1$, and $\text{up}^2(\gamma) \subset \Lambda_2$ then for all γ' with $\gamma \subset \gamma' \subset \Lambda_0$, either $\text{up}^2(\gamma)$ and $\text{up}^2(\gamma')$ are comparable, or $\text{up}^2(\gamma) \supseteq \bar{\alpha}^\frown\langle\infty\rangle$ and $\text{up}^2(\gamma') \supseteq \bar{\alpha}^\frown\langle\bar{\beta}\rangle$ for some $\bar{\alpha}$ and $\bar{\beta}$ with $\text{lev}(\bar{\alpha}) = 2$. Suppose not, and say $\text{up}^2(\gamma)$ and $\text{up}^2(\gamma')$ split at $\bar{\alpha}$. First assume $\text{lev}(\bar{\alpha}) = 1$. Then $\bar{\beta} \subset \text{up}(\gamma)$ for some $\bar{\beta}$ with $\text{up}(\bar{\beta}) = \bar{\alpha}$, and so $\Lambda_1 \restriction (|\bar{\beta}|+1) \subseteq \lambda(\gamma')$ by $\text{up}(\gamma) \subset \Lambda_1$, contradicting $\text{up}^2(\gamma)$ and $\text{up}^2(\gamma')$ splitting at $\bar{\alpha}$. So assume $\bar{\alpha}^\frown\langle\bar{\beta}\rangle \subseteq \text{up}^2(\gamma)$ for some $\bar{\beta} \in T_1$. Then $\bar{\beta}^\frown\langle\infty\rangle \subseteq \text{up}(\gamma) \subset \Lambda_1$, and so again $\Lambda_1 \restriction (|\bar{\beta}| + 1) \subseteq \lambda^2(\gamma')$, contradicting $\text{up}^2(\gamma)$ and $\text{up}^2(\gamma')$ splitting at $\bar{\alpha}$.

Now suppose $x_0 \in J$. Then $\alpha^\frown\langle\beta\rangle \subset \Lambda_2$ (for some β) for the unique $\alpha \subset \Lambda_2$ working on (15) with this x_0. Thus $\beta^\frown\langle\infty\rangle \subset \Lambda_1$, and $\neg\forall y \leq \text{par}(\beta)\exists s R(x_0, y, s)$. Let $\gamma \subset \Lambda_0$ be minimal with $\text{up}(\gamma) = \beta$. Then for all γ' with $\gamma \subseteq \gamma' \subset \Lambda_0$, we have $\lambda(\gamma') \supseteq \beta$. Any such γ' with $\text{up}(\gamma') \supseteq \beta$ working on (15) with x_0 will measure (17) negatively (since $\text{par}(\text{up}(\gamma')) \geq \text{par}(\beta)$) and thus set $\Gamma^A(x_0, y) = 1$. For any $\gamma' \subset \Lambda_0$ with $\text{up}(\gamma') \subset \beta$ working on (15) with this x_0, if γ' sets $\Gamma^A(x_0, y) = 0$ then $\text{up}(\gamma')^\frown\langle\gamma'\rangle \subset \beta$ by $\beta \subset \Lambda_1$, and thus $\gamma' \subset \gamma$. Therefore $\Gamma^A(x_0, y) = 1$ for all $y \geq \text{par}(\gamma)$. (Note that $\Gamma^A(x_0, y)$ is defined for all y since all $\gamma' \subset \Lambda_0$ with $\text{up}(\gamma') \subset \Lambda_1$ and $\text{up}^2(\gamma') = \alpha$ define $\Gamma^A(x_0, y)$

with the same use $\text{par}(\gamma_*)$ for the fixed string γ_* that γ' uses for y if $y \leq \text{par}(\gamma')$ unless already set to 0 with use -1.)

On the other hand, suppose $x_0 \notin J$. Then $\alpha^\frown\langle\infty\rangle \subset \Lambda_2$ for the unique $\alpha \subset \Lambda_2$ working on (15) with this x_0. Thus $\beta^\frown\langle\gamma_\beta\rangle \subset \Lambda_1$ (for some γ_β) for all $\beta \subset \Lambda_1$ with $\text{up}(\beta) = \alpha$; and for all these β (except for β_0, the least of them), γ_β will try to ensure $\Gamma^A(x_0, y) = 0$ with use -1 for all $y \in (\text{par}(\gamma_{\beta^-}), \text{par}(\gamma_\beta)]$ (where β^- is the maximal $\beta' \subset \beta$ with $\text{up}(\beta') = \text{up}(\beta)$). This will establish $\Gamma^A(x_0, y) = 0$ with use -1 for all $y > \text{par}(\gamma_{\beta_0})$ (and $\Gamma^A(x_0, y)$ is defined for $y \leq \text{par}(\gamma_{\beta_0})$ as above).

Thus suppose $\Gamma^A(x_0, y) \downarrow = 1$ for some (least) $y > \text{par}(\gamma_{\beta_0})$. Then some (minimal) γ_β would like to set it to 0. Thus $\text{par}(\gamma_{\beta^-}) < y \leq \text{par}(\gamma_\beta)$, and some $\bar{\gamma}$ with $\gamma_{\beta^-} \subseteq \bar{\gamma} \subset \gamma_\beta$ set $\Gamma^A(x_0, y) = 1$. Set $\bar{\alpha} = \text{up}^2(\bar{\gamma})$ and $\bar{\beta} = \text{up}(\bar{\gamma})$.

First suppose $\alpha \neq \bar{\alpha}$. Since both work on the same x_0, they must be incomparable, say they split at $\tilde{\alpha}$.

By our observation, $\tilde{\alpha}^\frown\langle\infty\rangle \subseteq \alpha$ and $\tilde{\alpha}^\frown\langle\tilde{\beta}\rangle \subseteq \bar{\alpha}$ for some $\tilde{\beta} \in T_1$, and so $\lambda(\bar{\gamma}) \supseteq \bar{\beta} \supseteq \tilde{\beta}^\frown\langle\infty\rangle$ but $\lambda(\gamma_\beta) \not\supseteq \tilde{\beta}^\frown\langle\infty\rangle$. Pick $\hat{\gamma}$ maximal with $\bar{\gamma} \subset \hat{\gamma} \subset \gamma_\beta$ and $\lambda(\hat{\gamma}) \supseteq \tilde{\beta}^\frown\langle\infty\rangle$. Set $\hat{\alpha} = \text{up}^2(\hat{\gamma})$ and $\hat{\beta} = \text{up}(\hat{\gamma})$. So $\hat{\beta} \subseteq \tilde{\beta}$. We will show $\hat{\beta} = \tilde{\beta}$ and thus $\hat{\alpha} = \tilde{\alpha}$. First suppose $\hat{\alpha} \neq \tilde{\alpha}$. Now $\hat{\alpha} \subset \tilde{\alpha}$ is impossible by $\tilde{\alpha} \subset \Lambda_2$ and $\hat{\beta} \subseteq \tilde{\beta}$. Also $\tilde{\alpha} \subset \hat{\alpha}$ implies $\tilde{\alpha}^\frown\langle\tilde{\beta}\rangle \subseteq \hat{\alpha}$ by $\lambda(\hat{\gamma}) \supseteq \tilde{\beta}^\frown\langle\infty\rangle$, contradicting $\hat{\beta} \subseteq \tilde{\beta}$. So $\tilde{\alpha}$ and $\hat{\alpha}$ must be incomparable, say they split at $\check{\alpha}$. By our observation, $\check{\alpha}^\frown\langle\check{\beta}\rangle \subseteq \hat{\alpha}$ and $\check{\alpha}^\frown\langle\infty\rangle \subseteq \tilde{\alpha}$ for some $\check{\beta} \subset \hat{\beta}$, contradicting $\hat{\beta} \subseteq \tilde{\beta}$. Thus $\hat{\alpha} = \tilde{\alpha}$. But then $\hat{\beta} \subset \tilde{\beta}$ contradicts $\text{up}(\hat{\beta}) = \text{up}(\tilde{\beta})$ (as $\hat{\beta}$ changes outcome), so $\hat{\beta} = \tilde{\beta}$. By $\gamma_{\beta^-} \subset \bar{\gamma}$ and $\beta^- \subset \Lambda_1$, we have $\beta^- \subset \tilde{\beta}$. By $\tilde{\alpha} \subset \alpha$, there is some $\tilde{\beta}_0 \subset \beta^-$ with $\text{up}(\tilde{\beta}_0) = \tilde{\alpha}$. Thus $\hat{\gamma}$ enumerates $\text{par}(\hat{\gamma}^-)$ into A for its node $\gamma^- = \hat{\gamma}^-$. By $\bar{\beta} \supseteq \tilde{\beta}^\frown\langle\infty\rangle$ and $\bar{\alpha} \supset \tilde{\alpha}$, necessarily $\hat{\gamma}^- \subset \bar{\gamma}$, and so $\text{par}(\hat{\gamma}^-) < \text{par}(\bar{\gamma})$. Thus $\hat{\gamma}^-$ would destroy the computation $\Gamma^A(x_0, y) = 1$ that was defined by $\bar{\gamma}$ as desired.

So α and $\bar{\alpha}$ must be equal. By $\beta^- \subset \Lambda_1$ and $\gamma_{\beta^-} \subset \bar{\gamma}$, we must have $\beta^{-\frown}\langle\gamma_{\beta^-}\rangle \subseteq \bar{\beta}$. But then γ_β would put a number $\leq \text{par}(\bar{\gamma})$ into A, destroying $\bar{\gamma}$'s definition of $\Gamma^A(x_0, y)$, yielding the desired contradiction. Thus $\Gamma^A(x_0, y) = 0$ for all $y > \text{par}(\gamma_{\beta_0})$.

As for the other type of requirements, suppose $\alpha \subset \Lambda_2$ is the unique strategy working on some fixed requirement (16). We will show first that any $\gamma \subset \Lambda_0$ with $\text{up}(\gamma) \subset \Lambda_1$ and $\text{up}^2(\gamma) = \alpha$ can define $\Delta^{J \oplus \Lambda_1}(x)$ for its x without being prevented from doing so by a previous definition. For the sake of a contradiction, suppose some $\gamma' \subset \Lambda_0$ sets $\Delta^{\sigma' \oplus \tau'}(x)$ to a different value for $\sigma' \oplus \tau'$ compatible with γ's intended oracle $\sigma \oplus \tau$. Since each requirement $\mathcal{R}_e$ is worked on exactly by all $\alpha' \in T_2$ with $|\alpha'| = e$, we must have $|\sigma| = |\sigma'|$ and thus $\sigma = \sigma'$. By the construction, $\beta = \text{up}(\gamma) \subset \tau$ and $\beta' = \text{up}(\gamma') \subset \tau'$; and by compatibility, $\beta' \subset \beta$ or $\beta \subset \beta'$. Furthermore, $|\text{up}(\beta)| = |\text{up}(\beta')|$, and again by the splitting-up from T_2 to T_1 and $\text{lev}(\beta) = \text{lev}(\beta') = 1$, $\text{up}(\beta)$ and $\text{up}(\beta')$ must be incomparable, say they split at $\bar{\alpha}$. Since β and β' are comparable, we must have $\text{up}(\beta) \supseteq \bar{\alpha}^\frown\langle\infty\rangle$ or $\text{up}(\beta') \supseteq \bar{\alpha}^\frown\langle\infty\rangle$. By $\sigma = \sigma'$, $\bar{\alpha}$ cannot work on a requirement (15), thus it works on a requirement (16). By $\text{lev}(\beta) = 1$, there is a unique $\bar{\beta}$ such that $\text{up}(\bar{\beta}) = \bar{\alpha}$, $\bar{\beta} \subset \beta$, and $\bar{\beta} \subset \beta'$. But β and β' make different predictions on the outcome of $\bar{\beta}$, a contradiction. This establishes that γ can define $\Delta^{J \oplus \Lambda_1}(x)$.

Now suppose first that $\alpha^\frown\langle\infty\rangle \subset \Lambda_2$ for the unique $\alpha \subset \Lambda_2$ working on a requirement (16). Then $\beta^\frown\langle\infty\rangle \subset \Lambda_1$ for the unique $\beta \subset \Lambda_1$ working on this requirement, so all $\gamma \subset \Lambda_0$ with $\mathrm{up}(\gamma) = \beta$ will measure (18′) negatively, i.e. $\Phi_x^A(x)\uparrow$. Furthermore, the least such γ, say γ_0, will set $\Delta^{\sigma\oplus(\beta^\frown\langle\infty\rangle)}(x) = 0$.

On the other hand, suppose $\alpha^\frown\langle\gamma\rangle \subset \Lambda_2$ (for some $\gamma \in T_0$) where α works on (16). Then $\beta^\frown\langle\gamma\rangle \subset \Lambda_1$ for the unique $\beta \subset \Lambda_1$ with $\mathrm{up}(\beta) = \alpha$, and $\gamma^\frown\langle\infty\rangle \subset \Lambda_0$. Thus γ measures (18′) positively and will set $\Delta^{\sigma\oplus(\beta^\frown\langle\gamma\rangle)}(x) = 1$. So we need to show $\Phi_x^A(x)\downarrow$.

It suffices to show that $A_{\mathrm{par}(\gamma)} \restriction (\varphi_{x,\mathrm{par}(\gamma)}(x)+1) = A \restriction (\varphi_{x,\mathrm{par}(\gamma)}(x)+1)$. Note that $\mathrm{par}(\gamma) > \varphi_{x,\mathrm{par}(\gamma)}(x)$. For the sake of contradiction, suppose some $\bar\gamma$ with $\gamma \subset \bar\gamma \subset \Lambda_0$ puts some $y \le \mathrm{par}(\gamma)$ into A where $y = \mathrm{par}(\bar\gamma^-)$ for $\bar\gamma$'s node $\gamma^- = \bar\gamma^-$ from the construction. We set $\bar\beta^- = \mathrm{up}(\bar\gamma^-)$. By $\mathrm{par}(\bar\gamma^-) \le \mathrm{par}(\gamma)$, we have $\bar\gamma^- \subset \gamma$. We set $\bar\alpha = \mathrm{up}^2(\bar\gamma)$ and $\bar\beta = \mathrm{up}(\bar\gamma)$. We denote by $\bar\beta_0$ and $\bar\gamma_0$ the nodes β_0 and γ_0 of $\bar\gamma$ from the construction, respectively. Then $\bar\beta_0{}^\frown\langle\bar\gamma_0\rangle \subseteq \bar\beta$. We proceed by comparing the positions of α and $\bar\alpha$ on T_2. As $\beta \subset \Lambda_1$, $\bar\beta_0 \subset \beta \subset \bar\beta$.

First, $\alpha \subset \bar\alpha$ is impossible since α changes outcome at γ, and thus $\lambda^2(\bar\gamma_0) \supseteq \alpha^\frown\langle\infty\rangle$ and $\lambda^2(\bar\gamma) \supseteq \alpha^\frown\langle\gamma\rangle$, contradicting $\bar\alpha \subseteq \lambda^2(\bar\gamma_0), \lambda^2(\bar\gamma)$.

Next, suppose α and $\bar\alpha$ are incomparable, say they split at $\tilde\alpha$. By our observation, $\alpha \supseteq \tilde\alpha^\frown\langle\infty\rangle$ and $\bar\alpha \supseteq \tilde\alpha^\frown\langle\tilde\beta\rangle$ for some $\tilde\beta \in T_1$, thus $\lambda(\bar\gamma_0) \supseteq \tilde\beta^\frown\langle\infty\rangle$ and $\lambda(\gamma) \not\supseteq \tilde\beta^\frown\langle\infty\rangle$, so $\lambda(\bar\gamma) \supseteq \tilde\beta^\frown\langle\infty\rangle$ is impossible by $\bar\gamma_0 \subset \gamma \subset \bar\gamma$.

Next, $\bar\alpha^\frown\langle\tilde\beta\rangle \subseteq \alpha$ (for some $\tilde\beta \in T_1$) is impossible since then $\tilde\beta^\frown\langle\infty\rangle \subseteq \beta$; thus, by $\gamma \subset \bar\gamma$ and $\lambda(\bar\gamma) \supseteq \tilde\beta^\frown\langle\infty\rangle$, we have $\mathrm{up}(\bar\gamma) = \tilde\beta$, and so $\bar\gamma$ will not enumerate any number by $\mathrm{up}(\bar\gamma)^\frown\langle\infty\rangle \subset \Lambda_1$.

Finally, assume $\bar\alpha^\frown\langle\infty\rangle \subseteq \alpha$. By $\bar\gamma_0{}^\frown\langle\infty\rangle \subseteq \bar\gamma^- \subset \gamma$, we must have $\bar\beta_0{}^\frown\langle\bar\gamma_0\rangle \subseteq \bar\beta^-$ and $\bar\beta_0{}^\frown\langle\bar\gamma_0\rangle \subseteq \beta$. By $\beta = \mathrm{up}(\gamma) \subset \Lambda_1$, we must have $\lambda(\gamma') \supseteq \beta$ for all γ' with $\gamma \subseteq \gamma' \subset \Lambda_0$; thus β and $\bar\beta$ must be comparable. If $\bar\beta \subset \beta$ then by $\beta \subset \Lambda_1$, we have $\bar\beta^\frown\langle\bar\gamma\rangle \subseteq \beta$, contradicting $\gamma \subset \bar\gamma$. Thus $\beta^\frown\langle\gamma\rangle \subseteq \bar\beta$ by $\gamma^\frown\langle\infty\rangle \subseteq \bar\gamma$. By our assumption on $\bar\gamma$ and $\bar\gamma_0$, there are no β' with $\bar\beta_0 \subset \beta' \subset \beta$ and $\mathrm{up}(\beta') = \mathrm{up}(\bar\beta_0)$, and by $\bar\alpha^\frown\langle\infty\rangle \subseteq \alpha$, there can be no $\gamma' \subset \gamma$ with $\mathrm{up}(\gamma') = \bar\beta$; so β and $\bar\beta^-$ must be incomparable, say they split at $\tilde\beta$. Note that $\tilde\beta \supseteq \bar\beta_0{}^\frown\langle\bar\gamma_0\rangle$ since $\beta, \bar\beta^- \supseteq \bar\beta_0{}^\frown\langle\bar\gamma_0\rangle$; and that $\bar\beta^- \supseteq \tilde\beta^\frown\langle\infty\rangle$ since $\bar\gamma^- \subset \gamma$.

We now set $\tilde\alpha = \mathrm{up}(\tilde\beta)$ and compare the positions of $\bar\alpha$ and $\tilde\alpha$ on T_2 (still under the assumption $\bar\alpha^\frown\langle\infty\rangle \subseteq \alpha$).

If $\tilde\alpha \subset \bar\alpha$ then, by our observation, $\mathrm{lev}(\alpha) = 2$ and $\tilde\alpha^\frown\langle\infty\rangle \subseteq \bar\alpha$ by $\bar\beta_0 \subset \tilde\beta$ and $\bar\alpha \subset \Lambda_2$. But this is impossible by $\bar\beta^- \supseteq \tilde\beta^\frown\langle\infty\rangle$.

If $\bar\alpha \subset \tilde\alpha$ then, by $\tilde\beta \subset \bar\beta$, we must have $\mathrm{lev}(\tilde\alpha) = 2$ and $\bar\alpha^\frown\langle\infty\rangle \subseteq \tilde\alpha$, contradicting $\bar\beta \supseteq \tilde\beta^\frown\langle\infty\rangle$.

If $\bar\alpha$ and $\tilde\alpha$ are incomparable, say they split at $\check\alpha$, then, by our observation, we must have $\bar\alpha \supseteq \check\alpha^\frown\langle\infty\rangle$ and $\tilde\alpha \supseteq \check\alpha^\frown\langle\check\beta\rangle$ for some $\check\beta \in T_1$, which in turn contradicts $\tilde\beta \subset \bar\beta$ and $\bar\alpha = \mathrm{up}(\bar\beta)$.

We have now established that the existence of a $\bar\gamma \supset \gamma$ destroying $\Phi_x^A(x)$ is impossible, so $\Phi_x^A(x)\downarrow$ as desired. ■

Sacks Density Theorem (Sacks [16]). *For any r.e. set $D <_T C$ there is an r.e. set A such that $D <_T A <_T C$.*

Proof. We will actually build an r.e. set A such that if $D <_T C \oplus D$ then $D <_T A \oplus D <_T C \oplus D$. So we need to build A satisfying

$$A \leq_T C \oplus D \tag{19}$$

and, for all p.r. functionals Φ and Ψ,

$$C = \Phi^{A \oplus D} \to C \leq_T D, \text{ and} \tag{20}$$

$$A = \Psi^D \to C \leq_T D. \tag{21}$$

We could ensure (19) by building an explicit reduction. Instead, we will show at the end that $A \leq_T C \oplus D$ essentially by showing that Λ_2 is co-r.e. in $C \oplus D$. We ensure (20) and (21) by strategies $\alpha \in T_3$ threatening to build a reduction Γ_α or Δ_α, respectively, showing $C \leq_T D$. (Of course, once we have established $C \leq_T D$ for some requirement, we do not have to ensure the lower-priority requirements since the hypothesis of the theorem fails.) Requirements (20) and (21) then take the form

$$\begin{aligned} &\forall x \exists s \forall y \leq x \forall t \geq s \big(C_s(y) = \Phi_s^{A_s \oplus D_s}(y)\downarrow \;\&\; (A_s \oplus D_s) \restriction (\varphi_s(y)+1) = \\ &\qquad (A_t \oplus D_t) \restriction (\varphi_s(y)+1)\big) \to \\ &\forall x > x_0 \exists s \forall y \forall t > s \big(x_0 < y \leq x \to C_s(y) = \Gamma_{\alpha,s+1}^{D_s}(y)\downarrow \;\&\; \\ &\qquad D_s \restriction (\gamma_{\alpha,s}(y)+1) = D_t \restriction (\gamma_{\alpha,s}(y)+1)\big), \end{aligned} \tag{22}$$

and

$$\begin{aligned} &\forall x \exists s \forall y \leq x \forall t \geq s \big(A_s(y) = \Psi_s^{D_s}(y)\downarrow \;\&\; \\ &\qquad D_s \restriction (\psi_s(y)+1) = D_t \restriction (\psi_s(y)+1)\big) \to \\ &\forall x \exists s \forall y \leq x \forall t > s \big(C_s(y) = \Delta_{\alpha,s+1}^{D_s}(y)\downarrow \;\&\; \\ &\qquad D_s \restriction (\delta_{\alpha,s}(y)+1) = D_t \restriction (\delta_{\alpha,s}(y)+1)\big). \end{aligned} \tag{23}$$

(Here x_0 will be a number determined by $\alpha \in T_3$. Note that (22) and (23) both only use $\neg\rho \to \tau$ in (7) whereas $\rho \to \sigma$ is vacuous.)

We fix an effective ω-ordering of the above requirements (22) and (23) and assign requirement $\mathcal{R}_e$ to all nodes in $\{\alpha \in T_3 \mid |\alpha| = e \;\&\; \forall i < e(\alpha(i) \neq \infty)\}$ since outcome ∞ of any requirement corresponds to showing $C \leq_T D$.

In going from T_3 to T_2, we split up α's requirement into the kth instance for $\beta \in T_2$ by bounding $x \leq k$, except in the left-hand side of (23) where we bound $x \leq \text{par}(\beta)$.

In going from T_2 to T_1, and from T_1 to T_0, we split up a subrequirement into its lth, or mth, instance for $\gamma \in T_1$, or $\delta \in T_0$, by bounding $s \leq l$, or $t \leq \text{par}(\delta)$, respectively.

The action of a $\delta \in T_0$ working on (the mth instance of the lth instance of the kth instance of) (22) is to measure if

$$\begin{aligned} &\forall x \leq k \exists s \leq l \forall y \leq x \forall t \in [s, \text{par}(\delta)] \big(C_s(y) = \Phi_s^{A_s \oplus D_s}(y)\downarrow \;\&\; \\ &\qquad (A_s \oplus D_s) \restriction (\varphi_s(y)+1) = (A_t \oplus D_t) \restriction (\varphi_s(y)+1)\big), \end{aligned} \tag{24}$$

which just means

$$\begin{aligned} C_l \upharpoonright (k+1) &= \Phi_l^{A_l \oplus D_l} \upharpoonright (k+1)\downarrow \ \& \\ (A_l \oplus D_l) \upharpoonright (\varphi_l(k)+1) &= (A_{\mathrm{par}(\delta)} \oplus D_{\mathrm{par}(\delta)}) \upharpoonright (\varphi_l(k)+1). \end{aligned} \tag{24'}$$

(This uses the tacit assumption that use functions are increasing in the argument and nondecreasing in the stage. Here and in (25), $A_{\mathrm{par}(\delta)}$ and $D_{\mathrm{par}(\delta)}$ are the sets A and D enumerated before δ's action.) If the answer is yes then δ has outcome 0 (recall that (24′) is an instance of $\neg\rho$ in (7)) and defines $\Gamma_\alpha^D(k) = C_l(k)$ with use $\gamma_\alpha(k) = \varphi_l(k)$ (unless already defined to a different value) where $\alpha = \mathrm{up}^3(\delta)$; otherwise δ has outcome ∞.

The action of a $\delta \in T_0$ working on (the mth instance of the lth instance of the kth instance of) (23) is to measure if

$$\begin{aligned} \forall x \le \mathrm{par}(\beta)\exists s \le l \forall y \le x \forall t \in [s, \mathrm{par}(\delta)]\big(A_s(y) &= \Psi_s^{D_s}(y)\downarrow \ \& \\ D_s \upharpoonright (\psi_s(y)+1) &= D_t \upharpoonright (\psi_s(y)+1)\big), \end{aligned} \tag{25}$$

which just means

$$\begin{aligned} A_l \upharpoonright (\mathrm{par}(\beta)+1) &= \Psi_l^{D_l} \upharpoonright (\mathrm{par}(\beta)+1)\downarrow \ \& \\ D_l \upharpoonright (\psi_l(\mathrm{par}(\beta))+1) &= D_{\mathrm{par}(\delta)} \upharpoonright (\psi_l(\mathrm{par}(\beta))+1). \end{aligned} \tag{25'}$$

Independent of the answer to (25′), δ enumerates $\mathrm{par}(\beta)$ into A if now $\Delta_\alpha^D(k)\downarrow \ne C(k)$. If the answer to (25′) is yes then δ has outcome 0 (again recall that (25′) is an instance of $\neg\rho$ in (7)) and defines $\Delta_\alpha^D(k) = C_l(k)$ (unless already defined to a different value) with use $\delta_\alpha(k) = \psi_l(\mathrm{par}(\beta))$. (Here $\alpha = \mathrm{up}^3(\delta)$ and $\beta = \mathrm{up}^2(\delta)$.) If the answer to (25′) is no then δ has outcome ∞.

We now verify that the construction satisfies all requirements (22) and (23) (up to the highest-priority one showing $C \le_T D$, if any), and that $A \le_T C \oplus D$.

We observe first that on T_3, $\alpha\hat{\ }\langle\infty\rangle \subseteq \alpha'$ implies that no requirement is assigned to α'; therefore on T_2, $\beta \subseteq \beta'$ (if subrequirements are assigned to β and β') implies $\mathrm{up}(\beta) \subseteq \mathrm{up}(\beta')$ as follows: If $\mathrm{up}(\beta) \not\subseteq \mathrm{up}(\beta')$, say $\mathrm{up}(\beta) \supseteq \alpha_0\hat{\ }\langle\beta_0\rangle$ and $\mathrm{up}(\beta') \not\supseteq \alpha_0\hat{\ }\langle\beta_0\rangle$ for some minimal α_0 and some β_0, then $\beta \supseteq \beta_0\hat{\ }\langle\infty\rangle$ but $\beta' \not\supseteq \beta_0\hat{\ }\langle\infty\rangle$, contradicting $\beta \subseteq \beta'$. Furthermore, if β_1 and β_2 are incomparable on T_2 and $\mathrm{up}(\beta_1) \subseteq \mathrm{up}(\beta_2)$ then β_1 and β_2 split at a $\beta \in T_2$ with $\mathrm{up}(\beta) = \mathrm{up}(\beta_1)$. This is because otherwise $\mathrm{up}(\beta) \subset \mathrm{up}(\beta_1)$, say $\mathrm{up}(\beta)\hat{\ }\langle\bar\beta\rangle \subseteq \mathrm{up}(\beta_1)$, and so $\bar\beta\hat{\ }\langle\infty\rangle \subseteq \beta_1, \beta_2$ and $\beta \subseteq \bar\beta$ by (3.3), contradicting β_1 and β_2 splitting at β.

First suppose that $C = \Phi^{A \oplus D}$ (and none of the higher-priority requirements has shown $C \le_T D$). Then $\alpha\hat{\ }\langle\infty\rangle = \Lambda_3$ for the unique $\alpha \subset \Lambda_3$ working on this requirement. Thus $\beta_k\hat{\ }\langle\gamma_k\rangle \subset \Lambda_2$ (for some $\gamma_k \in T_1$) for all k where β_k is the unique $\beta \subset \Lambda_2$ working on the kth instance of α's requirement. Since $\gamma_k\hat{\ }\langle\infty\rangle \subset \Lambda_1$, all $\delta \subset \Lambda_0$ with $\mathrm{up}(\delta) = \gamma_k$ answer (24′) positively, and so one of them defines $\Gamma_\alpha^D(k)$ D-correctly (unless some other strategy does so). So Γ_α^D is total. Suppose $\Gamma_\alpha^D(k) \ne C(k)$ for some k. Then some $\delta \subset \Lambda_0$ last defines $\Gamma_\alpha^D(k) = 0$ D-correctly, but $k \in C$ and some (least) $\bar\delta \supset \delta$ enumerates some $y \le \frac{1}{2}\varphi_{\mathrm{par}(\delta)}(k)$ into A (since $\Phi^{A\oplus D}(k) = C(k)$). We set $\beta = \mathrm{up}^2(\delta)$, $\gamma = \mathrm{up}(\delta)$, $\bar\alpha = \mathrm{up}^3(\bar\delta)$, $\bar\beta = \mathrm{up}^2(\bar\delta)$, and $\bar\gamma = \mathrm{up}(\bar\delta)$. Then $y = \mathrm{par}(\bar\beta)$. Notice that $\varphi_l(k) < l \le \mathrm{par}(\gamma)$ (where l bounds s for γ above) and $\varphi_{\mathrm{par}(\delta)}(k) < \mathrm{par}(\delta)$ by the usual convention of stage bounding use. Below, we will typically show $\mathrm{par}(\gamma) < \mathrm{par}(\bar\beta)$ or $\mathrm{par}(\delta) < \mathrm{par}(\bar\beta)$ for a contradiction.

We argue by cases, comparing the positions of α and $\bar\alpha$ on T_3. First assume that α and $\bar\alpha$ are incomparable, say $\alpha \supseteq \alpha_0\hat{}\langle\beta_0\rangle$ and $\bar\alpha \supseteq \alpha_0\hat{}\langle\bar\beta_0\rangle$ for $\beta_0 \neq \bar\beta_0$. (Recall that there is no outcome ∞ on T_3.) Thus $\beta \supseteq \beta_0\hat{}\langle\infty\rangle$ and $\bar\beta \supseteq \bar\beta_0\hat{}\langle\infty\rangle$. We distinguish subcases, comparing the positions of β_0 and $\bar\beta_0$.

First assume $\beta_0 \subset \bar\beta_0$. Then $\beta_0\hat{}\langle\gamma_0\rangle \subseteq \bar\beta_0$ for some γ_0 since $\mathrm{up}(\beta_0) = \mathrm{up}(\bar\beta_0)$. Thus $\bar\gamma \supseteq \gamma_0\hat{}\langle\infty\rangle$ but $\gamma \not\supseteq \gamma_0\hat{}\langle\infty\rangle$. If $\gamma \subseteq \gamma_0$ then $\mathrm{par}(\gamma) \leq \mathrm{par}(\gamma_0) < \mathrm{par}(\bar\beta_0) < \mathrm{par}(\bar\beta)$, a contradiction. Thus γ and $\gamma_0\hat{}\langle\infty\rangle$ are incomparable. Since $\delta \subset \bar\delta$ we must have $\gamma \supseteq \tilde\gamma\hat{}\langle\infty\rangle$ and $\bar\gamma \supseteq \gamma_0\hat{}\langle\infty\rangle \supseteq \tilde\gamma\hat{}\langle\tilde\delta\rangle$ for some $\tilde\gamma$ and some $\tilde\delta \supset \delta$, again a contradiction by $\mathrm{par}(\bar\beta) > \mathrm{par}(\bar\beta_0) > \mathrm{par}(\gamma_0) > \mathrm{par}(\tilde\delta) > \mathrm{par}(\delta)$.

Next assume $\bar\beta_0 \subset \beta_0$. Then $\bar\beta_0\hat{}\langle\bar\gamma_0\rangle \subseteq \beta_0$ for some $\bar\gamma_0$ since $\mathrm{up}(\bar\beta_0) = \mathrm{up}(\beta_0)$. Thus $\gamma \supseteq \bar\gamma_0\hat{}\langle\infty\rangle$ but $\bar\gamma \not\supseteq \bar\gamma_0\hat{}\langle\infty\rangle$. Now $\alpha_0\hat{}\langle\beta_0\rangle \subseteq \alpha \subset \Lambda_3$, so $\bar\beta_0\hat{}\langle\bar\gamma_0\rangle \subseteq \beta_0 \subset \Lambda_2$ and $\bar\gamma_0\hat{}\langle\infty\rangle \subset \Lambda_1$. Thus $\bar\gamma \subseteq \bar\gamma_0$ is impossible by $\bar\delta \supset \delta$. Also $\bar\gamma_0\hat{}\langle\infty\rangle \subseteq \gamma \subseteq \lambda(\delta)$, so $\bar\gamma_0\hat{}\langle\infty\rangle \subseteq \lambda(\delta')$ for all δ' with $\delta \subseteq \delta' \subset \Lambda_0$, contradicting $\bar\gamma \subseteq \lambda(\bar\delta)$ and $\delta \subset \bar\delta \subset \Lambda_0$.

Finally assume that β_0 and $\bar\beta_0$ are incomparable, say they split at β_1. By our observation above, $\mathrm{up}(\beta_1) = \alpha_0$, and there are distinct γ_1 and $\bar\gamma_1$ such that $\beta_1\hat{}\langle\gamma_1\rangle \subseteq \beta_0$ and $\beta_1\hat{}\langle\bar\gamma_1\rangle \subseteq \bar\beta_0$. So $\gamma_1\hat{}\langle\infty\rangle \subseteq \gamma$ and $\bar\gamma_1\hat{}\langle\infty\rangle \subseteq \bar\gamma$. Thus γ and $\bar\gamma_1\hat{}\langle\infty\rangle$ are incomparable. Since $\delta \subset \bar\delta$ we must have $\gamma \supseteq \tilde\gamma\hat{}\langle\infty\rangle$ and $\bar\gamma \supseteq \bar\gamma_1\hat{}\langle\infty\rangle \supseteq \tilde\gamma\hat{}\langle\tilde\delta\rangle$ for some $\tilde\gamma$ and some $\tilde\delta \supset \delta$, again a contradiction by $\mathrm{par}(\bar\beta) > \mathrm{par}(\bar\beta_0) > \mathrm{par}(\bar\gamma_1) > \mathrm{par}(\tilde\delta) > \mathrm{par}(\delta)$.

We have thus established that α and $\bar\alpha$ must be comparable. So assume next that $\alpha \subset \bar\alpha$, say $\alpha\hat{}\langle\beta_0\rangle \subseteq \bar\alpha$ for some β_0. Then $\beta_0\hat{}\langle\infty\rangle \subseteq \bar\beta$. We distinguish subcases, comparing the positions of β and β_0.

First assume $\beta \subset \beta_0$. Then $\beta\hat{}\langle\gamma_0\rangle \subseteq \beta_0$ for some γ_0 since $\mathrm{up}(\beta) = \mathrm{up}(\beta_0)$, and so $\bar\gamma \supseteq \gamma_0\hat{}\langle\infty\rangle$ but $\gamma \not\supseteq \gamma_0\hat{}\langle\infty\rangle$. If $\gamma \subseteq \gamma_0$ then $\mathrm{par}(\gamma) \leq \mathrm{par}(\gamma_0) < \mathrm{par}(\beta_0) < \mathrm{par}(\bar\beta)$, a contradiction. Thus γ and $\gamma_0\hat{}\langle\infty\rangle$ are incomparable. Since $\delta \subset \bar\delta$ we must have $\gamma \supseteq \tilde\gamma\hat{}\langle\infty\rangle$ and $\bar\gamma \supseteq \gamma_0\hat{}\langle\infty\rangle \supseteq \tilde\gamma\hat{}\langle\tilde\delta\rangle$ for some $\tilde\gamma$ and some $\tilde\delta \supset \delta$, again a contradiction by $\mathrm{par}(\bar\beta) > \mathrm{par}(\beta_0) > \mathrm{par}(\gamma_0) > \mathrm{par}(\tilde\delta) > \mathrm{par}(\delta)$.

Next assume $\beta \supseteq \beta_0$, so $\beta \not\supseteq \beta_0\hat{}\langle\infty\rangle$ but $\bar\beta \supseteq \beta_0\hat{}\langle\infty\rangle$. By $\lambda^2(\bar\delta) \supseteq \bar\beta$, $\lambda(\bar\delta) \supseteq \gamma\hat{}\langle\infty\rangle$ is impossible; so let $\tilde\delta \subset \bar\delta$ be maximal such that for all δ' with $\delta \subset \delta' \subseteq \tilde\delta$, $\lambda(\delta') \supseteq \gamma\hat{}\langle\infty\rangle$. Then $\tilde\gamma = \mathrm{up}(\tilde\delta) \subseteq \gamma$, $\tilde\gamma\hat{}\langle\infty\rangle \subseteq \gamma\hat{}\langle\infty\rangle$, and $\tilde\gamma\hat{}\langle\tilde\delta\rangle \subseteq \lambda(\tilde\delta\hat{}\langle\infty\rangle)$. We set $\tilde\beta = \mathrm{up}(\tilde\gamma)$ and $\tilde\alpha = \mathrm{up}^2(\tilde\gamma)$. We will show $\tilde\alpha = \alpha$ and $\tilde\beta \subseteq \beta$. For the sake of a contradiction, first suppose $\alpha \not\subseteq \tilde\alpha$, say $\alpha_1\hat{}\langle\beta_1\rangle \subseteq \alpha$ but $\alpha_1\hat{}\langle\beta_1\rangle \not\subseteq \tilde\alpha$ for minimal such α_1 and some β_1. Thus $\beta_1\hat{}\langle\infty\rangle \subseteq \beta$ but $\beta_1\hat{}\langle\infty\rangle \not\subseteq \tilde\beta$. Assume $\beta_1 \not\subseteq \tilde\beta$, say $\hat\beta \subseteq \tilde\beta$ is maximal with $\hat\beta \subset \beta_1$. Then, by our observation, $\mathrm{up}(\hat\beta) = \alpha_1$, and so $\hat\beta\hat{}\langle\hat\gamma\rangle \subseteq \beta_1$ for some $\hat\gamma$. By $\alpha \subset \Lambda_3$ we have $\beta_1\hat{}\langle\infty\rangle \subset \Lambda_2$ and so $\hat\gamma\hat{}\langle\infty\rangle \subset \Lambda_1$. Now $\delta \subset \tilde\delta \subset \Lambda_0$ implies $\hat\gamma\hat{}\langle\infty\rangle \subseteq \lambda(\tilde\delta)$ and thus $\hat\beta\hat{}\langle\hat\gamma\rangle \subseteq \lambda^2(\tilde\delta)$ by $\tilde\beta \subseteq \lambda^2(\tilde\delta)$ and $\hat\beta \subseteq \tilde\beta$. By $\tilde\beta \not\supseteq \hat\beta\hat{}\langle\hat\gamma\rangle$ and $\tilde\beta \subseteq \lambda^2(\tilde\delta)$, we have $\tilde\beta = \hat\beta$, and thus $\tilde\gamma = \hat\gamma$ by $\hat\gamma\hat{}\langle\infty\rangle \subseteq \lambda(\tilde\delta)$. But we have, $\tilde\gamma\hat{}\langle\infty\rangle \subset \Lambda_1$, contradicting the choice of $\tilde\gamma = \hat\gamma$. Thus $\beta_1 \subseteq \tilde\beta$. Now $\beta_1\hat{}\langle\gamma_1\rangle \subseteq \tilde\beta$ for some γ_1 is impossible by $\tilde\gamma \subseteq \gamma$ and $\beta_1\hat{}\langle\infty\rangle \subseteq \beta$; so $\tilde\beta = \beta_1$ and thus $\tilde\beta\hat{}\langle\infty\rangle \subseteq \beta$, contradicting $\tilde\gamma\hat{}\langle\infty\rangle \subseteq \gamma$. Thus $\alpha \subseteq \tilde\alpha$, so suppose, again for the sake of a contradiction, $\alpha \subset \tilde\alpha$, say $\alpha\hat{}\langle\beta_1\rangle \subseteq \tilde\alpha$, and so $\tilde\beta \supseteq \beta_1\hat{}\langle\infty\rangle$. Now $\beta_1 \not\subseteq \beta$ is impossible since then

$\beta_2{}^\frown\langle\gamma_2\rangle \subseteq \beta_1$ but $\beta_2{}^\frown\langle\gamma_2\rangle \not\subseteq \beta$ for minimal such β_2 and some γ_2 (by $\mathrm{up}(\beta_1) = \alpha$ and our observation), and so $\tilde\gamma \supseteq \gamma_2{}^\frown\langle\infty\rangle$ but $\gamma \not\supseteq \gamma_2{}^\frown\langle\infty\rangle$, contradicting $\tilde\gamma \subseteq \gamma$. Thus $\beta_1 \subseteq \beta$. Now $\lambda(\tilde\delta) \supseteq \gamma^\frown\langle\infty\rangle$, and so $\lambda^2(\tilde\delta) \supseteq \beta^\frown\langle\gamma\rangle$ or $\lambda^2(\tilde\delta) \supseteq \beta_2{}^\frown\langle\gamma_2\rangle$ for some $\beta_2 \subset \beta$ and some $\gamma_2 \supseteq \gamma^\frown\langle\infty\rangle$. The latter implies $\tilde\gamma \supseteq \gamma_2{}^\frown\langle\infty\rangle$ by $\alpha \subset \tilde\alpha$, contradicting $\tilde\gamma \subseteq \gamma$. Thus $\lambda^2(\tilde\delta) \supseteq \beta^\frown\langle\gamma\rangle$, contradicting $\lambda^2(\tilde\delta) \supseteq \tilde\beta$. Thus $\alpha = \tilde\alpha$. We next show that $\tilde\beta \subseteq \beta$. Suppose not; then $\tilde\beta \supseteq \beta_1{}^\frown\langle\gamma_1\rangle$ but $\beta \not\supseteq \beta_1{}^\frown\langle\gamma_1\rangle$ for minimal such β_1 and some γ_1. Then $\tilde\gamma \supseteq \gamma_1{}^\frown\langle\infty\rangle$, contradicting $\tilde\gamma \subseteq \gamma$. We have thus established $\tilde\alpha = \alpha$ and $\tilde\beta \subseteq \beta$, and so $D_{\tilde l} \upharpoonright (\varphi_{\tilde l}(\tilde k)+1) \neq D_{\mathrm{par}(\tilde\delta)} \upharpoonright (\varphi_{\tilde l}(\tilde k)+1)$ for $\tilde\delta$'s numbers $\tilde k$ and $\tilde l$.

We will now show that this implies $D_l \upharpoonright (\varphi_l(k)+1) \neq D \upharpoonright (\varphi_l(k)+1)$, contradicting the D-correctness of δ's definition of $\Gamma^D_\alpha(k)$. First of all, there is some maximal $\tilde\delta_0 \subset \delta$ with $\mathrm{up}(\tilde\delta_0) = \tilde\gamma$ since $\tilde\gamma \subseteq \gamma$. Furthermore, $\tilde\delta_0{}^\frown\langle 0\rangle \subseteq \delta$, and so $D_{\tilde l} \upharpoonright (\varphi_{\tilde l}(\tilde k)+1) = D_{\mathrm{par}(\tilde\delta_0)} \upharpoonright (\varphi_{\tilde l}(\tilde k)+1)$. By $\mathrm{up}^i(\tilde\delta_0) \subseteq \mathrm{up}^i(\delta)$ for all $i \in \{0,1,2,3\}$ and by (5), we must have $\tilde m_0 \geq m + \mathrm{par}(\gamma)$ (where $\tilde\delta_0$ and δ work on the $\tilde m_0$th and the mth instance of $\tilde\gamma$'s and γ's subrequirement, respectively) since otherwise the $(\tilde m_0+1)$th instance of $\tilde\gamma$'s subrequirement would have been assigned to δ rather than the mth instance of γ's subrequirement. Thus $\mathrm{par}(\tilde\delta_0) \geq \tilde m_0 \geq \mathrm{par}(\gamma) \geq l$, and so $D_{\tilde l} \upharpoonright (\varphi_{\tilde l}(\tilde k)+1) = D_l \upharpoonright (\varphi_{\tilde l}(\tilde k)+1)$ and $\varphi_{\tilde l}(\tilde k) = \varphi_l(\tilde k)$. Therefore $D_l \upharpoonright (\varphi_{\tilde l}(\tilde k)+1) \neq D \upharpoonright (\varphi_{\tilde l}(\tilde k)+1)$, which gives the desired implication above, using $\tilde k \leq k$ and $\varphi_{\tilde l}(\tilde k) = \varphi_l(\tilde k) \leq \varphi_l(k)$.

Finally assume β and β_0 are incomparable, say they split at β'. Then, by our observation, $\mathrm{up}(\beta') = \alpha$, and there are distinct γ_1 and γ_2 such that $\beta'{}^\frown\langle\gamma_1\rangle \subseteq \beta$ and $\beta'{}^\frown\langle\gamma_2\rangle \subseteq \beta_0$. So $\gamma_1{}^\frown\langle\infty\rangle \subseteq \gamma$ and $\gamma_2{}^\frown\langle\infty\rangle \subseteq \bar\gamma$. Thus γ and $\gamma_2{}^\frown\langle\infty\rangle$ are incomparable. Since $\delta \subset \bar\delta$ we must have $\gamma \supseteq \tilde\gamma^\frown\langle\infty\rangle$ and $\bar\gamma \supseteq \gamma_2{}^\frown\langle\infty\rangle \supseteq \tilde\gamma^\frown\langle\tilde\delta\rangle$ for some $\tilde\gamma$ and some $\tilde\delta \supset \delta$, again a contradiction by $\mathrm{par}(\bar\beta) > \mathrm{par}(\beta_0) > \mathrm{par}(\gamma_2) > \mathrm{par}(\tilde\delta) > \mathrm{par}(\delta)$.

We have thus established that $\bar\alpha \subseteq \alpha$, in fact $\bar\alpha \subset \alpha$ since they work on different requirements, say $\bar\alpha^\frown\langle\bar\beta_0\rangle \subseteq \alpha$. Then $\bar\beta_0{}^\frown\langle\infty\rangle \subseteq \beta$, and $\bar\beta_0{}^\frown\langle\infty\rangle \subset \Lambda_2$ by $\alpha \subset \Lambda_3$. We distinguish subcases, comparing the positions of $\bar\beta_0$ and $\bar\beta$. First assume $\bar\beta \subset \bar\beta_0$, say $\bar\beta^\frown\langle\bar\gamma_0\rangle \subseteq \bar\beta_0$ for some $\bar\gamma_0$. Now $\bar\beta^\frown\langle\bar\gamma_0\rangle \subseteq \bar\beta_0 \subset \beta$ and so $\bar\gamma_0{}^\frown\langle\infty\rangle \subseteq \gamma$; also $\bar\beta^\frown\langle\bar\gamma_0\rangle \subset \Lambda_2$ and so $\bar\gamma_0{}^\frown\langle\infty\rangle \subset \Lambda_1$. Thus $\lambda(\delta') \supseteq \bar\gamma_0{}^\frown\langle\infty\rangle$ for all δ' with $\delta \subset \delta' \subset \Lambda_0$. This implies $\mathrm{up}(\bar\delta) = \bar\gamma_0$, by (3.3), contradicting $\bar\delta^\frown\langle\infty\rangle \subset \Lambda_0$.

Next assume $\bar\beta \not\subseteq \bar\beta_0$, say $\tilde\beta \subset \bar\beta$ is maximal with $\tilde\beta \subseteq \bar\beta_0$. By our observation, $\mathrm{up}(\tilde\beta) = \bar\alpha$, and so $\tilde\beta^\frown\langle\tilde\gamma\rangle \subseteq \bar\beta$ for some $\tilde\gamma$ but $\tilde\beta^\frown\langle\tilde\gamma\rangle \not\subseteq \beta$. Thus $\tilde\gamma^\frown\langle\infty\rangle \subseteq \bar\gamma$ but $\tilde\gamma^\frown\langle\infty\rangle \not\subseteq \gamma$. If γ and $\tilde\gamma$ are incomparable, say they split at $\hat\gamma$, then $\hat\gamma^\frown\langle\infty\rangle \subseteq \gamma$ and $\hat\gamma^\frown\langle\hat\delta\rangle \subseteq \tilde\gamma$ for some $\hat\delta \supset \delta$ by $\delta \subset \bar\delta$. But then $\mathrm{par}(\bar\beta) > \mathrm{par}(\tilde\gamma) > \mathrm{par}(\hat\delta) > \mathrm{par}(\delta)$, a contradiction. So necessarily $\gamma \subset \tilde\gamma$, and thus $\mathrm{par}(\gamma) < \mathrm{par}(\tilde\gamma) < \mathrm{par}(\bar\beta)$, again a contradiction.

We are thus left with only one subcase, namely $\bar\beta = \bar\beta_0$. Since $\bar\alpha^\frown\langle\bar\beta\rangle \subseteq \alpha$, this subcase can generate at most one injury for each $\bar\alpha \subset \alpha$. We have thus established $C =^* \Gamma^D_\alpha$ as desired. (We remark that this finite injury to the Γ's can be eliminated if we rephrase requirement (21) as

$$A = \Psi^D \to \bar C \text{ r.e. in } D. \tag{21'}$$

Now the witness $\bar{\beta}_0 \subset \Lambda_2$ for (21′) will not be the least $\beta \subset \Lambda_2$ working on (21′) with $A \restriction (\mathrm{par}(\beta)+1) \neq \Psi^D \restriction (\mathrm{par}(\beta)+1)$, but the least such for which also $k_\beta \notin C$ (unless C is cofinite in which case certainly $C \leq_{\mathrm{T}} D$). The trick is that for this $\bar{\beta}_0$, $\mathrm{par}(\bar{\beta}_0)$ will not be enumerated into A (which causes the finite injury in the above construction). This also implies that $A \leq_{\mathrm{T}} \Lambda_1 \leq_{\mathrm{T}} C \oplus D$, giving an easier proof of (19).)

Next suppose $A = \Psi^D$ (and none of the higher-priority requirements has shown $C \leq_{\mathrm{T}} D$). Then $\alpha^\frown\langle\infty\rangle = \Lambda_3$ for the unique $\alpha \subset \Lambda_3$ working on this requirement. Thus $\beta_k{}^\frown\langle\gamma_k\rangle \subset \Lambda_2$ (for some $\gamma_k \subset \Lambda_1$) for all k where β_k is the unique $\beta \subset \Lambda_2$ working on the kth instance of α's requirement. Since $\gamma_k{}^\frown\langle\infty\rangle \subset \Lambda_1$, all $\delta \subset \Lambda_0$ with $\mathrm{up}(\delta) = \gamma_k$ answer (25′) positively, and one of these δ defines $\Delta_\alpha^D(k)$ D-correctly (unless some other strategy does so). So Δ_α^D is total. Suppose $\Delta_\alpha^D(k) \neq C(k)$ for some k. Then one of the $\delta \subset \Lambda_0$ with $\mathrm{up}(\delta) = \gamma_k$, say $\bar{\delta}$, will enumerate $\mathrm{par}(\beta_k)$ into A, trying to destroy the D-correct computation $\Delta_\alpha^D(k)$ defined by some $\hat{\delta} \subset \bar{\delta}$. We set $\hat{\beta} = \mathrm{up}^2(\hat{\delta})$ and $\hat{\gamma} = \mathrm{up}(\hat{\delta})$. (Note $\alpha = \mathrm{up}^3(\hat{\delta})$.) We will show $\hat{\beta} = \beta_k$, so $D_l \restriction (\psi_l(\mathrm{par}(\beta_k))+1) \neq D \restriction (\psi_l(\mathrm{par}(\beta_k))+1)$ (since $A(\mathrm{par}(\beta_k)) = \Psi^D(\mathrm{par}(\beta_k))$), and $\hat{\delta}$ did therefore not define $\Delta_\alpha^D(k)$ D-correctly by $\delta_\alpha(k) = \psi_l(\mathrm{par}(\beta_k))$, yielding the desired contradiction. Now since β_k and $\hat{\beta}$ work on the same k, $\beta_k \neq \hat{\beta}$ would imply that they are incomparable, say they split at β_0. Then, by our observation, we have $\mathrm{up}(\beta_0) = \alpha$; and so $\beta_0{}^\frown\langle\gamma_0\rangle \subseteq \beta_k$ and $\beta_0{}^\frown\langle\hat{\gamma}_0\rangle \subseteq \hat{\beta}$ for some γ_0 and $\hat{\gamma}_0$. Thus $\gamma_0{}^\frown\langle\infty\rangle \subseteq \gamma_k$ and $\hat{\gamma}_0{}^\frown\langle\infty\rangle \subseteq \hat{\gamma}$, and so $\gamma_0{}^\frown\langle\infty\rangle$ and $\hat{\gamma}_0{}^\frown\langle\infty\rangle$ must be incomparable. Therefore there is some δ' with $\hat{\delta} \subset \delta' \subseteq \bar{\delta}$ such that $\lambda(\delta') \not\supseteq \hat{\gamma}_0{}^\frown\langle\infty\rangle$. Let $\tilde{\delta} \subset \bar{\delta}$ be maximal such that for all δ' with $\hat{\delta} \subseteq \delta' \subseteq \tilde{\delta}$, $\lambda(\delta') \supseteq \hat{\gamma}_0{}^\frown\langle\infty\rangle$. Then $\tilde{\gamma} = \mathrm{up}(\tilde{\delta}) \subseteq \hat{\gamma}$, $\tilde{\gamma}^\frown\langle\infty\rangle \subseteq \hat{\gamma}^\frown\langle\infty\rangle$, and $\tilde{\gamma}^\frown\langle\tilde{\delta}\rangle \subseteq \lambda(\tilde{\delta}^\frown\langle\infty\rangle)$. We set $\tilde{\beta} = \mathrm{up}(\tilde{\gamma})$ and $\tilde{\alpha} = \mathrm{up}^2(\tilde{\gamma})$. We will show $\tilde{\alpha} = \alpha$ and $\tilde{\beta} \subseteq \beta_0$. For the sake of a contradiction, first suppose $\alpha \not\subseteq \tilde{\alpha}$, say $\alpha_1{}^\frown\langle\beta_1\rangle \subseteq \alpha$ but $\alpha_1{}^\frown\langle\beta_1\rangle \not\subseteq \tilde{\alpha}$ for minimal such α_1 and some β_1. Thus $\beta_1{}^\frown\langle\infty\rangle \subseteq \hat{\beta}$ but $\beta_1{}^\frown\langle\infty\rangle \not\subseteq \tilde{\beta}$. Assume $\beta_1 \not\subseteq \tilde{\beta}$, say $\breve{\beta} \subseteq \tilde{\beta}$ is maximal with $\breve{\beta} \subset \beta_1$. Then, by our observation, $\mathrm{up}(\breve{\beta}) = \alpha_1$, and so $\breve{\beta}^\frown\langle\breve{\gamma}\rangle \subseteq \beta_1$ for some $\breve{\gamma}$. By $\alpha \subset \Lambda_3$ we have $\beta_1{}^\frown\langle\infty\rangle \subset \Lambda_2$ and so $\breve{\gamma}^\frown\langle\infty\rangle \subset \Lambda_1$. Now $\hat{\delta} \subset \tilde{\delta} \subset \Lambda_0$ implies $\breve{\gamma}^\frown\langle\infty\rangle \subseteq \lambda(\tilde{\delta})$ and thus $\breve{\beta}^\frown\langle\breve{\gamma}\rangle \subseteq \lambda^2(\tilde{\delta})$ by $\tilde{\beta} \subseteq \lambda^2(\tilde{\delta})$ and $\breve{\beta} \subseteq \tilde{\beta}$. By $\tilde{\beta} \not\supseteq \breve{\beta}^\frown\langle\breve{\gamma}\rangle$ we have $\tilde{\beta} = \breve{\beta}$, and thus $\tilde{\gamma} = \breve{\gamma}$ by $\breve{\gamma}^\frown\langle\infty\rangle \subseteq \lambda(\tilde{\delta})$. But we have $\tilde{\gamma}^\frown\langle\infty\rangle \subset \Lambda_1$, contradicting the choice of $\tilde{\gamma} = \breve{\gamma}$. Thus $\beta_1 \subseteq \tilde{\beta}$. Now $\beta_1{}^\frown\langle\gamma_1\rangle \subseteq \tilde{\beta}$ for some γ_1 is impossible by $\tilde{\gamma} \subseteq \hat{\gamma}$; so $\tilde{\beta} = \beta_1$ and thus $\tilde{\beta}^\frown\langle\infty\rangle \subseteq \hat{\beta}$, contradicting $\tilde{\gamma}^\frown\langle\infty\rangle \subseteq \hat{\gamma}_0$. Thus $\alpha \subseteq \tilde{\alpha}$, so suppose, again for the sake of a contradiction, $\alpha \subset \tilde{\alpha}$, say $\alpha^\frown\langle\beta_1\rangle \subseteq \tilde{\alpha}$ for some β_1, and so $\beta_1{}^\frown\langle\infty\rangle \subseteq \tilde{\beta}$. Now $\beta_1 \not\subseteq \beta_0$ is impossible since then $\beta_2{}^\frown\langle\gamma_2\rangle \subseteq \beta_1$ but $\beta_2{}^\frown\langle\gamma_2\rangle \not\subseteq \beta_0$ for minimal such β_2 and some γ_2 (by $\mathrm{up}(\beta_1) = \alpha$ and by our observation above), and so $\tilde{\gamma} \supseteq \gamma_2{}^\frown\langle\infty\rangle$ but $\hat{\gamma} \not\supseteq \gamma_2{}^\frown\langle\infty\rangle$, contradicting $\tilde{\gamma} \subseteq \hat{\gamma}_0$. Thus $\beta_1 \subseteq \beta_0$. Now $\lambda(\tilde{\delta}) \supseteq \hat{\gamma}_0{}^\frown\langle\infty\rangle$, and so $\lambda^2(\tilde{\delta}) \supseteq \beta_0{}^\frown\langle\hat{\gamma}_0\rangle$ or $\lambda^2(\tilde{\delta}) \supseteq \beta_2{}^\frown\langle\gamma_2\rangle$ for some $\beta_2 \subset \beta_0$ and some $\gamma_2 \supseteq \hat{\gamma}_0{}^\frown\langle\infty\rangle$. The latter implies $\tilde{\gamma} \supseteq \gamma_2{}^\frown\langle\infty\rangle$ by $\alpha \subset \tilde{\alpha}$, contradicting $\tilde{\gamma} \subseteq \hat{\gamma}_0$. Thus $\lambda^2(\tilde{\delta}) \supseteq \beta_0{}^\frown\langle\hat{\gamma}_0\rangle$, contradicting $\lambda^2(\tilde{\delta}) \supseteq \tilde{\beta}$. Thus $\tilde{\alpha} = \alpha$. We next show $\tilde{\beta} \subseteq \beta_0$. Suppose not; then $\tilde{\beta} \supseteq \beta_1{}^\frown\langle\gamma_1\rangle$ but $\beta_0 \not\supseteq \beta_1{}^\frown\langle\gamma_1\rangle$ for minimal such β_1 and some γ_1. Then $\tilde{\gamma} \supseteq \gamma_1{}^\frown\langle\infty\rangle$ but $\hat{\gamma}_0 \not\supseteq \gamma_1{}^\frown\langle\infty\rangle$, contradicting $\tilde{\gamma} \subseteq \hat{\gamma}_0$. We have thus established $\tilde{\alpha} = \alpha$ and $\tilde{\beta} \subseteq \beta_0$. Therefore, $D_{\tilde{l}} \restriction (\psi_{\tilde{l}}(\mathrm{par}(\tilde{\beta}))+1) \neq D_{\mathrm{par}(\bar{\delta})} \restriction (\psi_{\tilde{l}}(\mathrm{par}(\tilde{\beta}))+1)$ for $\tilde{\delta}$'s number $l = \tilde{l}$.

We will show that this implies $D_{\hat{l}} \restriction (\psi_{\hat{l}}(\mathrm{par}(\hat{\beta}))+1) \neq D_{\mathrm{par}(\hat{\delta})} \restriction (\psi_{\hat{l}}(\mathrm{par}(\hat{\beta}))+1)$ for $\hat{\delta}$'s number $l = \hat{l}$, contradicting the D-correctness of $\hat{\delta}$'s definition of $\Delta^A_\alpha(k)$. First of all, there is some maximal $\tilde{\delta}_0 \subset \hat{\delta}$ with $\mathrm{up}(\tilde{\delta}_0) = \tilde{\gamma}$ since $\tilde{\gamma} \subset \hat{\gamma}$. Also $\tilde{\delta}_0{}^\frown\langle 0\rangle \subseteq \hat{\delta}$ and so $D_{\tilde{l}} \restriction (\psi_{\tilde{l}}(\mathrm{par}(\tilde{\beta}))+1) = D_{\mathrm{par}(\tilde{\delta}_0)} \restriction (\psi_{\tilde{l}}(\mathrm{par}(\tilde{\beta}))+1)$. By $\mathrm{up}^i(\tilde{\delta}_0) \subseteq \mathrm{up}^i(\hat{\delta})$ for all $i \in \{0,1,2,3\}$ and by (5), we must have $\tilde{m}_0 \geq \hat{m} + \mathrm{par}(\hat{\gamma})$ (where $\tilde{\delta}_0$ and $\hat{\delta}$ work on the $\tilde{m}_0$th and the $\hat{m}$th instance of $\tilde{\gamma}$'s and $\hat{\gamma}$'s subrequirement, respectively) since otherwise the $(\tilde{m}_0+1)$th instance of $\tilde{\gamma}$'s subrequirement would have been assigned to $\hat{\delta}$ rather than the $\hat{m}$th instance of $\hat{\gamma}$'s subrequirement. Thus $\mathrm{par}(\tilde{\delta}_0) \geq \tilde{m}_0 \geq \mathrm{par}(\hat{\gamma}) \geq \hat{l}$, and so $D_{\tilde{l}} \restriction (\psi_{\tilde{l}}(\mathrm{par}(\tilde{\beta}))+1) = D_{\hat{l}} \restriction (\psi_{\tilde{l}}(\mathrm{par}(\tilde{\beta}))+1)$ and $\psi_{\tilde{l}}(\mathrm{par}(\tilde{\beta})) = \psi_{\hat{l}}(\mathrm{par}(\tilde{\beta}))$. Therefore $D_{\hat{l}} \restriction (\psi_{\hat{l}}(\mathrm{par}(\tilde{\beta}))+1) \neq D_{\mathrm{par}(\tilde{\delta})} \restriction (\psi_{\hat{l}}(\mathrm{par}(\tilde{\beta}))+1)$, which gives the desired implication above, using $\mathrm{par}(\tilde{\beta}) < \mathrm{par}(\hat{\beta})$ and $\mathrm{par}(\tilde{\delta}) < \mathrm{par}(\hat{\delta})$. Thus $\hat{\beta} = \beta_k$ as desired.

This establishes the satisfaction of requirements (20) and (21) if $C \not\leq_T D$ since then clearly Λ_3 must be an infinite path. It remains to prove $A \leq_T C \oplus D$. We first exhibit a $(C \oplus D)$-recursive set $M \subseteq T_2 \times \omega$ and a $(C \oplus D)$-r.e. set $M_0 \subseteq T_2$ such that

$$M_0 = \{\, \beta \mid \exists t\,(\langle \beta, t\rangle \in M)\,\} = T_2 - \Lambda_2, \text{ and} \tag{26.1}$$

$$\forall \langle \beta, t\rangle \in M\, \forall t' \geq t (\beta \not\subseteq \lambda^2(\Lambda_0 \restriction t')). \tag{26.2}$$

We enumerate M as follows: The pair $\langle \beta^\frown\langle\infty\rangle, t\rangle$ is enumerated into M iff

$$\langle \beta, t\rangle \in M, \text{ or} \tag{27.1}$$

$$\begin{aligned} \exists\gamma_0\exists\delta_0\Big(&\mathrm{up}(\delta_0) = \gamma_0 \;\&\; \mathrm{up}(\gamma_0) = \beta \;\&\; \delta_0{}^\frown\langle 0\rangle \subset \Lambda_0 \;\&\; |\delta_0{}^\frown\langle 0\rangle| = t \;\& \\ &\delta_0\text{'s condition } (24') \text{ or } (25') \text{ is } D\text{-correct } \;\& \\ &\forall\bar{\beta} \subseteq \beta (k_{\bar{\beta}} \in C \to k_{\bar{\beta}} \in C_t \;\&\; \neg(\Delta^{D_t}_{\mathrm{up}(\bar{\beta}),t}(k_{\bar{\beta}})\downarrow = 0))\Big). \end{aligned} \tag{27.2}$$

The pair $\langle \beta^\frown\langle\gamma\rangle, t\rangle$ is enumerated into M iff

$$\langle \beta, t\rangle \in M, \text{ or} \tag{28.1}$$

$$\mathrm{up}(\gamma) \neq \beta, \text{ or} \tag{28.2}$$

$$\langle \beta^\frown\langle\infty\rangle, t\rangle \in M \;\&\; \gamma^\frown\langle\infty\rangle \not\subseteq \lambda(\Lambda_0 \restriction t), \text{ or} \tag{28.3}$$

$$\exists\delta_0\,(\mathrm{up}^2(\delta_0) = \beta \;\&\; \delta_0{}^\frown\langle\infty\rangle \subset \Lambda_0 \;\&\; |\delta_0{}^\frown\langle\infty\rangle| = t) \tag{28.4}$$

We first establish (26.2). By induction, and by inspection (for (28.4)), we have to verify (26.2) only for pairs enumerated into M through (27.2) or (28.3). So suppose $\langle \beta^\frown\langle\infty\rangle, t\rangle \in M$ via (27.2), or $\langle \beta^\frown\langle\gamma\rangle, t\rangle \in M$ via (28.3) (where the latter implies the former). We will show

$$\forall t' \geq t\,(\beta \not\subseteq \lambda^2(\Lambda_0 \restriction t') \text{ or } \beta^\frown\langle\gamma_0\rangle \subseteq \lambda^2(\Lambda_0 \restriction t')) \tag{29}$$

for the γ_0 in (27.2). Suppose this fails for some (least) $t' \geq t$. Then $\beta \subseteq \lambda^2(\Lambda_0 \restriction t')$ but $\gamma_0{}^\frown\langle\infty\rangle \not\subseteq \lambda(\Lambda_0 \restriction t')$. Let $\bar{\delta}$ be maximal with $\lambda(\bar{\delta}) \supseteq \gamma_0{}^\frown\langle\infty\rangle$ and $\delta \subset \bar{\delta} \subset \Lambda_0$.

Thus $\bar{\gamma}^\frown\langle\infty\rangle \subseteq \gamma_0{}^\frown\langle\infty\rangle$ and $\bar{\gamma}^\frown\langle\bar{\delta}\rangle = \lambda(\bar{\delta}^\frown\langle\infty\rangle)$ for $\bar{\gamma} = \mathrm{up}(\bar{\delta})$. We distinguish cases, comparing the positions of β and $\bar{\beta} = \mathrm{up}(\bar{\gamma})$. If β and $\bar{\beta}$ are incomparable then by $\bar{\gamma} \subset \gamma_0$, $\lambda(\gamma_0) \supseteq \beta$ and $\lambda(\bar{\gamma}) \supseteq \bar{\beta}$, we have $\hat{\beta}^\frown\langle\infty\rangle \subseteq \bar{\beta}$ and $\hat{\beta}^\frown\langle\hat{\gamma}\rangle \subseteq \beta$ for some $\hat{\beta}$ and $\hat{\gamma}$, which implies $\bar{\gamma}^\frown\langle\infty\rangle \subset \hat{\gamma}^\frown\langle\infty\rangle \subseteq \gamma_0$, contradicting $\lambda(\bar{\delta}) \supseteq \gamma_0$ and $\lambda^2(\bar{\delta}) \supseteq \bar{\beta}$. If $\bar{\beta} \subset \beta$ then $\bar{\gamma}^\frown\langle\infty\rangle \not\subseteq \lambda(\delta')$ and so $\beta \not\subseteq \lambda^2(\delta')$ for all δ' with $\bar{\delta} \subset \delta' \subset \Lambda_0$, a contradiction. If $\beta \subset \bar{\beta}$ then $\beta^\frown\langle\gamma_0\rangle \subseteq \bar{\beta}$ by (29) and $\bar{\beta} \subseteq \lambda^2(\bar{\delta})$, contradicting $\bar{\gamma} \subset \gamma_0$. So $\beta = \bar{\beta}$, and, by (3.3) and $\bar{\gamma}^\frown\langle\infty\rangle \subseteq \gamma_0{}^\frown\langle\infty\rangle$, we must have $\bar{\gamma} = \gamma_0$. Now $\bar{\delta}$'s condition (24′) or (25′) was D-correct by (27.2). Thus $\bar{\delta}$ works on a requirement (22), and $(A_l \oplus D_l) \upharpoonright (\varphi_l(k)+1) \neq (A_{\mathrm{par}(\bar{\delta})} \oplus D_{\mathrm{par}(\bar{\delta})}) \upharpoonright (\varphi_l(k)+1)$ by an A-change.

The y at which A changed was enumerated into A by some $\hat{\delta}$ with $\delta_0 \subset \hat{\delta} \subset \bar{\delta}$. We set $\hat{\beta} = \mathrm{up}^2(\hat{\delta})$ and $\hat{\gamma} = \mathrm{up}(\hat{\delta})$, so $y = \mathrm{par}(\hat{\beta})$. We show that such a $\hat{\delta}$ cannot exist, distinguishing subcases by comparing the positions of β and $\hat{\beta}$. If $\beta \subset \hat{\beta}$ then, by $\hat{\beta} \subseteq \lambda^2(\hat{\delta})$ and $\hat{\delta} \subset \bar{\delta}$, we have $\beta^\frown\langle\gamma_0\rangle \subseteq \hat{\beta}$; so $y = \mathrm{par}(\hat{\beta}) > \mathrm{par}(\gamma_0) \geq l > \varphi_l(k)$, a contradiction. Also $\hat{\beta} \subseteq \beta$ is impossible by the last conjunct of (27.2). So β and $\hat{\beta}$ must be incomparable, say they split at $\tilde{\beta}$. If $\tilde{\beta}^\frown\langle\tilde{\gamma}\rangle \subseteq \beta$ for some $\tilde{\gamma}$ then, by $\tilde{\beta}^\frown\langle\tilde{\gamma}\rangle \not\subseteq \hat{\beta}$, $\tilde{\gamma}^\frown\langle\infty\rangle \not\subseteq \lambda(\delta')$ for any δ' with $\hat{\delta} \subseteq \delta' \subset \Lambda_0$, contradicting $|\delta| < t'$. So $\tilde{\beta}^\frown\langle\infty\rangle \subseteq \beta$ and $\tilde{\beta}^\frown\langle\tilde{\gamma}\rangle \subseteq \hat{\beta}$ for some $\tilde{\gamma}$, and thus $\tilde{\gamma}^\frown\langle\infty\rangle \not\subseteq \gamma_0$ but $\tilde{\gamma}^\frown\langle\infty\rangle \subseteq \hat{\gamma}$. If $\gamma_0 \subset \tilde{\gamma}$ then $\mathrm{par}(\hat{\beta}) > \mathrm{par}(\tilde{\gamma}) > \mathrm{par}(\gamma_0) \geq l > \varphi_l(k)$, a contradiction. So γ_0 and $\tilde{\gamma}$ must be incomparable, say they split at $\check{\gamma}$. By $\delta_0 \subset \hat{\delta}$ and by $\gamma_0 \subseteq \lambda(\delta_0)$, we must have $\check{\gamma}^\frown\langle\infty\rangle \subseteq \gamma_0$ and $\check{\gamma}^\frown\langle\check{\delta}\rangle \subseteq \tilde{\gamma}$ for some $\check{\delta} \supset \delta_0$. Thus $\mathrm{par}(\hat{\beta}) > \mathrm{par}(\tilde{\gamma}) > \mathrm{par}(\check{\delta}) > \mathrm{par}(\delta_0) \geq l > \varphi_l(k)$, again a contradiction. This shows that no A-change $\leq \varphi_l(k)$ can have occurred; thus $\bar{\delta}$ cannot exist, establishing (29) and (26.2).

By Lemma 1(ii) and (26.2), we can conclude $M_0 \cap \Lambda_2 = \emptyset$. We establish (26.1) by showing that $T_2 - M_0$ does not contain incomparable nodes. (Note that $T_2 - M_0$ is a tree by (27.1) and (28.1).) For the sake of a contradiction, suppose β_1 and β_2 are incomparable nodes in $T_2 - M_0$ of minimal length. By minimality, any node in $T_2 - M_0$ of length $< |\beta_1|$ must be on Λ_2; also $|\beta_1| = |\beta_2|$ and $\beta_1 \upharpoonright (|\beta_1|-1) = \beta_2 \upharpoonright (|\beta_1|-1) = \beta$, say. By (26.2), we can assume $\beta_1 \subset \Lambda_2$.

First assume $\beta^\frown\langle\gamma\rangle = \beta_1$ for some γ. Then $\gamma^\frown\langle\infty\rangle \subset \Lambda_1$, and for all $\delta \subset \Lambda_0$ with $\mathrm{up}(\delta) = \gamma$, the first five conjuncts of (27.2) hold true for $\gamma = \gamma_0$ and $\delta = \delta_0$. By $\mathrm{up}^j(\gamma) \subset \Lambda_{1+j}$ for $j \in \{0,1,2\}$, there infinitely many $\delta \subset \Lambda_0$ with $\mathrm{up}(\delta) = \gamma$, and by the correctness of the functionals $\Gamma_{\bar{\alpha}}$ and $\Delta_{\bar{\alpha}}$ (for $\bar{\alpha} \subseteq \mathrm{up}(\beta)$), all conjuncts of (27.2) must hold for some $\delta \subset \Lambda_0$ with $\mathrm{up}(\delta) = \gamma$. Thus β_2 must eventually enter M_0, a contradiction.

So assume $\beta^\frown\langle\infty\rangle = \beta_1$. Then $\mathrm{up}(\beta) \subset \Lambda_3$, and so any $\beta^\frown\langle\gamma\rangle$ must eventually enter M_1 via (28.2) or (28.4), again a contradiction. We have thus established (26.1) and (26.2). It is now easy to see why $A \leq_T C \oplus D$ as follows: Fix x. By the surjectivity of par, determine if $x = \mathrm{par}(\beta)$ for some $\beta \in T_2$. If not then $x \notin A$. Otherwise check if β works on a requirement (23) and if $k_\beta \in C$. If not then $x \notin A$. Otherwise simultaneously enumerate (recursively in $C \oplus D$) the sets M and A. If $\langle\beta, t\rangle \in M$ for some t then $x \in A$ iff $x \in A_t$ by (26.2). Otherwise $k_\beta \in C_s$, say, so the first $\delta \subset \Lambda_0$ with $\mathrm{up}^2(\delta) = \beta$ and $s < \mathrm{par}(\delta)$ will enumerate $\mathrm{par}(\beta)$ into A if then $\Delta^D_{\mathrm{up}(\beta)}(k_\beta)\downarrow = 0$, or else $\mathrm{par}(\beta) \notin A$. This establishes $A \leq_T C \oplus D$ and thus concludes the proof of the Sacks Density Theorem. ■

REFERENCES

1. C.J. Ash, Stability of recursive structures in arithmetical degrees, Ann. Pure Appl. Logic *32* (1986), 113-135.
2. C.J. Ash, Labelling systems and r.e. structures, to appear.
3. R.M. Friedberg, Two recursively enumerable sets of incomparable degree of unsolvability, Proc. Natl. Acad. Sci. USA *43* (1957), 236-238.
4. M.J. Groszek, T.A. Slaman, Foundations of the priority method I: Finite and infinite injury, preprint.
5. A. Kučera, An alternative priority-free solution to Post's problem, in: Twelfth Symposium held in Bratislava, August 25-29, 1986, edited by J. Gruska, B. Rovan, and J. Wiederman, Lecture Notes in Computer Science No. 233, Proceedings, Mathematical Foundations of Computer Science '86, Springer-Verlag, Berlin, 1986.
6. A. Kučera, On the use of diagonally nonrecursive functions, Logic Colloquium '87 (Granada, 1987), Studies in Logic and Foundations of Mathematics, 129, North-Holland, Amsterdam, 1989, 219-239.
7. A.H. Lachlan, On some games relevant to the theory of recursively enumerable sets, Ann. of Math. *(2) 91* (1970), 291-310.
8. A.H. Lachlan, The priority method for the construction of recursively enumerable sets, in **[12]**, 299-310.
9. S. Lempp, M. Lerman, The decidability of the existential theory of the poset of recursively enumerable degrees with jump relations, in preparation.
10. S. Lempp, M. Lerman, The decidability of the existential theory of the poset of the Turing degrees with jump and 0, in preparation.
11. M. Lerman, Admissible ordinals and priority arguments, in **[12]**, 311-344.
12. A.R.D. Mathias, H. Rogers, Jr., Cambridge Summer School in Mathematical Logic (editors), held in Cambridge (England), August 1-21, 1971, Lecture Notes in Mathematics No. 337, Springer-Verlag, Berlin, 1973.
13. A.A. Muchnik, On the unsolvability of the problem of reducibility in the theory of algorithms, Dokl. Akad. Nauk SSSR, N.S. 108 (1956), 194-197.
14. G.E. Sacks, On the degrees less than $\mathbf{0}'$, Ann. of Math. *(2) 77* (1963), 211-231.
15. G.E. Sacks, Recursive enumerability and the jump operator, Trans. Amer. Math. Soc. *108* (1963), 223-239.
16. G.E. Sacks, The recursively enumerable degrees are dense, Ann. of Math. *(2) 80* (1964), 300-312.
17. R.A. Shore, A non-inversion theorem for the jump operator, Ann. Pure Appl. Logic *40* (1988), 277-303.
18. R.I. Soare, The infinite injury priority method, J. Symbolic Logic *41* (1976), 513-530.
19. R.I. Soare, Recursively enumerable sets and degrees, Perspectives in Mathematical Logic, Omega Series, Springer-Verlag, Berlin, 1987.

On the Relationship between the Complexity, the Degree, and the Extension of a Computable Set

Wolfgang Maass*
Dept. of Math., Stat., and Comp. Sci.
University of Illinois at Chicago
Chicago, IL. 60680

Theodore A. Slaman**
Dept. of Mathematics
University of Chicago
Chicago, IL. 60637

ABSTRACT

We consider the equivalence relation $A =_C B$ ("A and B have the same time complexity") $\Leftrightarrow$ (for all time constructible $f: A \in DTIME(f) \Leftrightarrow B \in DTIME(f)$). In this paper we give a survey of the known relationships between this equivalence relation and degree theoretic and extensional properties of sets. Furthermore we illustrate the proof techniques that have been used for this analysis, with emphasis on those arguments that are of interest from the point of view of recursion theory. Finally we will discuss in the last section some open problems and directions for further research on this topic.

1. Introduction.

In the subsequent analysis of the fine structure of time complexity classes we consider the following set of time bounds:

$$T := \{f : \mathbf{N} \to \mathbf{N} \mid f(n) \geq n \text{ and } f \text{ is time constructible on a RAM}\}.$$

f is called time constructible on a RAM if some RAM can compute the function $1^n \mapsto 1^{f(n)}$ in $O(f(n))$ steps. We do not allow arbitrary recursive functions as time bounds in our approach in order to avoid pathological phenomena (e.g. gap theorems [HU], [6]). In this way we can focus on those aspects of complexity classes that are relevant for concrete complexity (note that all functions that are actually used as time bounds in the analysis of algorithms are time constructible).

We use the random access machine (RAM) with uniform cost criterion as machine model (see [3], [1], [15], [19]) because this is the most frequently considered model in

*Written under partial support by NSF-Grant CCR 8903398. Part of this research was carried out during a visit of the first author at the University. The first author would like to thank the Department of Computer Science at the University of Chicago for its hospitality.

**Written under partial support by Presidential Young Investigator Award DMS-8451748 and NSF-Grant DMS-8601856.

algorithm design, and because a RAM allows more sophisticated diagonalization - constructions than a Turing machine. It does not matter for the following which of the common versions of the RAM-model with instructions for ADDITION and SUBTRACTION of integers is chosen (note that it is common to exclude MULTIPLICATION of integers from the instruction set in order to ensure that the computation time of the RAM is polynomially related to time on a Turing machine). In order to be specific we consider the RAM model as it was defined by Cook and Reckhow [3] (we use the common "uniform cost criterion" [1], i.e. $l(n) = 1$ in the notation of [3]). This model consists of a finite program, an infinite array $X_0, X_1, \ldots$ of registers (each capable of holding an arbitrary integer), and separate one-way input- and output-tapes. The program consists of instructions for ADDITION and SUBTRACTION of two register contents, the conditional jump "TRA m if $X_i > 0$" which causes control to be transferred to line m of the program if the current content of register X_i is positive, instructions for the transfer of register contents with indirect addressing, instructions for storing a constant, and the instruction "READ X_i" (transfer the content of the next input cell on the input-tape to register X_i) and "PRINT X_i" (print the content of register X_i on the next cell of the output-tape).

The computation time of a RAM for input x is the number of instructions that it executes for this input. One says that a RAM is of time complexity f if for every $n \in \mathbf{N}$ and every $x \in \{0,1\}^n$ its computation time for input x is $\leq f(n)$. The relationship between computation time on RAM's and Turing machines is discussed in [3] (Theorem 2), [1] (section 1.7), and [19] (chapter 3). It is obvious that a multi-tape Turing machine of time complexity $t(n)$ can be simulated by a RAM of time complexity $O(t(n))$. With a little bit more work (see [19]) one can construct a simulating RAM of time complexity $O(n + (t(n)/\log t(n)))$ (assuming that the output has length $O(n + (t(n)/\log t(n)))$). We define

$$DTIME(f) := \left\{ A \subseteq \{0,1\}^* \;\middle|\; \begin{array}{l} \text{there is a RAM of time complexity} \\ O(f) \text{ that computes } A \end{array} \right\}.$$

For A and B contained in $\{0,1\}^*$ (the set of all finite binary sequences) we define

$$A \leq_C B :\Leftrightarrow \forall f \in T(B \in DTIME(f) \Rightarrow A \in DTIME(f))$$

and

$$A =_C B :\Leftrightarrow \forall f \in T(A \in DTIME(f) \Leftrightarrow B \in DTIME(f)).$$

Intuitively, $A =_C B$ if A and B have the same deterministic time complexity. The $=_C$-equivalence classes are called *complexity types*. We write $\mathbf{0}$ for $DTIME(n)$, which is the smallest complexity type. Note that for every complexity type $\mathcal{C}$ and every $f \in T$ one has either $\mathcal{C} \subseteq DTIME(f)$ or $\mathcal{C} \cap DTIME(f) = \emptyset$.

Remark. Geske, Huynh and Selman [5] have considered the related partial order "the complexity of A is polynomially related to the complexity of B", and the associated equivalence classes ("polynomial complexity degrees"). Our results for complexity types (e.g. Theorems 5 and 6 below) provide corresponding results also for polynomial complexity degrees.

In order to prove results about the structure of complexity types one needs a technique to construct a set within a given time complexity while simultaneously controlling its other properties. It is less difficult to ensure that a constructed set is of complexity type $\mathcal{C}$ if one can associate with $\mathcal{C}$ an "optimal" time bound $f_{\mathcal{C}} \in T$ such that for all sets $X \in \mathcal{C}$

$$\{f \in T \mid X \in DTIME(f)\} = \{f \in T \mid f = \Omega(f_{\mathcal{C}})\}.$$

In this case, we call $\mathcal{C}$ a *principal* complexity type. Blum's speed-up theorem [2] asserts that there are complexity types that are not principal. For example, there is a complexity type $\mathcal{C}$ such that for every $X \in \mathcal{C}$,

$$\{f \in T \mid X \in DTIME(f)\} = \left\{f \in T \mid \exists i \in \mathbf{N}\left(f(n) = \Omega\left(\frac{n^2}{(\log n)^i}\right)\right)\right\}.$$

Note that this effect occurs even if one is only interested in time constructible time bounds (and sets X of "low" complexity).

In order to prove our results also for complexity types which are non-principal, we show that in some sense the situation of Blum's speed- up theorem (where we can characterize the functions f with $X \in DTIME(f)$ with the help of a "cofinal" sequence of functions) is already the worst that can happen (unfortunately this is not quite true, since we cannot always get a cofinal sequence of functions f_i with the same nice properties as in Blum's speed-up theorem). More precisely: we will show that each complexity type can be characterized by a cofinal sequence of time bounds with the following properties.

Definition. $(t_i)_{i\in\mathbf{N}} \subseteq \mathbf{N}$ is called a *characteristic sequence* if $t : i \mapsto t_i$ is recursive and
(1) $\forall i \in \mathbf{N}(\{t_i\} \in T$ and program t_i is a witness for the time-constructibility of $\{t_i\}$);
(2) $\forall i, n \in \mathbf{N}(\{t_{i+1}\}(n) \leq \{t_i\}(n))$.

Definition. Assume that $A \subseteq \{0,1\}^*$ is an arbitrary set and $\mathcal{C}$ is an arbitrary complexity type. One says that $(t_i)_{i\in\mathbf{N}}$ is *characteristic for* A if $(t_i)_{i\in\mathbf{N}}$ is a characteristic sequence and

$$\forall f \in T(A \in DTIME(f) \Leftrightarrow \exists i \in \mathbf{N}(f(n)) = \Omega(\{t_i\}(n)))).$$

One says that $(t_i)_{i\in\mathbf{N}}$ is *characteristic for* $\mathcal{C}$ if $(t_i)_{i\in\mathbf{N}}$ is characteristic for some $A \in \mathcal{C}$ (or equivalently: for all $A \in \mathcal{C}$).

Remark. The idea of characterizing the complexity of a recursive set by a sequence of "cofinal" complexity bounds is rather old (see e.g. [17], [10], [LY], [20], [MW]). However none of our predecessors exactly characterized the time complexity of a recursive set in terms of a uniform cofinal sequence of time constructible time bounds. This is the form of characterization which we employ in later proofs. [10] and [MW] give corresponding results for space complexity of Turing-machines. Their results exploit the linear speed-up theorem for space complexity on Turing-machines, which is not available for time complexity on RAM's. Time complexity on RAM's has been considered in [20], but only sufficient conditions are given for the cofinal sequence of time bounds (these conditions are stronger than ours). The more general results on complexity sequences in axiomatic complexity theory ([17], [20]) involve "overhead functions", or deal with nonuniform sequences, which makes the specialization to the notions considered here impossible. Because of the lack of a fine time hierarchy theorem for multi-tape Turing-machines, it is an open problem whether one can give a similar characterization for the Turing machine time complexity of a recursive set.

The relationship between complexity types and characteristic sequences is clarified in Section 2 of this paper. Theorem 1 states that one can associate with every complexity type $\mathcal{C}$ of recursive sets a sequence $(t_i)_{i\in\mathbf{N}}$ that is characteristic for $\mathcal{C}$. This fact will be used in most of the subsequent results. We show in Theorem 2 that the converse of Theorem 1 is also true: for every characteristic sequence $(t_i)_{i\in\mathbf{N}}$ there exists a complexity type $\mathcal{C}$ such that $(t_i)_{i\in\mathbf{N}}$ is characteristic for $\mathcal{C}$. We give a complete proof of Theorem 2, since this proof provides the simplest example of the new finite injury priority argument that occurs as sub-strategy in the proofs of most of our subsequent results.

As an immediate consequence of the proofs of these results we show in Theorem 3 that every complexity type contains a sparse set. As another consequence we get in Theorem 4 that the complexity types of computable sets form a lattice under the partial order $\leq_C$.

In Section 3 we show that with the help of these tools one is able to prove sharper versions of various familiar results about polynomial time degrees. It is shown in Theorem 5 that every complexity type outside of P contains sets of incomparable polynomial time Turing degree. In Theorem 6 we construct a complexity type that contains a minimal pair of polynomial time Turing degrees. A comparison of the proofs of these results with the proofs of the related results by Ladner [7] and Landweber, Lipton, Robertson [8] shows that the sharper versions which are considered here pose a more serious challenge to our construction techniques (we apply finite injury priority arguments and a constructive version of Cohen forcing).

In Section 4 we use the concepts and techniques that are introduced in this paper for an investigation of the fine structure of P. We show in Theorem 7 that in P each complexity type $\mathcal{C} \neq \mathbf{0}$ contains a rich structure of linear time degrees, and we show in Theorem 8 that these degree structures are not all isomorphic (in particular we characterize those C that have a maximal linear time degree). The proofs of these two theorems provide evidence that finite injury priority arguments are not only relevant for the investigation of sets "higher up", but also for the analysis of the finite structure of P.

In Section 5 we study the partial order of sets in a given complexity type under inclusion (modulo finite sets). We show that this partial order is dense, and we prove a splitting theorem for arbitrary sets in any given complexity type.

Finally, in Section 6 we will discuss some open problems.

The main results of this paper have previously been reported in the extended abstracts [12], [13].

2. The Relationship between Complexity Types and Characteristic Sequences.

Theorem 1. ("inverse of the speed-up-theorem for time on RAM's"). *For every recursive set A there exists a sequence $(t_i)_{i\in\mathbf{N}}$ that is characteristic for A.*

We refer to [14] for a proof of this result (the technique of this proof is not needed for the proofs of the subsequent results.)

In the proof of Theorem 8 we will need the following stronger version of Theorem 1.

Corollary 1. For every recursive set A there exist recursive sequences $(t_i)_{i\in\mathbf{N}}$, $(c_i)_{i\in\mathbf{N}}$, $(d_i)_{i\in\mathbf{N}}$, $(e_i)_{i\in\mathbf{N}}$ such that

$(t_i)_{i\in\mathbf{N}}$ is characteristic for A

$\{t_i\}(n)$ converges in $\leq c_i \cdot \{t_i\}(n)$ steps for all $i, n \in \mathbf{N}$

$\{e_i\} =^* A$ and $\{e_i\}(x)$ converges in $\leq d_i \cdot \{t_i\}(|x|)$ steps, for all $i \in \mathbf{N}$, $x \in \{0,1\}^*$.

□

Theorem 2. ("refinement of the speed-up-theorem for time on RAM's"). *For every characteristic sequence $(t_i)_{i\in\mathbf{N}}$ there is a set A such that $(t_i)_{i\in\mathbf{N}}$ is characteristic for A.*

PROOF: Fix an arbitrary characteristic sequence $(t_i)_{i\in\mathbf{N}}$, and a sequence $(K_i)_{i\in\mathbf{N}}$ of numbers such that $\{t_i\}(n)$ converges in $\leq K_i \cdot \{t_i\}(n)$ steps, for all $i, n \in \mathbf{N}$.

The claim of the theorem does not follow from any of the customary versions of Blum's speed-up-theorem [2], because it need not be the case that $\{t_i\} = o(\{t_{i-1}\})$. In fact it may occur that $\{t_i\} = \{t_{i-1}\}$ for many (or even for all) $i \in \mathbf{N}$. Even worse, one may have that $K_i \cdot \{t_i\}(n) > K_{i-1} \cdot \{t_{i-1}\}(n)$ for many $i, n \in \mathbf{N}$ (this may occur for example in the characteristic sequences that arise from the construction of Theorem 1 if i does not encode a "faster" algorithms $(i)_0$ of time complexity $\{(i)_1\}(n) < \{t_{i-1}\}(n)$; in this case one may have that $\{t_i\}(n) = \{t_{i-1}\}(n)$ and the computation of $\{t_i\}(n)$ takes longer than the computation of $\{t_{i-1}\}(n)$ because the former involves simulations of both $\{t_{i-1}\}(n)$ and $\{(i)_1\}(n)$). Therefore it is more difficult than in Blum's speed-up-theorem to ensure in the following proof that the constructed set A satisfies $A \in DTIME(\{t_i\})$ for every $i \in \mathbf{N}$. To achieve $A \in DTIME(\{t_i\})$ it is no longer enough to halt for all $j \geq i$ the attempts towards making $A(x) \neq \{j\}(x)$ after $\{t_j\}(|x|)$ steps, because the value of $\{t_j\}(|x|)$ may only be known after $K_j \cdot \{t_j\}(|x|)$ steps and it is possible that

$$\lim_{j,n\to\infty} \frac{K_j \cdot \{t_j\}(n)}{\{t_i\}(n)} = \infty.$$

Instead, in the following construction one simulates together with $\{j\}(x)$ and $\{t_j\}(|x|)$ also the computations $\{t_0\}(|x|), \ldots, \{t_{j-1}\}(|x|)$. That one of these $j+2$ computations which converges first will halt the computation of $A(x)$. It is somewhat delicate to implement these simultaneous simulations of $j+2$ computations (where j grows with $|x|$) in such a way that for each fixed $i \in \mathbf{N}$ the *portion* of the total computation time for $A(x)$ that is devoted to the simulation of $\{t_i\}(|x|)$ does not shrink when j grows to infinity (obviously this property is needed to prove that $A \in DTIME(\{t_i\})$).

In the following we construct a RAM R that computes a set A for which $(t_i)_{i\in\mathbf{N}}$ is a characteristic sequence. We fix some coding of RAM's by binary strings analogously as in [3]. We arrange that each code for a RAM is a binary string without leading zeros, so that one can also view it as binary notation for some number $j \in \mathbf{N}$. We assume that the empty string codes the "empty" RAM (which has no instructions). Thus one can associate with each binary string x the longest initial segment of x that is a code for a RAM, which will be denoted by j_x in the following.

The RAM R with input $x \in \{0,1\}^*$ acts as follows. It first checks via "looking back" for $|x|$ steps whether the requirement "$A \neq \{j\}$" for $j := j_x$ has already been satisfied during the first $|x|$ steps of the construction (for shorter inputs). If the answer is "yes", R decides without any further computation that $x \notin A$. If the answer is "no", R on input x dovetails the simulations of $t_0, \ldots, t_j$ on input $|x|$ and of j on input x in the following manner. The simultaneous simulation proceeds in phases $p = 0, 1, 2, \ldots$ At the beginning of each phase p the RAM R simulates one more step of j on input x. If m_j is the number of steps that R has used to simulate this step of j on input x, then R simulates subsequently m_j steps of t_j on input $|x|$. If m_{j-1} is the number of steps that R has used so far in phase p (for the simulation of one step of j on x and m_j steps of t_j on $|x|$), then R simulates subsequently m_{j-1} steps of t_{j-1} on input $|x|$; etc. If the number of steps that R spends in phase p exceeds $|x|$, R immediately halts and rejects x.

If R has finished phase p (by simulating m_0 steps of t_0 on input $|x|$, where m_0 is the total number of steps which R has spent in phase p for the simulation of $t_1, \ldots, t_j$ on $|x|$ and j on x) and none of the $j+2$ simulated computations has reached a halting state during phase p, then R proceeds to phase $p+1$.

If the computation of j on input x has converged during phase p then R puts x into A if and only if $\{j\}(x) = 0$. With this action the requirement "$A \neq \{j\}$" becomes satisfied and the computation of R on input x is finished.

If j on input x has not converged during phase p, but one of the other $j+1$ simulated computations (of $t_0, \ldots, t_j$ on $|x|$) converges during phase p, then R decides that $x \notin A$ and halts.

We now describe some further details of the program of R in order to make a precise time analysis possible.

In order to avoid undesirable interactions R reserves for each simulated RAM a separate infinite sequence ARRAY$_k$ of registers, where ARRAY$_k$ consists of the registers

$$X_{2^{k+1}\cdot(2j+1)}, j = 0, 1, 2, \ldots$$

Furthermore we specify for each $k \in \mathbf{N}$ a sub-array MEMORY$_k$ of ARRAY$_k$, consisting of the registers $X_{2^{k+1}\cdot(2j+1)}$ for $j = 0, 2, 4, \ldots$ in ARRAY$_k$. For each $k \in \mathbf{N}$ the constructed RAM R uses ARRAY$_k$ for the simulation of the RAM coded by t_k (more precisely: R uses MEMORY$_k$ to simulate the registers of the RAM t_k, and it uses the remaining registers in ARRAY$_k$ to store the program that is coded by t_k). Furthermore R uses ARRAY$_{j+1}$ to simulate the RAM coded by $j = j_x \in \mathbf{N}$ (this does not cause a conflict since j and t_{j+1} are never simultaneously simulated).

We have made these arrays explicit because in order to simulate a step of the k-th one of these machines that involves, say, the i-th register of that machine, R has to compute on the side (via iterated additions) the "real" address $2^{k+1} \cdot (2 \cdot 2i + 1)$ of the corresponding register in MEMORY$_k$. The number of steps required by this side-computation depends on k, but is independent of i. Therefore there exist constants $c(k), k \in \mathbf{N}$, that bound the number of steps that R has to spend to simulate a single step of the k-th one of these machines. It is essential that for any fixed $k \leq j_x + 1$ this constant $c(k)$ is independent of the number $j_x + 2$ of simulated machines for input x (this follows from the fact that ARRAY$_k$ is independent of j_x).

At the beginning of its computation on input x the RAM R computes $j := j_x$. Then R "looks back" for $|x|$ steps to check whether "$A \neq \{j\}$" has already been satisfied at some argument x' such that j_x is a prefix of x' and x' is a prefix of x. Let $x_1, \ldots, x_k$ be a list of all such binary strings x' (ordered according to their length). We implement the "looking back" for argument x as follows. For each of the shorter inputs $x_1, x_2, \ldots$ (one after the other) R carries out the same computation as discussed below for x. If it turns out after $|x|$ steps of this subcomputation that for one of these arguments x_i ($i \in \{1, \ldots, k\}$) the construction satisfied the requirement $A \neq \{j\}$, then R immediately halts and rejects x (note that for the first such x_i we actually have then $A(x_i) \neq \{j\}(x_i)$). Otherwise R continues the computation on input x as follows. R calls a program for the recursive function $i \mapsto t_i$ and uses it to compute the numbers $t_0, \ldots, t_j$. Afterwards

R "decodes" each of the programs t_i ($i \in \{0, \dots, j\}$), which are given as binary string (the binary representations of the number $t_i \in \mathbf{N}$). R uses for each instruction S of program t_i 4 registers in ARRAY$_i$ (that are not in MEMORY$_i$) to store the opcode and the up to three operands of S (similarly as in the proof of Theorem 3 in [3]). In the same way R "decodes" the program j.

If this preprocessing phase of R on input x takes more than $100 \cdot |x|$ steps (in addition to the steps spend on "looking back") then R immediately halts and rejects x.

During the main part of its computation (while it simulates $t_0, \dots, t_j, j$) R maintains in its odd-numbered registers (which do not belong to any of the arrays ARRAY$_k$) two counters that count the total number of steps that have been spent so far in the current phase p, as well as the number of steps of the currently considered program that have already been simulated during phase p. In order to allow enough time for the regular updating of both counters, as well as for the regular comparison of the values of both counters with the preset thresholds $(|x|, m_i)$, it is convenient to add every c steps the number c to each counter (where $c \in \mathbf{N}$ is a sufficiently large constant in the program of R).

In addition R stores in its odd-numbered registers the input x, j_x, the number of the currently simulated program, and for each of the simulated programs the address of the register that contains the opcode for the next instruction that has to be executed for that program. With this information R can resume the simulation of an earlier started program without any further "overhead steps" because each simulated program only acts on the registers in its "own" array ARRAY$_k$. Of course R always has to spend several steps to simulate a single instruction of any of the stored programs $t_0, \dots, t_j, j$. It has to apply a series of branching instructions in order to go from the stored (numerical) value of the opcode in ARRAY$_k$ to the actual instruction in its own program that corresponds to it, and it has to calculate the "real" addresses of the involved registers in MEMORY$_k$. However it is obvious that for each simulated program k there exists a constant $c(k)$ (independent from $j = j_x$ and x) that bounds the number of steps that R has to spend to simulate a single instruction of program k.

We now verify that $(t_i)_{i \in \mathbf{N}}$ is characteristic for A.

Claim 1. $A \in DTIME(\{t_i\})$ for every $i \in \mathbf{N}$.

PROOF: Fix i and set $S_i := \{x \in \{0,1\}^* \mid j_x < i\}$. Every $\tilde{x} \in S_i \cap A$ is placed into A in order to satisfy the requirement "$A \neq \{j\}$" for some $j < i$. We then have $A(\tilde{x}) \neq \{j\}(\tilde{x})$, and for sufficiently longer inputs x with $j_x = j$ the constructed RAM R finds out during the first part of its computation on input x (while "looking back" for $|x|$ steps) that the requirement "$A \neq \{j\}$" has already been satisfied. This implies that $x \notin A$. Thus for each $j < i$ only finitely many $\tilde{x}$ are placed into A in order to satisfy the requirement "$A \neq \{j\}$". Therefore $S_i \cap A$ is finite. Since $S_i \in DTIME(n)$ it only remains to prove that $\overline{S}_i \cap A \in DTIME(\{t_i\})$.

We show that R uses for every input $x \in \overline{S}_i$ at most $O(\{t_i\}(|x|))$ computation steps. By assumption we have $\{t_i\}(|x|) \geq |x|$, and therefore R uses only $O(\{t_i\}(|x|))$ steps in its preprocessing phase for input x.

We had fixed constants $K_i \in \mathbf{N}$ such that the computation of program t_i on input n consists of $\leq K_i \cdot \{t_i\}(n)$ steps (for all $n \in \mathbf{N}$). By construction the total number of steps that R spends for input x on the simulation of $j := j_x$ on x and of $t_{i+1}, \ldots, t_j$ on $|x|$ is bounded by $|x| + T_{i,x}$, where $T_{i,x}$ is the number of computation steps of program t_i on input $|x|$ that are simulated by R on input x. By construction we have $T_{i,x} \leq K_i \cdot \{t_i\}(|x|) + |x|$ (because the computation of R on x is halted at the latest at the end of the first phase p where $\{t_i\}(|x|)$ is seen to converge, and no phase p consists of more than $|x|$ steps of R).

Furthermore we know that there is a constant $c(t_i)$ that bounds the number of steps that R needs to simulate a single instruction of t_i. Thus R spends $\leq c(t_i) \cdot T_{i,x} = O(\{t_i\}(|x|))$ steps on the simulation of $t_{i+1}, \ldots, t_j$ on $|x|$ and j on x. Furthermore the number of "overhead steps" of R for the updating of counters, the comparison of their values with preset thresholds, and the switching of programs can be bounded by a constant times the number of steps that R spends on the actual simulations. Thus it just remains to be shown by induction on $i - k$ that R on input $x \in \overline{S}_i$ spends for every $k < i$ altogether at most $O(\{t_i\}(|x|))$ steps on the simulation of t_k on $|x|$. However this follows immediately from the construction, using the definition of the parameters m_k and the observation that the constants $c(t_k)$ do not depend on x.

Claim 2. Let $U(j, n)$ be the maximal number of steps that program j uses on an input of length n. Then:

$$A = \{j\} \Rightarrow U(j,n) = \Omega(\{t_j\}(n)).$$

PROOF: Fix any $j \in \mathbf{N}$ such that $A = \{j\}$. Assume for a contradiction that it is not the case that $U(j,n) = \Omega(\{t_j\}(n))$, i.e. we have $\sup_n \dfrac{\{t_j\}(n)}{U(j,n)} = \infty$. We show that then the requirement "$A \neq \{j\}$" gets satisfied at some argument x with $j_x = j$.

Obviously we have for all sufficiently long x with $j_x = j$ that the computation of R on x does not get halted prematurely because the preprocessing phase takes too many steps, or because a phase p in the main part of the computation requires more than $|x|$ steps. Furthermore for each $i \leq j$ there exists by construction a constant c_i such that for all x with $j_x = j$ R simulates for each step in the computation of j on x at most c_i steps in the computation of t_i on input $|x|$. Therefore our assumption $\sup_n \dfrac{\{t_j\}(n)}{U(j,n)} = \infty$ together with the fact that $\{t_0\}(n) \geq \ldots \geq \{t_j\}(n)$ for all $n \in \mathbf{N}$ implies that there is some x with $j_x = j$ so that R on input x does not halt prematurely because some $\{t_i\}(|x|)$ with $i \leq j$ is seen to converge before $\{j\}(x)$ is seen to converge. For such input x the RAM R succeeds in satisfying the requirement "$A \neq \{j\}$" by setting $A(x) \neq \{j\}(x)$. This contradiction completes the proof of Claim 2.

Claim 1 and Claim 2 together imply the claim of the theorem. □

As an immediate consequence we get from the preceding two theorems the following result (recall that $S \subseteq \{0,1\}^*$ is called **sparse** if there is a polynomial p such that $\forall n(|\{x \in S \mid |x| = n\}| \leq p(n))$; see [16] for a contrasting result about sparse sets).

Theorem 3. *Every complexity type contains a sparse set.*

PROOF: Let C be an arbitrary complexity type. By Theorem 1 there exists some sequence $(t_i)_{i\in\mathbf{N}}$ which is characteristic for C. In order to get a sparse set S such that $(t_i)_{i\in\mathbf{N}}$ is characteristic for S we use a variation of the proof of Theorem 2. In this variation of the construction one never places x into the constructed set unless $|j_x| \leq \log|x|$.

□

Remarks.

1. It is an open problem whether every complexity type contains a tally set.
2. Further results about the relationship between extensional properties of a set and its complexity type can be found in Section 5.

It is obvious that the partial order $\leq_C$ on sets (which was defined in Section 1) induces a partial order $\leq_C$ on complexity types. In this paper we are not concerned with the structure of this partial order $\leq_C$, however we want to mention the following immediate consequence of Theorems 1 and 2.

Theorem 4. *The complexity types of computable sets with the partial order $\leq_C$ form a lattice. Furthermore, if the characteristic sequence $(t'_i)_{i\in\mathbf{N}}$ is characteristic for the complexity type C' and the characteristic sequence $(t''_i)_{i\in\mathbf{N}}$ is characteristic for the complexity type C'', then some sequence $(t_i^{\min})_{i\in\mathbf{N}}$ with $\{t_i^{\min}(n)\} = \min(\{t'_i\}(n), \{t''_i\}(n))$ is characteristic for the infimum $C' \wedge C''$ of C', C'', and some sequence $(t_i^{\max})_{i\in\mathbf{N}}$ with $\{t_i^{\max}\}(n) = \max(\{t'_i\}(n), \{t''_i\}(n))$ is characteristic for the supremum $C' \vee C''$ of C', C''.*

PROOF: The key fact for the proof is the following elementary observation. Consider any two sets $T, S \subseteq \{0,1\}^*$ and characteristic sequences $(t_i)_{i\in\mathbf{N}}$ and $(s_i)_{i\in\mathbf{N}}$ such that $(t_i)_{i\in\mathbf{N}}$ is characteristic for T and $(s_i)_{i\in\mathbf{N}}$ is characteristic for S. Then

$$T \geq_C S \Leftrightarrow \forall i \exists j(\{t_i\} = \Omega(\{s_j\})).$$

In order to prove the claim of the theorem one first has to verify that one can find programs $t_i^{\min}$, $t_i^{\max}$ for $\min(\{t'_i\}, \{t''_i\})$, respectively $\max(\{t'_i\}, \{t''_i\})$, so that $(t_i^{\min})_{i\in\mathbf{N}}$ and $(t_i^{\max})_{i\in\mathbf{N}}$ are characteristic sequences. The only nontrivial point is the requirement to define the recursive function $i \mapsto t_i^{\min}$ in such a way that for all $i \in \mathbf{N}$, $\{t^{\min}\}(n) = \min(\{t'_i\}(n), \{t''_i\}(n))$ and $t_i^{\min}$ is a witness for the time constructibility of this function. In order to achieve this, it is essential that program $t_i^{\min}$ "knows" time constructibility factors c'_i, c''_i such that $\{t'_j\}(n)$ converges in $\leq c'_i \cdot \{t'_i\}(n)$ steps and $\{t''_i\}(n)$ converges in $\leq c_i'' \cdot \{t_i''\}(n)$ steps. Since program $t_i^{\min}$ has to compute $\min(\{t_i'\}(n), \{t_i''\}(n))$ in a time constructible fashion, it needs c_i', c_i'' in order to know when it is "safe" to abandon the simulation of the longer one of the two computations $\{t_i'\}(n)$, $\{t_i''\}(n)$ (after the shorter one has converged). But this is no problem, since the proof of Theorem 1 shows (as in the corollary to Theorem 1) that one can assume without loss of generality that recursive sequences $(c_i')_{i\in\mathbf{N}}$, $(c_i'')_{i\in\mathbf{N}}$ of time constructibility factors are given together with the characteristic sequences $(t_i')_{i\in\mathbf{N}}$ and $(t_i'')_{i\in\mathbf{N}}$.

Since $(t_i^{\min})_{i\in\mathbf{N}}$ is a characteristic sequence, there exists by Theorem 2 a set $T^{\min} \subseteq \{0,1\}^*$ such that $(t_i^{\min})_{i\in\mathbf{N}}$ is characteristic for $T^{\min}$. Let $\mathcal{C}^{\min}$ be the complexity type of $T^{\min}$. It is obvious that $\mathcal{C}^{\min} \leq_C \mathcal{C}'$ and $\mathcal{C}^{\min} \leq_C \mathcal{C}''$. In order to show that $\mathcal{C}^{\min}$ is the infimum of $\mathcal{C}'$ and $\mathcal{C}''$, one has to verify for an arbitrary complexity type $\mathcal{C}$ with $\mathcal{C} \leq_C \mathcal{C}'$ and $\mathcal{C} \leq_C \mathcal{C}''$ that $\mathcal{C}^{\min} \geq_C \mathcal{C}$. Let the sequence $(t_i)_{i\in\mathbf{N}}$ be characteristic for $\mathcal{C}$ ($(t_i)_{i\in\mathbf{N}}$ exists by Theorem 1). The key fact at the beginning of this proof implies that $\forall i \exists j(\{t_i'\} = \Omega(\{t_j\}))$ and $\forall i \exists j(\{t_i''\} = \Omega(\{t_j\}))$. Hence $\forall i \exists j(\{t_i^{\min}\} = \Omega(\{t_j\}))$, which implies that $\mathcal{C}^{\min} \geq_C \mathcal{C}$.

One verifies analogously that the complexity type $\mathcal{C}^{\max}$ which is defined by $(t_i^{\max})_{i\in\mathbf{N}}$ is the supremum of $\mathcal{C}'$, $\mathcal{C}''$ with regard to $\leq_C$. □

3. Time Complexity versus Polynomial Time Reducibility.

In this section we study the relationship between the complexity type of a set and its polynomial time Turing degree. We first introduce the customary recursion theoretic vocabulary for the discussion of priority constructions. One important part of the following constructions is the construction technique of Theorem 2, which allows us to control the complexity type of a set that is constructed to meet various other requirements. Therefore we will review briefly the construction of Theorem 2 in recursion theoretic terms.

A *stage* is just an integer s viewed in the context of a definition by recursion. A *strategy* is just an algorithm to determine the action taken during stage n, recursively based on various parameters of the construction. Typically, a strategy is used to show that the sets being constructed have some specific property; we call this a *requirement*. We organize our attempts to satisfy the requirements in some order which we call the *priority ordering*. If requirement Q comes before requirement R, we say that Q has *higher priority* and R has *lower priority*.

In the construction of Theorem 2 we built A to satisfy two families of requirements. For each i, A had to be computable on a RAM, running in time $O(\{t_i\})$. Second, any RAM that computed A had to operate with time $\Omega(\{t_i\})$, for some i. We assigned priority by interleaving the two types of requirements in order of their indices.

For an element of the first family of requirements, we used a strategy which imposed a sequence of time controls on the construction. The behavior of the ith strategy was to terminate all action of lower priority in determining $A(x)$ once the construction had executed sufficiently many phases to exceed $\{t_i\}(|x|)$ many steps. Suppose that strategies of higher priority only act finitely often to cause A to accept strings. Then, using the finite data describing the higher priority activity we can correctly compute A at x in time $O(\{t_i\})$ by first checking the data for an answer at x. If the value of A at x is not included in the data then we run the construction until the ith time control strategy calls a halt to lower priority activity. We then read off the answer.

For each of the second family of requirements, we used a diagonalization strategy. Namely, if $\{i\}$ ever converges at an argument x before a strategy of higher priority terminates our action then we define A at x to make $A(x) \neq \{i\}(x)$. By our association of at most one diagonalization strategy to each string, the two possibilities were for $\{i\}$ to disagree with A or to have running time $\Omega(\{t_i\})$. The latter being the case when diagonalization is impossible since the time control associated with higher priority strategies terminates the diagonalization attempt at every string.

We collectively refer to the time control and diagonalization strategies as the *$\mathcal{C}$-strategies.*

The proof that the sequence $(t_j)_{j\in\mathbf{N}}$ is characteristic for A has two essential features. The first is that each of the constituent strategies is injured finitely often. The jth time control strategy s_j is injured when the value of A at x is determined differently from the one assumed by s_j. In the proof of Theorem 2, this occurs when the value of A is determined by a diagonalization strategy with index less than j. The second feature, which occurs on a higher level, is that no single move in the construction prohibits the subsequent application and complete implementation of further time control and diagonalization strategies.

With this in mind, we can look for other families of strategies, which are compatible with the $\mathcal{C}$-strategies, to produce interesting examples within a given complexity type. Our next result compares complexity with relative computability.

A Turing reduction is given by a RAM M, augmented with the ability to query an oracle as to whether it accepts the query string. We evaluate M relative to A by answering all queries with the value of A on the query string. M specifies a polynomial time Turing reduction if there is a polynomial g such that for every oracle A and every string x, the evaluation of M with input x halts in less than $g(|x|)$ steps. Say that B is polynomial time Turing reducible to A, if there is a polynomial time Turing reduction that, when evaluated on a string x relative to A returns value $B(x)$. We use the term *Turing* reduction to avoid confusion with the related notion of many-one reduction. Note, the choice of RAM's in the definition of polynomial time Turing reduction is not important; for example, the same class is obtained using any of a variety of machine models.

Theorem 5. *A complexity type C contains sets A and B that are incomparable with regard to polynomial time Turing reductions if and only if $C \not\subseteq P$.*

PROOF: Let C be fixed and let $(t_i)_{i \in \mathbf{N}}$ be characteristic for C. We build A and B and use the C-strategies to ensure that A and B belong to C. In addition, we ensure for each polynomial time Turing reduction $\{e\}$,

$$(3.1) \qquad \begin{aligned} \{e\}(A) = B &\Longrightarrow C \subseteq P \\ \{e\}(B) = A &\Longrightarrow C \subseteq P. \end{aligned}$$

We will describe the strategies for the new requirement, the *inequality strategies.* In fact, since they are symmetric, we only describe the strategy to ensure the first of the two implications in 3.1. These strategies are combined with the earlier ones using the same combinatorial pattern as before.

We describe the eth new strategy. We first arrange, by a variation of looking back, that no strategy of lower priority is implemented until we have established $\{e\}(A, x) \neq B(x)$ for some specific string x. Let l_0 be the greatest length of a string which is accepted into A or B by the effect of a strategy of higher priority. Let A_0 and B_0 be A and B restricted to strings of length less than or equal to l_0. Again, by looking back, we may assume that A_0, B_0 and l_0 are known. For each length l, we use a string x_l of length l to attempt to establish the inequality. (The choice of x_l is made to ensure that x_l is will not even potentially be used by some strategy of higher priority.) We simulate the computation relative to the oracle that is equal to A_0 on strings of length less than l_0 and empty elsewhere. If we are able to complete the simulation without being canceled by a time control strategy of higher priority, then we define $B(x_l)$ to disagree with the answer returned by the simulation.

There are two possible outcomes for this strategy. We could succeed in establishing the inequality between $\{e\}(A)$ and B. In this case, the inequality strategy is compatible with the C-strategies. It only requires that finitely may strings are in B and it places no permanent impediment to the implementation of all of the C-strategies. These are the two features we already isolated as determining compatibility.

On the other hand, the inequality between $\{e\}(A)$ and B might never be established. This can only occur if for every l, the inequality strategy is terminated before completion of its simulation of $\{e\}(A, x_l)$ by some time control strategy of higher priority. But then the time control imposed by that higher priority strategy must be polynomial, since it is bounded by a constant times the running time of $\{e\}$. In this case, C is contained in P.

Thus, either the inequality strategies are compatible with the C-strategies and we may use the framework developed earlier to build A and B in C of incomparable polynomial time Turing degree or C is contained in P. □

The inequality strategies are, in some ways, simpler than the C-strategies. In the style of Ladner's early constructions [7], they act to establish an inequality during the first opportunity to do so. If no opportunity arises, we conclude that C is contained in P. On the other hand, these strategies have a feature not appearing in the earlier argument. If an inequality strategy never finds a string at which it can establish the desired inequality, then it completely halts the implementation of lower priority strategies. This behavior is acceptable in the context of our construction, since if it occurs then we may conclude that the theorem is true for a fairly trivial reason. Of course, these types of strategies appear in basic looking back constructions. Here, we interleaved them with other finite injury strategies.

Minimal pairs. In Theorem 5, we gave a construction showing that in every complexity type there is a pair of sets which are polynomial time Turing incomparable. Thus, in the sense of Theorem 5, complexity type is never directly tied to relative computability. In this section, we ask whether there is any correlation between informational content as expressed by polynomial time Turing degree and complexity type. We give a partial negative answer in Theorem 6.

Theorem 6. *There is a complexity type C which contains sets A and B that form a minimal pair with regard to polynomial time Turing reductions (i.e. $A, B \notin P$, but for all X, $X \leq_p A$ and $X \leq_p B$ implies that $X \in P$).*

PROOF: We build A and B by recursion. A *condition* is a specification of an oracle on all strings whose lengths are bounded by a fixed integer; i.e. a finite approximation to an oracle. For U a condition, let $domain(U)$ denote the set of strings on which U is defined. During stage n, we specify conditions A_n and B_n on A and B with domains at least including $\{0,1\}^{\leq n}$ so that $A_{n-1} \subseteq A_n$ and $B_{n-1} \subseteq B_n$.

Readers familiar with recursion theoretic forcing and priority methods (see [9]) will recognize our building a pair of Cohen generic subsets of $\{0,1\}^*$. Their mutual genericity with respect to polynomial time Turing reductions implies that they form a minimal pair of P-degrees. We use the priority method to arrange that A and B meet enough dense sets for genericity to apply.

The fact that A and B are recursive follows from the observation that the recursion step in the generation of A_n and B_n is computable. In fact, there is a RAM that implements this recursion. When we speak of a step in the construction we are referring to a step in the execution of this RAM.

In the construction, we take steps to ensure that A and B satisfy the claims of the theorem by ruling out each possible counter example. Thus, we individually satisfy the following individual requirements.

G_d. If $\{d\} = A$ then there is a RAM that runs at least as fast as $\{d\}$ and computes B. Similarly, for A and B with roles reversed.

$H_{e,f}$. If $\{e\}$ and $\{f\}$ are polynomial time oracle RAM's such that $\{e\}(A) = \{f\}(B)$, then their common value X is in P. In fact, in satisfying $H_{e,f}$ we will exhibit the polynomial time algorithm to compute X.

We ensure each of these requirements by use of an associated strategy. We will shortly sketch how the strategies operate. First, we give some indication of their context, since it is somewhat different than that in the earlier constructions. As before, we assign a priority ranking to the requirements. We invoke the first n strategies during stage n, and thereby arrange that every strategy is in use during all but finitely many stages. We determine the action taken in the main recursion during stage n by means of a nested (minor) recursion of length n in which we calculate the effects of strategies. During stage n, we extend A and B so as to agree with the common value chosen by the maximum possible initial segment of strategies. Provided that for each strategy Σ,

(1) for all but finitely many stages, the conditions on A and B chosen by Σ are the same as those chosen by all higher priority strategies and
(2) for any sets A and B which are produced by a construction whose operation during all but finitely many stages is determined by Σ, A and B satisfy the requirement associated with Σ,

then A and B constructed as above will satisfy all of the requirements. We will first sketch the operation of the strategies and then the way by which they are combined in the minor recursion.

We turn now to our specific strategies. We view a strategy Σ as a *procedure*. It is called with arguments A_{n-1}, B_{n-1}, $A_n^{default}$, $B_n^{default}$ and $default\_time$. These arguments have the following types: the first four are conditions and the final one is an integer. Their intended roles in the construction are to have A_{n-1} and B_{n-1} as the conditions on A and B determined in the previous stage; the next two, $A_n^{default}$ and $B_n^{default}$, are the default conditions on A and B which will be used at the end of stage n if Σ does not disallow their use; $default\_time$ is the number of steps needed to run the construction up to the point of calling Σ, which is enough to compute $A_n^{default}$ and $B_n^{default}$. The strategy returns two conditions A_n and B_n which extend A_{n-1} and B_{n-1}. In the construction, they indicate the conditions that Σ intends be used as the stage n computation of A and B.

G_d. The strategy g_d to ensure the satisfaction of the requirement G_d acts as follows.

(1) First, check for a string y in $domain(A_{n-1})$ such that $\{d\}(y)$ converges in less than n steps and gives a value that is different from $A_{n-1}(y)$. If there is such a y, return $A_n^{default}$ and $B_n^{default}$. *(In step (1), we look back to see whether G_d is already satisfied by an inequality between $\{d\}$ and A.)*
(2) Otherwise, check for a string y in $domain(A_n^{default}) - domain(A_{n-1})$ such that $\{d\}$ converges at y in less than $default\_time$ many steps. If there is such a y then for A_n, return the extension A^* of A_{n-1} that is identically equal to 0 at every string

in $domain(A_n^{default}) - domain(A_{n-1})$ other than y. At y, A^* is defined to disagree with $\{d\}(y)$. For B_n, return the extension B^* of B_{n-1} that is identically equal to 0 on every string in $domain(B_n^{default}) - domain(B_{n-1})$. *(We look for an argument where it is faster to compute $\{d\}$ than it is to run the default computation. If we find one then we define A_n to establish $A \neq \{d\}$.)*

(3) If neither of these cases apply, then return $A_n^{default}$ and $B_n^{default}$. *(If no action is required, return the default values.)*

If there is a y such that the evaluation of $\{d\}(y)$ takes less time then the evaluation of the default value for $B(y)$, then g_d ensures that $\{d\} \neq A$. Otherwise, g_d ensures that B is always given by the default calculation and so can be computed in less time than the evaluation of $\{d\}$. Thus, g_d ensures that if $\{d\} = A$ then there is an algorithm to compute B that runs in less time.

In addition, the values for A and B returned by g_d only deviate from the input default values finitely often. Once g_d's outputs are identical with the ultimate values of A_n and B_n, if g_d ever returns a value other than its input default value then, by (1), it automatically returns the default during every stage large enough to verify $\{d\} \neq A$.

$H_{e,f}$. Let $\{e\}$ and $\{f\}$ be polynomial time oracle RAM's. Let p be a polynomial that bounds their running times. The strategy $h_{e,f}$ to ensure the satisfaction of $H_{e,f}$ acts as follows.

(1) First, check for a string y such that $\{e\}(A_{n-1}, y) \neq \{f\}(B_{n-1}, y)$ is verified by a computation of length less than n. If there is such a y, return $A_n^{default}$ and $B_n^{default}$. *(Look back to see whether the requirement is already satisfied by the inequality $\{e\}(A) \neq \{f\}(B)$.)*

(2) Otherwise, check for a y of length less than or equal to $default\_time$ such that there are two extensions A' and A'' of A_{n-1} such that $\{e\}(A', y)$ and $\{e\}(A'', y)$ are defined by computations with queries only to the domains of A' and A'' and have different values. If there is such a y, then return the value B^* for B_n that extends B_{n-1} and is identically equal to 0 on every string in $\{0,1\}^{\leq p(default\_time)} - domain(B_{n-1})$. Return as value for A_n whichever of A' and A'' establishes $\{e\}(A_n, y) \neq \{f\}(B_n, y)$, for the (returned) value $B_n = B^*$. *(If the condition A_{n-1} does not already decide the value of $\{e\}(A, y)$ for all y's of length less than or equal to $default\_time$ then use the split in $\{e\}$ to make $\{e\}(A) \neq \{f\}(B)$. We attempt to meet a set of conditions associated with mutual genericity.)*

(3) If neither of these cases apply, then return $A_n^{default}$ and $B_n^{default}$. *(As in g_d, if no action is required then return the default values.)*

Assume that $h_{e,f}$ is respected during all but finitely many stages. If the inequality between $\{e\}(A)$ and $\{f\}(B)$ is established in (2), then the requirement is trivially satisfied.

Assume that $\{e\}(A)$ and $\{f\}(B)$ are equal and let X be their common value. Let y be a string such that $h_{e,f}$ is respected during the stage when $A(y)$ is defined. We compute $X(y)$ as follows. First, compute the largest m so that the evaluation of $A_{m-1}^{default}$ and $B_{m-1}^{default}$ involves less than $|y|$ steps. Then, evaluate $\{e\}(A^*, y)$, where A^* is equal to $A_{m-1}^{default}$ on $domain(A_{m-1}^{default})$ and is identically equal to 0 elsewhere.

Since its operations are explicitly evaluated in polynomial time, it is clear that this procedure can be implemented on a RAM in polynomial time. To see that it correctly computes X, note that since $h_{e,f}$ is respected during almost every stage and does not establish the inequality between $\{e\}(A)$ and $\{f\}(B)$, the default values for A and B are the ones actually used in the construction. Further, the m computed by y is the stage when $h_{e,f}$ examines all conditions extending A_{m-1} $(= A_{m-1}^{default})$ to find a pair of conditions that gives a pair of incompatible values for $\{e\}$ at y. By assumption, $h_{e,f}$ could not split the values of $\{e\}$, so every extension of A_{m-1} gives the same answer to $\{e\}(-, y)$ as A gives. In particular, $\{e\}(A^*, y) = \{e\}(A, y) = X(y)$, as desired.

Note, that this strategy also only returns conditions different from the input default values finitely often by the same argument that applied to g_d.

The construction. This method of combining strategies also appears in [SSl, Shinoda-Slaman]. We let A_0 and B_0 be the trivial conditions with empty domain.

During stage n, we set up the minor recursion to invoke the first n strategies in the priority ordering. First, we execute the nth strategy with arguments A_{n-1}, B_{n-1}; $A_n^{default}$ and $B_n^{default}$, given by the trivial extensions of A_{n-1} and B_{n-1} which are are identically 0 on every string of length less than or equal to n not in the domains of A_{n-1} and B_{n-1}; and *default_time*, equal to the number of steps needed to compute these quantities. By recursion, in decreasing order of priority, we execute the next strategy with arguments A_{n-1}, B_{n-1}; $A_n^{default}$ and $B_n^{default}$, obtained from the previous strategy's returned values; and *default_time*, equal to the number of steps needed to compute the construction through the point of executing the previous strategy. The output of the highest priority strategy (namely, the last strategy executed) gives the values for A_n and B_n.

Suppose that Σ is one of the above strategies. Then, Σ is in operation for all but finitely many stages. Hence, it determines the default value given to the strategies of higher priority for all but finitely many stages. Further, each strategy of higher priority than Σ only returns conditions different from the its input default values at most finitely often. So, the values returned by Σ will be the ones used by the construction during all but finitely many stages. By the above remarks, this is enough to conclude that the requirement associated with Σ is satisfied. □

Remark. It is an open problem whether *every* complexity type $C \not\subseteq P$ contains a minimal pair.

4. On the Fine Structure of P.

In contrast to many results in structural complexity theory that are only relevant for sets outside of P, the investigation of complexity types also leads to some challenging questions about the fine structure of P itself. One may argue that the exploration of the possibilities and limitations of construction techniques for sets in P may potentially be useful in order to distinguish P from larger complexity classes (e.g., $PSPACE$).

Those time bounds f that are commonly used in the analysis of algorithms for problems in P have the property that

$$\sup\{f(m) \mid m \leq c \cdot n\} = O(f(n))$$

for every constant $c \in N$, and that f agrees almost everywhere with some concave function g (i.e. $\forall k > n(g(n)) \leq \frac{n}{k}g(k))$). These two properties together entail the useful fact that $DTIME(f)$ is closed under linear time Turing-reductions (we assume that the query tape is erased after each query). Note also that the first property alone guarantees already that $DTIME(f)$ is closed under linear time many-one reductions.

In view of the preceding fact it is of interest to analyze for sets in P a slightly different notion of complexity type, where the underlying set T of time bounds is replaced by the class T_L of those f in T that satisfy the two additional properties above. This version has the advantage that each linear time Turing degree is contained in a single complexity type (in other words: each complexity type is closed under the equivalence relation $=_{\text{lin}}$). Therefore we assume in this section that T has been replaced by T_L.

Linear time reductions have provided the only successful means to show that certain concrete sets have exactly the same time complexity (e.g. Dewdney [4] proved that BIPARTITE MATCHING $=_{\text{lin}}$ VERTEX CONNECTIVITY ("are there $\geq k$ disjoint uv-paths in G, for u, v, k given")). The following result implies that this method is not general.

Theorem 7. *Every complexity type $C \not\subseteq DTIME(n)$ of polynomial time computable sets contains infinitely many different linear time degrees, and the linear time degrees in C are dense. Furthermore C contains incomparable linear time degrees, but no smallest linear time degree.*

PROOF: Assume that $C \not\subseteq DTIME(n)$. The construction of sets A, B in C that are incomparable with regard to linear time reductions proceeds as in the proof of Theorem 5.

In order to show that C contains no smallest linear time degree we fix some arbitrary set $A \in C$. We construct a set $B \in C$ with $A \not\leq_{\text{lin}} B$ by deleting from A all elements that lie in certain "intervals" $I_{n,m} := \{x \in \{0,1\}^* \mid n \leq |x| \leq m\}$. Since $A \notin DTIME(n)$ one can falsify each possible linear time reduction from A to B by choosing the length $m - n$ of the removed interval $I_{n,m}$ sufficiently large (for given n define m via "looking back"). In order to guarantee that in addition $B \in \mathcal{C}$, we combine these strategies in a finite injury priority argument with the "$\mathcal{C}$-strategies" from the proof of Theorem 2 (see the discussion at the beginning of Section 3).

In order to show that the linear time degrees in $\mathcal{C}$ are dense we assume that sets $A, B \in \mathcal{C}$ are given with $B <_{\text{lin}} A$. Let $\widetilde{B} := \{0^\frown x \mid x \in B\}$ and $\widetilde{A} := \{1^\frown x \mid x \in A\}$. Then we have $A =_{\text{lin}} \widetilde{A} \cup \widetilde{B}$ (thus $\widetilde{A} \cup \widetilde{B} \in \mathcal{C}$ and $B <_{\text{lin}} \widetilde{A} \cup \widetilde{B}$). One constructs in the usual manner (with "looking back" as in [7]) a linear time computable set L such that for $D := (\widetilde{A} \cap L) \cup \widetilde{B}$ one has $D \not\leq_{\text{lin}} B$ and $\widetilde{A} \cup \widetilde{B} \not\leq_{\text{lin}} D$. Furthermore it is obvious from the definition of D that $B \leq_{\text{lin}} D \leq_{\text{lin}} \widetilde{A} \cup \widetilde{B}$. Thus $D \in \mathcal{C}$ and $B <_{\text{lin}} D <_{\text{lin}} \widetilde{A} \cup \widetilde{B} =_{\text{lin}} A$.

Finally we observe that the existence of sets A, B in $\mathcal{C}$ with $A \not\leq_{\text{lin}} B$ (see the beginning of the proof) implies that there is a set $G \in \mathcal{C}$ with $B <_{\text{lin}} G$: set

$$G := \{0^\frown x \mid x \in A\} \cup \{1^\frown x \mid x \in B\}.$$

Thus an iterative application of the preceding density result implies that $\mathcal{C}$ contains infinitely many linear time degrees. □

It is tempting to conjecture that for every polynomial time computable set $A \notin DTIME(n)$ there is a set B of the same complexity type with $A \not\leq_{\text{lin}} B$ and $B \not\leq_{\text{lin}} A$, furthermore that the structure of linear time degrees of sets in a complexity type is the same for every polynomial time computable complexity type $\mathcal{C} \not\subseteq DTIME(n)$. The following result implies that both conjectures are false (see section 1 for the definition of a principal complexity type).

Theorem 8. *A complexity type $\mathcal{C} \subseteq P$ has a largest linear time degree if and only if $\mathcal{C}$ is principal (in fact if $\mathcal{C}$ is non-principal then it does not even contain a maximal linear time degree).*

Idea behind the proof of Theorem 8. If $\mathcal{C}$ is a principal complexity type then it contains a set U which is almost (i.e. up to padding) universal for $\mathcal{C}$. One has then $X \leq_{\text{lin}} U$ for every $X \in \mathcal{C}$.

Assume now that $\mathcal{C}$ is non-principal, and that X is an arbitrary set in $\mathcal{C}$. One then constructs a set $A \in \mathcal{C}$ with $A \not\leq_{\text{lin}} X$. This will imply the claim, since one has then $(A \vee X) \in \mathcal{C}$ and $X <_{\text{lin}} (A \vee X)$, where $(A \vee X) := \{y0 \mid y \in A\} \cup \{y1 \mid y \in X\}$.

To achieve $A \not\leq_{\text{lin}} X$ it is sufficient to satisfy each instance of the following requirement.

R_e. $A \neq \{e\}(X)$ or $\{e\}(X, y)$ uses more than $e \cdot |y|$ steps for some string y, where as usual each oracle query counts as one step.

The corollary of Theorem 1 shows that there exist recursive sequences $(t_i)_{i\in\mathbf{N}}$ and $(e_i)_{i\in\mathbf{N}}$ such that $(t_i)_{i\in\mathbf{N}}$ is a characteristic sequence for $\mathcal{C}$ and $\forall i(\{e_i\} =^* X$ and $\{e_i\}$ is of time complexity $O(\{t_i\}))$, where $=^*$ denotes equality modulo finite sets. The constructed set A will lie in $\mathcal{C}$, if we can satisfy each of the following requirements.

S_e. A belongs to $DTIME(\{t_e\}$.

T_j. If $\{j\} \in T$ and $A \in DTIME(\{j\})$ then for some $i \in \mathbf{N}$, $\{j\} = \Omega(\{t_i\})$.

In the case of a conflict between different requirements the one with the smaller index (i.e. higher priority) wins. In particular in order to satisfy S_e one has to make sure that all attempts to satisfy a requirement $R_{e'}$ with $e' > e$ at some argument $y \in \{0,1\}^*$ (where one tries to achieve that $A(y) \neq \{e'\}(X,y)$) are halted after $C_e \cdot \{t_e\}(|y|)$ steps (where C_e is some constant). On the other hand, in order to satisfy requirement $R_{e'}$ at argument y one first has to know the value of $\{e'\}(X,y)$. The number of computation steps needed for that depends on the algorithm $\{e_i\}(z)$ that one uses to simulate oracle queries "$z \in X$?" We ignore in this sketch that we have only $\{e_i\} =^* X$ instead of $\{e_i\} = X$. This will only cause finitely many additional "injuries" that force us to repeat the attempt to satisfy $R_{e'}$ (each time when a new discrepancy between $\{e_i\}$ and X is detected by "looking back"). It is not a priori clear which attempt to satisfy $R_{e'}$ (via some y, e_i as above) will succeed within the negative restraint imposed by S_e, because we know very little about the behaviour of the given sequence $(t_i)_{i\in\mathbf{N}}$. In particular we may have that $\{t_i\} = \{t_{i+1}\}$ for many i, and therefore the number of steps needed to evaluate $\{e'\}(\{e_i\},y)$ need not be bounded by $C_e \cdot \{t_e\}(|y|)$ (if $e' > e$ is sufficiently large). However, since $\mathcal{C}$ is non-principal there is an infinite set $W \subseteq \mathbf{N}$ such that there exists for every $i \in W$ an infinite set $H_i \subseteq \mathbf{N}$ with $\{t_i\}|_{H_i} = o(\{t_{i-1}\}|_{H_i})$. This set W supplies a "dense" set of "windows" through which at least one attempt for each requirement $R_{e'}$ can be carried out without interference by requirements S_e with $e < e'$.

Additional priority conflicts (and finite injuries) occur between the requirements S_e and T_j, and between the requirements R_e and T_j. In order to satisfy T_j one has to make A different from any set that is computed by an algorithm that is "too fast" (this requires that certain y are placed into A, or kept out of A). The interaction between the requirements S_e and T_j is handled in the same way as in the proof of Theorem 2.

5. Extensional Properties of Sets that have the Same Time Complexity.

In this section we investigate some basic properties of the partial order

$$PO(C) := \langle \{X | X \in C\}, \subseteq^* \rangle,$$

where C is an arbitrary complexity type and $\subseteq^*$ denotes inclusion modulo finite sets (i.e. $X \subseteq^* Y :\Leftrightarrow X - Y$ is finite). This investigation is part of the long range project to study the relationship between extensional properties of a set and its computational complexity. Among other work in this direction we would like to mention in particular the study of the complexity of sparse sets (see e.g. [16]), and the investigation of the relationship between properties of recursively enumerable sets under $\subseteq^*$ and their degree of computability. Our approach differs from this preceding work insofar as it also applies to "actually computable" sets (i.e. sets in P). Therefore it provides an opportunity to develop finer construction tools that can be used to examine also the structure of sets of small complexity.

Theorem 9. Every set X can be split into two sets A, B of the same complexity type as X (i.e. $X = A \cup B$, $A \cap B = \emptyset$, $X =_C A =_C B$).

Idea of the Proof of Theorem 9. Associate with the given set X a characteristic sequence $(t_i)_{i \in N}$ as in Theorem 1. For every $e, n \in \mathbf{N}$ and $x \in \{0,1\}^*$ define

$$TIME(e, x) := (\text{number of steps in the computation of } \{e\} \text{ on input } x)$$

and

$$MAXTIME(e, n) := \max\{TIME(e, x) \mid |x| = n\}.$$

It is sufficient to partition X into sets A and B in such a way that for every $e \in \mathbf{N}$ the following requirements R_e^A, R_e^B, S_e^A, S_e^B are satisfied:

$$R_e^A :\Leftrightarrow (A = \{e\} \Rightarrow \forall f \in T(\forall n(MAXTIME(e, n) \le f(n)) \Rightarrow \exists j \in \mathbf{N}(f = \Omega(\{t_j\}))))$$

$$S_e^A :\Leftrightarrow A \in DTIME(\{t_e\}(n)).$$

R_e^B, S_e^B are defined analogously.

Note that it is not possible to satisfy R_e^A by simply setting $A(x) := 1 - \{e\}(x)$ for some x: in order to achieve that $A \subseteq X$ we can only place x into A if $x \in X$.

Instead, we adopt the following strategy to satisfy R_e^A (the strategy for R_e^B is analogous): For input $x \in \{0,1\}^*$ compute $\{e\}(x)$.

Case I. If $\{e\}(x) = 0$, then this strategy issues the constraint "$x \in A \Leftrightarrow x \in X$".

Case II. If $\{e\}(x) = 1$, then this strategy issues the constraint "$x \notin A$" (which forces x into B if $x \in X$).

In the case of a conflict for some input x between strategies for different requirements one lets the requirement with the highest priority (i.e. the smallest index e) succeed (this causes in general an "injury" to the other competing requirements).

The interaction between the described strategies is further complicated by the fact that in the case where R_e^A is never satisfied via Case II, or via Case I for some $x \in X$, we have to be sure that Case I issues a constraint *for almost every input* x (provided that the simulation of $\{e\}(x)$ is not prematurely halted by some requirement S_i^A with $i \leq e$, see below). Consequently the number of requirements whose strategies act on the same input x grows with $|x|$ (only those R_i^A, R_i^B with $i < |x|$ can be ignored where one can see by "looking back" for $|x|$ steps that they are already satisfied).

The strategy for requirement $S_e^A(S_e^B)$ is as follows: it issues the constraint that for all inputs x with $|x| \geq e$ the *sum* of all steps that are spent on simulations for the sake of requirements $R_i^A, R_i^B, S_i^A, S_i^B$ with $i \geq e$ has to be bounded by $O(\{t_e\}(|x|))$. One can prove that in this way $S_e^A(S_e^B)$ becomes satisfied (because only finitely many inputs are placed into A or B for the sake of requirements of higher priority). One also has to prove that the constraint of S_e^A does not hamper the requirements of lower priority in a serious manner.

In order to verify that this construction succeeds, one has to show that each requirement R_e^A, R_e^B is "injured" at most finitely often. This is not obvious, because we may have for example that R_{e-1}^B (which has higher priority) issues overriding constraints for infinitely many arguments x according to Case I. However in this case we know that only finitely many of these x are elements of X (otherwise R_{e-1}^B would have been seen to be satisfied from some point of the construction on), and all of its other constraints are "compatible" with the strategies of lower priority (since we make $A, B \subseteq X$).

Finally we verify that each requirement $R_e^A(R_e^B)$ is satisfied. This is obvious if Case II occurs in the strategy for R_e^A for some input x where R_e^A is no longer injured; or if Case I occurs for such input x with $x \in X$ (in both cases we can make $A \neq \{e\}$). However it is also possible that $x \notin X$ for each such x (and that $\{e\} = A$), in which case R_e^A becomes satisfied for a different reason. In this case we have $\{e\}(x) = 0 = X(x)$ for each such x. Therefore we can use $\{e\}$ to design a new algorithm for X that is (for every input) at least as fast as the algorithm $\{e\}$ for A (it uses $\{e\}$ for those inputs where $\{e\}$ is faster than the "old" algorithm for X of time complexity $\{t_e\}$). Therefore one can prove that $X \in DTIME(f)$ for every $f \in T$ that bounds the running time of algorithm $\{e\}$ for A. This implies that $f(n) = \Omega(\{t_j\}(n))$ for some $j \in \mathbf{N}$ (by construction of the characteristic sequence $(t_i)_{i \in \mathbf{N}}$). □

Corollary 2. For every complexity type $\mathcal{C} \neq \mathbf{0}$ the partial order $PO(\mathcal{C})$ of sets in $\mathcal{C}$ has neither minimal nor maximal elements.

Let $PO_{0,1}(\mathcal{C})$ be the partial order

$$\langle \{X | X \in \mathcal{C} \vee X = \{0,1\}^* \vee X = \phi\}, \subseteq^* \rangle$$

(thus $PO_{0,1}(\mathcal{C})$ results from $PO(\mathcal{C})$ by adding the smallest set ϕ and the largest set $\{0,1\}^*$).

Corollary 3. For every complexity type $\mathcal{C}$ there is an embedding E of the partial order of the countable atomless Boolean algebra AB into $PO_{0,1}(\mathcal{C})$ with $E(1) = \{0,1\}^*$, $E(0) = \phi$, and $E(a \vee b) = E(a) \cup E(b)$ as well as $E(a \wedge b) = E(a) \cap E(b)$ for all elements $a, b \in AB$.

Remark. In order to define this embedding E one starts with any two sets X, $\{0,1\}^* - X$ in $\mathcal{C}$, and applies the splitting theorem iteratively. The idea is to represent the elements of AB by arbitrary finite unions of those sets in $\mathcal{C}$ that are constructed in this way. In order to guarantee that these finite unions U are also in $\mathcal{C}$ (unless $U =^* \{0,1\}^*$) one has to prove a slightly stronger version of the splitting theorem. The following additional property of A, B is needed: For every $f \in T$ for which there exists some $U \in DTIME(f)$ with $U \cap X = A$ or $U \cap X = B$ one has $X \in DTIME(f)$. One can prove this stronger version with a small variation in the strategy for requirement R_e^A (R_e^A can now be satisfied via Case II at a single input x only if $\underline{x \in X}$; if R_e^A never gets satisfied at any input x via Case I or Case II one can argue that $x \notin X$ whenever the simulation of $\{e\}(x)$ can be finished in an attempt for R_e^A, independently of the output of $\{e\}(x)$).

In the following we write $Y \underset{\infty}{\subset} X$ if $Y \subseteq X$ and $X - Y$ is infinite.

Theorem 10. (Density Theorem)

Assume that $Y \underset{\infty}{\subset} X$ and $Y \leq_{\mathcal{C}} X$ (i.e. $\forall f \in T(X \in DTIME(f) \Rightarrow Y \in DTIME(f))$). Then there is a set A such that $Y \underset{\infty}{\subset} A \underset{\infty}{\subset} X$ and $A =_{\mathcal{C}} X$.

Idea of the proof of Theorem 10. Let $(t_i)_{i \in N}$ be a characteristic sequence for X. It is sufficient to construct A such that $Y \subseteq A \subseteq X$ and for all $e \in N$ the requirements R_e, S_e, T_e, U_e are satisfied, where R_e, S_e are identical with the requirements R_e^A, S_e^A in the proof of Theorem 9 (together they ensure that $A =_{\mathcal{C}} X$) and

$$T_e: \; |A - Y| \geq e$$

$$U_e: \; |X - A| \geq e.$$

The strategy to satisfy R_e is similar to the strategy for satisfying R_e^A. However in Case II (where $\{e\}(x) = 1$), unlike in the splitting theorem, R_e does not have the power to keep x out of A (even if R_e has the highest priority) because x may later enter Y. Instead, R_e issues in Case II the constraint "$x \notin A \Leftrightarrow x \notin Y$" (i.e. R_e wants to keep x out of A if it turns out that $x \notin Y$).

It is easy to see that R_e becomes satisfied if Case I occurs for some $x \in X$, or if Case II occurs for some $x \notin Y$ (provided that R_e is not "injured" at x by requirements of higher priority). If neither of these events occurs, then we can conclude that $\{e\}(x) = X(x)$ whenever the simulation of $\{e\}(x)$ can be finished before it is halted for the sake of some requirement S_i with $i \leq e$. This information can be used (as in the proof of Theorem 9) to design an algorithm for X that converges for every input x "at least as fast" as the computation $\{e\}(x)$. □

Corollary 4. For every complexity type $\mathcal{C}$ the partial order $PO(\mathcal{C})$ is dense.

It is easy to see that $PO(\mathcal{C})$ is isomorphic to the countable atomless Boolean algebra AB if $\mathcal{C} = \mathbf{0}$. Furthermore it was shown that AB can be embedded into $PO_{0,1}(\mathcal{C})$ for every complexity type $\mathcal{C}$. However the following corollary suggests that the structure of the partial order $PO(\mathcal{C})$ is substantially more complicated than that of AB if $\mathcal{C} \neq \mathbf{0}$. Obviously any complexity type $\mathcal{C} \neq \mathbf{0}$ is closed under complementation, but not under union or intersection. However, it could still be the case that any two sets $A, B \in \mathcal{C}$ have a least upper bound in the partial order $PO(\mathcal{C})$. This is ruled out by the following result.

Corollary 5. Consider an arbitrary complexity type $\mathcal{C} \neq \mathbf{0}$. Then any two sets $A, B \in \mathcal{C}$ have a least upper bound in the partial order $PO(\mathcal{C})$ if and only if $A \cup B \in \mathcal{C}$. In particular one can define with a first order formula over $PO(\mathcal{C})$ whether $A \cup B \in \mathcal{C}$ (respectively $A \cap B \in \mathcal{C}$) for $A, B \in \mathcal{C}$.

Proof. Assume that $A, B \in \mathcal{C}$, $A \cup B \notin \mathcal{C}$, $A \cup B \subseteq D$ and $D \in \mathcal{C}$. Then $(A \cup B) \leq_{\mathcal{C}} D$ and $(A \cup B) \underset{\infty}{\subset} D$. Thus there exists by Theorem 10 a set $D' \in \mathcal{C}$ with $(A \cup B) \underset{\infty}{\subset} D' \underset{\infty}{\subset} D$. Therefore D is not a least upper bound for A and B in $PO(\mathcal{C})$. □

Remark.
This result suggests that the first order theory of the partial order $PO(\mathcal{C})$ is non-trivial for $\mathcal{C} \neq \mathbf{0}$.

6. Open Problems.

There are various results in structural complexity theory which state that there exist in some complexity class K sets with a certain property Q. Each such result gives rise to the more precise question which complexity types in K contain sets with property Q. As examples we mention the open questions whether every complexity type contains a tally set (i.e., a subset of $\{0\}^*$), and whether every complexity type $\mathcal{C} \not\subseteq P$ contains a minimal pair of polynomial time Turing degrees. The answers to a number of open problems of this type appear to be of interest on their own. Furthermore their solutions may help to enlarge our reservoir of construction techniques for computable sets (in particular also for sets of "low" complexity).

Another problem area is the characterization of the structure of the (time-bounded) degrees of computability of the sets in a complexity type $\mathcal{C}$. For example one would like to know whether the sets in $\mathcal{C}$ realize infinitely many different types in the first order language of the partial order of the degrees in $\mathcal{C}$, and whether this theory is decidable. Other open problems arise if one compares the degree structures of different complexity types. For example we do not know whether the structure of polynomial time Turing degrees of sets in a complexity type $\mathcal{C} \not\subseteq P$ is the same for each such complexity type, and we do not know whether Theorem 8 specifies the only difference between the structure of linear time degrees within a non-zero complexity type $\mathcal{C} \subseteq P$.

An interesting open question about the lattice of all complexity types under $\leq_C$ is whether there is an automorphism of the partial order of complexity types that moves some complexity type in P to a complexity type that is not in P.

With regard to the partial order $PO(\mathcal{C})$ of sets in a complexity type $\mathcal{C}$ under inclusion, it would be of interest to know whether the structure of $PO(\mathcal{C})$ depends on $\mathcal{C}$ for $\mathcal{C} \neq 0$.

Finally, we would like to point out that all other resource-bounds for computations (e.g. nondeterministic time, or deterministic space) give also rise to the consideration of corresponding equivalence classes (or "complexity types") of those sets that are equivalent with regard to these complexity measures. Many questions that relate complexity types for deterministic computations with complexity types for nondeterministic computations or space bounded computations are obviously very hard. However some of these may turn out to be easier to answer than the related "global" questions about inclusions among the corresponding complexity classes. As an example we would like to mention the problem whether there are sets A, B such that $A \neq_C B$ (i.e. A and B have different deterministic time complexity), but for all space constructible space bounds f we have $A \in DSPACE(f) \Leftrightarrow B \in DSPACE(f)$.

References

[1] A.V. AHO, J.E. HOPCROFT, J.D. ULLMAN, The Design and Analysis of Computer Algorithms, Addison-Wesley (Reading, 1974).

[2] M. BLUM, A machine-independent theory of the complexity of recursive functions, *J. ACM,* **14**(1967), 322-336.

[3] S.A. COOK, R.A. RECKHOW, Time-bounded random access machines, *J. Comp. Syst. Sc.,* **7**(1973), 354-375.

[4] A.K. DEWDNEY, Linear time transformations between combinatorial problems, *Internat. J. Computer Math.,* **11**(1982), 91-110.

[5] J.G. GESKE, D.T. HUYNH, A.L. SELMAN, A hierarchy theorem for almost everywhere complex sets with applications to polynomial complexity degrees, *Proc. of the 4th Annual Symposium on Theoretical Aspects of Computer Science*, Lecture Notes in Computer Science, vol. 247 (Springer, 1987), 125-135.

[6] J. HARTMANIS, J.E. HOPCROFT, An overview of the theory of computational complexity, *J. ACM*, **18**(1971), 444-475.

[7] R. LADNER, On the structure of polynomial-time reducibility, *J. ACM*, **22**(1975), 155-171.

[8] L. LANDWEBER, R. LIPTON, AND E. ROBERTSON, On the structure of sets in NP and other classes, *TCS*, **15**(1981), 181-200.

[9] M. LERMAN, Degrees of Unsolvability, Springer (1983).

[10] L.A. LEVIN, On storage capacity for algorithms, *Soviet Math. Dokl.*, **14**(1973), 1464-1466.

[11] N. LYNCH, Helping: several formalizations, *J. of Symbolic Logic*, **40**(1975), 555-566.

[12] W. MAASS, T. A. SLAMAN, The complexity types of computable sets (extended abstract), *Proc. of the Structure in Complexity Theory Conference*, 1989.

[13] W. MAASS, T.A. SLAMAN, Extensional properties of sets of time bounded complexity (extended abstract), *Proc. of the 7th Int. Conf. on Fundamentals of Computation Theory*, J. Csirik, J. Demetrovics, F. Gésceg, eds., Lecture Notes in Computer Science vol. 380, Springer Verlag (Berlin 1989), 318-326.

[14] W. MAASS, T. A. SLAMAN, The complexity types of computable sets, in preparation.

[15] M. MACHTEY, P. YOUNG, An Introduction to the General Theory of Algorithms, North-Holland (Amsterdam, 1978).

[16] S. MAHANEY, Sparse complete sets for NP: solution of a conjecture of Berman and Hartmanis, *J. Comp. Syst. Sc.*, **25**(1982), 130-143.

[17] A.R. MEYER, P.C. FISCHER, Computational speed-up by effective operators, *J. of Symbolic Logic*, **37**(1972), 55-68.

[18] A.R. MEYER, K. WINKLMANN, The fundamental theorem of complexity theory, *Math. Centre Tracts*, **108**(1979), 97-112.

[19] W.J. PAUL, Komplexitaetstheorie, Teubner (Stuttgart, 1978).

[20] C.P. SCHNORR, G. STUMPF, A characterization of complexity sequences, *Zeitschr. f. math. Logik u. Grundlagen d. Math.*, **21**(1975), 47-56.

[21] J. SHINODA, T.A. SLAMAN, On the theory of the PTIME degrees of the recursive set, *Proc. of Structure in Complexity Theory Conference*, 1988, 252-257.

POLYNOMIALLY ISOLATED SETS

A. Nerode[1]
J. B. Remmel[2]

§1. **Preface.** Polynomial time equivalence types (PET's), and their operations of addition "+" and multiplication "·", were introduced by the authors in [30]. The PET's are a resource–bounded refinement of the recursive equivalence types (RET's) of Dekker [5], obtained by replacing 1–1 partial recursive functions with 1–1 honest p–time partial functions with p–time domains (definition 2 below). The authors proved in [30], by a long argument, the analogue for PET's of Friedberg's cancellation theorem for RET's of [12]: for positive integers n and PET's X, Y, nX = nY implies X = Y. This indicates the tractability of developing a general theory of PET's, but with longer and harder proofs than in RET's.

The most completely studied RET's are the isols, the RET's of recursively isolated sets – that is, sets with no infinite recursively enumerable subset. The isols are closed under "+", "·" and obey the cancellation laws for "+", "·". Is there a good analogue of isols in the PET's? In §2, theorem 14 we show that the PET's of recursively isolated sets do not obey cancellation laws for "+", "·". PET's of recursively isolated sets will therefore certainly not have a theory like that of the isols.

Is there a definition of "isolated set", different from that for "recursive isolated set", which is more appropriate for PET's and also such that "isolated sets" obey cancellation for "+", "·"? To seek an answer we examine Dekker's proof of multiplicative cancellation for isols in §3 to find out upon what property of sets the cancellation law for "·" really depends. Having found this out, we identify in §4, definition 5, a new class of "isolated sets" A, characterized by polynomial time termination of iteration of p–time monotone set maps of finite sets mapping finite subsets of A to finite subsets of A. These we call "polynomial isolated sets", or "p–isolated sets". We show in theorem 17, §4, that, as intended, p–isolated sets obey cancellation laws for "+", "·".

Among the p–isolated sets are the p–finite sets of definition 6, §4. These p–finite sets A are characterized by polynomial time termination of iteration of p–time maps (monotone or not) of finite sets mapping finite subsets of A to finite subsets of A. In §5, theorems 19, 20, we construct p–finite sets which are infinite p–time sets. (p–time sets are sets with polynomial time characteristic functions). Infinite p–time p–finite sets are infinite recursive sets and are certainly not recursively isolated, so with them we are entirely outside the domain of recursively isolated sets. In §6, theorem 21, we show that p–time, p–isolated sets are closed under "+", "·". In §6, theorem 22, we show that every subset of a p–time p–isolated set is itself p–isolated. It is open whether the p–time p–finite sets are likewise closed under "+", "·". Now we begin the paper proper with a review of RET's, isols and PET's.

RET's. A recursive equivalence is a partial function of non–negative integers to non–negative integers which is extendible to a 1–1 partial recursive function. Sets α, β of non–negative integers are recursively equivalent if there is a recursive equivalence mapping α onto β. The equivalence class $<\alpha>$ of a set α of non–negative integers is called a recursive equivalence type, or an RET. Embed the non–negative integers into the RET's by mapping the non–negative integer n into the RET $<\{0, ..., n-1\}>$. If J is a recursive pairing function for the non–negative integers, then addition and multiplication for RET's are induced by operations of effective disjoint union and effective cartesian product, all within the non–negative integers.

$$<\alpha> + <\beta> = <\{J(0, x) \mid x \in \alpha)\} \cup \{J(1, x) \mid x \in \beta\}>$$

$$<\alpha> \times <\beta> = <J(x, y) \mid x \in \alpha \wedge y \in \beta>.$$

Also, $<\alpha> \leq <\beta>$ if there is a γ, $<\alpha> + <\gamma> = <\beta>$.

For the theory of RET's in general, see [3, 5, 11, 18, 19, 23, 24, 26].

Isols. An RET x is called an isol if $x \neq x + 1$, or equivalently if for any representative α of x, α cannot be mapped by a recursive equivalence onto a proper subset of α. A recursively isolated set is one with no infinite recursively enumerable subset. The isols x are precisely the recursive equivalence types of recursively isolated sets. Finite sets are obviously recursively isolated, infinite recursively isolated sets are called recursively immune. Recursively isolated sets and recursively immune sets are usually just called isolated sets and immune sets respectively. We retain the adjective "recursively" to distinguish them from the polynomially isolated and polynomially immune sets introduced in this paper.

Isol theory has been very extensively developed. For the beginnings of isolic arithmetic see Dekker [3], Dekker and Myhill [5]. For Myhill's fundamental notion of combinatorial function, see Myhill [8] and Dekker [4]. For the Myhill–Nerode theory of combinatorial functions and the Nerode theory of frames, see Nerode [20, 21, 22, 25], McLaughlin [17]. For the theory of structures with recursive operations and isolated domains, see Crossley–Nerode [2]. For the theory of RET's of co–simple sets (= recursively isolated sets with r. e. complements) see Hay [13, 14]. For the theory of co–simple structures see Remmel [32]. For the theorem that those countable models of Peano first order logic arithmetic which have the same solvable Diophantine equations as the standard model are isomorphic to substructures of the isols, see Nerode [26]. For the analogous theory of Dedekind finite cardinals (those with $x \neq x+1$) in Mostowski or Zermelo–Fraenkel set theory without the axiom of choice, see Ellentuck [6, 7, 8, 9, 10, 11]. Ellentuck's work is done using full classical logic as ordinary classical mathematics. But developing the subject in intuitionistic logic gives a very clean connection between isols and Dedekind finite cardinals. McCarty [1980] showed that the first order theory of isols can be identified with the theory of certain stable Dedekind finite cardinals in a recursive realizability model of Intuitionistic Zermelo–Fraenkel set theory (IZF). Bibliographies of recursive equivalence, isols, and recursive algebra papers will be found in [1, 28].

Polynomial time equivalence types. We restate without proof needed material from Nerode and Remmel [30]. The present paper continues the numbering of definitions, propositions, lemmas, and theorems of [30]. As in [30], propositions, lemmas, and theorems are consecutively numbered. Non–negative integers y are thought of as given in the tally representation 0^y from $\{0\}^*$, that is, as a string of zeros. Let $\Sigma = \{0, 1\}$, and let Σ^* be the set of all finite sequences from Σ. For the definition of polynomial time functions and their elementary properties, see Hopcroft and Ullman [15]. Here is the standard definition of honest p–time (total) function.

DEFINITION 1 (from [30]). A polynomial time (total) function f is <u>honest</u> if there is a polynomial q such that for all $x \in \Sigma^*$, $q(|f(x)|) \geq |x|$. (Here $|x|$ is the length of string x.)

More generally, we need to define partial p–time functions and "honest" partial p–time functions.

DEFINITION 2 (from [30]).

(i) A partial function $f: \Sigma^* \to \Sigma^*$ is polynomial time (p–time) if

the domain of f, dom(f), is a polynomial time set (i.e, has a polynomial time characteristic function),

there is a deterministic transducer M such that

$$\forall x \in dom(f)\ (M(x) = f(x)),$$

and there is a polynomial p such that for all x in dom(f),

M(x) is computed in fewer than $p(|x|)$ steps.

(Here $|x|$ is the length of string x.)

(ii) We say a partial function $f: \Sigma^* \to \Sigma^*$ is honest if f is p–time and there exists a polynomial q such that

$$\forall x \in dom(f)\ (q(|f(x)|) \geq |x|).$$

We can always extend such an f to a total function $\hat{f}$ by letting $\hat{f}(x) = x$ if $x \notin dom(f)$. Then $\hat{f}$ is also an honest p–time total function.

Our basic objects, to be compared by polynomial time equivalence (p–equivalence), will be tally sets, that is sets of strings of zeros from $\{0\}^*$. A p–equivalence is a 1–1 function extendible to a 1–1 honest partial p–time function..

DEFINITION 3 (from [30]). Given $X, Y \subsetneq \{0\}^*$, we say that X is polynomial time equivalent to Y (X is p–equivalent to Y) if there exists a p–equivalence defined on X and mapping X onto Y. Write $X \sim_p Y$ if X is p–equivalent to Y.

Remark. Let f be a 1–1 honest partial p–time function, let $X, Y \subsetneq \{0\}^*$, and suppose that the restriction of f to X maps X onto Y. It is easy to see that

$$\bar{f} = \{\langle x, y\rangle \mid x \in \{0\}^*, y \in \{0\}^* \wedge f(x) = y\}$$

is also a 1–1 honest partial p–time function such that f and $\bar{f}$ agree on X. Thus there is no loss of generality in assuming that all p–time equivalences are extendible to 1–1 honest partial p–time functions f such that rgn(f), dom(f) $\subsetneq \{0\}^*$. We shall make these additional assumptions on f implicitly throughout the paper. Also for the rest of this paper we shall implicitly identify the natural numbers $N = \{0, 1, 2, ...\}$ with $\{0\}^*$ so that we can think of RET's as collections of subsets of $\{0\}^*$. Via the mapping $f: N \to \{0\}^*$ where $f(n) = 0^n$, we identify subsets of natural numbers with subsets of strings wherever possible. Hence we can say $A \subseteq \{0\}^*$ is recursively immune (recursively isolated, r. e., respectively) if and only if $f^{-1}(A)$ is recursively immune (recursively isolated, r. e., respectively). If X, $Y \subseteq \{0\}^*$, we write $X \sim Y$ if X is recursively equivalent to Y. We write $<X>$ for the recursive equivalence type of X, i.e., $<X> = \{Y \subseteq \{0\}^* \mid X \sim Y\}$.

PROPOSITION 1 (from [30]). "$\sim_p$" is an equivalence relation on $\mathscr{P}(\{0\}^*)$ (where $\mathscr{P}(X)$ denotes the power set of X).

LEMMA 1 (from [30]). Suppose that f and g are partial 1–1 honest p–time functions such that dom(f), rng(f), dom(g), and rng(g) are contained in $\{0\}^*$. Then f^{-1} and fog are partial 1:1 honest p–time functions.

Given proposition 1, we define the PET's as follows.

DEFINITION 4 (from [30]). Suppose that $X \subseteq \{0\}^*$. Then we define the <u>polynomial time equivalence type of</u> X by

$$<X>_p = \{Y \mid Y \subseteq \{0\}^* \wedge X \sim_p Y\}.$$

$<X>_p$ is also called the PET of X. Here is the pairing function J that we use on integers in the tally representation.

$$J(0^n, 0^m) = 0^{\frac{1}{2}[(m+n)^2+3m+n]}.$$

Given $X, Y \subseteq \{0\}^*$, we let

$X \oplus Y = \{0^{2n} \mid 0^n \in X\} \cup \{0^{2n+1} \mid 0^n \in Y\}$, and

$X \times Y = \{J(0^m, 0^n) \mid 0^m \in X \;\&\; 0^n \in Y\}$.

Then given PET's α and β, we define

$\alpha + \beta = <X \oplus Y>_p$, where $X \in \alpha$ and $Y \in \beta$

$\alpha \cdot \beta = <X \times Y>_p$, where $X \in \alpha$ and $Y \in \beta$.

Define a partial ordering relation on PET's by

$\alpha \leq \beta$ iff there exists a PET γ such that $\alpha + \gamma = \beta$.

It is easy to check that if $\alpha \leq \beta$ and $A \in \alpha$ and $B \in \beta$, then there is a p–time set S such that $A \sim_p B \cap S$.

LEMMA 9 (from [30]). Suppose φ is a 1:1 partial p–time function such that both $dom(\varphi)$ and $rng(\varphi)$ are contained in $\{0\}^*$. Then φ is honest iff φ^{-1} is a partial polynomial time function.

§2. **No cancellation for "+", "·" in recursively isolated PET's.** It was shown in [30] that additive and multiplicative cancellation fail for PET's, but the PET's exhibited in [30] were not the PET's of recursively isolated sets. Here we fill in this gap with co–simple sets, that is, infinite recursively isolated sets with r. e. complements.

THEOREM 14 (following the numbering of [30]). There are co–simple recursively isolated sets X, Y, and Z such that $X \oplus Z \sim_p Y \oplus Z$, but $X \not\sim_p Y$.

PROOF. We recall our non–recursively isolated counterexample to additive cancellation [30]. Let

$$A = \{0^{2n+1} \mid n > 0\},\ B = \{0^{(2n+1)2^{2n+1}} \mid n > 0\}, \text{ and}$$
$$C = \{0^{(2k+1)2^m} \mid k \geq 0 \text{ and } 1 \leq m \leq 2k\}.$$

Define a function f such that for fixed $k > 0$,

$$f(0^{2(2k+1)}) = 0^{2(2k+1)2+1},$$

$$f(0^{2(2k+1)2^m+1}) = 0^{2(2k+1)2^{m+1}+1} \quad \text{if } 1 \leq m < 2k,$$

$$f(0^{2(2k+1)2^{2k}+1}) = 0^{2(2k+1)2^{2k+1}},$$

and for $k = 0$, let $f(0^2) = 0^4$.

Then f is a p–time function and A, B, and C are p–time sets. For each k, f maps the set in $A \oplus C$,

$$0^{2(2k+1)}, 0^{2[(2k+1)2]+1}, 0^{2[(2k+1)4]+1}, \ldots, 0^{2[(2k+1)2^{2k}]+1},$$

onto the set in $B \oplus C$.

$$0^{2[(2k+1)2]+1}, 0^{2[(2k+1)4]+1}, \ldots, 0^{2[(2k+1)2^{2k}]+1}, 0^{2[(2k+1)2^{2k+1}]}$$

Thus we are using the set $\{0^{(2k+1)2^m} \mid 1 \leq m \leq 2k\}$ in C to form a bridge from 0^{2k+1} in A to $0^{(2k+1)2^{2k+1}}$ in B which the function $f: A \oplus C \to B \oplus C$ traverses essentially by doubling. Now for any set $W \subseteq N-\{0\}$, let

$$A_W = \{0^{2k+1} \mid k \in W\},$$

$$B_W = \{0^{(2k+1)2^{2k+1}} \mid k \in W\}, \text{ and}$$

$$C_W = \{0^{(2k+1)2^m} \mid k \in W \,\&\, 1 \leq m \leq 2^k\}.$$

Then it is easy to check that f is a bijection between $A_W \oplus C_W$ and $B_W \oplus C_W$ and hence f witnesses that

$$A_W \oplus C_W \sim_p B_W \oplus C_W.$$

Observe that A_W and B_W are recursively equivalent to $\{0^k \mid k \in W\}$. Thus if W is recursively immune, then A_W and B_W are recursively immune. Note also that if C_W contains an infinite r.e. set S, then

$$S' = \{k \mid \exists m(1 \leq m \leq 2k \,\&\, 0^{(2k+1)2^m} \in S\}$$

must also be an infinite r.e. set. Thus if W is recursively immune, then C_W is recursively immune. Moreover if N–W is r.e., it is easy to see that $\{0, 1\}^* - A_W$, $\{0, 1\}^* - B_W$, and $\{0, 1\}^* - C_W$ are r.e. Thus if W is any co–simple subset of N–{0}, then A_W, B_W, and C_W are co–simple subsets of $\{0, 1\}^*$.

We claim that for any infinite set $W \subseteq N-\{0\}$, $A_W \not\sim_p B_W$. That is, suppose φ is a 1:1 honest partial p–time function such that $\text{dom}(\varphi) \supseteq A_W$ and $\varphi(A_W) = B_W$. Let p(x) be a polynomial with positive coefficients which bounds the running time of φ. Then note that there is an n_0 such that $p(n) < 2^n$ for $n \geq n_0$. Because W is infinite, there is an $m_0 > n_0$ such that

$$m_0 \in W \text{ and } |\varphi(0^{2k+1})| < 2^{2m_0+1} \text{ for all } k \leq n_0.$$

But then if $E = W \cap \{n | n \leq m_0\}$ and $F = E-\{m_0\}$, we know that $|F| = |E|-1$ and $\varphi(A_E) \subseteq B_F$. That is, our choice of m_0 ensures that for each $x \in E$,

$|\varphi(0^{2x+1})| < (2m_0+1)2^{2m_0+1}$ and B_F represents the set of all $y \in B_W$ such that

$|y| < (2m_0+1)2^{2m_0+1}$. But since $|A_E| > |B_F|$, φ cannot be one to one which violates

our choice of φ. Thus there can be no such φ and hence $A_W \not\sim_p B_W$.

Thus if W is any co-simple subset of $N-\{0\}$, then letting $X = A_W$, $Y = B_W$, and $Z = C_W$ will produce $X, Y,$ and Z which satisfy the properties claimed in the theorem.
□

The proof of theorem 14 actually establishes the following result.

THEOREM 15. Let α be any infinite RET. Then there exists $X, Y \in \alpha$ and a set $C \subseteq \{0\}^*$ such that $X \oplus C \sim_p Y \oplus C$ but $X \not\sim_p Y$. Moreover there exists a set C such that if α is an isol, then C is recursively isolated and if α is a co-simple isol, then C is a co-simple set.

Let P be the property of being an r. e. set. In RET's, α is recursively isolated iff there is no $X \in \alpha$ which contains an infinite set with property P. The counterexamples to additive cancellation in PET's above indicate that there is no property Q of sets of integers which can be used as a definition of "polynomially isolated set" and which is both of the form

"X is a 'polynomially isolated set' iff X contains no infinite subset with property Q",

and is also such that the PET's of "polynomially isolated sets" obey additive cancellation. So what definition of "polynomially isolated" can we adopt to get additive and multiplicative cancellation? To find out, we go over Dekker's proof of multiplicative cancellation for isols.

§3. Dekker's proof of multiplicative cancellation for isols.   Assume that A, B, and C are recursively isolated sets with C non-empty, and that f is a 1:1 partial recursive function such that $A \times C \subseteq \mathrm{dom}(f)$ and $f(A \times C) = B \times C$. Let $c_0 \in C$. The cancellation law is proved if we construct a 1:1 partial recursive function g such that $A \subseteq \mathrm{dom}(g)$ and $g(A) = B$. Such a g is constructed in stages. At any given stage s, we assume we have defined g on some finite set D_s and we attempt to extend our domain of definition of g. For any $x \notin D_s$ such that $(x, c_0) \in \mathrm{dom}(f)$, we construct a sequence of pairs of sets

$$(A_1, C_1), (A_2, C_2), (A_3, C_3), \ldots$$

as follows.

Step 1. Compute $f(<x, c_0>) = <y, d_0>$. If $d_0 = c_0$ and $y \notin g(D_s)$, stop. Otherwise let

$C_1 = \{c_0, d_0\}$,

$A_1 = \{x\}$ if $y \notin g(D_s)$,

$A_1 = \{x, g^{-1}(y)\}$ if $y \in g(D_s)$.

Step s. Having defined A_{s-1} and C_{s-1}, compute $f(A_{s-1} \times C_{s-1})$. If $f(A_{s-1} \times C_{s-1})$ is of the form $B_{s-1} \times C_{s-1}$, stop. Otherwise let

$C_s = C_{s-1} \cup \{d \mid \exists y\ (<y, d> \in f(A_{s-1} \times C_{s-1}))\}$ and

$A_s = A_{s-1} \cup \{g^{-1}(y) \mid y \in g(D_s) \wedge \exists d\ (<y, d> \in f(A_{s-1} \times C_{s-1}))\}$.

It is easy to prove by induction that if $x \in A$, and we have defined g up to stage s so that for all x, $x \in A$ iff $g(x) \in B$, then $A_i \subseteq A$ and $C_i \subseteq C$ for all i.

Since A and C are isolated and

$A_1 \subseteq A_2 \subseteq A_3 \subseteq \ldots$ and

$C_1 \subseteq C_2 \subseteq C_3 \subseteq \ldots$

are increasing r.e. sequences of sets within A and C respectively, it follows that these sequences must be finite. Hence there is some step t such that $f(A_t \times C_t) = B_t \times C_t$. It then follows that $A_t \subseteq A$, $B_t \subseteq B$, and $|A_t| = |B_t|$. Then the idea is to show that the construction ensures $|A_t - D_s| = |B_t - g(D_s)|$, so that we can extend g to $A_t - D_s$ by matching the elements of $A_t - D_s$ to $B_t - f(D_s)$ in some specified manner.

At the same time, we must ensure that g maps A onto B, so for each $y \notin g(D_s)$, we build a similar sequence of sets starting from $<y, c_0>$ but using f^{-1} and g^{-1} in place of f and g in the previous definitions.

As the stages progress, we gain more and more knowledge about f. We will be able to compute longer and longer initial sements of the sequence

$$(A_0, C_0), (A_1, C_1), (A_2, C_2), \dots$$

and hence will be able to extend the definition of g. (Of course, we have to show that one can match such sets in a consistent manner, which Dekker did by a lemma similar to the one which we shall use in theorem 17 to follow.) Let us reformulate the construction above using partial recursive monotone set maps. Partially order pairs of sets (A', C') by

$$(A', C') \leq (A'', C'') \text{ iff } A' \subseteq A'', C' \subseteq C''.$$

The only place Dekker used the fact that $A \times C$ and $B \times C$ are recursively isolated was to conclude that the monotone increasing sequence of pairs of finite sets we constructed

$$(A_1, C_1) \leq (A_2, C_2) \leq (A_3, C_3) \dots,$$

starting off with any pair of finite sets (A_1, C_1) which is $\leq (A, C)$, closes off to become a fixed finite set

$$(A_n, C_n) = (A_{n+1}, C_{n+1}) = (A_{n+2}, C_{n+2}) \dots$$

in a finite number of steps. But the transition from $A_i \times C_i$ to $A_{i+1} \times C_{i+1}$ is uniformly recursive, so in mapping (A_n, C_n) to (A_{n+1}, C_{n+1}) we are iterating a partial recursive monotone increasing set map φ that maps pairs of finite sets $(A_1, C_1) \leq (A, C)$ to larger pairs of finite sets $(A_2, C_2) \leq (A, C)$. What recursive isolation of $A \times C$ provided to Dekker for his proof of multiplicative cancellation was this.

* For partial recursive monotone increasing map φ on pairs of finite sets, mapping pairs of finite sets $\leq (A, C)$ to pairs of finite sets $\leq (A, C)$, and for any pair of finite sets $F = (A', C') \leq (A, C)$, the sequence $\varphi(F) \leq \varphi^2(F) \leq \varphi^3(F) \leq \dots$ terminates in a finite number of steps with a fixed finite set $\varphi^n(x)$, which can be recognized effectively.

Now fix a finite $C' \subseteq C$. For finite A', define a partial recursive monotone increasing set map by $\varphi_1(A')$ = the first coordinate of $\varphi(A', C')$.

$*_1$ For any finite $A' \subseteq A$, the monotone increasing sequence of finite sets $\varphi_1(A') \subseteq \varphi_1^2(A') \subseteq \varphi_1^3(A')$.. terminates in a finite number of steps in a fixed finite $\varphi_1^m(C')$, which can be effectively recognized.

Now fix a finite set $A' \subseteq A$. For finite C', define a partial recursive monotone increasing set map by $\varphi_2(C')$ = the second coordinate of $\varphi(A', C')$.

$*_2$ For any finite $C' \subseteq C$, the monotone increasing sequence of finite sets $\varphi_2(C') \subseteq \varphi_2^2(C') \subseteq \varphi_2^3(C')$... terminates in a finite number of steps in a fixed finite set $\varphi_2^n(C')$ which can be effectively recognized.

But

$$\varphi(A', C') = ((\varphi_1^m(A'), \varphi_2^n(C')) \text{ for } (A', C') \leq (A, C).$$

So if for i = 1, 2, $*_i$ is assumed for all possible recursive monotone set maps φ_i of finite subsets of A, C, respectively, this will imply *. This reduces the property of termination for iteration of recursive monotone partial recursive maps of pairs of finite sets $\leq (A, C)$ to a corresponding property $*_i$ for the individual sets A, C. Since $\cup_k \varphi_1^k(A')$, $\cup_k \varphi^k(C')$ are r. e. subsets of recursively isolated sets A, C, respectively, recursive isolation of A, C guarantees $\cup_k \varphi_1^k(A')$, $\cup_k \varphi^k(C')$ are finite, and therefore verifies $*_1$, $*_2$. This is the only place recursive isolation enters.

As far as we know all constructions of recursive equivalences in all isol papers can be naturally rewritten in the same manner using partial recursive monotone increasing set maps on finite sets. Indeed, this is the import of the frame method of [20], which is based on monotone set maps.

To prove cancellation in RET's we started out with a recursive equivalence f and had to produce a recursive equivalence g. In PET's we will start out with a stronger hypothesis, namely that f is an honest p–time partial function. But we have to prove a much stronger conclusion, namely that g is an honest p–time partial function. To guarantee this for g, we need to know that $\varphi(A', C')$ is a "p–time map" closing off, when iterated, at a stage k in such a way that the total computation of $\varphi^k(A', C')$ is polynomial time in the sum of the lengths of A' and C'. We will be able to conclude this if we assume that $\varphi_1(A')$, $\varphi_2(C')$ are "p–time maps", each closing off, when iterated, at stages m and n in such a way that the total computation of $\varphi_1^m(A')$ and $\varphi_2^n(C')$ is polynomial time in the data too. This suggests we should use as a definition of polynomially isolated A, that iterating

$$\psi(A'), \psi^2(A'), \psi^3(A'), \ldots$$

any p–time monotone map ψ of finite subsets A into finite subsets of A terminates in polynomial time in the length of A'. A more exact definition along these lines follows.

§4. Polynomially isolated sets and cancellation. We need some notation for strings.

(1) Given a finite set $X \subseteq \{0\}^*$, we identify X with the string $\overline{X} = x_1 1 x_2 1 \ldots x_n 1$ where $X = \{x_1, \ldots, x_n\}$ and $|x_1| < \ldots < |x_n|$.

If X and Y are finite subsets of $\{0\}^*$, we write $\overline{X} \cup \overline{Y}$ for $\overline{X \cup Y}$ and $\overline{X} \cap \overline{Y}$ for $\overline{X \cap Y}$. Similarly we write $\overline{X} \subseteq \overline{Y}$ if $X \subseteq Y$, etc.

(2) Let $A \subseteq \{0\}^*$. Define fin(A) as $\{\overline{X} \mid X \subseteq A \wedge X \text{ is finite}\}$. Note that for any p–time set $A \subseteq \{0\}^*$, fin(A) is also a p–time set.

(3) Let $\overline{X}_i \in \text{fin}(\{0\}^*)$. Then write $|\bigcup_{i \geq 0} X_i| \leq n$ if $\bigcup_{i \geq 0} X_i = X$, where X is a finite set and $|X| \leq n$. We write $|\bigcup_{i \geq 0} X_i| > n$ if not $|\bigcup_{i \geq 0} X_i| \leq n$, i.e., if either $\bigcup_{i \geq 0} X_i$ is infinite or $\bigcup_{i \geq 0} X_i = X$ is finite but $|X| > n$.

(4) Let $p: \text{fin}(\{0\}^*) \to \text{fin}(\{0\}^*)$. Then p is a <u>p–time monotone set map</u> if p is a partial p–time function such that

(i) $\text{dom}(p) = \text{fin}(\{0\}^*)$,

(ii) $\text{rng}(p) \subseteq \text{fin}(\{0\}^*)$, and

(iii) for all $\overline{X}, \overline{Y} \in \text{fin}(\{0\}^*)$, $\overline{X} \subseteq \overline{Y} \to p(\overline{X}) \subseteq p(\overline{Y})$.

DEFINITION 5. A set $A \subseteq \{0\}^*$ is <u>polynomially isolated</u> (<u>p–isolated</u>) iff for every p–time monotone set map $p: \mathrm{fin}(\{0\}^*) \to \mathrm{fin}(\{0\}^*)$ such that

$$\text{for all } \overline{X} \in \mathrm{fin}(A),\ p(\overline{X}) \in \mathrm{fin}(A),$$

there is a polynomial q such that

$$\text{for all } \overline{X} \in \mathrm{fin}(A),\ \left|\bigcup_{n=1}^{\infty} p^n(\overline{X})\right| \le q(|\overline{X}|).$$

DEFINITION 6. A set $A \subseteq \{0\}^*$ is <u>p–finite</u> if for any partial p–time map p such that $\mathrm{dom}(p) \supseteq \mathrm{fin}(A)$ and p maps $\mathrm{fin}(A)$ into $\mathrm{fin}(A)$, there is a polynomial q such that for all $\overline{X} \in \mathrm{fin}(A)$,

$$\left|\bigcup_{i=1}^{\infty} p^n(\overline{X})\right| \le q(|\overline{X}|).$$

In our existence proof for p–isolated sets, we will in fact produce p–finite sets. We postpone proofs of existence until after we prove that p–isolated sets satisfy additive and multiplicative cancellation in the PET's. First observe that p–finiteness and p–isolation are preserved under polynomial time equivalence.

PROPOSITION 16. Suppose $X, Y \subseteq \{0\}^*$ and $X \sim_p Y$. Then X is p–finite iff Y is p–finite. Also X is p–isolated iff Y is p–isolated.

PROOF. Let f be the 1:1 partial p–time honest function such that $\mathrm{dom}(f)$, $\mathrm{rng}(f) \subseteq \{0\}^*$, $X \subseteq \mathrm{dom}(f)$, and $f(X) = Y$. By lemma 9 we know that f^{-1} is also a 1:1 partial p–time map.

Suppose g is any 1:1 partial p–time map whose domain and range are contained in $\{0\}^*$. Let $\overline{g}$ be the partial p–time set map whose domain and range is contained in $\mathrm{fin}(\{0\}^*)$ such that for any $\overline{X} \in \mathrm{fin}(\{0\}^*)$,

$$\overline{g}(\overline{X}) = \begin{cases} \overline{g(X)} & \text{if } X \subseteq \mathrm{dom}(g), \text{ else} \\ \overline{X}. \end{cases}$$

By our definition of partial p–time map, we know that $\mathrm{dom}(g)$ is a p–time set and hence we can decide in polynomial time in $|\overline{X}|$ whether $X \subseteq \mathrm{dom}(g)$. It follows that $\overline{g}$ is a partial p–time function with $\mathrm{dom}(\overline{g}) = \mathrm{fin}(\{0\}^*)$. Let $r(x)$ be a polynomial with positive coefficients which bounds the running time of g. Then for any $X = \{x_1,\dots,x_n\} \subseteq \{0\}^*$, we have that $|\bar{X}| = n+|x_1|+\dots+|x_n|$ and $|\bar{g}(X)| = n+|g(x_1)|+\dots+|g(x_n)|$. It follows that

$$(4.1) \quad |\bar{g}(X)| = n+|g(x_1)|+\dots+|g(x_n)| \le n+r(|x_1|)+\dots+r(|x_n|) \le r(n+|x_1|+\dots+|x_n|) = r(|\bar{X}|).$$

Suppose that X is p–finite (p–isolated) and that p is a partial p–time map (p–time monotone set map) such that dom(p), rng(p) $\subseteq$ fin($\{0\}^*$), fin(Y) $\subseteq$ dom(p), and $p \restriction Y$: fin(Y) $\to$ fin(Y). Let p_X be defined by $p_X = \overline{f^{-1}} \circ p \circ \bar{f}$. It is then easy to see that p_X is a partial p–time map (p–time monotone set map) such that dom(p_X), rng(p_X) $\subseteq \{0\}^*$, fin(X) $\subseteq$ dom(p_X), and $p_X \restriction$ fin(X): fin(X) $\to$ fin(X). Moreover note that for any $\bar{B} \in$ fin(Y), $\bar{f} \circ \overline{f^{-1}}(\bar{B}) = \bar{B}$ so that $p_X^n(\bar{A}) = \overline{f^{-1}} \circ p^n \circ \bar{f}(\bar{A})$ for any $\bar{A} \in$ fin(X). We claim that for any $\bar{B} \in$ fin(Y)

$$(4.2) \quad \bigcup_{n\ge 0} p^n(\bar{B}) = \bar{f}(\bigcup_{n\ge 0} p_X^n(\overline{f^{-1}}(\bar{B}))).$$

That is, since X is p–finite (p–isolated), we know that there is a fixed polynomial h(x) such that for any $\bar{A} \in$ fin(X),

$$(4.3) \quad |\bigcup_{n\ge 0} p_X^n(\bar{A})| \le h(|\bar{A}|).$$

In particular, $\bigcup_{n\ge 0} p_X^n(\bar{A}) \in$ fin(X). Note also for any $\bar{C} \in$ fin(X), $\overline{f^{-1}} \circ \bar{f}(\bar{C}) = \bar{C}$. Thus if $\bar{A} = \overline{f^{-1}}(\bar{B})$,

$$\bar{f}(\bigcup_{n\ge 0} p_X^n(\bar{A})) = \bigcup_{n\ge 0} \bar{f} \circ p_X^n(\bar{A}) = \bigcup_{n\ge 0} \bar{f} \circ \overline{f^{-1}} \circ p^n \circ \bar{f} \circ \overline{f^{-1}}(\bar{B}) = \bigcup_{n\ge 0} p^n(\bar{B}).$$

Now let q_1 and q_2 be polynomials with positive coefficients which bound the running times of f and f^{-1} respectively. Then by (4.1) and (4.3), we have

$$|\bigcup_{n\ge 0} p^n(\bar{B})| = |\bar{f}(\bigcup_{n\ge 0} p_X^n(\overline{f^{-1}}(\bar{B}))| \le q_1(|\bigcup_{n\ge 0} p_X^n(\overline{f^{-1}}(\bar{B}))|)$$

$$\leq q_1(h(|\overline{f^{-1}(\bar{B})}|)) \leq q_1(h(q_2(|\bar{B}|)).$$

Thus we see that $q_1 \circ h \circ q_2(x)$ is the polynomial which witness that Y is p–finite (p–isolated). □

Given proposition 16, we see that proving additive and multiplicative cancellation for PET's of p–isolated sets reduces to proving the following theorem.

THEOREM 17. Let $A, B, C \subseteq \{0\}^*$.
(i) Suppose $A \oplus C \sim_p B \oplus C$ and $A \oplus C$ and $B \oplus C$ are p–isolated. Then $A \sim_p B$.
(ii) Suppose $A \times C \sim_p B \times C$ and $A \times C$ and $B \times C$ are p–isolated. Then $A \sim_p B$ if $C \neq \phi$.

PROOF OF ADDITIVE CANCELLATION.

Let φ be a 1:1 partial honest p–time function such that $\mathrm{dom}(\varphi) \supseteq A \oplus C$ and $\varphi(A \oplus C) = B \oplus C$. By lemma 9, φ^{-1} is also a 1:1 partial honest p–time function. We shall use φ to construct a 1:1 partial honest p–time function f whose domain and range are contained in $\{0\}^*$ and for which $f(A) = B$.

We noted previously that there is no loss in generality in assuming $\mathrm{dom}(\varphi), \mathrm{rng}(\varphi) \subseteq \{0\}^*$. First we use φ and φ^{-1} to define a pair of p–time monotone set maps. Given $\bar{X} \in \mathrm{fin}(\{0\}^*)$, let

$$p(\bar{X}) = \bar{X} \cup \overline{\{0^{2x+1} \mid \exists y \in X(\varphi(y) = 0^{2x+1})\}},$$

$$q(\bar{X}) = \bar{X} \cup \overline{\{0^{2x+1} \mid \exists y \in X(\varphi^{-1}(y) = 0^{2x+1})\}}.$$

It is easy to see that both p and q are p–time monotone set maps and that for all $\bar{X} \in \mathrm{fin}(A \oplus C)$, $p(\bar{X}) \in \mathrm{fin}(A \oplus C)$, and for all $\bar{Y} \in \mathrm{fin}(B \oplus C)$, $q(\bar{Y}) \in \mathrm{fin}(B \oplus C)$. Note if $0^{2x} \in A \oplus C$, i.e., if $0^x \in A$, then we have the following picture for the computation of $\bigcup_{n \geq 1} p^n(\bar{X})$, where $\bar{X} = \overline{\{0^{2x}\}}$.

for all $k \geq 3$.

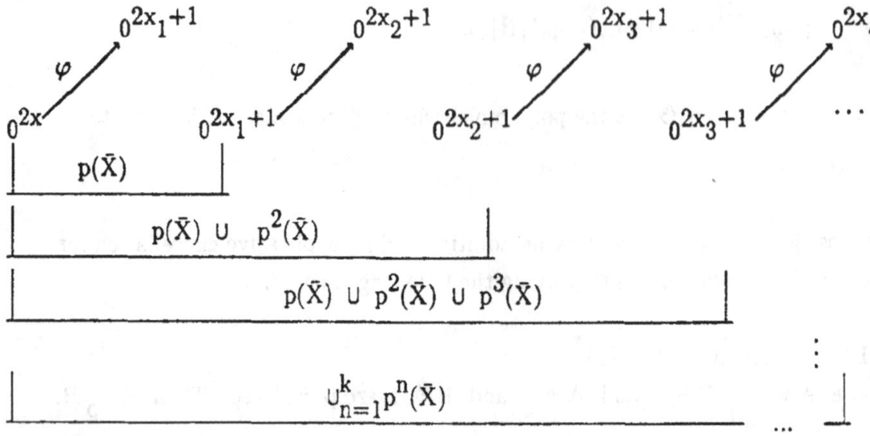

That is, as we iterate $p^n(\bar{X})$, we keep picking up more and more odd length elements until we finally hit an $x' \in p^n(\bar{X})$ such that $\varphi(x')$ has even length. (In the situation pictured above $0^x \in A$ iff $0^{x_4} \in B$.) Note that our assumption that $A \oplus C$ is p–isolated ensures that if $\bar{X} = \overline{\{0^{2x}\}}$ where $0^x \in A$, then there is an $n \geq 1$ such that $\bar{X} \subset p(\bar{X}) \subset p^2(\bar{X}) \subset \ldots \subset p^{n-1}(\bar{X}) = p^n(\bar{X}) = p^{n+1}(\bar{X}) = \ldots$ and

$$p^n(\bar{X}) = \bigcup_{k \geq 0} p^k(\bar{X}).$$

In fact, for any $\bar{Y} \in \text{fin}(A \oplus C)$, there must be an $n(Y)$ such that

$$p^{n(Y)-1}(\bar{Y}) = p^{n(Y)}(\bar{Y}) = \bigcup_{k \geq 0} p^k(\bar{Y}).$$

Moreover there is a fixed polynomial r such that $|p^{n(Y)}(\bar{Y})| \leq r(|\bar{Y}|)$ for all $\bar{Y} \in \text{fin}(A \oplus C)$. Thus we know that

$$|\bar{X}| < |p(\bar{X})| < |p^2(\bar{X})| < \ldots < |p^n(\bar{X})| \leq r(2x+1),$$

so that to find n, we need only calculate p on at most $r(2x+1)$ different values. Now if h is the polynomial which bounds the running time of the deterministic transducer which computes p, then one can see that to compute $p^k(\bar{X})$ for $k \leq n$, requires at most

$$\sum_{i=0}^{r(2x+1)} h(i) \leq \sum_{i=0}^{h(r(2x+1))} i = \frac{((h(r(2x+1))+1))(h(r(2x+1)))}{2}$$

steps. Since $\frac{1}{2}(h(r(2x+1))+1)(h(r(2x+1)))$ is a polynomial, it follows that there is some fixed polynomial $b(x)$, such that if $0^x \in A$, then we can find an n and $p^n(\{\overline{0^{2x}}\})$ such that $p^{n-1}(\{\overline{0^{2x}}\}) = p^n(\{\overline{0^{2x}}\})$ in fewer than $b(x)$ steps. Now if $0^x \notin A$, then we have no guarantee that the sequence

$$\{\overline{0^{2x}}\} = X \subset p(X) \subset p^2(X) \subset \dots$$

might not go on forever. But we can start computing the sequence $p(X), p^2(X), \dots$ until we use $b(x)$ steps, at which point we know that if the computation has not stopped, then $0^x \notin A$.

By an entirely similar argument there is a polynomial $c(x)$ such that if $0^{2x} \in B \oplus C$, then within $c(x)$ steps, we can find n and $q^n(\{\overline{0^{2x}}\})$ such that

$$q^{n-1}(\{\overline{0^{2x}}\}) = q^n(\{\overline{0^{2x}}\}) = \bigcup_{k \geq 0} q^k(\{\overline{0^{2x}}\}).$$

Moreover if we start with $X = \{\overline{0^{2x}}\}$ and compute the sequence $q(X) \subset q^2(X) \subset \dots$ and we do not find an n such that $q^{n-1}(X) = q^n(X)$ within $c(x)$ steps, then we know $0^x \notin B$.

This given, it is now easy to see that we can define our desired functions f and f^{-1} inductively as follows.

Stage 2x. We define $f(0^x)$ as follows. Begin the computation of the sequence

$$p(X) \subseteq p^2(X) \subseteq p^3(X) \subseteq \dots$$

for $b(x)$ steps where $X = \{\overline{0^{2x}}\}$. If within $b(x)$ steps we find n such that $p^{n-1}(X) = p^n(X)$, then compute $\varphi(z)$ for each $z \in p^n(X) = \bigcup_{k \geq 0} p^k(X)$. Now if $p^n(X) - \{\overline{0^{2x}}\} = Z$, then every element of Z must be of odd length and there can be at most one $z' \in p^n(X)$ such that $\varphi(z')$ is of even length. If there is such a $z' \in p^n(X)$, then let y be such that $0^{2y} = \varphi(z')$. (Note we will try to compute $f^{-1}(0^y)$ at stage $2y+1$ and it is crucial that its computation is consistent.) Then compute

$$q(Y) \subseteq q^2(Y) \subseteq q^3(Y) \subseteq \dots,$$

(where $Y = \{\overline{0^{2y}}\}$) for $c(y)$ steps. (Note since $|y|$ is bounded by a polynomial in x, this computation will also be bounded by a polynomial in x.) If within $c(y)$ steps we find m such that $q^m(Y) = q^{m-1}(Y)$, then it is easy to check that $q^m(Y) = \{\overline{0^{2y}}\} \cup \overline{Z}$ so we can define $f(0^x) = 0^y$. If we find no such m within $c(y)$ steps, then $0^y \notin B$ and we would declare 0^y not in $\mathrm{dom}(f^{-1})$ so we declare $0^x \notin \mathrm{dom}(f)$. Finally, if we cannot find such an n within $b(x)$ steps or there is no such $z' \in p^n(X)$, then declare $0^x \notin \mathrm{dom}(f)$.

<u>Stage 2y+1</u>. Define $f^{-1}(0^y)$ as follows. Begin the computation of the sequence

$$q(Y) \subseteq q^2(Y) \subseteq q^3(Y) \subseteq \ldots,$$

(where $Y = \{\overline{0^{2y}}\}$) for $c(y)$ steps. If within $c(y)$ steps we find m such that $q^{m-1}(Y) = q^m(Y)$, then compute $\varphi^{-1}(z)$ for each $z \in q^m(Y) = \bigcup_{k \geq 0} q^k(Y)$. Note if

$q^m(Y) - \{\overline{0^{2y}}\} = W$, then every element of W must be of odd length. Moreover there can be at most one $w' \in W$ such that $\varphi^{-1}(w')$ is of even length. If there is such a $w' \in W$, then let $0^{2x} = \varphi^{-1}(w')$. (Again $|x|$ will be bounded by a polynomial in y so we can run the computation of $f(0^x)$ to check that there are no consistency problems.) Then compute

$$p(X) \subseteq p^2(X) \subseteq p^3(X) \subseteq \ldots,$$

(where $X = \{\overline{0^{2x}}\}$) for $b(x)$ steps. If within $b(x)$ steps we find an n such that $p^n(X) = p^{n-1}(X)$, then $p^n(X) = \overline{W} \cup \{\overline{0^{2x}}\}$. We then let $f^{-1}(0^y) = 0^x$. Otherwise $0^x \notin \mathrm{dom}(f)$ so declare $0^y \notin \mathrm{dom}(f^{-1})$.

Finally if we cannot find such an m within $c(y)$ steps or there is no such $z' \in p^n(X)$, then declare $0^y \notin \mathrm{dom}(f^{-1})$. This completes our construction.

It is now routine, given our discussion preceding the construction, to see that both f and f^{-1} are partial p–time functions whose domain and range are contained in $\{0\}^*$. Moreover, since φ is 1:1 and $\varphi(A \oplus C) = B \oplus C$, it is easy to prove that the definitions of f and f^{-1} are consistent, $\mathrm{dom}(f) \supseteq A$, $\mathrm{dom}(f^{-1}) \supseteq B$, and $0^x \in A$ iff $f(0^x) \in B$. Thus f is a 1:1 partial honest p–time function such that $A \subseteq \mathrm{dom}(f)$ and $f(A) = B$. Hence $A \sim_p B$.

PROOF OF MULTIPLICATIVE CANCELLATION.

There is no loss in generality in assuming that

$$A \subseteq \{0^{2x} \mid x \geq 0\} \text{ and } B \subseteq \{0^{2x+1} \mid x \geq 0\}.$$

Let ψ be a 1:1 partial honest p–time function which witnesses that $A \times C \sim_p B \times C$. Note that

$$A \times C \subseteq \{<0^{2y}, 0^x> \mid y \geq 0, x \geq 0\},$$

$$B \times C \subseteq \{<0^{2y+1}, 0^x> \mid y \geq 0, x \geq 0\},$$
$$\{<0^{2y}, 0^x> \mid y \geq 0, x \geq 0\},$$
$$\{<0^{2y+1}, 0^x> \mid x \geq 0, y \geq 0\}$$

are all p–time sets.

Thus there is no loss in generality in assuming
$$A \times C \subseteq \text{dom}(\psi) \subseteq \{<0^{2y}, 0^x> \mid y > 0, x \geq 0\},$$
$$B \times C \subseteq \text{rng}(\psi) \subseteq \{<0^{2y+1}, 0^x> \mid y \geq 0, x \geq 0\}, \text{ and}$$
$$\psi(A \times C) = B \times C.$$

In particular $\text{dom}(\psi)$, $\text{rng}(\psi) \subseteq \{0\}^*$ so by lemma 9, ψ^{-1} is a 1:1 partial honest p–time function. We shall use ψ to construct a 1:1 honest p–time function f such that $\text{dom}(f) \subseteq \{0^{2x} \mid x \geq 0\}$, $\text{rng}(f) \subseteq \{0^{2x+1} \mid x \geq 0\}$, $A \subseteq \text{dom}(f)$, and $f(A) = B$. First we use ψ and ψ^{-1} to define a pair of p–time monotone set maps α and β. Given $X \in \text{fin}(\{0\}^*)$, let
$$X_1 = \{0^y \mid \exists 0^z(<0^y, 0^z> \in X)\} \text{ and}$$
$$X_2 = \{0^z \mid \exists 0^y(<0^y, 0^z> \in X\}.$$

It is easy to see that given X, we can find $\overline{X_1 \times X_2}$ in polynomial time in the length of X. Then let $\alpha(X) = s(X) \times t(X)$ where $s(X)$ and $t(X)$ are constructed by the following three step process.

Step 1. Form $X_1 \times X_2$.

Step 2. Form $Y_1 \times Y_2$, where
$$Y_1 = \{0^y \mid \exists x \in X_1 \times X_2 \ \exists w \ (\psi(x) = <0^y, 0^w>)\}, \text{ and}$$
$$Y_2 = X_2 \cup \{0^w \mid \exists x \in X_1 \times X_2 \ \exists y \ (\psi(x) = <0^y, 0^w>)\}.$$

Step 3. $s(X) = X_1 \cup \{0^x \mid \exists y \in Y_1 \times Y_2 \ \exists z \ (\psi^{-1}(y) = <0^x, 0^z>)\}$ and
$$t(X) = Y_2 \cup \{0^z \mid \exists y \in Y_1 \times Y_2 \ (\psi^{-1}(y) = <0^x, 0^z>)\}.$$

Similarly, we let $\beta(X) = u(X) \times v(X)$ where $u(X)$ and $v(X)$ are constructed by the following three step process.

Step 1. Form $X_1 \times X_2$.

Step 2. Form $Z_1 \times Z_2$ where

$Z_1 = \{0^y \mid \exists x \in X_1 \times X_2\ \exists w\ (\psi^{-1}(x) = <0^y, 0^w>)\}$ and

$Z_2 = X_2 \cup \{0^w \mid \exists x \in X_1 \times X_2\ \exists y\ (\psi^{-1}(x) = <0^y, 0^w>)\}$.

Step 3. $u(\overline{X}) = X_1 \cup \{0^x \mid \exists y \in Z_1 \times Z_2\ \exists z\ (\psi(y) = <0^x, 0^z>)\}$ and

$v(\overline{X}) = Z_2 \cup \{0^z \mid \exists y \in Y_1 \times Y_2\ \exists z\ (\psi(y) = <0^x, 0^z>)\}$.

It is routine to check that α and β are p–time monotone set maps. Now let c_0 be the minimal element of C which exists since we assume C is non–empty. Then suppose that $0^x \in A$ and consider the sequence

$$\alpha(\overline{X}) \subseteq \alpha^2(\overline{X}) \subseteq \alpha^3(\overline{X}) \subseteq \ldots,$$

where $\overline{X} = \overline{\{<0^x, c_0>\}}$. Because $A \times C$ is p–isolated there must be an $n \geq 1$ such that $\overline{X} \subset \alpha(\overline{X}) \subset \alpha^2(\overline{X}) \subset \ldots \subset \alpha^{n-1}(\overline{X}) = \alpha^n(\overline{X}) = \alpha^{n+1}(\overline{X}) = \ldots$ and $\alpha^n(\overline{X}) = \bigcup_{k \geq 0} \alpha^k(\overline{X})$.

Since $A \times C$ is p–isolated, it must be the case that if $\overline{Y} \in \text{fin}(A \times C)$, there is an $n(Y)$ such that $\alpha^{n(Y)-1}(\overline{Y}) = \alpha^{n(Y)}(\overline{Y}) = \bigcup_{k \geq 0} \alpha^k(\overline{Y})$. Moreover there is a fixed polynomial r such that $|\overline{\alpha^{n(Y)}(\overline{Y})}| \leq r(|\overline{Y}|)$ for all $\overline{Y} \in \text{fin}(A \times C)$. Thus in our case, we know that

$$|\overline{X}| < |\alpha(\overline{X})| < |\alpha^2(\overline{X})| < \ldots < |\alpha^n(\overline{X})| \leq r(|<0^x, c_0>|)$$
$$= r(\tfrac{1}{2}(x+|c_0|)^2+3x+|c_0|+2).$$

Hence to find n, we need only calculate p on at most $r(\frac{1}{2}(x+|c_0|)^2+3x+|c_0|+2)$ different values. Let h be the polynomial which bounds the running time of the deterministic transducer which computes p, then one can see that to compute $\alpha^k(\overline{X})$ for $k \leq n$, requires at most

$$\sum_{i=0}^{r(\frac{1}{2}(x+|c_0|)^2+3x+|c_0|+2)} h(i) \leq \sum_{i=0}^{h(r(\frac{1}{2}(x+|c_0|)^2+3x+|c_0|+2))} i$$

$$= \frac{[h(r(\frac{1}{2}(x+|c_0|)^2+3x+|c_0|+2))+1][h(r(\frac{1}{2}(x+|c_0|)^2+3x+|c_0|+2))]}{2}$$

steps. Thus there is some fixed polynomial $a(x)$ such that if $0^x \in A$, then we can find n and $\alpha^n(\overline{\{<0^x, c_0>\}})$, where $\alpha^{n-1}(\overline{\{<0^x, c_0>\}}) = \alpha^n(\overline{\{<0^x, c_0>\}})$, in fewer than $a(x)$ steps. It is easy to see from our definition of α that the only way $\alpha(\overline{X}) = \overline{X}$ is if $\overline{X} = \overline{X_1 \times X_2}$ and $\psi(X_1 \times X_2) = Y_1 \times X_2$. Thus if $\overline{X} = \overline{\{<0^x, c_0>\}}$, where $0^x \in A$ and $\alpha^{n-1}(\overline{X}) = \alpha^n(\overline{X})$, then we must have $\alpha^n(\overline{X}) = \overline{A_x \times C_x}$ and $\psi(A_x \times C_x) = B_x \times C_x$, where $A_x \subsetneq A$, $B_x \subsetneq B$, and $C_x \subsetneq C$. It follows that under such circumstances that there is a polynomial b such that we can find $\overline{A}_x$, $\overline{B}_x$ and $\overline{C}_x$ in fewer than $b(x)$ steps and $\text{card}(A_x) = \text{card}(B_x)$.

In case $0^x \notin A$, then $\overline{\{<0^x, c_0>\}} \notin \text{fin}(A \times C)$. We have no guarantee that $\bigcup_{n \geq 0} \alpha^n(\overline{\{<0^x, c_0>\}})$ is even finite. Nevertheless, we can start the computation of

$$\alpha(\overline{X}) \subsetneq \alpha^2(\overline{X}) \subsetneq \alpha^3(\overline{X}) \subsetneq \ldots,$$

where $\overline{X} = \overline{\{<0^x, c_0>\}}$. If in fewer than $a(x)$ steps we find n and $\alpha^n(\overline{X})$ such that $\alpha^{n-1}(\overline{X}) = \alpha^n(\overline{X})$, then as before $\alpha^n(\overline{X})$ must be of the form $\overline{A_x \times C_x}$ where $\psi(A_x \times C_x) = B_x \times C_x$, and we can find A_x, B_x, and C_x in fewer than $b(x)$ steps. In such a case, we will say A_x, B_x, and C_x are defined. However if within $a(x)$ steps, we find no such n, then we say A_x, B_x, and C_x are undefined. The key point to note is that within $a(x)+b(x)$ we can either find A_x, B_x, and C_x or determine that A_x, B_x, and C_x are undefined.

By a completely symmetric argument there are polynomials $c(x)$ and $d(x)$ such that if $0^y \in B$, then we can find n and $\beta^n(\overline{\{<0^y, c_0>\}})$ within $c(y)$ steps where

$$\beta^{n-1}(\overline{\{<0^y, c_0>\}}) = \beta^n(\overline{\{<0^y, c_0>\}}) = \bigcup_{k \geq 0} \beta^k(\overline{<0^y, c_0>}).$$

Moreover it must be the case that

$$\beta^n(\overline{\{<0^y, c_0>\}}) = D_y \times C_y, \psi^{-1}(D_y \times C_y) = E_y \times C_y,$$

where $D_y \subsetneq A$, $E_y \subsetneq B$, and $C_y \subsetneq C$, and we can find $\overline{D}_y$, E_y, and $\overline{C}_y$ in fewer than $d(y)$ steps. As before if $0^y \notin B$, we will start the computation of

$$\beta(\Upsilon) \subsetneq \beta^2(\Upsilon) \subsetneq \beta^3(\Upsilon) \subsetneq ...,$$

where $\Upsilon = \overline{\{<0^y, c_0>\}}$. If within $c(y)$ steps, we find n such that $\beta^n(\Upsilon) = \beta^{n-1}(\Upsilon)$, then $\beta^n(\Upsilon) = \overline{D_y \times C_y}$, where $\psi(D_y \times C_y) = E_y \times C_y$ and we can find D_y, E_y, and C_y in fewer than $d(y)$ steps. In such a case we say D_y, E_y, and C_y are defined. If we find no such n within $c(y)$ steps, then we say D_y, E_y, and C_y are undefined. Thus for any y, we will either find D_y, E_y, and C_y in fewer than $c(y)+d(y)$ steps or determine if D_y, E_y, and C_y are undefined.

Given $Z \subsetneq \{0\}^*$, let

$$Z_e = \{0^x \in Z \mid x \text{ is even}\} \text{ and } Z_o = \{0^x \in Z \mid x \text{ odd}\}.$$

We say a finite set $Z \subsetneq \{0\}^*$ is a block if there is a finite set $F \subsetneq \{0\}^*$ such that F is non–empty, $Z_e \times F \subsetneq \mathrm{dom}(\psi)$, and $\psi(Z_e \times F) = Z_o \times F$. Note that if Z is a block, then $\mathrm{card}(Z_e) = \mathrm{card}(Z_o)$. Moreover if Z and W are blocks, then $Z \cap W$ is a block. That is, let E and F be such that $Z_e \times F$, $W_e \times E \subsetneq \mathrm{dom}(\psi)$, $\psi(Z_e \times F) = Z_o \times F$, and $\psi(W_e \times E) = W_o \times E$. Then it is easy to see that

$$\psi((Z_e \cap W_e) \times (F \cap E)) = (Z_o \cap W_o) \times (F \cap E), \text{ so that}$$

$$Z \cap W = (Z_e \cap W_e) \cup (Z_o \cap W_o) \text{ is a block.}$$

We use the following lemma of Manaster.

LEMMA 18. Suppose we have a collection of finite subsets of $\{0\}^*$, which we call blocks, satisfying the following two properties.

(i) Any intersection of two blocks is a finite union of blocks.

(ii) If B is any block, then $\mathrm{card}(B_e) = \mathrm{card}(B_o)$.

Let A be any finite union of blocks. Then $\mathrm{card}(A_e) = \mathrm{card}(A_o)$.

PROOF. We proceed by induction on card(A). If card(A) = 2, then clearly A must be a block by itself and hence $\mathrm{card}(A_e) = \mathrm{card}(A_o)$ by property (ii). For an induction, suppose $A = \bigcup_{i=1}^{n} B_i$, where B_i is a block for i = 1,...,n and B is a block such that B–A is non–empty. Clearly $(A \cup B)_e = A_e + (B-A)_e$, where + denotes disjoint union, and $(A \cup B)_o = A_o + (B-A)_o$. Since by induction $\mathrm{card}(A_e) = \mathrm{card}(A_o)$, we need only show $\mathrm{card}((B-A)_e) = \mathrm{card}((B-A)_o)$. Now $B = (B-A)+(B \cap A)$. Hence $\mathrm{card}((B-A)_e) = \mathrm{card}(B_e) - \mathrm{card}((B \cap A)_e)$, and $\mathrm{card}((B-A)_o) = \mathrm{card}(B_o) - \mathrm{card}((B \cap A)_o)$.

But since B is a block, $\mathrm{card}(B_e) = \mathrm{card}(B_o)$. Hence we need only show $\mathrm{card}((B \cap A)_e) = \mathrm{card}((B \cap A)_o)$. But $B \cap A = B \cap (\bigcup_{i=1}^{n} B_i) = \bigcup_{i=1}^{n} (B \cap B_i)$. Since for each i, $B \cap B_i$ is a finite union of blocks of property (i), it follows that $B \cap A$ is a finite union of blocks. But $\mathrm{card}(B \cap A) \leq \mathrm{card}(A)$ so by our induction hypothesis $\mathrm{card}((B \cap A)_e) = \mathrm{card}((B \cap A)_o)$. □

Next we need to study the sets $A_x \times C_x$ and $B_x \times C_x$ with $0^x \in \{0^{2x} \mid z \geq 0\}$, and also the sets $D_y \times C_y$ and $E_y \times C_y$ with $0^y \in \{0^{2z+1} \mid z \geq 0\}$. Our basic idea is to match off the elements in A_x with the elements of B_x and to match off the elements in D_y with the elements of E_y to build up our desired function f. We can appeal to lemma 18 to show that we can always extend our matchings, however we must be careful if we are going to ensure that f(A) = B. It is easy to see that

(i) $0^x \in A$ implies $A_x \subseteq A$, $B_x \subseteq B$, $C_x \subseteq C$, and

(ii) $0^y \in B$ implies $D_y \subseteq A$, $E_y \subseteq B$, $C_y \subseteq C$.

However if $0^x \notin A$, it may be that $\psi(<0^x, c_0>) = <m', c'>$, where $m' \in B$ but $c' \notin C$. Then $\psi^{-1}(<m', c_0>) = <n'', c''>$, where $n'' \in A$ and $c'' \in C$. In this way we get $n'' \in A_x \cap A$ and $m' \in B_x \cap B$. Thus if $0^x \notin A$, it may be that $A_x \cap A$, $B_x \cap B$ are non-empty, and $C_x \cap C$ is non-empty. Similarly if $0^y \notin B$, it may be that $D_y \cap A$ is non-empty, $E_y \cap B$ is non-empty, and $C_y \cap C$ is non-empty. Thus in matching elements in A_x to those in B_x and elements in D_y to those in E_y we must be careful that we match elements of A with elements of B and vice versa. However, even in the case where $0^x \notin A$, we still have that $\psi(A_x \times C_x) = B_x \times C_x$. Since $\psi(A \times C) = B \times C$ and ψ is 1:1, it is easy to see that we also have

(iii) $\quad \psi((A_x \cap A) \times (C_x \cap C)) = (B_x \cap B) \times (C_x \cap C).$

Similarly, even if $0^y \notin B$, we still have

(iv) $\quad \psi((D_y \cap A) \times (C_y \cap C)) = (E_y \cap B) \times (C_y \cap C).$

Moreover, it is also the case that for any x, $A_x \times C_x$ is the minimal set of the form $A' \times C'$ such that $<0^x, c_0> \in A' \times C'$ and $\varphi(A' \times C') = B' \times C'$. That is, we get the following.

(v) if $A' \times C'$ is such that $\psi(A' \times C') = B' \times C'$ and $<0^x, c_0> \in A' \times C'$, then $A_x \times C_x \subseteq A' \times C'$ and $B_x \times C_x \subseteq B' \times C'$.

Similarly, we have the following.

(vi) If $A' \times C'$ is such that $\psi(A' \times C') = B' \times C'$ and $<0^y, c_0> \in B' \times C'$, then $D_y \times C_y \subseteq A' \times C'$ and $E_y \times C_y \subseteq B' \times C'$.

Now let $0^x \in \{0^{2z} \mid z \geq 0\}$, and assume A_x, B_x and C_x are defined. Then we know that $c_0 \in C_x$. It follows from (v) and (vi) that for all $0^z \in A_x$ and all $0^w \in B_x$,

$A_z \times C_z \subseteq A_x \times C_x$ and $B_z \times C_z \subseteq B_x \times C_x$ if A_z, B_z, & C_z are defined, and

$D_w \times C_w \subseteq A_x \times C_x$ and $E_w \times C_w \subseteq B_x \times C_x$ if D_w, E_w, and C_w are defined.

We let $\mathscr{P}_x$ be the partially ordered set consisting of

$$\{A_z \cup B_z \mid 0^z \in A_x\} \cup \{D_w \cup E_w \mid 0^w \in B_x\},$$

ordered by inclusion. Let $\mathscr{L}_x$ be some linear order extending this partial order on the same domain. Let $\mathscr{B}_1, \mathscr{B}_2, \dots, \mathscr{B}_p$ be a listing of the domain of $\mathscr{L}_x$ under inclusion. The $\mathscr{B}_i$'s are all blocks in the sense of our earlier definition. Then, given our list of blocks, we can define $f \upharpoonright A_x$ via the following procedure.

Procedure Match. Given a list of blocks $\mathscr{B}_1, \dots, \mathscr{B}_n$, find $f \upharpoonright (\cup \mathscr{B}_i)_e$.

Step 1. Let $\{a_0^0, \dots, a_{n_0}^0\} = (\mathscr{B}_1)_e$, where $|a_0| < \dots < |a_{n_0}|$ and $\{b_0^0, \dots, b_{m_0}^0\} = (\mathscr{B}_1)_o$, where $|b_0^0| < \dots < |b_{m_0}^0|$.

Since $\mathscr{B}_1$ is a block, we know $n_0 = m_0$ so we can define $f(a_i^0) = b_i^0$.

Step $k \geq 1$. Assume we have defined f on $\bigcup_{i=1}^{k-1} (\mathscr{B})_e$. Then let $\{a_0^k, \dots, a_{n_k}^k\} = (\mathscr{B}_k - \bigcup_{i=1}^{k-1} \mathscr{B}_i)_e$, where $|a_0^k| < \dots < |a_{n_k}^k|$, and let $\{b_0^k, \dots, b_{m_k}^k\} = (\mathscr{B}_k - \bigcup_{i=1}^{k-1} \mathscr{B}_i)_o$, where $|b_0^k| < \dots < |b_{m_k}^k|$. Then by lemma 15, we know $n_k = m_k$ so that we can let $f(a_i^k) = b_i^k$ for $i \leq n_k$.

Now suppose in addition that $0^x \in A$, then A_x, B_x, and C_x are defined and, in fact, $|\overline{A}_x| + |\overline{B}_x| + |\overline{C}_x| \leq b(x)$. Because $A_x \subseteq A$ and $B_x \subset B$, it follows that if $z \in A_x$, then A_z, B_z, and C_z are defined and

$$|\overline{A}_z| + |\overline{B}_z| + |\overline{C}_z| \leq b(|z|) \leq b(b(x))$$

since $|z| \leq b(x)$. A similar argument works in case $w \in B_x$. Then D_w, E_w, and C_w are defined and

$$|\overline{D}_w|+|E_w|+|\overline{C}_w| \le b(|w|) \le b(b(x)).$$

Moreover, the total time required to find A_z, B_z, C_z for all $z \in A_x$ and to find D_y, E_y, C_y for all $w \in B_x$ is no more than

$$\sum_{i=0}^{b(x)} (a(i)+b(i)+c(i)+d(i)) \le \sum_{i=0}^{a(b(x))+b(b(x))+c(b(x))+d(b(x))} i = \binom{a(b(x))+b(b(x))+c(b(x))+d(b(x))+1}{2}.$$

Thus it requires only polynomially many steps in x to find

$$\{\overline{A_z \cup B_z} \mid 0^z \in A_z\} \cup \{\overline{D_w \cup E_w} \mid 0^w \in B_x\} = \mathscr{P}_x.$$

Since $\text{card}(A_x \cup B_x) \le |\overline{A}_x|+|\overline{B}_x|+|\overline{C}_x| \le b(x)$, it follows that we have at most $2b(x)$ blocks in $\mathscr{P}_x$ and the size of any block is bounded by $b(b(x))$.

Since each block has a size bounded by the polynomial $b(b(x))$, it follows that the time required to compare two blocks in $\mathscr{P}_x$ for containment is also bounded by a polynomial in x. Finding the linear extension $\mathscr{L}_x$ can certainly be done with order $(\text{card}(\mathscr{P}_x))^2$ comparisons, so we can find the list of $\mathscr{L}_x$ in increasing order, $\mathscr{B}_1, \ldots, \mathscr{B}_p$, in polynomial time in x. Finally, given the list $\mathscr{B}_1, \ldots, \mathscr{B}_p$, it is clear that (in polynomial time in $|\overline{\mathscr{B}}_1|+\ldots+|\overline{\mathscr{B}}_p|$) we can find $f \restriction A_x$. Thus if $0^x \in A$, then the entire process of finding $f \restriction A_x$ takes only polynomial time in x. Similarly, if $0^x \notin A$, then the process of finding $f \restriction A_x$ takes polynomial time in x (This assumes that A_x, B_x, and C_x are defined for each $0^z \in A_x$, all of A_z, B_z, and that C_z is defined, and that for each $0^w \in B_x$, all of D_w, E_w, C_w are defined). Thus our formal procedure for computing $f(0^x)$ is as follows.

COMPUTATION OF $f(0^x)$ FOR x EVEN.

First, determine if A_x, B_x, and C_x are defined. (This takes at most $a(x)$ steps.) If not, then declare $0^x \notin \text{dom } f$. If A_x, B_x, and C_x are defined, then check whether for all $0^z \in A_x$, all of A_z, B_z, and C_z are defined and check whether for all $0^w \in B_x$, all of D_w,

E_w, and C_w are defined. If not, declare $0^x \notin \text{dom}(f)$. Otherwise, let $\mathcal{B}_1, \ldots, \mathcal{B}_n$ be a list in increasing order of $\mathcal{L}_x$ where $\mathcal{L}_x$ is a linear extension of $\mathcal{P}_x$. Of course here $\mathcal{P}_x$ is the partially ordered set with domain $\{A_z \cup B_z \mid 0^z \in A_x\} \cup \{D_w \cup E_w \mid 0^w \in B_x\}$ ordered by inclusion. Then use Match to define $f \restriction (\bigcup_{i=1}^{n} \mathcal{B}_i)_e$. Since $0^x \in (\bigcup_{i=1}^{n} \mathcal{B}_i)_e$, we have defined $f(0^x)$.

Our remarks preceding the formal definition of $f(0^x)$ make it clear that we can determine if $0^x \in \text{dom}(f)$ in polynomial time in x. If $0^x \in \text{dom}(f)$, then we can compute $f \restriction A_x$ in polynomial time in x. By symmetry, we can compute $f^{-1}(0^y)$ for $0^y \in \{0^{2z+1} \mid z \geq 0\}$. That is, we formally compute $f^{-1}(0^y)$ as follows.

COMPUTATION OF $f^{-1}(0^y)$ FOR y ODD.

First determine if all of D_y, E_y, and C_y are defined. (This takes at most $c(x)$ steps.) If not, declare $0^y \notin \text{dom}(f^{-1})$. If all of D_y, E_y, and C_y are defined then check whether for all $0^z \in D_y$, all of A_z, B_z, and C_z are defined and check whether for all $0^w \in E_y$, all of D_w, E_w, and C_w are defined. If not, declare $0^y \notin \text{dom}(f^{-1})$. Otherwise, let $\mathcal{B}_1, \ldots, \mathcal{B}_n$ be a list in increasing order of $\mathcal{L}_y$ where $\mathcal{L}_y$ is a linear extension of $\mathcal{P}_y$. Here $\mathcal{P}_y$ is the partially ordered set with domain

$$\{A_z \cup B_z \mid 0^z \in D_y\} \cup \{D_w \cup E_w \mid 0^w \in E_y\}$$

ordered by inclusion. Then use Match to define $f \restriction (\bigcup_{i=1}^{n} \mathcal{B}_i)_e$. Since $0^y \in (\bigcup_{i=1}^{n} \mathcal{B}_i)_0$, we have defined $f^{-1}(0^y)$.

By arguments entirely similar to the ones we used for computing $f(0^x)$, we can show that it requires only polynomially many steps in y to determine if $0^y \in \text{dom}(f^{-1})$. Moreover, if $0^y \in \text{dom}(f^{-1})$, then we can also compute $f \restriction D_y$ in polynomially many steps in y. Thus f and f^{-1} will be partial polynomial time functions. Hence f will be a 1:1 partial honest polynomial time function provided that the definitions of f and f^{-1} are consistent.

CONSISTENCY OF DEFINITIONS

First, we must show that if

$$0^{x_1}, 0^{x_2} \in \{0^{2z} \mid z \geq 0\}, \text{ and } 0^{y_1}, 0^{y_2} \in \{0^{2z+1} \mid z \geq 0\}, \text{ and}$$
$$0^{x_1}, 0^{x_2} \in \text{dom}(f), \text{ and } 0^{y_1}, 0^{y_2} \in \text{dom}(f^{-1}),$$

then (a), (b), (c) below all hold.

(a) $f \restriction A_{x_1}$ and $f \restriction A_{x_2}$ agree on $A_{x_1} \cap A_{x_2}$.

(b) $f \restriction A_{x_1}$ and $f \restriction D_{y_1}$ agree on $A_{x_1} \cap D_{y_1}$.

(c) $f \restriction D_{y_1}$ and $f \restriction D_{y_2}$ agree on $D_{y_1} \cap D_{y_2}$.

Second, we must show that if 0^x is not in dom(f), then $0^x \notin A_z$ for any $0^z \in$ dom(f) and $0^x \notin D_w$ for any $0^w \in \text{dom}(f^{-1})$.

Third, we must show that if 0^y is not in $\text{dom}(f^{-1})$, then $0^y \notin A_z$ for any $0^z \in$ dom(f) and $0^y \notin D_w$ for any $0^w \in \text{dom}(f^{-1})$.

The second and third conditions are easy to deal with. For example, if A_x, B_x, and C_x are not defined, then whenever $0^x \in A_z$ or $0^x \in D_w$, then neither z nor w can meet the criterion to have $0^z \in$ dom(f) or $0^w \in \text{dom}(f^{-1})$. Similarly if A_x, B_x, and C_x are defined, then by property (v), $0^x \in A_z$ implies that $A_x \subseteq A_z$ and $B_x \subseteq B_z$ and $C_x \subseteq C_z$. Thus if $0^x \notin$ dom(f), then either

(I) there is some $0^u \in A_x \subseteq A_z$ such that A_u, B_u, and C_u are not defined, or

(II) there is some $0^w \in B_x \subseteq B_z$ such that D_w, E_w, and C_w are not defined.

Thus in either case $z \notin$ dom(f). Similarly, if $0^x \in D_w$, then we know $A_x \subseteq D_w$, $B_x \subseteq E_w$, and $C_x \subseteq C_w$, so that 0^w cannot meet the criterion to be in $\text{dom}(f^{-1})$. By an entirely symmetric argument, if $0^y \notin \text{dom}(f^{-1})$ then $0^y \in A_z$ implies 0^z cannot meet the criterion to be in dom(f) and $0^y \in D_w$ implies 0^w cannot meet the criterion to be in

$\mathrm{dom}(f^{-1})$.

We verify condition (a). Note that if

$$\psi(A_{x_1} \times C_{x_1}) = B_{x_1} \times C_{x_1} \text{ and } \psi(A_{x_2} \times C_{x_2}) = B_{x_2} \times C_{x_2}, \text{ then}$$

$$\psi((A_{x_1} \cap A_{x_2}) \times (C_{x_1} \cap C_{x_2})) = (B_{x_1} \cap B_{x_2}) \times (C_{x_1} \cap C_{x_2}).$$

But then if $0^x \in A_{x_1} \cap A_{x_2}$, condition (v) ensures

$$A_x \subseteq A_{x_1} \cap A_{x_2},\ B_x \subseteq B_{x_1} \cap B_{x_2}, \text{ and } C_x \subseteq C_{x_1} \times C_{x_2}.$$

Similarly if $0^y \in B_{x_1} \cap B_{x_2}$, then condition (vi) ensures that

$$D_y \subseteq A_{x_1} \cap A_{x_2},\ E_y \subseteq B_{x_1} \cap B_{x_2}, \text{ and } C_x \subseteq C_{x_1} \cap C_{x_2}.$$

This means that if $0^x \in A_{x_1} \cap A_{x_2}$, then the definition of $f(0^x)$ in either our definition of $f \upharpoonright A_{x_1}$ or our definition of $f \upharpoonright A_{x_2}$ depends only on the blocks of the form $A_z \cup B_z$ with $z \in A_{x_1} \cap A_{x_2}$ or $D_w \cup E_w$ with $w \in B_{x_1} \cap B_{x_2}$. Since such blocks are common to both $\mathcal{P}_{x_1}$ and $\mathcal{P}_{x_2}$, it is easy to see that $f(0^x)$ must get the same value in either our definition of $f \upharpoonright A_{x_1}$ or $f \upharpoonright A_{x_2}$. A similar argument will establish conditions (b) and (c).

Having established that the definition of $f(0^x)$ is consistent no matter what rectangles we use to define it, it then easily follows that $f(A) = B$. That is, we have already argued that if $0^x \in A$, then $f \upharpoonright A_x$ can be defined. But since under such circumstances, $A_x \subseteq A$ and $B_x \subseteq B$, it follows that $f(x) \in B$. Similarly, it is easy to show that if $0^y \in B$, then $f^{-1} \upharpoonright D_y$ must be defined, $D_y \subseteq A$, and $E_y \subseteq B$. Thus $f^{-1}(0^y) \in A$. Thus we have shown that f is a 1:1 partial honest polynomial function such that $A \subseteq \mathrm{dom}(f)$ and $f(A) = B$. Hence $A \sim_p B$. □

§5. Existence of p–time, p–finite sets.

THEOREM 19. There exist infinite p–time, p–finite sets.

PROOF. Let z be any integer such that $z \geq 2$. Define $z^{(n)}$ by the inductive definition

$$z^{(0)} = z, \; z^{(n+1)} = z^{z^{(n)}}.$$

It easy to check that for each such z, $E^z = \{0^{z^{(n)}} \mid n \geq 0\}$ is a p–time set.
We claim that E^z is p–finite. To prove this, suppose that p is a partial p–time map such that $\mathrm{dom}(p) \supseteq \mathrm{fin}(E^z)$ and $p \upharpoonright \mathrm{fin}(E^z)\colon \mathrm{fin}(E^z) \to \mathrm{fin}(E^z)$. If X is a finite set contained in E^z and $m_X = z^{(n)}$ is the maximum element of X, then $X \subseteq E^z_n = \{z^{(m)} \mid m \leq n\}$.

But we have

$$|\overline{E^z_n}| = \sum_{m=0}^{n} (z^{(m)}+1) \leq n + \sum_{i=0}^{z^{(n)}+1} i = n + \binom{z^{(n)}+2}{2} \leq (z^{(n)})^3.$$

So $|\overline{E^z_n}|$ is bounded by a polynomial in $|m_X|$. Now let r be the polynomial which bounds the running time of the p–time tranducer which computes p. Then certainly $|p(X)| \leq r(|X|)$ for all $X \in \mathrm{fin}(A)$. Thus for all $X \in \mathrm{fin}(A)$,
$|p(X)| \leq r(|X|) \leq r(|m_X|^3)$. Since $r(n^3)$ is a polynomial in n, there is a k so large that $r(n^3) < z^n$ for all $n \geq k$. If $n > k$ and $m_X = z^{(n)}$, then

$$|p(X)| \leq r(|m_X|^3) \leq r((z^{(n)})^3) < z^{z^{(n)}} = z^{(n+1)}.$$

That is, for sufficiently large n, if $m_X = z^{(n)}$, then $|p(X)|$ is not large enough to have $p(X)$ contain any $z^{(m)}$ where $m > n$. So if $n > k$ and $X \in \mathrm{fin}(E^z)$ is such that $X \subseteq \overline{E^z_n}$, then $p(X) \subseteq \overline{E^z_n}$. Thus for such X, $\bigcup_{m \geq 0} p^m(X) \subseteq \overline{E^z_n}$. But this means that if $X \in \mathrm{fin}(E^z)$ and $m_X = z^{(n)}$ with $n \geq k$, then

$$\Big|\bigcup_{m\geq 0} p^m(X)\Big| \leq |\overline{E^z_n}| \leq |m_X|^3 \leq |X|^3.$$

Moreover if $m_X = z^{(n)}$ where $n < k$, then we still have that $\big|\bigcup_{m \geq 0} p^m(X)\big| \leq |\overline{E^z_k}| = c_k$ where c_k is a fixed constant. Thus for any $X \in \mathrm{fin}(E^z)$, we have

$$\Big|\bigcup_{m\geq 0} p^m(X)\Big| \leq |X|^3 + c_k$$

so that E^z is p–finite as claimed. □

THEOREM 20. There exist infinitely many PET's which contain infinite p–time, p–finite sets.

PROOF. In [30] we showed that if f is defined by $f(0^n) = 0^{2^{2^n}}$ and A is any infinite set contained in $\{0\}^*$, then $A \not\equiv_p f(A)$. Actually the proof in [30] only used the fact that $|f(0^n)| \geq 2^n$.

Fix any integer $z \geq 2$. Consider the map g_z such that $g_z(0^{z^{(n)}}) = 0^{(z^z)^{(n)}}$. It is easy to prove by induction that

(5.1) $$z^{(z^{(n)})} \leq (z^z)^{(n)}.$$

Thus since $|g_z(0^{z^{(n)}})| \geq z^{(z^{(n)})} \geq 2^{2^{z^{(n)}}}$, we can conclude that $g_z(E_z) = E_{z^z} \not\equiv_p E_z$. It then follows that if we define $2^{[n]}$ inductively by $2^{[0]} = 2$, $2^{[n+1]} = (2^{[n]})^{2^{[n]}}$, then $E_{2^{[0]}}, E_{2^{[1]}}, E_{2^{[2]}}, \ldots$ are all in distinct PET's. □

It is an open problem as to whether p–finite sets are closed under $\oplus$ and $\times$. We prove these closure properties for p–time, p–isolated sets in §6.

§6. p–time, p–finite sets are closed under addition and multiplication.

THEOREM 21. Suppose A and B are p–time, p–isolated subsets of $\{0\}^*$. Then $A \oplus B$ and $A \times B$ are p–isolated.

PROOF. Since A and B are p–time sets,
$$A \oplus B = \{0^{2n} \mid 0^n \in A\} \cup \{0^{2n+1} \mid 0^n \in B\}$$
is a p–time set. Now suppose $A \oplus B$ is not p–isolated. Then there exists a p–time monotone set map $p\colon \mathrm{fin}(\{0\}^*) \to \mathrm{fin}(\{0\}^*)$ such that both

for all X in $\mathrm{fin}(A \oplus B)$, $p(X)$ is in $\mathrm{fin}(A \oplus B)$)

and yet

for any polynomial q, there is an $X_q \in \mathrm{fin}(A \oplus B)$ such that
$$\Big|\bigcup_{n=1}^{\infty} p^n(X_q)\Big| > q(|X_q|).$$

For any n, let

$$A_n = \{0^x \mid 0^x \in A \ \&\ x \le n\} \text{ and } B_n = \{0^x \mid 0^x \in B \ \&\ x \le n\}.$$

Because A and B are p–time sets, it easily follows that we can find $\overline{A}_n$ and $\overline{B}_n$ in polynomial time in n. That is, suppose a(x) is a polynomial which bounds the running time of the deterministic Turing machine M which accepts A. Note that the total time required to compute $M(\phi),\ldots,M(0^n)$ is no more than

$$\sum_{i=0}^{n} a(i) \le \sum_{i=0}^{a(n)} i = \frac{(a(n)+1)(a(n))}{2}.$$

Hence it easily follows that we can compute $\overline{A}_n$ in polynomial time in n. This given, we can now define two polynomial time monotone set maps p_A and p_B from p.

(6.1) $p_A(\overline{X}) = \overline{X} \cup \overline{A}_n$, where $n = |p(\overline{A_{|X|} \oplus B_{|X|}})|$.

(6.2) $p_B(\overline{X}) = \overline{X} \cup \overline{B}_n$, where $n = |p(\overline{A_{|X|} \oplus B_{|X|}})|$.

Then p_A and p_B are polynomial time monotone set maps because A and B are polynomial time sets. It is also easy to see that

(6.3) $\forall \overline{X} \in \text{fin}(A)(p_A(\overline{X}) \in \text{fin}(A))$, and

(6.4) $\forall X \in \text{fin}(B)(p_B(\overline{X}) \in \text{fin}(B))$.

We claim that either p_A shows that A is not p–isolated or p_B shows that B is not p–isolated. For fix any polynomial q with positive coefficients. It is easy to see that there is a fixed polynomial r such that if $\overline{A_1 \oplus B_1} \in \text{fin}(A \oplus B)$, then

$$|\overline{A_{|A_1 \oplus B_1|} \oplus B_{|A_1 \oplus B_1|}}| \le r(|\overline{A_1 \oplus B_1}|).$$

Moreover,

$$|\overline{A_1 \oplus B_1}| \le 2|\overline{A}_1|+2|\overline{B}_1|+\text{card}(B_1) \le 3|\overline{A}_1|+3|\overline{B}_1| \le \max(6|\overline{A}_1|, 6|\overline{B}_1|).$$

So consider the polynomial 6q(r(x)). We know that there is an $X \in \text{fin}(A \oplus B)$ such that $|\bigcup_{n\ge1} p^n(X)| > 6q(r(|X|))$. Now suppose $\overline{X} = \overline{U_0 \oplus V_0}$ where $\overline{U}_0 \in \text{fin}(A)$ and $\overline{V}_0 \in \text{fin}(B)$. Let $p^n(\overline{X}) = \overline{U_n \oplus V_n}$, where $\overline{U}_n \in \text{fin}(A)$ and $\overline{V}_n \in \text{fin}(B)$. It is easy to see by the monotonicity of p that if $A' = \overline{A_{|X|}}$ and $B' = \overline{B_{|X|}}$ then $p_A^n(A') \supseteq \overline{U}_n$ and $p_B^n(B') \supseteq \overline{V}_n$. Note by our notation $p^n(A') \in \text{fin}(A)$ so we let $\underline{p^n(A')}$ be the subset of A such that $\underline{p^n(A')} = p^n(A')$. We define $\underline{p^n(B')}$ similarly. Then we have the following.

(6.5) $$|\bigcup_{n\geq 1} p^n(X)| = |\overline{\bigcup_{n\geq 1} U_n \oplus \bigcup_{n\geq 1} V_n}|$$

$$\leq |\overline{\bigcup_{n\geq 1} \underline{p^n(A')} \oplus \bigcup_{n\geq 1} \underline{p^n(B')}}|$$

$$\leq 6\max(|\bigcup_{n\geq 1} p^n(A')|, |\bigcup_{n\geq 1} p^n(B')|).$$

On the other hand,

$$|\bigcup_{n\geq 1} p^n(X)| > 6q(r(|X|)$$

$$\geq 6q(|\overline{A_{|X|} \oplus B_{|X|}}|) \geq 6\max(q(|A'|), q(|B'|)).$$

Thus we must have that

(6.6) $$|\bigcup_{n\geq 1} p_A^n(A')| > q(|A'|), \text{ or}$$

(6.7) $$|\bigcup_{n\geq 1} p_B^n(B')| > q(|B'|).$$

Now consider the sequence of polynomials $q_n(x) = \sum_{k=0}^{n} nx^k$. It follows that either

(1) there are infinitely many n such that there exists $\overline{A_{h(n)}} \in \text{fin}(A)$ and $|\bigcup_{m\geq 1} p_A^m(\overline{A}_{h(n)})| > q_n(|\overline{A}_{h(n)}|)$, or

(2) there are infinitely many n such that there exists $\overline{B}_{k(n)} \in \text{fin}(B)$ and $|\bigcup_{m\geq 1} p_B^m(\overline{B}_{k(n)})| > q_n(|\overline{B}_{k(n)}|)$.

For any polynomial q there is an n such that $|q(x)| \leq |q_n(x)|$ for all $x \geq 0$. If (1) holds, then A is not p-isolated. If (2) holds, then B is not p-isolated. Thus if both A and B are p-time and p-isolated, then $A \oplus B$ must be p-time and p-isolated. □

Next assume that $A \times B$ is not p–isolated. Thus there is a polynomial time monotone set map f such that

$$f \restriction \text{fin}(A \times B): \text{fin}(A \times B) \to \text{fin}(A \times B),$$

and for any polynomial $q(x)$, there is an $X_q \in \text{fin}(A \times B)$, such that

$$|\bigcup_{n\geq 0} f^n(X_q)| > q(|X_q|).$$

We then define two polynomial time monotone set maps f_A and f_B from f.

(6.8) $\quad f_A(X) = X \cup A_n$, where $n = |f(\overline{A_{|X|} \times B_{|X|}})|$,

(6.9) $\quad f_B(X) = X \cup B_n$ where $n = |f(\overline{A_{|X|} \times B_{|X|}})|$.

Again it is easy to check that f_A and f_B are polynomial time monotone set maps and that

(6.10) $\quad \forall X \in \text{fin}(A)\ (f_A(X) \in \text{fin}(A))$ and

(6.11) $\quad \forall X \in \text{fin}(B)\ (f_B(X) \in \text{fin}(B))$.

We claim that either f_A shows that A is not p–isolated or f_B show that B is not p–isolated. By the same argument as we used for $\oplus$, we need only show that for any polynomial q with positive coefficients, there exist $A' \in \text{fin}(A)$ and $B' \in \text{fin}(B)$ such that

(6.12) $$|\bigcup_{n\geq 0} f_A^n(A')| > q(|A'|) \text{ or}$$

(6.13) $$|\bigcup_{n\geq 0} f_B^n(B')| > q(|B'|).$$

Again it is easy to see that there is a fixed polynomial $s(x)$ such that if $\overline{A_1 \times B_1} \in \text{fin}(A \times B)$, then

$$|\overline{A_{|A_1\times B_1|} \times B_{|A_1\times B_1|}}| \leq s(|\overline{A_1 \times B_1}|).$$

Moreover,

$$|\overline{A_1 \times B_1}| = \mathrm{card}(A_1)\cdot\mathrm{card}(B_1) + \sum_{0^n \in A_1} \sum_{0^m \in B_1} |<0^n, 0^m>|$$

$$= \mathrm{card}(A_1)\cdot\mathrm{card}(B_1) + \sum_{0^n \in A_1} \sum_{0^m \in B_1} \tfrac{1}{2}m^2+m\cdot n+\tfrac{1}{2}n^2+3m+n$$

$$\leq \mathrm{card}(A_1)\cdot\mathrm{card}(B_1) + \sum_{0^n \in A_1} \sum_{0^m \in B_1} (m^2+3m+1)(n^2+3n+1)$$

$$= \mathrm{card}(A_1)\cdot\mathrm{card}(B_1) + \Big[\sum_{0^n \in A_1} (n^2+3m+1)\Big]\Big[\sum_{0^m \in B_1} (m^2+3m+1)\Big]$$

$$\leq \mathrm{card}(A_1)\cdot\mathrm{card}(B_1) + (|\overline{A}_1|^2+3|\overline{A}_1|+1)(|\overline{B}_1|^2+3|\overline{B}_1|+1)$$

$$\leq 2\max((|\overline{A}_1|^2+3|\overline{A}_1|+1)^2, (|B_1|^2+3|B_1|+1)^2).$$

So let $t(x)$ be the polynomial $2(x^2+3x+1)^2$ and consider the polynomial $t\circ q\circ s(x)$. We know there is a $Y \in \mathrm{fin}(A \times B)$ such that $|\bigcup_{n\geq 0} f^n(Y)| \geq t\circ q\circ r(|Y|)$. Let π_1 and π_2 be the corresponding projection functions for our polynomial time pairing function J. For $n \geq 0$, let

$$U_n = \pi_1(\underline{f^n(Y)}) \text{ and } V_n = \pi_2(\underline{f^n(Y)}).$$

It is straightforward to prove by induction using the monotonicity of f, that if $A' = \overline{A_{|X|}}$ and $B' = \overline{B_{|X|}}$, then for all n,

(6.14) $\quad f^n(X) \subseteq \overline{U_n \times V_n} \subseteq \overline{\underline{f_A^n(A')} \times \underline{f_B^n(B')}}.$

Thus

(6.15) $\quad |\bigcup_{n\geq 0} f^n(X)| \leq |\bigcup_{n\geq 0} \overline{U_n \times V_n}| \leq |\overline{(\bigcup_{n\geq 0} U_n) \times (\bigcup_{n\geq 1} V_n)}|$

$$\le |\overline{\bigcup_{n\geq 0} f_A^n(A') \times \bigcup_{n\geq 0} f_B^n(B')}| \le t(\max|\bigcup_{n\geq 0} f_A^n(A')|, |\bigcup_{n\geq 0} f_B^n(B')|).$$

On the other hand, using our definition of s and the fact that q has positive coefficients, we have

(6.16) $|\bigcup_{n\geq 1} f^n(X)| > t{\circ}q{\circ}s(|X|) \geq t{\circ}q(|\overline{A_{|X|} \times B_{|X|}}|) \geq t(\max(q(|A'|), q(|B'|)))$.

Combining (6.15) and (6.16), we conclude that

(6.17) $t(\max(|\bigcup_{n\geq 1} f_A^n(A')|, |\bigcup_{n\geq 1} f_B^n(B')|)) \geq t(\max(q(|A'|), q(|B'|)))$.

Now because t has positive coefficients, we know

(6.18) $\max(|\bigcup_{n\geq 1} f_A^n(A')|, |\bigcup_{n\geq 1} f_B^n(B')|) \geq \max(q(|A'|), q(|B'|))$.

But (6.18) implies that either (6.12) or (6.13) holds. This is what we needed to prove. □

THEOREM 22. Suppose A is a p–time, p–isolated subset of $\{0\}^*$ and $C \subsetneq A$. Then C is p–isolated.

PROOF. Suppose that C is a subset of p–time, p–isolated A and C is not p–isolated. Let p be a monotone set map which witnesses that C is not p–isolated. That is, we have $\forall X \in \mathrm{fin}(C)\ (p(X) \in \mathrm{fin}(C))$. But for any polynomial q, there is an $X_q \in \mathrm{fin}(C)$ such that $|\bigcup_n p^n(X_q)| > q(|X_q|)$. Now the only reason that we cannot use p to show that A is not p–isolated is that it may not be the case that $\forall X \in \mathrm{fin}(A)\ (p(X) \in \mathrm{fin}(A))$. So define an alternative monotone set map $p_A(X)$ by

$$p_A(X) = A_n, \text{ where } n = |p(X)|.$$

Then because A is p–time, it is easy to see that p_A is a p–time monotone set map from $\mathrm{fin}(\{0\}^*)$ into $\mathrm{fin}(A)$. Moreover, for any $X \in \mathrm{fin}(C)$, $p_A(X) \supseteq p(X)$. Thus for any polynomial q,

$$\bigcup_n p^n(X_q) \subseteq \bigcup_n p_A^n(X_q),$$

and hence

$$|\bigcup_{n\geq 0} p_A^n(X_q)| \geq |\bigcup_{n\geq 0} p^n(X_q)| > q(|X_q|).$$

Thus p_A^n shows that A is not p–isolated. □

By combining theorems 20 and 21, we have the following.

THEOREM 23. Suppose A and B are p–isolated subsets of $\{0\}^*$ such that there exist p–time, p–isolated subsets A' and B' of $\{0\}^*$ with $A \subsetneq A'$ and $B \subsetneq B'$. Then $A \oplus B$ and $A \times B$ are p–isolated.

PROOF. Note by Theorem 7, $A' \oplus B'$ and $A' \times B'$ are p–time, p–isolated sets. But $A \oplus B \subsetneq A' \oplus B'$ and $A \times B \subsetneq A' \times B'$ so by Theorem 8, $A \oplus B$ and $A \times B$ are p–isolated. □

§7. **Remarks.** There are two well investigated subclasses of the RET's, namely Λ, the set of isols, and Λ_z, the set of co–simple isols. Based on the results of this paper, there are at least three natural classes of PET's which should have interesting theories with respect to $\oplus$, $\times$, and $\leq$. Let us define

$P\Lambda = \{\alpha \mid \alpha$ is a PET and $\exists x \in \alpha$ (x is p–isolated set contained in a p–time, p–isolated subset of $\{0\}^*$)$\}$,

$P\Lambda_z = \{\alpha \mid \alpha$ is a PET and $\exists x \in \alpha$ (x is a cosimple, p–isolated set contained in a p–time, p–isolated subset of $\{0\}^*$)$\}$,

$P\Lambda_p = \{\alpha \mid \alpha$ is a PET and $\exists x \in \alpha$ (x is a p–time, p–isolated set)$\}$.

Based on known theorems about isols and results of this paper, each of $P\Lambda$, $P\Lambda_z$, and $P\Lambda_p$ is closed under $\oplus$ and $\times$, and obeys additive and multiplicative cancellation.

Like Λ, $P\Lambda$ has a continuum of elements. We expect $(P\Lambda, \oplus, \times, \leq)$ to have a theory somewhat like that of $(\Lambda, \oplus, \times, \leq)$, developed in [2, 3, 5, 9, 10, 11, 13, 14, 17, 20, 21, 22, 25].

Like Λ_z, $P\Lambda_z$ has only countably many elements. We expect the theory of $(P\Lambda_z, \oplus, \times, \leq)$

to be somewhat like that of $(\Lambda_z, \oplus, \times, \leq)$ developed in [13, 14, 32].

What corresponds in the isols to $P\Lambda_p$ is the set of isols of recursive sets, that is, the set of non–negative integers. So we get no hint from the isols as to what the non–integer members of $P\Lambda_p$ are like. The theory of $P\Lambda_p$ is an interesting new theory arising due to the polynomial time bounding of computational resources. We expect the theory of $P\Lambda_p$ to be closer to recursive algebra [1, 28] and p–time algebra [29, 30] than it is to isols.

Even in the case of $P\Lambda$ and $P\Lambda_z$, we cannot expect very complete parallels. Certainly many of the arguments which produce counterexamples in isols to various laws of arithmetic will lift easily to $P\Lambda$ and $P\Lambda_z$, since it is easier to diagonalize over all p–time functions than to diagonalize over all partial recursive functions. But we cannot expect that positive results about laws of arithmetic holding in Λ or Λ_z, proved for the isols by using unbounded computational resources, will lift easily. For in $P\Lambda$ and $P\Lambda_z$ we have to produce 1–1 honest p–time partial functions, while in Λ or Λ_z, we merely had to produce 1–1 partial recursive functions. All three classes $P\Lambda$, $P\Lambda_z$, and $P\Lambda_p$ present interesting objects of study.

We thank J. N. Crossley for many valuable conversations.

[1]Supported by NSF grant MCS–83–01850 and by ARO contract DAAG29–85–C–0018.

[2]Supported by NSF grant DMS–85–05004 and by ARO contract DAAG29–85–C–0018.

BIBLIOGRAPHY

1. Crossley, J. N., Aspects of Effective Algebra, Proc. of a Conference at Monash University (1–4 August, 1979), Upside Down A Book Company, Steele's Creek, Box 226, Yarra Glenn, Victoria, Australia, 1981.

2. Crossley, J. N., and Nerode, A., Combinatorial Functors, (monograph), Ergebnisse der Mathematik und ihrer Grenzgebiete, vol. 81, Springer–Verlag, New York, 1974.

3. Dekker, J. C. E., "A Non–Constructive Extension of the Number System" (two abstracts), J. Symb. Logic 20 (1955), 204–205.

4. Dekker, J. C. E., Les Functions Combinatoires et Les Isols, Collection de Logique Mathematique ser. A. No. 22, Gauthier–Villars, Paris, 1966.

5. Dekker, J. C. E., and Myhill, J., "Recursive Equivalence Types", Univ. of California Publications in Mathematics n.s. 3 (1959), 67–214.

6. Ellentuck, E., "The Universal Properties of Dedekind Finite Cardinals", Ann. Math. 82 (1965), 225–248.

7. Ellentuck, E., "Generalized Idempotence in Cardinal Arithmetic", Fund. Math. 58 (1966),225–248.

8. Ellentuck, E., "The First Order Properties of Dedekind Finite Integers", Fund. Math. 63, (1966), 7–25.

9. Ellentuck, E., "Almost Combinatorial Skolem Functions", J. Symb. Logic 35 (1970), 378–382.

10. Ellentuck, E., "Non–Recursive Combinatorial Functions", Jour. Symb. Logic 37 (1972), 90–95.

11. Ellentuck, E., "The Positive Properties of Isolic Integers", J. Symbolic Logic 37, (1972), 114–132.

12. Friedberg, R. M., "The Uniqueness of Finite Division for Recursive Equivalence Types", Math. Z. 75 (1961), 3–7.

13. Hay, L., "The Co–Simple Isols", Ann. Math. 83 (1966), 231–256.

14. Hay, L., "Elementary Differences Between the Isols and the Co–Simple Isols", Trans. Amer. Math. Soc. 127 (1967), 427–441.

15. Hopcroft, J. E., and Ullman, J. D., Introduction to Automata Theory, Languages, and Computation, Addison–Wesley, Reading, Mass, 1979.

16. McCarty, D. C., Realizability and Recursive Mathematics, D. Phil dissertation, Oxford, 1980.

17. McLaughlin, T., Regressive Sets and the Theory of Isols, Marcel Dekker, New York, 1983.

18. Myhill, J., "Recursive Equivalence Types and Combinatorial Functions", Bull. Amer. Math. Soc. 64 (1958), 373–376.

19. Myhill, J., "Absorption Laws in the Arithmetic of Recursive Equivalence Types", Notices. Amer. Math Soc. 6 (1959), 526–527.

20. Nerode, A., "Extensions to Isols", Ann. of Math. 73 (1961), 362–403.

21. Nerode, A., "Extensions to Isolic Integers", Ann. of Math. 75 (1962), 419–448.

22. Nerode, A., "Non–Linear Combinatorial Functions of Isols", Math. Z. 86 (1965), 410–424.

23. Nerode, A., "Additive Relations Among Recursive Equivalence Types", Math. Ann. 159 (1965), 329–343.

24. Nerode, A., "Combinatorial Series and Recursive Equivalence Types", Fund. Math. 58 (1966), 133–141.

25. Nerode, A., "Additive Relations among Recursive Equivalence Types", Math. Ann. 159 (1965), 329–343.

26. Nerode, A., "Diophantine Correct Non–Standard Models in the Isols", Ann. of Math. 84 (1966), 421–432.

27. Nerode, A., "Combinatorial Series and Recursive Equivalence Types", Fund. Math. 58 (1966), 133–141.

28. Nerode, A., and Remmel, J. B., "A Survey of R. E. Substructures", Proc. Symp. in Pure Mathematics vol. 42, Amer. Math. Soc., 1985, 323–373.

29. Nerode, A. and Remmel, J. B., "Complexity–Theoretic Algebra 1: Vector Spaces over Finite Fields", Structure in Complexity (2nd Annual Conference), Computer Society of IEEE, 1987, 218–241.

30. Nerode, A., and Remmel, J. B., "Complexity–Theoretic Algebra II: Boolean Algebras", Proc. Third Asian Logic and Computer Science Symposium, Beijing 1987, Ann. of Pure and Applied Logic, to appear.

31. Nerode, A. and Remmel, J. B., "Polynomial Time Equivalence Types", CMU 1987 Logic and Computation Workshop volume, Contemporary Mathematics series, Amer Math. Soc., 1989, to appear.

32. Remmel, J. B., "Combinatorial Functors on Co–R. E. Structures", Ann. Math. Logic 11 (1976), 281–287.

Mathematical Sciences Institute
Cornell University
Ithaca, N. Y., 14853, U.S.A.

Mathematics Department
University of California at San Diego
La Jolla, CA 92093, U.S.A.

A Characterization of Effective Topological Spaces

Dieter Spreen
Fachbereich Mathematik, Theoretische Informatik
Universität-GH Siegen
Hölderlinstr. 3, D-5900 Siegen
West Germany

Abstract. Starting with D. Scott's work on the mathematical foundations of programming language semantics, interest in topology has grown up in theoretical computer science, under the slogan "open sets are semidecidable properties". But whereas on Scott domains all such properties are also open, this is no longer true in general. In this paper we present a characterization of effectively given topological spaces that says which semidecidable sets are open. We consider countable topological T_0-spaces that satisfy certain additional topological and computational requirements which can be verified for a general class of Scott domains and metric spaces, and we show that the given topology is the recursively finest topology generated by semidecidable sets which is compatible with it. From this general result we derive the above mentioned theorem about the correspondence of the semidecidable properties with the Scott open sets. This theorem, in its turn, is a generalization of the Rice/Shapiro theorem on index sets of classes of recursively enumerable sets. Moreover, characterizations of the canonical topology of a recursively separable recursive metric space are derived. It is shown that it is the recursively finest effective T_3-topology that can be generated by semidecidable sets the topological complement of which is also semidecidable.

1. Introduction

Topological spaces that satisfy certain natural effectivity requirements have been studied by various authors (cf. [5,13,14,16-24,35-39,60,61,65]). Among these investigations two main directions of interest can be observed, which are related to two different schools, the Russian and the North American. In both schools one is interested in studying which topological constructions can be done effectively, but while in the Russian school the main interest is to find out which important results of classical analysis hold effectively (constructively), the North American approach is a more recursion theoretic one and follows Nerode's program for studying recursively enumerable (r.e.) substructures of a recursively presented structure, in this case r.e. classes of basic open sets and the open sets generated by them.

The present paper has been written under a computer science point of view. In theoretical computer science interest in topology has started with Scott's pioneering work on models of the λ-calculus and the semantics of programming languages (cf. [46-53]). Many important notions in this theory can be interpreted in terms of the canonical topology of a Scott domain, i.e. the Scott topology. A similar approach has independently been developed by Eršov [7-9]. Scott domains are not the only kind of spaces that is successfully applied in computer science. Several authors used metric spaces to give a mathematical meaning to concurrent programs (cf. [1,2,12,34,40,41,43]). In [56,57] Smyth showed that both types of spaces are special quasi-uniform spaces. He also showed that there are interesting computational examples of the latter kind of spaces arising in studies of "observational preorder" for CCS and related systems (cf. [15,30]) which are neither domains nor metric spaces. As has been pointed out by him [55], the use of topology in computing theory is not merely a technical trick. The topology captures an essential computational notion, under the slogan "open sets are semidecidable properties" (for an introduction into this program cf. [62]).

In the case of Scott domains the (effectively) open sets are the only semidecidable properties. This follows from a generalization of the Rice/Shapiro theorem on index sets of classes of r.e. sets (cf. [42]), saying that under certain natural effectivity requirements the restriction of the Scott topology to the computable domain elements is equivalent to the topology generated by all completely enumerable subsets of this set. The latter topology is usually called Eršov topology.

Topologies that are generated by a basis of completely enumerable subsets can be defined on each indexed set. They have first been considered by Mal'cev [29]. Therefore, we call any such topology a Mal'cev topology. Since the Eršov topology is generated by all completely enumerable subsets of a given indexed set, it is the finest Mal'cev topology on this set. For recursive metric spaces it has been proved in [59] that in general the Eršov topology on such a space is strictly finer than the metric topology. This shows that for some topological spaces there are semidecidable properties which are incompatible with the topological structure. In the present paper we deal with the question, which of these properties are compatible with the given topology, i.e., which generate a Mal'cev topology that is equivalent to the given topology.

We consider countable topological T_0-spaces with a countable basis, on which a relation of strong inclusion is defined such that for any two basic open sets and any point belonging to both of them, one can effectively find a further basic open set, which contains the given point and is strongly included in each of the given basic open sets. Under some additional effectivity assumptions, which among others imply that every basic open set is completely enumerable, i.e. semidecidable, we then show that each such topology is the recursively finest Mal'cev topology being compatible with it. Here, by saying that a topology η is compatible with a topology τ we roughly mean that, if some basic open set of topology τ is *not* included in a given basic open set of topology η, then we must be able to effectively find a witness for this, i.e. an element of the basic open set of topology τ not contained in the basic open set of topology η. Moreover, topology τ is recursively finer than topology η, if given a basic open set in τ and a point contained in it, we can effectively find a basic open set in η which contains the given point and is contained in the given basic open set of τ. If the space is also recursively separable with respect to τ, which means that it has an effectively enumerable dense base, this implies that given a basic open set in τ one can compute its representation as a union of basic open sets in η.

The effectivity conditions that have to be satisfied are very natural. First of all the topology has to be compatible with itself. Furthermore, we require that the elements of each basic open set can uniformly be enumerated as well as all basic open sets containing a given point. By the last condition we can enumerate a base of the neighbourhood filter of each point. Since we are dealing with T_0-spaces, we may demand that we can also do the converse, i.e. from a (normed) enumeration of a base of basic open sets of such a filter compute the point determined by it. These conditions are discussed in Section 2.

In Section 3 Mal'cev topologies are studied, and in Section 4 the above mentioned characterization theorem is proved. As we shall see in the remaining sections, the assumptions of this result are easily verified in the case of the spaces mentioned in the beginning. In Section 5 we apply the results of Section 4 to topological spaces consisting of the computable elements of an effective complete partial order. The characterization theorem for the Scott topology on such general domains has been introduced by K. Weihrauch (cf. [63]). As a special case we obtain the Rice/Shapiro theorem.

In Section 6, finally, we consider recursively separable recursive metric spaces. By applying the results of Section 4 again, it follows that the metric topology is recursively equivalent to the effectively semi-regular topology associated with the Eršov topology, i.e. the topology generated by all regular basic open sets in the Eršov topology with completely enumerable exterior. Note that these are just the regular elements in the lattice of all completely enumerable subsets of the metric space. Moreover, the metric topology is the recursively finest Mal'cev topology that effectively satisfies the T_3-axiom and is complemented, which means that the exterior of each basic open set is also completely enumerable.

Observe that the well-known results on the effective continuity of effective operators such as the theorems by Myhill and Shepherdson [33], Kreisel, Lacombe and Shoenfield [27], Ceĭtin [3], and Moschovakis [31,32], and also the uniform generalization of these theorems given by the author and P. Young [58] can easily be obtained from the characterizations in this paper by considering for an effective operator the inverse image of the range topology under this operator.

2. Strongly Effective Spaces

In what follows, let $<\ ,\ >: \omega^2 \to \omega$ be a recursive pairing function with corresponding projections π_1 and π_2 $(\pi_i(<a_1,a_2>) = a_i)$, let $P^{(n)}$ $(R^{(n)})$ denote the set of all n-ary partial (total) recursive functions, and let W_i be the domain of the i-th partial recursive function φ_i with respect to some Gödel numbering φ. We let $\varphi_i(a)\downarrow$ mean that the computation of $\varphi_i(a)$ stops, and $\varphi_i(a)\downarrow_n$ mean that it stops within n steps. In the opposite cases we write $\varphi_i(a)\uparrow$ and $\varphi_i(a)\uparrow_n$ respectively.

Now, let $\mathcal{T} = (T, \tau)$ be a countable topological T_0-space with a countable basis $\mathcal{B}$. If η is any topology on T, then we also write $\eta = < C >$ to express that C is a countable basis of η. Moreover, for any subset X of T, $\mathrm{int}_\tau(X)$, $\mathrm{cl}_\tau(X)$, $\mathrm{ext}_\tau(X)$ and $\mathrm{bnd}_\tau(X)$ respectively are the interior, the closure, the exterior and the boundary of X. In the special cases we have in mind, a relation between the basic open sets can be defined which is stronger than the usual set inclusion, and one has to use this relation in order to derive the characterization result we talked about in the introduction. We call a relation $\prec$ on $\mathcal{B}$ *strong inclusion*, if for all $X, Y \in \mathcal{B}$, from $X \prec Y$ it follows that $X \subseteq Y$. Furthermore, we say that $\mathcal{B}$ is a *strong basis*, if for all $z \in T$ and $X, Y \in \mathcal{B}$ with $z \in X \cap Y$ there is a $V \in \mathcal{B}$ such that $z \in V$, $V \prec X$ and $V \prec Y$.

If one considers basic open sets as vague descriptions, then strong inclusion relations can be considered as "definite refinement" relations. Strong inclusion relations that satisfy much stronger requirements appear very naturally in the study of quasi-proximities (cf. [10]). Moreover, such relations have been used in Császár's approach to general topology (cf. [4]) and in Smyth's work on topological foundations of programming language semantics (cf. [56, 57]). Compared with the conditions used in these papers, the above requirements seem to be rather week, but as we go along, we shall meet a further requirement, and it is this condition which in applications prevents us from choosing $\prec$ to be ordinary set inclusion. For what follows we assume that $\prec$ is a strong inclusion on $\mathcal{B}$ and $\mathcal{B}$ is a strong basis.

Let $x: \omega \rightarrowtail T$ (onto) and $B: \omega \rightarrowtail \mathcal{B}$ (onto) respectively be (partial) indexings of T and $\mathcal{B}$ with domains dom(x) and dom($\mathcal{B}$). The value of x at $i \in \mathrm{dom}(x)$ is denoted, interchangeably, by x_i or $x(i)$. The same holds for the indexing B. A subset X of T is *completely enumerable*, if there is a recursively enumerable (r.e.) set A such that $x_i \in X$ iff $i \in A$, for all $i \in \mathrm{dom}(x)$. We say that B is *computable*, if there is some r.e. set L such that for all $i \in \mathrm{dom}(x)$ and $n \in \mathrm{dom}(B)$, $<i,n> \in L$ iff $x_i \in B_n$. Furthermore, the space $\mathcal{T}$ is called *strongly effective*, if B is a total indexing and the property of being a strong basis holds effectively, which means that there exists a function $b \in P^{(3)}$ such that for $i \in \mathrm{dom}(x)$ and $n,m \in \omega$ with $x_i \in B_m \cap B_n$, $b(i,m,n)\downarrow$, $x_i \in B_{b(i,m,n)}$, $B_{b(i,m,n)} \prec B_m$, and $B_{b(i,m,n)} \prec B_n$. (Note that, if $\prec$ is only set inclusion and B is not required to be total, then the space is effective in the sense of Nogina [36].) The following lemma presents a natural sufficient condition for a space to be strongly effective.

LEMMA 2.1. *Let B be computable and total such that $\{<m,n> \mid B_m \prec B_n\}$ is r.e. Then $\mathcal{T}$ is strongly effective.*

PROOF. Let $L \subseteq \omega$ witness that B is computable and $A = \{<m,n> \mid B_m \prec B_n\}$. Moreover, for $i \in \mathrm{dom}(x)$ and $n,m \in \omega$ define $Z_{i,m,n} = \{a \mid <i,a> \in L \wedge <a,m>, <a,n> \in A\}$. Then $Z_{i,m,n}$ is r.e., uniformly in i, m, n. Thus, there is some function $f \in R^{(3)}$ such that $\varphi_{f(i,n,m)}$ is a total enumeration of $Z_{i,m,n}$, if this set is nonempty. Since $\mathcal{B}$ is a strong basis, it follows that $Z_{i,m,n} \neq \emptyset$, for all $i \in \mathrm{dom}(x)$ and $n,m \in \omega$ with $x_i \in B_m \cap B_n$. Furthermore, $a \in Z_{i,m,n}$ iff $x_i \in B_a$, $B_a \prec B_m$ and $B_a \prec B_n$. Hence, $b = \lambda i,m,n.\varphi_{f(i,m,n)}(0)$ witnesses that $\mathcal{T}$ is strongly effective.

As it is well known, each point y of a T_0-space is uniquely determined by its neighbourhood filter $\mathfrak{N}(y)$ and/or a base of it. If B is computable, a base of basic open sets can effectively be enumerated for each such filter. The next result shows that for strongly effective spaces this can be done in a normed way. An enumeration $\left(B_{f(a)}\right)_{a\in\omega}$ with $f: \omega \to \omega$ such that range(f) $\subseteq$ dom(B) is said to be *normed*, if it is decreasing with respect to $\prec$. If f is recursive, it is also called *recursive* and any Gödel number of f is said to be an *index* of it. In case $\left(B_{f(a)}\right)$ enumerates a base of the neighbourhood filter of some point, we say it *converges* to that point.

LEMMA 2.2. *Let $\mathcal{T}$ be strongly effective and B be computable. Then there are functions $q \in P^{(1)}$ and $p \in P^{(2)}$ such that for all $i \in \mathrm{dom}(x)$ and all $n \in \omega$ with $x_i \in B_n$, $q(i)$ and $p(i,n)$ are indices of normed recursive enumerations of basic open sets which converge to x_i. Moreover, $B_{\varphi_{p(i,n)}(a)} \prec B_n$, for all $a \in \omega$.*

PROOF. Since B is computable, there is some $v \in R^{(1)}$ such that for all $i \in \mathrm{dom}(x)$, $a \in W_{v(i)}$ iff $x_i \in B_a$. Thus, $\{B_a | a \in W_{v(i)}\}$ is a filter base of $\mathfrak{N}(x_i)$. Let $b \in P^{(3)}$ witness that $\mathcal{T}$ is strongly effective and $s \in R^{(1)}$ be such that $\varphi_{s(a)}$ is a total enumeration of W_a, if $W_a \neq \emptyset$. Then define $g \in P^{(3)}$ by

$$g(i,n,a) = b(i,n,\varphi_{s(v(i))}(0))$$
$$g(i,n,a+1) = b(i,g(i,n,a),\varphi_{s(v(i))}(a+1)).$$

Now, let $i \in \mathrm{dom}(x)$ and all $n \in \omega$ with $x_i \in B_n$. Since b witnesses that $\mathcal{T}$ is strongly effective, it follows that $\lambda\, a.\; g(i,n,a)$ is a total function and $x_i \in B_{g(i,n,a+1)} \prec B_{g(i,n,a)} \prec B_n$, for all $a \in \omega$ Moreover, $B_{g(i,n,a)} \subseteq B_m$ where m is the a-th element of $W_{v(i)}$ in the enumeration $\varphi_{s(v(i))}$. Hence, $\{B_{g(i,n,a)} | a \in \omega\}$ is also a filter base of $\mathfrak{N}(x_i)$. Define $p \in P^{(2)}$ by $\varphi_{p(i,n)}(a) = g(i,n,a)$ and $q \in P^{(1)}$ by $q(i) = p(i,\varphi_{s(v(i))}(0))$.

In what follows, we not only want to be able to generate normed recursive enumerations of basic open sets that converge to a given point, but conversely, we need also to be able to pass effectively from such convergent enumerations to the point they converge to. We say that B is *acceptable*, if it is computable and there is a function $pt \in P^{(1)}$ such that, if m is an index of a normed recursive enumeration of basic open sets which converges to some point $y \in T$, then $pt(m)\downarrow$, $pt(m) \in \mathrm{dom}(x)$ and $x_{pt(m)} = y$.

In [58] an approach to effective topological spaces has been presented in which similar conditions are used. In that approach most of the concepts are based on effective point sequences. Working with filters instead, now seems to us to be more appropriate and natural.

As it is well known, each open set is the union of certain basic open sets. In the context of effective topology one is only interested in such open sets where the union is taken over an enumerable class of basic open sets. We call an open set $O \in \tau$ a *Lacombe* set, if there is an r.e. set $A \subseteq \mathrm{dom}(B)$ such that $O = \bigcup\{B_a | a \in A\}$. Set $L_n^\tau = \bigcup\{B_a | a \in W_n\}$, if $W_n \subseteq \mathrm{dom}(B)$, and let

L_n^τ be undefined, otherwise. Then L^τ is an indexing of the Lacombe sets of τ. Obviously, $B \leq_m L^\tau$, i.e., there is some function $f \in P^{(1)}$ such that $f(n)\downarrow$, $f(n) \in \mathrm{dom}(L^\tau)$ and $B_n = L^\tau_{f(n)}$, for all $n \in \mathrm{dom}(B)$.

We can now effectively compare effectively given topologies. Let $\eta = \langle \mathcal{C} \rangle$ be a further topology on T and $C: \omega \rightarrowtail\!\!\rightarrow \mathcal{C}$ (onto) be an indexing of $\mathcal{C}$. Then τ is *Lacombe finer* than η ($\eta \subseteq_L \tau$) and η *Lacombe coarser* than τ, if $C \leq_m L^\tau$. If both $\eta \subseteq_L \tau$ and $\tau \subseteq_L \eta$, then η and τ are called *Lacombe equivalent.*

There is also another possibility to effectively compare η and τ. τ is said to be *recursively finer* than η ($\eta \subseteq_r \tau$) and η *recursively coarser* than τ, if there is some function $g \in P^{(2)}$ such that $g(i,m)\downarrow$, $g(i,m) \in \mathrm{dom}(B)$ and $x_i \in B_{g(i,m)} \subseteq C_m$, for all $i \in \mathrm{dom}(x)$ and $m \in \mathrm{dom}(B)$ with $x_i \in C_m$.

LEMMA 2.3. *Let B be computable. Then, if η is Lacombe coarser than τ, it is also recursively coarser than τ.*

PROOF. Let $C \leq_m L^\tau$ via $f \in P^{(1)}$, and let L witness that B is computable. Then $\{\langle a,i,n\rangle \mid a \in W_{f(n)} \wedge \langle i,a\rangle \in L\}$ is r.e. For $i,n \in \omega$ let $\langle a',i',n'\rangle$ be the first element in some fixed enumeration of this set with $i' = i$ and $n' = n$. It follows that $g \in P^{(2)}$ defined by $g(i,n) = a'$ witnesses that $\eta \subseteq_r \tau$.

3. Mal'cev Topologies

A topology η on T is a *Mal'cev topology*, if it has a basis $\mathcal{C}$ of completely enumerable subsets of T. Any such basis is called a *Mal'cev basis*. Let CE be the class of all completely enumerable subsets of T and $\mathcal{E} = \langle CE \rangle$. $\mathcal{E}$ is called *Eršov topology*. All Mal'cev bases on T can be indexed in a uniform canonical way. Let M_n^η be the set of all x_i with $i \in W_n \cap \mathrm{dom}(x)$, if this set is contained in $\mathcal{C}$, and let M_n^η be undefined, otherwise. Then M^η is a computable numbering of $\mathcal{C}$, and for any other numbering C of $\mathcal{C}$, C is computable iff $C \leq_m M^\eta$. We will assume in this paper that any Mal'cev basis is indexed in a computable way and that CE is indexed by $M^{\mathcal{E}}$. Then $\mathcal{E}$ is the Lacombe finest Mal'cev topology on T.

For Mal'cev topologies the converse of Lemma 2.3 is also true, provided our space $\mathcal{T}$ is *recursively separable*, which means that there is some r.e. set $D \subseteq \mathrm{dom}(x)$ such that $\{ x_i \mid i \in D \}$ is dense in T, i.e., it intersects every basic open set.

LEMMA 3.1. *Let $\mathcal{T}$ be recursively separable. Then any Mal'cev topology that is recursively coarser than τ, is also Lacombe coarser than τ.*

PROOF. Let $DB = \{x_i \mid i \in D\}$ be the dense enumerable subset of T. Moreover, let $\eta = < C >$ be a Mal'cev topology on T and $C: \omega \rightarrowtail C$ (onto) be a computable indexing of C. Let the computability of C be witnessed by $L \subseteq \omega$, and let $g \in P^{(2)}$ witness that $\eta \subseteq_r \tau$. Then for $i \in \mathrm{dom}(x)$ and $m \in \mathrm{dom}(C)$ with $x_i \in C_m$, $g(i,m)\downarrow$, $g(i,m) \in \mathrm{dom}(B)$ and $x_i \in B_{g(i,m)} \subseteq C_m$. Hence,

$$C_m = \bigcup\{B_{g(i,m)} \mid x_i \in C_m \cap DB\} = \bigcup\{B_{g(i,m)} \mid i \in D \ \wedge \ < i,m > \in L\},$$

which implies that $\eta \subseteq_L \tau$.

There are some important classes of Mal'cev topologies which we shall consider now. As it is easy to see, the class CE of all completely enumerable subsets of T is a distributive lattice with respect to union and intersection. For $U \in CE$, let U^* denote its pseudocomplement, i.e. the greatest completely enumerable subset of $T \setminus U$, if it exists. U is called *regular*, if U^* and U^{**} both exist and $U^{**} = U$. We say that a Mal'cev topology is *regular based*, if it has a basis of regular sets. Any such basis is called a *regular* basis. Since the class REG of all regular subsets of T is closed under intersection, it also generates a regular based Mal'cev topology on T, which we denote by $\mathcal{R}$.

Let η be a regular based Mal'cev topology on T with regular basis C and $C: \omega \rightarrowtail C$ (onto) be a numbering of C. We say that C is *$*$-computable*, if there is some r.e. set L' such that for all $i \in \mathrm{dom}(x)$ and $m \in \mathrm{dom}(C)$, $< i,m > \in L'$ iff $x_i \in C_m^*$. Just as in the general case of all Mal'cev bases, also all regular bases on T can be indexed in a uniform way. Let to this end $R^\eta_{<m,n>} = M^\eta_m$, if $m \in \mathrm{dom}(M^\eta)$, $n \in \mathrm{dom}(M^\tau)$ and $M^{\eta}_m{}^* = M^\tau_n$, and let $R^\eta_{<m,n>}$ be undefined, otherwise. Then R^η is a computable and $*$-computable indexing of C, and for any other numbering C of C, C is both computable and $*$-computable, iff $C \leq_m R^\eta$. We assume in this paper that regular bases are always indexed, both computably and $*$-computably, and that REG is indexed by $R^{\mathcal{R}}$. Then $\mathcal{R}$ is the Lacombe finest regular based Mal'cev topology on T.

If η is a topology on T, then a subset X of T is called *weakly decidable*, if its interior and its exterior are both completely enumerable. η is called *effectively semi-regular*, if it is generated by those of its regular open sets which are also weakly decidable. Recall that an open set X is regular open, if $X = \mathrm{int}_\eta(\mathrm{cl}_\eta(X))$. Since the exterior of a regular open set is also regular open, any class of such sets which with each set also contains its exterior generates a topology in which these sets are again regular open. Moreover, the exterior of any of them is the same in both topologies. Hence, the class of all weakly decidable regular open sets of a topology η on T generates a topology η^* which is coarser than η and effectively semi-regular; it is said to be the effectively semi-regular topology *associated* with η.

Since every regular subset of T is regular open and weakly decidable with respect to the Eršov topology, and these are the only such sets, it follows that

LEMMA 3.2 *$\mathcal{R}$ is the effectively semi-regular topology associated with the Eršov topology, i.e., $\mathcal{R} = \mathcal{E}^*$.*

4. A Characterization

Let B be computable. Then all basic open sets are completely enumerable, which means that τ is a Mal'cev topology and hence Lacombe coarser than the Eršov topology on T. In this section we are interested in which Mal'cev topologies are recursively coarser than τ.

Let to this end for any $X \subseteq T$

$$\mathrm{hl}(X) = \bigcap\{O \in \mathcal{B} \mid \exists O' \in \mathcal{B}\ X \subseteq O' \prec O\},$$

let $\eta = \langle C \rangle$ be some further topology on T, and $C: \omega \rightarrowtail C$ (onto) be an indexing of C. Then η is said to be *compatible* with τ, if there are functions $s \in P^{(2)}$ and $r \in P^{(3)}$ such that for all $i \in \mathrm{dom}(x)$, $n \in \mathrm{dom}(B)$ and $m \in \mathrm{dom}(C)$ the following hold:

(i) If $x_i \in C_m$, then $s(i,m)\downarrow$, $s(i,m) \in \mathrm{dom}(C)$ and $x_i \in C_{s(i,m)} \subseteq C_m$.

(ii) If moreover $B_n \not\subseteq C_m$, then also $r(i,n,m)\downarrow$, $r(i,n,m) \in \mathrm{dom}(x)$ and $x_{r(i,n,m)} \in \mathrm{hl}(B_n) \setminus C_{s(i,m)}$.

In order to see how this will help us to answer the above question, let us suppose that $x_i \in C_m$ and that we have a normed recursive enumeration $\left(B_{f(c)}\right)_{c \in \omega}$ which converges to x_i. We want to force some neighbourhood $B_{f(a)}$ of x_i to be contained in C_m. Let us therefore assume that $B_{f(a)}$ is not contained in C_m. Then we can effectively find a (possibly) smaller neighbourhood $C_{s(i,m)}$ with $x_i \in C_{s(i,m)} \subseteq C_m$, and a point $x_{r(i,n,m)}$ which lies, not necessarily outside of C_m, but at least outside of $C_{s(i,m)}$. Now, we modify the given enumeration $\left(B_{f(c)}\right)$ to look like $B_{f(c)}$ for $c < a$, but to follow a normed recursive enumeration of basic open sets which converges to $x_{r(i,n,m)}$, for $c \geq a$. Then, if we choose a to be the number of steps necessary to find the point this enumeration converges to in $C_{s(i,m)}$, the modified enumeration will "look" like it is going to converge to a point in $C_{s(i,m)}$. But once this point has been found in $C_{s(i,m)}$, it changes into an enumeration that converges to a point outside of $C_{s(i,m)}$.

The next result is the essential step in the proof of our general characterization theorem.

THEOREM 4.1. *Let $\mathcal{T}$ be strongly effective and B be acceptable. Then any Mal'cev topology that is compatible with τ is recursively coarser than τ. If $\mathcal{T}$ is also recursively separable, than any such topology is even Lacombe coarser than τ.*

PROOF. Because of Lemma 3.1 we only have to show the first part of this theorem. Let to this end $\eta = < \mathcal{C} >$ be a Mal'cev topology which is compatible with τ, and $C: \omega \rightarrowtail \mathcal{C}$ (onto) be a computable indexing of $\mathcal{C}$. Moreover, let $L \subseteq \omega$, $pt \in P^{(1)}$, $s \in P^{(2)}$ and $r \in P^{(3)}$ respectively witness that C is computable, B is acceptable and η is compatible with τ. Finally, let $p \in P^{(2)}$ and $q \in P^{(1)}$ be as in Lemma 2.2 and

$$E = \left\{ < n,i,m > \mid < pt(n), s(i,m) > \in L \right\}.$$

Then E is r.e. Hence, there is some recursive set A such that E is the projection of A, i.e., $E = \left\{ a \mid \exists c < a,c > \in A \right\}$. Now, set

$$\hat{g}(n,i,m) = \mu\, c: << n,i,m >, c > \in A$$

and let $h \in P^{(3)}$ be such that

$$\varphi_{h(n,i,m)}(a) = \begin{cases} \varphi_{q(i)}(a), & \text{if } a \leq \hat{g}(n,i,m) \\ \varphi_{p(r(i,\varphi_{q(i)}(\hat{g}(n,i,m)+1),m),\varphi_{q(i)}(\hat{g}(n,i,m)))}(a - \hat{g}(n,i,m)), & \text{otherwise.} \end{cases}$$

By the recursion theorem there is then a function $d \in R^{(2)}$ with

$$\varphi_{h(d(i,m),i,m)} = \varphi_{d(i,m)}.$$

Let $g(i,m) = \hat{g}(d(i,m),i,m)$, and suppose that $g(i,m)\uparrow$ for some $i \in \text{dom}(x)$ and $m \in \text{dom}(C)$ with $x_i \in C_m$. Then $d(i,m)$ is an index of a normed recursive enumeration of basic open sets which converges to x_i. By the acceptability of B it follows that $pt(d(i,m))\downarrow$ and $x_{pt(d(i,m))} = x_i$. Moreover, since $x_i \in C_m$, we furthermore obtain that $s(i,m)\downarrow$ and $x_i \in C_{s(i,m)}$. Thus $x_{pt(d(i,m))} \in C_{s(i,m)}$, which implies that $g(i,m)\downarrow$. This contradicts our assumption. Therefore $g(i,m)\downarrow$ for all $i \in \text{dom}(x)$ and $m \in \text{dom}(C)$ with $x_i \in C_m$.

Assume next that $B(\varphi_{q(i)}(g(i,m)+1)) \not\subseteq C_m$, for some $i \in \text{dom}(x)$ and some $m \in \text{dom}(C)$ with $x_i \in C_m$. Then $r(i, \varphi_{q(i)}(g(i,m)+1), m)\downarrow$. Moreover

$$x_{r(i,\varphi_{q(i)}(g(i,m)+1),m)} \in \text{hl}(B_{\varphi_{q(i)}(g(i,m)+1)}) \setminus C_{s(i,m)} \subseteq B_{\varphi_{q(i)}(g(i,m))} \setminus C_{s(i,m)} \qquad (1)$$

The last inclusion holds, since $B(\varphi_{q(i)}(g(i,m)+1)) \prec B(\varphi_{q(i)}(g(i,m)))$. According to Lemma 2.2, $p(r(i,\varphi_{q(i)}(g(i,m)+1),m), \varphi_{q(i)}(g(i,m)))$ is an index of a normed recursive enumeration of basic open sets which converges to $x(r(i,\varphi_{q(i)}(g(i,m)+1),m))$. In addition

$$B_{\varphi_{p(r(i,\varphi_{q(i)}(g(i,m)+1),m),\varphi_{q(i)}(g(i,m)))}(0)} \prec B_{\varphi_{q(i)}(g(i,m))},$$

which implies that $d(i,m)$ is also an index of a normed recursive enumeration of basic open sets which converges to $x(r(i,\varphi_{q(i)}(g(i,m)+1),m))$. As above we obtain from this that $pt(d(i,m))\downarrow$ and $x(r(i,\varphi_{q(i)}(g(i,m)+1),m)) = x(pt(d(i,m)))$. Because of (1) it now follows that $x_{pt(d(i,m))} \notin C_{s(i,m)}$. Hence $g(i,m)\uparrow$, which contradicts what we have shown above. Thus

$$x_i \in B_{\varphi_{q(i)}(g(i,m)+1)} \subseteq C_m,$$

for all $i \in \text{dom}(x)$ and $m \in \text{dom}(C)$ with $x_i \in C_m$, showing that $\eta \subseteq_r \tau$.

Since τ is a Mal'cev topology, if B is computable, as a consequence we obtain our general characterization theorem: The topology of a strongly effective T_0-space that is compatible with itself, in which the elements of all basic open sets can effectively be enumerated, and in which one can effectively pass from a normed, recursive and convergent enumeration of basic open sets to the point it converges to is the recursively finest Mal'cev topology on that space which is compatible with it:

THEOREM 4.2. *Let $\mathcal{T}$ be strongly effective, B be acceptable, and τ be compatible with itself. Then τ is the recursively finest Mal'cev topology on T that is compatible with τ. If, in addition, $\mathcal{T}$ is recursively separable, then τ is even the Lacombe finest Mal'cev topology on T which is compatible with it.*

The requirements of this theorem seem to us to be very natural, and in practice it seems always easy to verify that the canonical topological bases of the spaces under consideration have an acceptable indexing. As we shall see, the compatibility requirement too holds in many natural situations, but we shall also see that there is a strongly effective space and a Mal'cev topology which is strictly finer than the topology of the space, showing that it is not compatible with this topology and thus that the restriction to compatible Mal'cev topologies is essential in Theorems 4.1 and 4.2.

In the next two sections we study two important types of spaces. Both of them will turn out to be strongly effective, and interesting Mal'cev topologies are compatible with their canonical topologies. First we consider spaces that consist of the computable elements of an effective complete partial order, and then we consider recursive metric spaces.

5. The Domain Case

Let $Q = (Q, \sqsubseteq)$ be a partial order. A subset S of Q is *directed*, if for all $y_1, y_2 \in S$ there is some $u \in S$ with $y_1, y_2 \sqsubseteq u$. Q is a *complete* partial order (cpo), if every directed subset S of Q has a least upper bound $\sup S$ in Q. Let $\bot = \sup\varnothing$. Moreover, let $\ll$ denote the *way-below relation* on Q, i.e., let $y_1 \ll y_2$ iff for directed subsets S of Q the relation $y_2 \sqsubseteq \sup S$ always implies the existence of a $u \in S$ with $y_1 \sqsubseteq u$. Note that $\ll$ is transitive.

A subset Z of Q is a *basis* of Q, if for any $y \in Q$ the set $Z_y = \{z \in Z \mid z \ll y\}$ is directed and $y = \sup Z_y$. If the cpo Q has a basis, then it is said to be *continuous*. For such cpo-s it is shown in [63, Lemma 2.3] that for $y, y_1, y_2, y_3 \in Q$

LEMMA 5.1. (i) $y_1 \ll y_2 \Rightarrow y_1 \sqsubseteq y_2$.

(ii) $y_1 \ll y_2 \sqsubseteq y_3 \Rightarrow y_1 \ll y_3$.

(iii) $y_1 \ll y_2 \Rightarrow \exists z \in Z\ y_1 \ll z \ll y_2$.

(iv) Z_y is directed with respect to $\ll$.

Moreover, for continuous cpo-s a canonical topology can be defined (cf. [11,47]): A subset X of Q is open, if $(O1)$ and $(O2)$ hold:

$(O1)$ $\forall u, y \in Q\ (u \in X \wedge u \sqsubseteq y \Rightarrow y \in X)$
$(O2)$ $\forall u \in X\ \exists y \in X\ y \ll u$.

The topology thus defined is called the *Scott topology* of Q. If Z is a basis of Q, then $\{O_z \mid z \in Z\}$ with $O_z = \{y \in Q \mid z \ll y\}$ is a basis for this topology.

There have been many suggestions in the literature as to which effectivity requirements a cpo should satisfy in order to develop a sufficiently rich computability theory for these structures (cf. [6,25,26,28,44,45,47,54]). Here we use a very general approach which is due to Weihrauch (cf. [63]).

Let Z be a basis of Q. If there exists an indexing $e: \omega \to Z$ (onto) of Z such that $\{<i,j> \mid e_i \ll e_j\}$ is r.e., then $Q = (Q, \sqsubseteq, Z, e)$ is called an *effective* cpo. An element $y \in Q$ is said to be *computable*, if $\{i \mid e_i \ll y\}$ $[= e^{-1}(Z_y)]$ is r.e. Let Q_c denote the set of all computable elements of Q, then $(Q_c, \sqsubseteq, Z, e)$ is called *constructive domain.* Q_c can be characterized as that subset D of Q for which, for every directed subset S of Z, $\sup S \in D$ iff $\{i \mid e_i \in S\}$ is r.e. Let σ be the relativization of the Scott topology to Q_c. Then $\mathcal{Q}_c = (Q_c, \sigma)$ is a countable T_0-space with basic open sets $B_n = O_{e(n)} \cap Q_c$ $(n \in \omega)$. Moreover, Z is dense in $\mathcal{Q}_c$.

An indexing $x: \omega \to Q_c$ (onto) of the computable cpo elements is called *admissible*, if it satisfies the following axioms $(A1)$ and $(A2)$:

$(A1)$ $\{<i,j> \mid e_i \ll x_j\}$ is r.e.
$(A2)$ There is a function $d \in R^{(1)}$ with $x_{d(i)} = \sup e(W_i)$, for all indices $i \in \omega$ such that $e(W_i)$ is directed.

As it is shown in [63, Satz 5] such indexings exist. Moreover, it is easy to verify that B is computable, if x satisfies $(A1)$. If x satisfies $(A2)$, then there is some $g \in R^{(1)}$ with $e = x \circ g$. Since Z is dense in $\mathcal{Q}_c$, it thus follows that $\mathcal{Q}_c$ is recursively separable in this case.

Now, for basic open sets B_m and B_n we define

$$B_m \prec B_n \Leftrightarrow e_n \ll e_m.$$

Then we obtain with Lemma 5.1 that $\prec$ is a strong inclusion and $\{B_n \mid n \in \omega\}$ is a strong basis.

The major theorem of this section guarantees that under rather general conditions, all Mal'cev topologies on Q_c are compatible with the relativization of the Scott topology to Q_c. It thus follows from Theorem 4.2 that this topology is the Lacombe finest Mal'cev topology on Q_c.

THEOREM 5.2. *Let $(Q_c, \sqsubseteq, Z, e)$ be a constructive domain and x an admissible indexing of Q_c. Then the following statements hold:*

(i) *Q_c is strongly effective.*

(ii) *B is acceptable.*

(iii) *Each completely enumerable subset of Q_c is upwards closed under $\sqsubseteq$.*

(iv) *Every Mal'cev topology on Q_c is compatible with σ.*

PROOF. (i) is a direct consequence of Lemma 2.1.

(ii) As we have already seen, B is computable under the above assumptions on x. To see that it is also acceptable, let m be an index of a normed recursive enumeration of basic open sets which converges to some point $y \in Q_c$. Then the sequence $(e(\varphi_m(a)))_{a\in\omega}$ is strictly increasing with respect to $\ll$ and $\sup e(\varphi_m(a)) = y$. Now, let $d \in R^{(1)}$ be as in $(A2)$ and $v \in R^{(1)}$ such that $W_{v(j)} = \text{range}(\varphi_j)$. It follows that $x_{d(v(m))} = y$.

(iii) Let Y be some completely enumerable subset of Q_c with $x_i \in Y$ and suppose that $x_i \sqsubseteq x_j$ but $x_j \notin Y$. Now, define $f \in R^{(1)}$ such that $\varphi_{f(n)}(a) = i$ if $\varphi_n(n)\uparrow_a$, and $\varphi_{f(n)}(a) = j$, otherwise. Then $\{x(\varphi_{f(n)}(a)) \mid a \in \omega\}$ is directed. As it is shown in [63, Lemma 11], there is some $t \in R^{(1)}$ with $x_{t(m)} = \sup\{x(\varphi_m(a)) \mid a \in \omega\}$, if $\{x(\varphi_m(a)) \mid a \in \omega\}$ is directed. Then $x_{t(f(n))} = x_i$ if $n \notin K = \{c \mid \varphi_c(c)\downarrow\}$, and $x_{t(f(n))} = x_j$, otherwise. Since Y is completely enumerable, there is some r.e. set A with $c \in A$ iff $x_c \in Y$, for all $c \in \omega$. Then we have that $t(f(n)) \in A$ iff $x_{t(f(n))} \in Y$. Because $x_{t(f(n))} = x_i \in Y$ iff $n \notin K$, we obtain that $t(f(n)) \in A$ iff $n \notin K$. This contradicts the fact that $\omega \setminus K$ is not r.e.

(iv) Let η be a Mal'cev topology on Q_c with Mal'cev basis C and let $C\colon \omega \rightarrowtail C$ (onto) be a computable indexing of C. Moreover, let $g \in R^{(1)}$ with $e = x \circ g$. Define $s \in P^{(2)}$ and $r \in P^{(3)}$ by $s(i,m) = m$ and $r(i,n,m) = g(n)$. Then (s,r) is a witness that η is compatible with

σ. In order to see this, assume that $B_n \not\subseteq C_m$. Then $e_n \notin C_m$, by (iii). But $e_n \in B_a$, for all B_a with $B_n \prec B_a$. Hence $e_n \in \text{hl}(B_n)$.

Since also the Eršov topology is the Lacombe finest Mal'cev topology on Q_c, we obtain

THEOREM 5.3. *Let $(Q_c, \sqsubseteq, Z, e)$ be a constructive domain with Q_c having an admissible indexing. Then the relativization of the Scott topology to Q_c is Lacombe equivalent to the Eršov topology on Q_c.*

This is the generalization of the Rice/Shapiro theorem to constructive domains mentioned earlier (cf. [63]). To see how this theorem follows for the r.e. sets, let P_ω be the class of all subsets of ω, let *FIN* be the subclass of the finite sets, and let $\rho: \omega \to FIN$ (onto) be a canonical indexing of *FIN*. Then $(P_\omega, \subseteq, FIN, \rho)$ is an effective cpo, and the numbering W is admissible (cf. [63]). Moreover, for pairs of sets such that at least one component is a finite set the way-below relation coincides with the cpo ordering. Hence we obtain (cf. [42])

COROLLARY 5.4 (Rice/Shapiro). *Let I be a subclass of the class RE of all r.e. sets. Then $W^{-1}(I)$ is r.e. iff there exists some r.e. set A such that*

$$I = \{X \in RE \mid \exists i \in A\ \rho(i) \subseteq f\}.$$

6. The Metric Case

Let $\mathbb{R}$ denote the set of all real numbers, and let ν be some canonical indexing of the rational numbers. Then a real number z is said to be *computable*, if there is a function $f \in R^{(1)}$ such that for all $m,n \in \omega$ with $m \leq n$, the inequality $|\nu_{f(m)} - \nu_{f(n)}| < 2^{-m}$ holds and $z = \lim \nu_{f(m)}$. Any Gödel number of the function f is called an *index* of z. This defines a partial indexing γ of the set $\mathbb{R}_c$ of all computable real numbers.

Now, let (M,δ) be a countable metric space with range$(\delta) \subseteq \mathbb{R}_c$, and let $x: \omega \rightarrowtail M$ (onto) be an indexing of M. Then (M,δ) is said to be *recursive*, if the distance function δ is effective, i.e., if there is some function $d \in P^{(2)}$ such that for all $i,j \in \text{dom}(x)$, $d(i,j)\downarrow$, $d(i,j) \in \text{dom}(\gamma)$ and $\delta(x_i,x_j) = \gamma_{d(i,j)}$. As it is well known, there is a canonical Hausdorff topology Δ on M. The collection of sets $H_{<i,m>} = \{y \in M \mid \delta(x_i, y) < 2^{-m}\}$ $(i \in \text{dom}(x),\ m \in \omega)$ is a basis of this topology.

Let $\mathcal{M} = (M, \Delta)$ be recursively separable and let $g \in R^{(1)}$ be such that $DB = \{x_{g(n)} \mid n \in \omega\}$ is dense in $\mathcal{M}$. Then the set of all $H_{<g(i),m>}$ $(i, m \in \omega)$ is also a basis of the metric topology on M. In what follows, we shall always use this basis and the numbering $B_{<i,m>} = H_{<g(i),m>}$. Since there is an r.e. set A with $<i,j> \in A$ iff $\gamma_i < \gamma_j$, for $i, j \in \text{dom}(\gamma)$ (cf. [31, Lemma 5]), it follows that B is computable (cf. [32, Lemma 1]). Moreover, it follows that there is some r.e. set L' with $<j,<i,m>> \in L'$ iff $\delta(x_i, x_j) > 2^{-m}$, for all $i, j \in \text{dom}(x)$ and $m \in \omega$. Thus, also the exterior of each basic open set is completely enumerable, uniformly in its index.

For basic open sets $B_{<i,m>}$ and $B_{<j,n>}$ let

$$B_{<i,m>} \prec B_{<j,n>} \Leftrightarrow \delta(x_{g(i)}, x_{g(j)}) + 2^{-m} < 2^{-n}.$$

Using the triangular inequation it is readily verified that $\prec$ is a strong inclusion and $\{B_a \mid a \in \omega\}$ is a strong basis.

As we have already seen, B is computable. We shall now state a condition on Cauchy sequences which ensures that B is also acceptable. Let $(y_a)_{a \in \omega}$ be a sequence with $y_a \in DB$, for all $a \in \omega$. Then (y_a) is said to be *normed*, if $\delta(y_m, y_n) < 2^{-m}$, for all $m, n \in \omega$ with $m \le n$. Moreover, (y_a) is *recursive*, if there is some function $f \in R^{(1)}$ with range$(f) \subseteq$ range(g) such that $y_a = x_{f(a)}$, for all $a \in \omega$. Any Gödel number of f is called an index of the sequence (y_a). We say that (y_a) *satisfies* (A), if the following holds (cf. [32]):

(A) There is a function $li \in P^{(1)}$ such that, if m is an index of a converging normed recursive sequence (y_a) of elements of the dense basis of $\mathcal{M}$, then $li(m)\downarrow$, $li(m) \in \text{dom}(x)$ and $x_{li(m)} = \lim y_a$.

LEMMA 6.1. *Let x satisfy (A). Then $\mathcal{M}$ is strongly effective and B is acceptable.*

PROOF. Since $\{<<i,m>,<j,n>> \mid \delta(x_{g(i)}, x_{g(j)}) + 2^{-m} < 2^{-n}\}$ is r.e., it follows from Lemma 2.1 that $\mathcal{M}$ is strongly effective.

As we have already seen, B is computable. To see that it is also acceptable, let m be an index of a normed recursive enumeration of basic open sets which converges to some point $z \in M$. Moreover, let $f \in R^{(1)}$ with

$$\varphi_{f(m)}(a) = g(\pi_1(\varphi_m(\mu c: a < \pi_2(\varphi_m(c))))).$$

Since the sequence $(B(\varphi_m(c)))$ is strictly decreasing with respect to $\prec$, it follows that $(\pi_2(\varphi_m(c)))$ is a strictly increasing sequence of natural numbers. Thus, $\varphi_{f(m)}$ is a total function. Now, define $y_a = x(\varphi_{f(m)}(a))$, for $a \in \omega$, then (y_a) is a normed recursive sequence of elements of the dense base of $\mathcal{M}$ which converges to z, and $f(m)$ is one of its indices. Because x satisfies (A), we obtain that $li(f(m))\downarrow$ and $x_{li(f(m))} = z$.

Since B is computable, it follows that all basic open sets are completely enumerable. In [59] it has been shown

THEOREM 6.2. *There exists a recursively separable recursive metric space such that the Eršov topology on this space is strictly finer than the metric topology.*

Thus, in general there are Mal'cev topologies which are not compatible with the given metric topology. This shows that the restriction to compatible Mal'cev topologies is essential in Theorems 4.1 and 4.2. Let us now look which Mal'cev topologies are compatible with Δ.

As has already been said, the exterior of each basic open set in the metric topology is completely enumerable, uniformly in its index. Moreover, we have

THEOREM 6.3. *Let x satisfy (A). Then the following statements hold:*

(i) *Each basic open set B_a is regular.*

(ii) *Every regular based Mal'cev topology on M is compatible with Δ.*

PROOF. (i) Since B_a and $\text{ext}_\Delta(B_a)$ are both completely enumerable, it suffices to show that for each completely enumerable open set X, $X^* = \text{ext}_\Delta(X)$.

Assume that there exists a completely enumerable subset S of $M \setminus X$ with $\text{ext}_\Delta(X) \subset S$. Then there is some point $z \in S \cap \text{bnd}_\Delta(X)$. We shall now construct a normed recursive sequence (y_a) of elements of the dense base of $\mathcal{M}$ such that $z = \lim y$ and $y_a \in X$, for all $a \in \omega$. As follows from above, there is some r.e. set E such that for all $j,m \in \omega$, $\langle j,m \rangle \in E$ iff

$\delta(z, x_{g(j)}) < 2^{-m}$ and $x_{g(j)} \in X$. For $a \in \omega$, let $<j',m'>$ be the first element in some enumeration of E with $m' = a+2$. Note that there is always such an element, since $z \in \mathrm{bnd}_\Delta(X)$, X is open and DB is dense in $\mathcal{M}$. Set $y_a = x_{g(j')}$. Let b be an index of this sequence and let $li \in P^{(1)}$ be as in condition (A). Moreover, let $n \in \omega$ be such that for all $a \in \mathrm{dom}(x)$, $a \in W_n$ iff $x_a \in S$. Then, by the recursion theorem there is some $i \in \omega$ with

$$\varphi_i(a) = \begin{cases} \varphi_b(a), & \text{if } \varphi_n(li(i))\uparrow_a \\ \varphi_b(\mu c\colon \varphi_n(li(i))\downarrow_c), & \text{otherwise.} \end{cases}$$

It is now easy to see that the assumption that $\varphi_n(li(i))\uparrow$ leads to a contradiction. Thus $\varphi_n(li(i))\downarrow$, which means that $\lim x(\varphi_i(a)) \in S$. Furthermore it follows for $a' = \mu c\colon \varphi_n(li(i))\downarrow_c$ that $\varphi_i(a) = \varphi_b(a')$ for all $a \geq a'$. Hence $\lim x(\varphi_i(a)) = x(\varphi_b(a')) = y_{a'}$. Since $y_{a'} \in X$ and $X \subseteq M \setminus S$, this contradicts what we have seen before. It follows that there is no completely enumerable superset of $\mathrm{ext}_\Delta(X)$ in $M \setminus X$, showing that $X^* = \mathrm{ext}_\Delta(X)$.

(ii) Let η be a regular based Mal'cev topology on M with regular basis $\mathcal{C}$, and let $C\colon \omega \rightarrowtail \mathcal{C}$ (onto) be a computable and $*$-computable indexing of $\mathcal{C}$. First we show for $n \in \omega$ and $m \in \mathrm{dom}(C)$ that, if B_n intersects $M \setminus C_m$, then B_n also intersects $C_m{}^*$. Assume to this end that $B_n \cap C_m^* = \emptyset$, but $B_n \cap (M \setminus C_m) \neq \emptyset$. Then $B_n \subseteq M \setminus C_m^*$ and $B_n \cap ((M \setminus C_m^*) \setminus C_m) \neq \emptyset$. It follows that $C_m \subset C_m \cup B_n \subseteq M \setminus C_m{}^*$. This contradicts the regularity of C_m.

Now, let $L, L' \subseteq \omega$ respectively witness that B is computable and C is $*$-computable. For $n,m \in \omega$ let $<a',n',m'>$ be the first element in some fixed enumeration of $\{<a,b,c> \mid <a,b> \in L \wedge <a,c> \in L'\}$ with $n' = n$ and $m' = m$. As we have just seen, there is always such a number a', if $n \in \omega$ and $m \in \mathrm{dom}(C)$ such that $B_n \not\subseteq C_m$. Define $r(i,n,m) = a'$ and $s(i,m) = m$. Then (s,r) is a witness that η is compatible with Δ.

Since B is also $*$-computable, it follows from Lemma 3.2 and Theorem 4.2

THEOREM 6.4. *Let (M,δ) be a recursively separable recursive metric space with indexing x satisfying (A). Then the canonical metric topology is Lacombe equivalent to the effectively semi-regular topology associated with the Eršov topology on M.*

It is well known that the metric topology satisfies the T_3-axiom. As we shall see next, it does so also effectively. This will lead to a further characterization of the metric topology which says that it is the Lacombe finest complemented Mal'cev topology which effectively satisfies the T_3-axiom.

Let $\eta = < C >$ be a topology on M and $C: \omega \rightarrowtail C$ (onto) be an indexing of C. We say that η *effectively satisfies the T_3-axiom*, if there is some function $s \in P^{(2)}$ such that $s(i,m)\downarrow$, $s(i,m) \in \mathrm{dom}(C)$ and

$$x_i \in C_{s(i,m)} \subseteq \mathrm{cl}_\eta(C_{s(i,m)}) \subseteq C_m,$$

for all $i \in \mathrm{dom}(x)$ and $m \in \mathrm{dom}(C)$ with $x_i \in C_m$.

If η is a Mal'cev topology and C a Mal'cev basis, then η is called *complemented*, if all of its basic open sets are weakly decidable. C is said to be *co-computable*, if there is some r.e. set L' such that for all $i \in \mathrm{dom}(x)$ and $m \in \mathrm{dom}(C)$, $< i,m > \in L'$ iff $x_i \in \mathrm{ext}_\eta(C_m)$. In what follows we assume that the bases of complemented Mal'cev topologies are indexed computably as well as co-computably. Just as in the case of regular based Mal'cev topologies one can show that such indexings always exist.

As we have already seen, Δ is complemented and B is co-computable. Moreover, we have

THEOREM 6.5. *(i) Δ effectively satisfies the T_3-axiom.*
(ii) Let x satisfy (A). Then each complemented Mal'cev topology on M which effectively satisfies the T_3-axiom is compatible with Δ.

PROOF. (i) From what has been said above it follows that there is some r.e. set E such that for $i \in \mathrm{dom}(x)$ and $j,m,a,c \in \omega$,

$$< i,j,m,a,c > \in E \text{ iff } \delta(x_{g(j)},x_i) + 2^{1-c} < 2^{-m} \text{ and } \delta(x_{g(a)},x_i) < 2^{-c}.$$

For $i,j,m \in \omega$ let $<i',j',m',a',c'>$ be the first element in some fixed enumeration of E with $i' = i$, $j' = j$ and $m' = m$. Define $s(i,< j,m >) = < a',c' >$. Then $s \in P^{(2)}$. Now, let $i \in \mathrm{dom}(x)$ and $j,m \in \omega$ be such that $x_i \in B_{<j,m>}$. Since DB is dense in $\mathcal{M}$, it follows that $s(i,< j,m >)\downarrow$. Moreover, $x_i \in B_{s(i,<j,m>)}$ and $\mathrm{cl}_\Delta(B_{s(i,<j,m>)}) = \{ y \in M \mid \delta(x_{g(a)},y) \le 2^{-c} \} \subseteq B_{<j,m>}$, where $< a,c > = s(i,< j,m >)$.

(ii) Let η be a complemented Mal'cev topology on M with Mal'cev basis C that effectively satisfies the T_3-axiom, and let $C: \omega \rightarrowtail C$ (onto) be a computable and co-computable indexing of C. Furthermore, let $s \in P^{(2)}$ witness that η effectively satisfies the T_3-axiom. Now, suppose that $B_n \not\subseteq C_m$, for some $n \in \omega$ and $m \in \mathrm{dom}(C)$. Then, for all $i \in \mathrm{dom}(x)$ with $x_i \in C_m$, we have that $B_n \setminus \mathrm{cl}_\eta(C_{s(i,m)}) \neq \emptyset$. We show that also $DB \cap (B_n \setminus \mathrm{cl}_\eta(C_{s(i,m)})) \neq \emptyset$.

Let to this end $z \in B_n \setminus \mathrm{cl}_\eta(C_{s(i,m)})$. As we have seen in the proof of Theorem 6.3 (i) there is a normed recursive sequence (y_a) which converges to z such that $y_a \in DB \cap B_n$, for all $a \in \omega$. Let b be an index of this sequence and let $li \in P^{(1)}$ be as in condition (A). Moreover, let $b' \in \omega$ be such that for $c \in \mathrm{dom}(x)$, $c \in W_{b'}$ iff $x_c \notin \mathrm{cl}_\eta(C_{s(i,m)})$. Then, by the recursion theorem there is some $a \in \omega$ with

$$\varphi_a(c) = \begin{cases} \varphi_b(c), & \text{if } \varphi_{b'}(li(a))\uparrow_c \\ \varphi_b(\mu c' : \varphi_{b'}(li(a))\downarrow_{c'}), & \text{otherwise.} \end{cases}$$

It is easy to see that the assumption that $\varphi_{b'}(li(a))\uparrow$ leads to a contradiction. Thus $\varphi_{b'}(li(a))\downarrow$, which means that $\lim x(\varphi_a(c)) \notin \mathrm{cl}_\eta(C_{s(i,m)})$. Furthermore it follows for $\hat{c} = \mu c' : \varphi_{b'}(li(a))\downarrow_{c'}$ that $\varphi_a(c) = \varphi_b(\hat{c})$ for all $c \geq \hat{c}$. Hence $x_{li(a)} = \lim x(\varphi_a(c)) = x(\varphi_b(\hat{c})) = y_{\hat{c}}$. Thus there is some $y \in DB \cap (B_n \setminus \mathrm{cl}_\eta(C_{s(i,m)}))$, namely $y = y_{\hat{c}}$.

Now, let L and L' respectively witness that B is computable and C is co-computable. Moreover, for $i,m,n \in \omega$ let $<i',m',n',a'>$ be the first element in some fixed enumeration of $E = \{<i,m,n,a> \mid <g(a),n> \in L \wedge <g(a),s(i,m)> \in L'\}$ with $i' = i$, $m' = m$ and $n' = n$. Define $r(i,n,m) = a'$. Then $r \in P^{(3)}$. In addition, it follows from what has just been shown that for $i \in \mathrm{dom}(x)$, $n \in \omega$ and $m \in \mathrm{dom}(C)$ with $x_i \in C_m$ and $B_n \not\subseteq C_m$, $r(i,n,m)\downarrow$ and $x_{r(i,n,m)} \in B_n \setminus C_{s(i,m)}$. Since always $B_n \subseteq \mathrm{hl}(B_n)$, we obtain that η is compatible with Δ.

With Theorem 4.2 we thus obtain

THEOREM 6.6. *Let (M,δ) be a recursively separable recursive metric space with indexing x satisfying (A). Then the canonical metric topology is the Lacombe finest complemented Mal'cev topology on M that effectively satisfies the T_3-axiom.*

Acknowledgement

The author is grateful to Ulrich Berger for suggesting improvements to an earlier version of the paper and fruitful discussions. Thanks for useful hints and discussions are also due to Professors Juri L. Eršov, Dana Scott, Michael B. Smyth, Klaus Weihrauch and Paul Young.

References

[1] America, P. and J. Rutten: Solving reflexive domain equations in a category of complete metric spaces. *Mathematical Foundations of Programming Language Semantics, 3rd Workshop* (Main, M. et al., eds.), 252-288. Lec. Notes Comp. Sci. 298. Berlin: Springer (1988).

[2] De Bakker, J.W. and J.I. Zucker: Processes and the denotational semantics of concurrency. *Inform. and Control* 54, 70-120 (1982).

[3] Ceĭtin, G.S.: Algorithmic operators in constructive metric spaces. *Trudy Mat. Inst. Steklov* 67, 295-361 (1962); English transl., *Amer. Math. Soc. Transl.* (2) 64, 1-80 (1967).

[4] Czászár, A.: *Foundations of General Topology*. New York: Pergamon (1963).

[5] Dyment, E.Z.: Recursive metrizability of numbered topological spaces and bases of effective linear topological spaces. *Izv. Vyssh. Uchebn. Zaved, Mat. (Kazan)*, no. 8, 59-61 (1984); English transl., *Sov. Math. (Iz. VUZ)* 28, 74-78 (1984).

[6] Egli, H. and R.L. Constable: Computability concepts for programming language semantics. *Theoret. Comp. Sci.* 2, 133-145 (1976).

[7] Eršov, Ju.L.: Computable functionals of finite type. *Algebra i Logika* 11, 367-437 (1972); English transl., *Algebra and Logic* 11, 203-242 (1972).

[8] Eršov, Ju.L.: The theory of *A*-spaces. *Algebra i Logika* 12, 369-416 (1973); English transl., *Algebra and Logic* 12, 209-232 (1973).

[9] Eršov, Ju.L.: Model $\mathfrak{C}$ of partial continuous functionals. *Logic Colloquium* 76 (Gandy, R. et al., eds.), 455-467. Amsterdam: North-Holland (1977).

[10] Fletcher, P. and W.F. Lindgren: *Quasi-Uniform Spaces*. New York: Dekker (1982).

[11] Gierz, G. et al.: *A Compendium of Continuous Lattices*. Berlin: Springer (1980).

[12] Golson, W.G. and W.C. Rounds: Connections between two theories of concurrency: Metric spaces and synchronisation trees. *Inform. and Control* 57, 102-124 (1983).

[13] Hauck, J.: Konstruktive Darstellungen in topologischen Räumen mit rekursiver Basis. *Zeitschr. f. math. Logik Grundl. d. Math.* 26, 565-576 (1980).

[14] Hauck, J.: Berechenbarkeit in topologischen Räumen mit rekursiver Basis. *Zeitschr. f. math. Logik Grundl. d. Math.* 27, 473-480 (1981).

[15] Hennessy, M. and G. Plotkin: A term model for CCS. *Mathematical Foundations of Computer Science* (Dembiński, E.P., ed.), 261-274. Lec. Notes Comp. Sci. 88. Berlin: Springer (1980).

[16] Hingston, Ph.: Non-complemented open sets in effective topology. *J. Austral. Math. Soc. (Series A)* 44, 129-137 (1988).

[17] Kalantari, I.: Major subsets in effective topology. *Patras Logic Symposium* (Metakides, G., ed.), 77-94. Amsterdam: North-Holland (1982).

[18] Kalantari, I. and A. Leggett: Simplicity in effective topology. *J. Symbolic Logic* 47, 169-183 (1982).

[19] Kalantari, I. and A. Leggett: Maximality in effective topology. *J. Symbolic Logic* 48, 100-112 (1983).

[20] Kalantari, I. and J.B. Remmel: Degrees of recursively enumerable topological spaces. *J. Symbolic Logic* 48, 610-622 (1983).

[21] Kalantari, I. and A. Retzlaff: Recursive constructions in topological spaces. *J. Symbolic Logic* 44, 609-625 (1979).

[22] Kalantari, I. and G. Weitkamp: Effective topological spaces I: A definability theory. *Ann. Pure Appl. Logic* 29, 1-27 (1985)

[23] Kalantari, I. and G. Weitkamp: Effective topological spaces II: A hierarchy. *Ann. Pure Appl. Logic* 29, 207-224 (1985).

[24] Kalantari, I. and G. Weitkamp: Effective topological spaces III: Forcing and definability. *Ann. Pure Appl. Logic* 36, 17-27 (1987).

[25] Kanda, A.: Gödel numbering of domain theoretic computable functions. Report no. 138, Dept. of Comp. Sci., Univ. of Leeds (1980).

[26] Kanda, A. and D. Park: When are two effectively given domains identical? *Theoretical Computer Science, 4th GI Conference* (Weihrauch, K., ed.), 170-181. Lec. Notes Comp. Sci. 67. Berlin: Springer (1979).

[27] Kreisel, G., D. Lacombe, and J. Shoenfield: Partial recursive functionals and effective operations. *Constructivity in Mathematics* (Heyting, A., ed.), 290-297. Amsterdam: North-Holland (1959).

[28] Kreitz, Ch.: Zulässige cpo-s, ein Entwurf für ein allgemeines Berechenbarkeitskonzept. Schriften zur Angew. Math. u. Informatik Nr. 76, Rheinisch-Westfälische Technische Hochschule Aachen (1982).

[29] Mal'cev, A.I.: *The Metamathematics of Algebraic Systems. Collected Papers:* 1936-1967 (Wells III, B.F., ed.). Amsterdam: North-Holland (1971).

[30] Milner, R.: *A Calculus of Communicating Sequences*. Lec. Notes Comp. Sci. 92. Berlin: Springer (1980).

[31] Moschovakis, Y.N.: Recursive analysis. Ph.D. Thesis, Univ. of Wisconsin, Madison, Wis. (1963).

[32] Moschovakis, Y.N.: Recursive metric spaces. *Fund. Math.* 55, 215-238 (1964).

[33] Myhill, J. and J.C. Shepherdson: Effective operators on partial recursive functions. *Zeitschr. f. math. Logik Grundl. d. Math.* 1, 310-317 (1955).

[34] Nivat, M.: Infinite words, infinite trees, infinite computations. *Foundations of Computer Science III, Part* 2 (de Bakker, J.W. et al., eds.), 1-52. Math. Centre Tracts 109. Amsterdam (1979).

[35] Nogina, E.Ju.: On effectively topological spaces. *Dokl. Akad. Nauk SSSR* 169, 28-31 (Russian) (1966); English transl., *Soviet Math. Dokl.* 7, 865-868 (1966).

[36] Nogina, E.Ju.: Relations between certain classes of effectively topological spaces. *Mat. Zametki* 5, 483-495 (Russian) (1969); English transl., *Math. Notes* 5, 288-294 (1969).

[37] Nogina, E.Ju.: Enumerable topological spaces. *Zeitschr. f. math. Logik Grundl. d. Math.* 24, 141-176 (Russian) (1978).

[38] Nogina, E. Ju.: On completely enumerable subsets of direct products of numbered sets. *Mathematical Linguistics and Theory of Algorithms*, 130-132 (Russian). Interuniv. thematic Collect., Kalinin Univ. (1978).

[39] Nogina, E.Ju.: The relation between separability and tracebility of sets. *Mathematical Logic and Mathematical Linguistics*, 135-144 (Russian). Kalinin Univ. (1981).

[40] Reed, G.M. and A.W. Roscoe: Metric spaces as models for real-time concurrency. *Mathematical Foundations of Programming Language Semantics, 3rd Workshop* (Main, M. et al., eds.), 330-343. Lec. Notes Comp. Sci. 298. Berlin: Springer (1988).

[41] Roscoe, A.W.: A mathematical theory of communicating processes. D.Phil. Thesis, Oxford Univ. (1982).

[42] Rogers, H., Jr.: *Theory of Recursive Functions and Effective Computability*. New York: McGraw-Hill (1967).

[43] Rounds, W.C.: Applications of topology to the semantics of communicating processes. *Seminar on Concurrency* (Brookes, S.D. et al., eds.), 360-372. Lec. Notes Comp. Sci. 197. Berlin: Springer (1985).

[44] Sciore, E. and A. Tang: Admissible coherent c.p.o.-s. *Automata, Languages and Programming* (Ausiello, G. et al., eds.), 440-456. Lec. Notes Comp. Sci. 62. Berlin: Springer (1978).

[45] Sciore, E. and A. Tang: Computability theory in admissible domains. *10th Annual ACM Symp. on Theory of Computing*, 95-104. New York: Ass. Comp. Mach. (1978).

[46] Scott, D.: Outline of a mathematical theory of computation. Techn. Monograph PRG-2, Oxford Univ. Comp. Lab. (1970).

[47] Scott, D.: Continuous lattices. *Toposes, Algebraic Geometry and Logic* (Bucur, I. et al., eds.), 97-136. Lec. Notes Math. 274. Berlin: Springer (1971).

[48] Scott, D.: Lattice theory, data types and semantics. *Formal Semantics of Programming Languages* (Rustin, R., ed.), 65-106. Englewood Cliffs, N.J.: Prentice-Hall (1972).

[49] Scott, D.: Models for various type-free calculi. *Logic, Methodology and Philosophy of Science IV* (Suppes, P. et al., eds.), 157-187. Amsterdam: North-Holland (1973).

[50] Scott, D.: Data types as lattices. *SIAM J. on Computing* 5, 522-587 (1976).

[51] Scott, D.: Lectures on a mathematical theory of computation. Techn. Monograph PRG-19, Oxford Univ. Comp. Lab. (1981).

[52] Scott, D.: Domains for denotational semantics. *Automata, Languages and Programming* (Nielsen, M. et al., eds.), 577-613. Lec. Notes Comp. Sci. 140. Berlin: Springer (1982).

[53] Scott, D. and Ch. Strachey: Towards a mathematical semantics for computer languages. *Computers and Automata* (Fox, J., ed.), 19-46. Brooklyn, N.Y.: Polytechnic Press (1971).

[54] Smyth, M.B.: Effectively given domains. *Theoret. Comp. Sci.* 5, 257-274 (1977).

[55] Smyth, M.B.: Power domains and predicate transformers. *Automata, Languages and Programming* (Diaz, J., ed.), 662-675. Lec. Notes Comp. Sci. 154. Berlin: Springer (1983).

[56] Smyth, M.B.: Completeness of quasi-uniform spaces in terms of filters. Manuscript (1987).

[57] Smyth, M.B.: Quasi-uniformities: reconciling domains with metric spaces. *Mathematical Foundations of Programming Language Semantics, 3rd Workshop* (Main, M. et al., eds.), 236-253. Lec. Notes Comp. Sci. 298. Berlin: Springer (1988).

[58] Spreen, D. and P. Young: Effective operators in a topological setting. *Computation and Proof Theory, Proc. Logic Colloquium '83* (Richter, M.M. et al., eds.), 437-451. Lec. Notes Math. 1104. Berlin: Springer (1984).

[59] Spreen, D.: Rekursionstheorie auf Teilmengen partieller Funktionen. Habilitationsschrift, Rheinisch-Westfälische Technische Hochschule Aachen (1985).

[60] Vaĭnberg, Ju.R. and E.Ju. Nogina: Categories of effectively topological spaces. *Studies in Formalized Languages and Nonclassical Logics*, 253-273 (Russian). Moscow: Izdat. "Nauka" (1974).

[61] Vaĭnberg, Ju.R. and E.Ju. Nogina: Two types of continuity of computable mappings of numerated topological spaces. *Studies in the Theory of Algorithms and Mathematical Logic, Vol.* 2, 84-99, 159 (Russian). Moscow: Vyčisl. Centr Akad. Nauk SSSR (1976).

[62] Vickers, St.: *Topology via Logic*. Cambridge: Cambridge Univ. Press (1989).

[63] Weihrauch, K. and Th. Deil: Berechenbarkeit auf cpo-s. Schriften zur Angew. Math. u. Informatik Nr. 63, Rheinisch-Westfälische Technische Hochschule Aachen (1980).

[64] Weihrauch, K.: *Computability*. Berlin: Springer (1987).

[65] Xiang, Li: Everywhere nonrecursive r.e. sets in recursively presented topological spaces. *J. Austral. Math. Soc. (Series A)* 44, 105-128 (1988).

INDEX OF AUTHORS

LIST OF PARTICIPANTS

(Participants marked with * have given a talk at the conference)

AMBOS-SPIES, K.*	Math. Inst., Univ. Heidelberg, Im Neuenheimer Feld 288, 6900 HEIDELBERG 1, FRG
ARSLANOV, M.M.*	Karbishev str. 13, apt. 109, 420087 KAZAN 87, USSR
ASSER, G.	Ludwig-Jahn-Str. 6, 2200 GREIFSWALD, DDR
BÖRGER, E.	Dipartimento di Informatica, Univ. di Pisa, Corso Italia 40, 56100 PISA, Italy
CLOTE, P.*	Dept. of Maths., Boston College, CHESTNUT HILL, MA 02167, USA
COOPER, S.B.	School of Maths., Univ. of Leeds, LEEDS, LS2 9JT, England
DEGTEV, A.N.*	Dept. of Maths., Tyuman State Univ., ul. Semakova 10, TYUMAN-3, USSR
DING, D.*	Dept. of Maths., Nanjing Univ., NANJING 210008, P.R. of China
DOMBROVSKY, M.	Computer Center of the Academy of Sciences, Vavilova, MOSKOW 117333, USSR *Home:* Anadyrsky Proezd, Dom 67 KV 70, MOSKOW 129336
DOWNEY, R.	Dept. of Maths., Victoria Univ. of Wellington, P.O. Box 600, WELLINGTON, New Zealand
EBBINGHAUS, H.-D.	Inst. f. Math. Logik u. Grundl. d. Math. d. Univ. Freiburg, Albertstr. 23b, 7800 FREIBURG, FRG
ERSHOV, Y. L.*	Inst. of Maths., Siberian Acad. of Sciences of the USSR, Universitetskii pr. 4, NOVOSIBIRSK 630090, USSR
FEJER, P.A.*	Dept. of Maths., Univ. of Mass. at Boston, Harbor Campus, BOSTON, MA 02125, USA
FRIEDMAN, S.D.*	Dept. of Maths., Massachusetts Inst. of Technology, CAMBRIDGE, MA 02139, USA
GANDY, R.O.*	9, Squitchey Lane, Summertown, OXFORD OX2 7LD, England

GRIFFOR, E.R. Dept. of Maths., Univ. of Uppsala, Thunbergsvägen 3,
752 38 UPPSALA, Sweden

GROSZEK, M.* Dept. of Maths., Dartmouth College, Bradley Hall,
HANOVER, NH 03755, USA

HARRINGTON, L.* Dept. of Maths., Univ. of California,
BERKELEY, CA 94720, USA

HAUGHT, C.* Dept. of Maths., Univ. of Chicago, 5734 University Ave.,
CHICAGO, IL 60637, USA

HINMAN, P.G.* Dept. of Maths., Univ. of Michigan,
ANN ARBOR, MI 48109-1003, USA

HOMER, S. Dept. of Comp. Sci., Boston Univ., 111 Cummington St.,
BOSTON, MA 02215, USA

JOKUSCH, C.G. Dept. of Maths., Univ. of Illinois, Altgeld Hall,
1409, West Green Str., URBANA, IL 61801, USA

KECHRIS, A.S.* Dept. of Maths., California Institute of Technology,
PASADENA, CA 91125, USA

KUČERA, A.* Dept. of Comp. Sci., Charles Univ., Malostranske nam 25,
11800 PRAHA, ČSSR

KUMMER, M.* Inst. f. Informatik d. Univ. Karlsruhe, Postfach 6980,
7500 KARLSRUHE 1, FRG

LEMPP, S.* Dept. of Maths., Univ. of Wisconsin-Madison, 480 Lincoln Drive
MADISON, Wisconsin 53706, USA

LOUVEAU, A.* Equipe d'Analyse, T.46, 4e etage, Univ. Pierre et Marie Curie,
(Univ. Paris VI), 4, Place Jussieu,
75252 PARIS-CEDEX 05, France

MAASS, W.* Dept. of Maths., Univ. of Illinois at Chicago, Box 4348,
CHICAGO, IL 60680, USA

MOHRHERR, J. P.O. BOX 9,
WILMETTE, IL 60091, USA

MOSCHOVAKIS, Y.N.* Dept. of Maths., Univ. of California, 405 Hilgard Avenue,
LOS ANGELES, CA 90024, USA

MÜLLER, G.H. Math. Inst. d. Univ. Heidelberg, Im Neuenheimer Feld 288,
6900 HEIDELBERG, FRG

NERODE, A.*	Dept. of Maths., Cornell Univ., 201 Caldwell Hall, ITHACA, NY 14853-2602, USA
ODIFREDDI, P.	Dip. di Informatica, Universita di Torina, Corso Svizzera 185, 10149 TORINO, Italy
POHLERS, W.*	Inst. f. Math. Logik u. Grundlagenforschung d. Univ. Münster, Einsteinstr. 62, 4400 MÜNSTER, FRG
RICHTER, M.M.	Fachbereich Informatik d. Univ. Kaiserslautern, Postfach 3049, 6750 KAISERSLAUTERN, FRG
SACKS, G.E.*	Dept. of Maths., Harvard Univ., One Oxford St., CAMBRIDGE, MA 02138, USA
SCHWICHTENBERG, H.*	Math. Inst. d. Univ. München, Theresienstr. 39, 8000 MÜNCHEN, FRG
SHORE, R.A.*	Dept. of Maths., Cornell Univ., White Hall, ITAKA, NY 14853-7901, USA
SLAMAN, T.A.*	Dept. of Maths., Univ. of Chicago, 5734 University Ave, CHICAGO, IL 60637, USA
SOARE, R.I.*	Dept. of Maths., Univ. of Chicago, 5734 University Ave, CHICAGO, IL 60637, USA
SPREEN, D.*	LG Theor. Informatik, FB 7 Physik, Univ. GHS Siegen, Postfach 101240, 5900 SIEGEN, FRG
STOB, M.*	Dept. of Maths., Calvin College, GRAND RAPIDS, Mich. 49506, USA
THIELE, E.-J.	Fachb. Informatik/FB3 d. Techn. Univ. Berlin, Straße des 17.Juni 135, 1000 BERLIN (WEST), FRG
WAINER, S.S.*	School of Maths., Univ. of Leeds, LEEDS, LS2 9JT, England
WECHSUNG, G.*	Sektion Mathematik, Friedrich-Schiller-Univ. Jena, Universitätshochhaus 17.OG., 6900 JENA, DDR
WEIHRAUCH, K.	FB Math. u. Inform., Fachrichtung Inform. d. Fernuniv. Hagen, Postfach 940, 5800 HAGEN, FRG
WOODIN, W.H.*	Dept. of Maths., Univ. of California, BERKELEY, CA 94720, USA
YANG, D.*	Inst. of Software, Academia Sinica, P.O. Box 8718, BEIJING, 100080, P.R. of China

LECTURE NOTES IN MATHEMATICS

Edited by A. Dold, B. Eckmann and F. Takens

Some general remarks on the publication of proceedings of congresses and symposia

Lecture Notes aim to report new developments - quickly, informally and at a high level. The following describes criteria and procedures which apply to proceedings volumes. The editors of a volume are strongly advised to inform contributors about these points at an early stage.

§1. One (or more) expert participant(s) of the meeting should act as the responsible editor(s) of the proceedings. They select the papers which are suitable (cf. §§ 2, 3) for inclusion in the proceedings, and have them individually refereed (as for a journal). It should not be assumed that the published proceedings must reflect conference events faithfully and in their entirety. Contributions to the meeting which are not included in the proceedings can be listed by title. The series editors will normally not interfere with the editing of a particular proceedings volume - except in fairly obvious cases, or on technical matters, such as described in §§ 2, 3. The names of the responsible editors appear on the title page of the volume.

§2. The proceedings should be reasonably homogeneous (concerned with a limited area). For instance, the proceedings of a congress on "Analysis" or "Mathematics in Wonderland" would normally not be sufficiently homogeneous.

One or two longer survey articles on recent developments in the field are often very useful additions to such proceedings - even if they do not correspond to actual lectures at the congress. An extensive introduction on the subject of the congress would be desirable.

§3. The contributions should be of a high mathematical standard and of current interest. Research articles should present new material and not duplicate other papers already published or due to be published. They should contain sufficient information and motivation and they should present proofs, or at least outlines of such, in sufficient detail to enable an expert to complete them. Thus resumes and mere announcements of papers appearing elsewhere cannot be included, although more detailed versions of a contribution may well be published in other places later.

Contributions in numerical mathematics may be acceptable without formal theorems resp. proofs if they present new algorithms solving problems (previously unsolved or less well solved) or develop innovative qualitative methods, not yet amenable to a more formal treatment.

Surveys, if included, should cover a sufficiently broad topic, and should in general not simply review the author's own recent research. In the case of such surveys, exceptionally, proofs of results may not be necessary.

§4. "Mathematical Reviews" and "Zentralblatt für Mathematik" recommend that papers in proceedings volumes carry an explicit statement that they are in final form and that no similar paper has been or is being submitted elsewhere, if these papers are to be considered for a review. Normally, papers that satisfy the criteria of the Lecture Notes in Mathematics series also satisfy this requirement, but we strongly recommend that the contributing authors be asked to give this guarantee explicitly at the beginning or end of their paper. There will occasionally be cases where this does not apply but where, for special reasons, the paper is still acceptable for LNM.

§5. Proceedings should appear soon after the meeeting. The publisher should, therefore, receive the complete manuscript (preferably in duplicate) within nine months of the date of the meeting at the latest.

§6. Plans or proposals for proceedings volumes should be sent to one of the editors of the series or to Springer-Verlag Heidelberg. They should give sufficient information on the conference or symposium, and on the proposed proceedings. In particular, they should contain a list of the expected contributions with their prospective length. Abstracts or early versions (drafts) of some of the contributions are helpful.

§7. Lecture Notes are printed by photo-offset from camera-ready typed copy provided by the editors. For this purpose Springer-Verlag provides editors with technical instructions for the preparation of manuscripts and these should be distributed to all contributing authors. Springer-Verlag can also, on request, supply stationery on which the prescribed typing area is outlined. Some homogeneity in the presentation of the contributions is desirable.

Careful preparation of manuscripts will help keep production time short and ensure a satisfactory appearance of the finished book. The actual production of a Lecture Notes volume normally takes 6 -8 weeks.

Manuscripts should be at least 100 pages long. The final version should include a table of contents.

§8. Editors receive a total of 50 free copies of their volume for distribution to the contributing authors, but no royalties. (Unfortunately, no reprints of individual contributions can be supplied.) They are entitled to purchase further copies of their book for their personal use at a discount of 33.3 %, other Springer mathematics books at a discount of 20 % directly from Springer-Verlag. Contributing authors may purchase the volume in which their article appears at a discount of 33.3 %.

Commitment to publish is made by letter of intent rather than by signing a formal contract. Springer-Verlag secures the copyright for each volume.

<u>Addresses:</u>

Professor A. Dold, Mathematisches Institut, Universität Heidelberg,
Im Neuenheimer Feld 288, 6900 Heidelberg, Federal Republic of Germany

Professor B. Eckmann, Mathematik, ETH-Zentrum
8092 Zürich, Switzerland

Prof. F. Takens, Mathematisch Instituut, Rijksuniversiteit Groningen,
Postbus 800, 9700 AV Groningen, The Netherlands

Springer-Verlag, Mathematics Editorial, Tiergartenstr. 17,
6900 Heidelberg, Federal Republic of Germany, Tel.: (06221) 487-410

Springer-Verlag, Mathematics Editorial, 175, Fifth Avenue,
New York, New York 10010, USA, Tel.: (212) 460-1596

Vol. 1259: F. Cano Torres, Desingularization Strategies for Three-Dimensional Vector Fields. IX, 189 pages. 1987.

Vol. 1260: N.H. Pavel, Nonlinear Evolution Operators and Semigroups. VI, 285 pages. 1987.

Vol. 1261: H. Abels, Finite Presentability of S-Arithmetic Groups. Compact Presentability of Solvable Groups. VI, 178 pages. 1987.

Vol. 1262: E. Hlawka (Hrsg.), Zahlentheoretische Analysis II. Seminar, 1984–86. V, 158 Seiten. 1987.

Vol. 1263: V.L. Hansen (Ed.), Differential Geometry. Proceedings, 1985. XI, 288 pages. 1987.

Vol. 1264: Wu Wen-tsün, Rational Homotopy Type. VIII, 219 pages. 1987.

Vol. 1265: W. Van Assche, Asymptotics for Orthogonal Polynomials. VI, 201 pages. 1987.

Vol. 1266: F. Ghione, C. Peskine, E. Sernesi (Eds.), Space Curves. Proceedings, 1985. VI, 272 pages. 1987.

Vol. 1267: J. Lindenstrauss, V.D. Milman (Eds.), Geometrical Aspects of Functional Analysis. Seminar. VII, 212 pages. 1987.

Vol. 1268: S.G. Krantz (Ed.), Complex Analysis. Seminar, 1986. VII, 195 pages. 1987.

Vol. 1269: M. Shiota, Nash Manifolds. VI, 223 pages. 1987.

Vol. 1270: C. Carasso, P.-A. Raviart, D. Serre (Eds.), Nonlinear Hyperbolic Problems. Proceedings, 1986. XV, 341 pages. 1987.

Vol. 1271: A.M. Cohen, W.H. Hesselink, W.L.J. van der Kallen, J.R. Strooker (Eds.), Algebraic Groups Utrecht 1986. Proceedings. XII, 284 pages. 1987.

Vol. 1272: M.S. Livšic, L.L. Waksman, Commuting Nonselfadjoint Operators in Hilbert Space. III, 115 pages. 1987.

Vol. 1273: G.-M. Greuel, G. Trautmann (Eds.), Singularities, Representation of Algebras, and Vector Bundles. Proceedings, 1985. XIV, 383 pages. 1987.

Vol. 1274: N. C. Phillips, Equivariant K-Theory and Freeness of Group Actions on C*-Algebras. VIII, 371 pages. 1987.

Vol. 1275: C.A. Berenstein (Ed.), Complex Analysis I. Proceedings, 1985–86. XV, 331 pages. 1987.

Vol. 1276: C.A. Berenstein (Ed.), Complex Analysis II. Proceedings, 1985–86. IX, 320 pages. 1987.

Vol. 1277: C.A. Berenstein (Ed.), Complex Analysis III. Proceedings, 1985–86. X, 350 pages. 1987.

Vol. 1278: S.S. Koh (Ed.), Invariant Theory. Proceedings, 1985. V, 102 pages. 1987.

Vol. 1279: D. Ieşan, Saint-Venant's Problem. VIII, 162 Seiten. 1987.

Vol. 1280: E. Neher, Jordan Triple Systems by the Grid Approach. XII, 193 pages. 1987.

Vol. 1281: O.H. Kegel, F. Menegazzo, G. Zacher (Eds.), Group Theory. Proceedings, 1986. VII, 179 pages. 1987.

Vol. 1282: D.E. Handelman, Positive Polynomials, Convex Integral Polytopes, and a Random Walk Problem. XI, 136 pages. 1987.

Vol. 1283: S. Mardešić, J. Segal (Eds.), Geometric Topology and Shape Theory. Proceedings, 1986. V, 261 pages. 1987.

Vol. 1284: B.H. Matzat, Konstruktive Galoistheorie. X, 286 pages. 1987.

Vol. 1285: I.W. Knowles, Y. Saitō (Eds.), Differential Equations and Mathematical Physics. Proceedings, 1986. XVI, 499 pages. 1987.

Vol. 1286: H.R. Miller, D.C. Ravenel (Eds.), Algebraic Topology. Proceedings, 1986. VII, 341 pages. 1987.

Vol. 1287: E.B. Saff (Ed.), Approximation Theory, Tampa. Proceedings, 1985–1986. V, 228 pages. 1987.

Vol. 1288: Yu. L. Rodin, Generalized Analytic Functions on Riemann Surfaces. V, 128 pages, 1987.

Vol. 1289: Yu. I. Manin (Ed.), K-Theory, Arithmetic and Geometry. Seminar, 1984–1986. V, 399 pages. 1987.

Vol. 1290: G. Wüstholz (Ed.), Diophantine Approximation and Transcendence Theory. Seminar, 1985. V, 243 pages. 1987.

Vol. 1291: C. Mœglin, M.-F. Vignéras, J.-L. Waldspurger, Correspondances de Howe sur un Corps p-adique. VII, 163 pages. 1987

Vol. 1292: J.T. Baldwin (Ed.), Classification Theory. Proceedings, 1985. VI, 500 pages. 1987.

Vol. 1293: W. Ebeling, The Monodromy Groups of Isolated Singularities of Complete Intersections. XIV, 153 pages. 1987.

Vol. 1294: M. Queffélec, Substitution Dynamical Systems – Spectral Analysis. XIII, 240 pages. 1987.

Vol. 1295: P. Lelong, P. Dolbeault, H. Skoda (Réd.), Séminaire d'Analyse P. Lelong – P. Dolbeault – H. Skoda. Seminar, 1985/1986. VII, 283 pages. 1987.

Vol. 1296: M.-P. Malliavin (Ed.), Séminaire d'Algèbre Paul Dubreil et Marie-Paule Malliavin. Proceedings, 1986. IV, 324 pages. 1987.

Vol. 1297: Zhu Y.-l., Guo B.-y. (Eds.), Numerical Methods for Partial Differential Equations. Proceedings. XI, 244 pages. 1987.

Vol. 1298: J. Aguadé, R. Kane (Eds.), Algebraic Topology, Barcelona 1986. Proceedings. X, 255 pages. 1987.

Vol. 1299: S. Watanabe, Yu.V. Prokhorov (Eds.), Probability Theory and Mathematical Statistics. Proceedings, 1986. VIII, 589 pages. 1988.

Vol. 1300: G.B. Seligman, Constructions of Lie Algebras and their Modules. VI, 190 pages. 1988.

Vol. 1301: N. Schappacher, Periods of Hecke Characters. XV, 160 pages. 1988.

Vol. 1302: M. Cwikel, J. Peetre, Y. Sagher, H. Wallin (Eds.), Function Spaces and Applications. Proceedings, 1986. VI, 445 pages. 1988.

Vol. 1303: L. Accardi, W. von Waldenfels (Eds.), Quantum Probability and Applications III. Proceedings, 1987. VI, 373 pages. 1988.

Vol. 1304: F.Q. Gouvêa, Arithmetic of p-adic Modular Forms. VIII, 121 pages. 1988.

Vol. 1305: D.S. Lubinsky, E.B. Saff, Strong Asymptotics for Extremal Polynomials Associated with Weights on $\mathbb{R}$. VII, 153 pages. 1988.

Vol. 1306: S.S. Chern (Ed.), Partial Differential Equations. Proceedings, 1986. VI, 294 pages. 1988.

Vol. 1307: T. Murai, A Real Variable Method for the Cauchy Transform, and Analytic Capacity. VIII, 133 pages. 1988.

Vol. 1308: P. Imkeller, Two-Parameter Martingales and Their Quadratic Variation. IV, 177 pages. 1988.

Vol. 1309: B. Fiedler, Global Bifurcation of Periodic Solutions with Symmetry. VIII, 144 pages. 1988.

Vol. 1310: O.A. Laudal, G. Pfister, Local Moduli and Singularities. V, 117 pages. 1988.

Vol. 1311: A. Holme, R. Speiser (Eds.), Algebraic Geometry, Sundance 1986. Proceedings. VI, 320 pages. 1988.

Vol. 1312: N.A. Shirokov, Analytic Functions Smooth up to the Boundary. III, 213 pages. 1988.

Vol. 1313: F. Colonius, Optimal Periodic Control. VI, 177 pages. 1988.

Vol. 1314: A. Futaki, Kähler-Einstein Metrics and Integral Invariants. IV, 140 pages. 1988.

Vol. 1315: R.A. McCoy, I. Ntantu, Topological Properties of Spaces of Continuous Functions. IV, 124 pages. 1988.

Vol. 1316: H. Korezlioglu, A.S. Ustunel (Eds.), Stochastic Analysis and Related Topics. Proceedings, 1986. V, 371 pages. 1988.

Vol. 1317: J. Lindenstrauss, V.D. Milman (Eds.), Geometric Aspects of Functional Analysis. Seminar, 1986–87. VII, 289 pages. 1988.

Vol. 1318: Y. Felix (Ed.), Algebraic Topology – Rational Homotopy. Proceedings, 1986. VIII, 245 pages. 1988

Vol. 1319: M. Vuorinen, Conformal Geometry and Quasiregular Mappings. XIX, 209 pages. 1988.

Vol. 1320: H. Jürgensen, G. Lallement, H.J. Weinert (Eds.), Semigroups, Theory and Applications. Proceedings, 1986. X, 416 pages. 1988.

Vol. 1321: J. Azéma, P.A. Meyer, M. Yor (Eds.), Séminaire de Probabilités XXII. Proceedings. IV, 600 pages. 1988.

Vol. 1322: M. Métivier, S. Watanabe (Eds.), Stochastic Analysis. Proceedings, 1987. VII, 197 pages. 1988.

Vol. 1323: D.R. Anderson, H.J. Munkholm, Boundedly Controlled Topology. XII, 309 pages. 1988.

Vol. 1324: F. Cardoso, D.G. de Figueiredo, R. Iório, O. Lopes (Eds.), Partial Differential Equations. Proceedings, 1986. VIII, 433 pages. 1988.

Vol. 1325: A. Truman, I.M. Davies (Eds.), Stochastic Mechanics and Stochastic Processes. Proceedings, 1986. V, 220 pages. 1988.

Vol. 1326: P.S. Landweber (Ed.), Elliptic Curves and Modular Forms in Algebraic Topology. Proceedings, 1986. V, 224 pages. 1988.

Vol. 1327: W. Bruns, U. Vetter, Determinantal Rings. VII,236 pages. 1988.

Vol. 1328: J.L. Bueso, P. Jara, B. Torrecillas (Eds.), Ring Theory. Proceedings, 1986. IX, 331 pages. 1988.

Vol. 1329: M. Alfaro, J.S. Dehesa, F.J. Marcellan, J.L. Rubio de Francia, J. Vinuesa (Eds.): Orthogonal Polynomials and their Applications. Proceedings, 1986. XV, 334 pages. 1988.

Vol. 1330: A. Ambrosetti, F. Gori, R. Lucchetti (Eds.), Mathematical Economics. Montecatini Terme 1986. Seminar. VII, 137 pages. 1988.

Vol. 1331: R. Bamón, R. Labarca, J. Palis Jr. (Eds.), Dynamical Systems, Valparaiso 1986. Proceedings. VI, 250 pages. 1988.

Vol. 1332: E. Odell, H. Rosenthal (Eds.), Functional Analysis. Proceedings, 1986–87. V, 202 pages. 1988.

Vol. 1333: A.S. Kechris, D.A. Martin, J.R. Steel (Eds.), Cabal Seminar 81–85. Proceedings, 1981–85. V, 224 pages. 1988.

Vol. 1334: Yu.G. Borisovich, Yu. E. Gliklikh (Eds.), Global Analysis – Studies and Applications III. V, 331 pages. 1988.

Vol. 1335: F. Guillén, V. Navarro Aznar, P. Pascual-Gainza, F. Puerta, Hyperrésolutions cubiques et descente cohomologique. XII, 192 pages. 1988.

Vol. 1336: B. Helffer, Semi-Classical Analysis for the Schrödinger Operator and Applications. V, 107 pages. 1988.

Vol. 1337: E. Sernesi (Ed.), Theory of Moduli. Seminar, 1985. VIII, 232 pages. 1988.

Vol. 1338: A.B. Mingarelli, S.G. Halvorsen, Non-Oscillation Domains of Differential Equations with Two Parameters. XI, 109 pages. 1988.

Vol. 1339: T. Sunada (Ed.), Geometry and Analysis of Manifolds. Procedings, 1987. IX, 277 pages. 1988.

Vol. 1340: S. Hildebrandt, D.S. Kinderlehrer, M. Miranda (Eds.), Calculus of Variations and Partial Differential Equations. Proceedings, 1986. IX, 301 pages. 1988.

Vol. 1341: M. Dauge, Elliptic Boundary Value Problems on Corner Domains. VIII, 259 pages. 1988.

Vol. 1342: J.C. Alexander (Ed.), Dynamical Systems. Proceedings, 1986–87. VIII, 726 pages. 1988.

Vol. 1343: H. Ulrich, Fixed Point Theory of Parametrized Equivariant Maps. VII, 147 pages. 1988.

Vol. 1344: J. Král, J. Lukeš, J. Netuka, J. Veselý (Eds.), Potential Theory – Surveys and Problems. Proceedings, 1987. VIII, 271 pages. 1988.

Vol. 1345: X. Gomez-Mont, J. Seade, A. Verjovski (Eds.), Holomorphic Dynamics. Proceedings, 1986. VII, 321 pages. 1988.

Vol. 1346: O. Ya. Viro (Ed.), Topology and Geometry – Rohlin Seminar. XI, 581 pages. 1988.

Vol. 1347: C. Preston, Iterates of Piecewise Monotone Mappings on an Interval. V, 166 pages. 1988.

Vol. 1348: F. Borceux (Ed.), Categorical Algebra and its Applications. Proceedings, 1987. VIII, 375 pages. 1988.

Vol. 1349: E. Novak, Deterministic and Stochastic Error Bounds in Numerical Analysis. V, 113 pages. 1988.

Vol. 1350: U. Koschorke (Ed.), Differential Topology. Proceedings, 1987. VI, 269 pages. 1988.

Vol. 1351: I. Laine, S. Rickman, T. Sorvali, (Eds.), Complex Analysis, Joensuu 1987. Proceedings. XV, 378 pages. 1988.

Vol. 1352: L.L. Avramov, K.B. Tchakerian (Eds.), Algebra – Some Current Trends. Proceedings, 1986. IX, 240 Seiten. 1988.

Vol. 1353: R.S. Palais, Ch.-l. Terng, Critical Point Theory and Submanifold Geometry. X, 272 pages. 1988.

Vol. 1354: A. Gómez, F. Guerra, M.A. Jiménez, G. López (Eds.), Approximation and Optimization. Proceedings, 1987. VI, 280 pages. 1988.

Vol. 1355: J. Bokowski, B. Sturmfels, Computational Synthetic Geometry. V, 168 pages. 1989.

Vol. 1356: H. Volkmer, Multiparameter Eigenvalue Problems and Expansion Theorems. VI, 157 pages. 1988.

Vol. 1357: S. Hildebrandt, R. Leis (Eds.), Partial Differential Equations and Calculus of Variations. VI, 423 pages. 1988.

Vol. 1358: D. Mumford, The Red Book of Varieties and Schemes. V, 309 pages. 1988.

Vol. 1359: P. Eymard, J.-P. Pier (Eds.), Harmonic Analysis. Proceedings, 1987. VIII, 287 pages. 1988.

Vol. 1360: G. Anderson, C. Greengard (Eds.), Vortex Methods. Proceedings, 1987. V, 141 pages. 1988.

Vol. 1361: T. tom Dieck (Ed.), Algebraic Topology and Transformation Groups. Proceedings, 1987. VI, 298 pages. 1988.

Vol. 1362: P. Diaconis, D. Elworthy, H. Föllmer, E. Nelson, G.C. Papanicolaou, S.R.S. Varadhan. École d'Été de Probabilités de Saint-Flour XV–XVII, 1985–87. Editor: P.L. Hennequin. V, 459 pages. 1988.

Vol. 1363: P.G. Casazza, T.J. Shura. Tsirelson's Space. VIII, 204 pages. 1988.

Vol. 1364: R.R. Phelps, Convex Functions, Monotone Operators and Differentiability. IX, 115 pages. 1989.

Vol. 1365: M. Giaquinta (Ed.), Topics in Calculus of Variations. Seminar, 1987. X, 196 pages. 1989.

Vol. 1366: N. Levitt, Grassmannians and Gauss Maps in PL-Topology. V, 203 pages. 1989.

Vol. 1367: M. Knebusch, Weakly Semialgebraic Spaces. XX, 376 pages. 1989.

Vol. 1368: R. Hübl, Traces of Differential Forms and Hochschild Homology. III, 111 pages. 1989.

Vol. 1369: B. Jiang, Ch.-K. Peng, Z. Hou (Eds.), Differential Geometry and Topology. Proceedings, 1986–87. VI, 366 pages. 1989.

Vol. 1370: G. Carlsson, R.L. Cohen, H.R. Miller, D.C. Ravenel (Eds.), Algebraic Topology. Proceedings, 1986. IX, 456 pages. 1989.

Vol. 1371: S. Glaz, Commutative Coherent Rings. XI, 347 pages. 1989.

Vol. 1372: J. Azéma, P.A. Meyer, M. Yor (Eds.), Séminaire de Probabilités XXIII. Proceedings. IV, 583 pages. 1989.

Vol. 1373: G. Benkart, J.M. Osborn (Eds.), Lie Algebras, Madison 1987. Proceedings. V, 145 pages. 1989.

Vol. 1374: R.C. Kirby, The Topology of 4-Manifolds. VI, 108 pages. 1989.

Vol. 1375: K. Kawakubo (Ed.), Transformation Groups. Proceedings, 1987. VIII, 394 pages, 1989.

Vol. 1376: J. Lindenstrauss, V.D. Milman (Eds.), Geometric Aspects of Functional Analysis. Seminar (GAFA) 1987–88. VII, 288 pages. 1989.

Vol. 1377: J.F. Pierce, Singularity Theory, Rod Theory, and Symmetry-Breaking Loads. IV, 177 pages. 1989.

Vol. 1378: R.S. Rumely, Capacity Theory on Algebraic Curves. III, 437 pages. 1989.

Vol. 1379: H. Heyer (Ed.), Probability Measures on Groups IX. Proceedings, 1988. VIII, 437 pages. 1989